MANUFACTURING PROCESSES

MANUFACTURING PROCESSES

HERBERT W. YANKEE

Associate Professor of Mechanical Engineering
Worcester Polytechnic Institute

PRENTICE-HALL, INC., *Englewood Cliffs, New Jersey 07632*

Library of Congress Cataloging in Publication Data

Yankee, Herbert W
    Manufacturing processes.

    Includes bibliographies and index.
    1. Manufacturing processes. I. Title.
TS183.Y36   670   78-13059
ISBN 0-13-555557-4

Editorial production supervision
and interior design by: *James M. Chege*

Page layout by: *Rita K. Schwartz*

Cover design by: *Edsal Enterprizes*

Manufacturing buyer: *Gordon Osbourne*

© 1979 by Prentice-Hall, Inc., Englewood Cliffs, N.J. 07632

Printed in the United States of America

10   9

PRENTICE-HALL INTERNATIONAL, INC., *London*
PRENTICE-HALL OF AUSTRALIA PTY. LIMITED, *Sydney*
PRENTICE-HALL OF CANADA, LTD., *Toronto*
PRENTICE-HALL OF INDIA PRIVATE LIMITED, *New Delhi*
PRENTICE-HALL OF JAPAN, INC., *Tokyo*
PRENTICE-HALL OF SOUTHEAST ASIA PTE. LTD., *Singapore*
WHITEHALL BOOKS LIMITED, *Wellington, New Zealand*

To
Tod, Jill, Joshua, Bethany,
and to
my parents

CONTENTS

35.   CONVENTIONAL DRAWING AND RELATED PROCESSES

PREFACE

Manufacturing Processes is a text book *and* a reference book. It has a clear, direct style designed to cut through a bewildering array of highly specialized technical information. Effective illustrations contribute to the clarity of the text.

The book gives insights into the "how" and "why" of production decisions. Important manufacturing concepts which relate to process selection are organized by the real-life constraints caused by cost, time, materials considerations, and by production capabilities. The reasons for certain trade-offs which invariably accompany process selection are discussed and justified. For each process, ample data is supplied, e.g., on the selection of appropriate materials, realistic and achievable production rates, available ranges of precision, and limitations often imposed in manufacturing by size, weight or factors of physical properties. All sizes are given in dual English-S1 units.

A leading objective is to provide information which serves as a base for evaluating the relative merits of one processing method with another. No attempt is made to describe any process in the depth and detail required by an engineer who is charged with the responsibility of setting up full manufacturing facilities for a particular process.

The text does offer, in condensed form, a vast accumulation of data covering virtually every significant commercial method of manufacturing used in industry today. Over one hundred different primary and secondary manufacturing processes are described in a brief, accurate form. Important principles of product design are stressed, particularly those influenced by the nature of the special requirements of a given process. An up-to-date listing of specific product applications for each process is provided.

Each chapter of the book contains review questions. These questions are carefully designed to reinforce the reading when used as a class text. In addition, a very complete bibliography is offered. This source material opens up excellent supplemental detail for the compacted data in each chapter.

Most of the problems at the end of the chapters are typical of the design and manufacturing decisions the engineer may be called upon to make. The open-ended problems have been organized to prevent routine "cranking-out" of answers and call for investigating many possible ways of solution. As in actual situations, there may be several acceptable solutions— some more acceptable than others. Few of the problems have answers that can be found in the text. In fact, the problems were developed with the intention of providing additional, supplementary experiences not possible within the confined scope of the text. Useful formulas have been included in the Appendix that may aid in the solution of some of the problems.

A technical book is never written alone. *Manufacturing Processes* is no exception. To a great extent, material has been developed from commercial literature and in-depth inquiries and correspondence with well-established manufacturers. A conscientious effort was made to ensure that the text reflects the countless technological developments of recent years, and the current practices and applications in modern industry. Every aspect of the book has been thoroughly reviewed to assure maximum accuracy and relevancy.

The author wishes to express his sincere appreciation for the generous contributions made by a vast number of industrial firms, both in the United States and abroad. The technical data and the outstanding illustrative material supplied by representatives of these industries is a particularly significant factor in this book.

Grateful acknowledgement for their technical assistance is expressed in particular to three of my colleagues at Worcester Polytechnic Institute: Professors Roger R. Borden, Roy F. Bourgault, and Bennett E. Gordon, Jr. I am particularly grateful for the competent work of Mrs. Robert A. Englund who typed the manuscript and to my wife, Rosemary, who labored countless hours at proofreading. Also, I am indebted to the following other individuals who significantly contributed to this book in various ways: Donald E. Cookson, Charles Wick, Melvin Long, Charles DesPlaines, John J. Racine, Dr. Mark S. MacDonald, Daniel B. Dallas, Kendall J.

Raslavsky, John A. Mock, Dr. Gordon L. Greene, Prof. S. K. Palmer, Anderson A. Osburn, C. E. Lampson, Dr. Walter F. Keil, Frank A. Kalus, Jr., Prof. Hugh K. Behn, David F. Bergeron, and Jane Marean.

Finally, special appreciation is expressed to the author's brother, Deane K. Yankee, Project Engineer, Packaging Industries, Hyannis, Massachusetts, who not only read the entire manuscript but made countless valuable suggestions for improvement. Many of his excellent recommendations have been incorporated in this text.

HERBERT W. YANKEE

Worcester, Mass.

MANUFACTURING
PROCESSES

1

PRODUCTION DESIGN AND PROCESS SELECTION

This chapter discusses some of the factors that influence the manufacture of a product and the role key personnel in various departments play in carrying out their respective functions. Also explained are the team and individual functions necessary for successful product development, elements of both functional and production design, and critical decisions to be made resulting in the most economical production process. A processing checklist is given for product designers to help them evaluate specific process capabilities as they design a part to be manufactured.

DEVELOPING THE PRODUCT SPECIFICATIONS

Management techniques and procedures used to initially conceive and develop a product vary widely from company to-company. Staff organizations ordinarily depend upon the size of the company, its position in the market, the nature of the product, and company policies regarding capitalization and planning techniques.

The responsibility for determining the nature of the product to be manufactured and related decisions regarding quantity, quality, and costs that will be competitive in the market place rests with management. The overall production plan usually evolves from a group achievement or team effort

1

consisting of "inputs" from expert key personnel from research and development, product development, sales, plant management, and from other departments. The team is generally charged with determining the feasibility of producing either an entirely new product or evaluating the relative merits of modifying or redesigning an existing product. In any case, product designers are ultimately presented with information regarding management's appraisal of the importance of a given production plan, specifications and special requirements of the product which relate to the basic function, cost limitations, time factors, customer requests, sales predictions, and thoughts on expected difficulties or suspected problem areas. Sound product design and economical process selection begin with a knowledge of as many related facts about the product as possible.

The Product Development Team

Product development responsibilities may fall within a manufacturing engineering department, or, in some firms, they may comprise a separate department. Manufacturing engineering departments vary widely in makeup, depending upon in-plant production capabilities, which are usually influenced by existing manpower, machines, equipment, tooling, materials, and buildings. In some companies the product design section is headed by a product development engineer or a manufacturing manager, who may also supervise the activities of process engineers, tool designers, plant layout, tool control and methods personnel, and, in some cases, a maintenance section.

Fundamentally, the actual work in product design starts with the designer, who is usually a member of a product development team. Most companies recognize the many benefits of group activity, which invariably lead to the generation of many useful ideas. At this stage in the design process, only a vague idea of what the end product may ultimately consist is known. Firm decisions are carefully avoided. The emphasis in these brainstorming or "blue-sky" sessions is upon developing ideas in quantity; quality evaluations are avoided. All ideas are recorded, since it is much more difficult to develop ideas than it is to reject them. It is often possible to develop useful ideas by uniquely combining what may at first have seemed to be a series of disassociated concepts. It is especially important that product development team members be highly creative and as knowledgeable as possible, since the initial idea-gathering stage should ultimately lead to an acceptable problem solution. It is recognized that not everyone possesses the same amount of creativity but that everyone has some.

The first phase of product development is to analyze thoroughly management's production plan. Such a plan generally consists of a written problem statement or directive which clearly outlines, in as precise terms as possible, the product specifications. The team may then begin by generating a list of many possible ways by which the product specifications can be met.

One member of the group acts as the recorder. After one or two such sessions, the list is carefully evaluated to determine which of the proposed ideas have the most merit, either singly or in combination with other ideas. Close examination is finally made of the most promising ideas from this list, usually in the form of freehand sketches which help to evaluate graphically the comparative merits.

Functional Design

Once one or more designers have reduced the proposed product development ideas to a few tentatively acceptable ideas, the product concepts are then ready for further evaluation by an accurately prepared layout drawing. It is at this point that the designer critically analyzes each component to assure that the product, mechanism, structure, or machine will function acceptably and will meet the necessary critical performance requirements. The layout drawing gives the designer a graphical model by which the various component parts and their relative locations may be studied. Care is taken to prevent the design from becoming needlessly complex. Designers strive for parts that are characterized by smooth, simple shapes. Strong, serviceable parts are generally those whose configurations are as straightforward and sturdy as possible. Most companies "prove out" a product by constructing an experimental mechanical model or prototype, which is thoroughly tested and evaluated. The layout drawing may be readily changed as design alterations become necessary.

Production Design

While both the layout drawing and the experimental model serve the designer in evaluating the functional requirements of a given product design, the alert designer has, from the start, been mentally "processing" each of the functional parts in an effort to simplify manufacture. The designer continually strives to accomplish the very most for the least. This process is known as *optimization*. It consists of making design decisions from as many points of view as possible. This ongoing process is vital in achieving successful results in production design.

Production design follows functional design. Basically, it is a study of the process of adapting component parts to the simplest method of manufacture. Careful design for production can achieve a variety of substantial economies in manufacture.

Actually, the drawings and specifications of the designer help to establish minimum manufacturing costs to a far greater degree than is often recognized. Regardless of process efficiency, tooling, or purchasing efforts, the designer's early influence on manufacturing cost often sets the limits beyond which very little improvement can be made. The total design effect upon manufacturing costs has been found to average as much as 35% of the major overall costs. Such costs apply to raw materials, equipment, direct and indirect labor, tooling, and engineering. The designer must know from experience the capabilities of the various automatic production machines in order to make full use of them. He must know what standard parts can be fitted to his requirements and whether or not certain parts can be obtained from vendors at lower costs than if produced in his shop.

Few things are as beneficial in designing high production products as a careful study of similar products already in production. Such a study acquaints designers with what has been done and has proved economical. It also informs them as to what is feasible and what, though feasible, may involve drawbacks, whose effect on overall costs and general suitability need to be watched. Comparative product studies suggest expedients and possible new design features of great merit, and often result in lowered costs.

Economical and practical design of machine components must be based upon, to a great extent, a wide knowledge of the various basic methods of manufacture. A fundamental objective in product design should be to achieve a minimum cost of production for a specific design using the least amount of material. A process must be selected that yields a high output per hour, low tooling costs, minimum set-up time, and a product having desirable physical properties. This can only emanate from a constant awareness of the need for simplification of physical design features and a thorough evaluation of all the manufacturing alternatives available.

CHOICE OF THE PRODUCTION PROCESS

Depending upon such factors as product configuration, material, size, weight, and the desired function, the choice of a particular production process for a given part may be a surprisingly simple decision to reach. Some parts, because of their unique basic requirements, seem to fall automatically to one or perhaps two processing methods. High-volume production of parts that are restricted to a single material specification often requires few, if any, processing decisions. Examples of parts in this classification may be zinc-alloy die casting, alloy castings from investment molds, or injection-molded plastic products. In other cases, the selection of an appropriate production process which is both practical and economical can be a very difficult decision. Usually, for most product requirements, the process can be efficiently narrowed down to just a few possibilities. The final decision is generally made only after a thorough evaluation of all the important competing factors.

Deciding upon a proper design *before* the part gets into production is an absolute necessity from the dual standpoint of quality and economy. Several people are frequently involved in determining how a part should be designed and, ultimately, how it should be manufactured. While the main responsibility usually rests with the engineering department of a manufacturing firm, the purchasing department can often contribute valuable information regarding material costs and suppliers. Tool engineers can often assist the designer in evaluating suspected problem areas pertaining to secondary operations on a given part. Sales engineers may be particularly helpful in expressing feedback from customers regarding the nature of a certain desirable design feature which may render the product more useful and salable. Metallurgists play an important role in the selection of suitable materials for product applications. Considerable assistance can also be obtained by consulting with experts from various manufacturing firms in the fields of metal and plastic molding, forging, stamping, machining, and in other services.

Most experienced designers would agree that final product designs often are the result of a gradual evolution from very sketchy original concepts. There is also little disagreement that all avenues of product development should be considered prior to selecting a final design before the part is released for production.

GENERAL RULES FOR DESIGN

The achievement of sound design which incorporates the advantages of maximum economy in manufacture and functional requirements of a part is dependent upon the designer's ability to apply certain basic rules. Although their relative importance need not be in the precise order shown, these rules

1. *Keep the functional and physical characteristics as simple as possible.* Complexity of the design has a direct bearing on production costs. As a general rule, the service life of a part can be extended considerably when the design of the parts is simple and sturdy. The load-carrying capacity of a part is more predictable when the design configuration is straightforward and uncomplicated. Because the work of computing stresses within a complex structure is often extremely difficult, coupled with the fact that actual stress distribution is often obscure, the value of design simplicity cannot be overemphasized. Designers are alert for ways to reduce costs by either totally eliminating the need for a given part or by combining two or more parts. On the other hand, some complicated parts may be more economical to produce and assemble by breaking them down into simpler parts.

2. *Design for the most economical production method.* It is particularly important to keep the scrap problem in mind. The ratio of the weight of original material required to the weight of the finished piece provides a clue to possible cost reductions. This is especially true for high-volume production of parts. Weight considerations often result in a reduction of scrap costs as well as a substantial cost savings of machining time. The inherent limitations in each process may present a far greater effect on costs than

ordinarily realized if a thorough survey is first made. For example, it may be unwise to use sand castings for pressure-tight parts or where high-quality surface finish or plating is required.

Designers strive to select appropriate materials which suit the intended production method as well as the functional design requirements. Material specification for a given part should occur at an early stage. For parts whose designs require considerable machining operations, the primary consideration must be to select a free machining material. In some cases, materials costs can be entirely or at least partially offset by a reduction or total elimination of machining time. As an example, aluminum alloy forgings cost more than steel parts of similar design, but the time required for machining aluminum alloys may amount to one-third of that required for machining corresponding steel parts. Magnesium alloys provide an even greater advantage, machining time being approximately one-sixth that of steel.

If a low-machinability material must be used for a given part, consideration should be given to ways of producing the part by some process where machinability is not a factor. Parts processed by powder metallurgy or by investment casting, for example, require little, and sometimes no machining. As another alternative, the designer may select a material that can be readily formed and easily machined, followed by heat treatment as a means of obtaining optimum physical properties.

3. *Design for a minimum number of machining operations.* The fewer operations necessary to complete a part to drawing specifications, the lower the cost. Needless fancy or nonfunctional configurations requiring extra operations and material should be omitted from the design. Actually, the greatest savings obtained in designing for production usually result from a reduction in the number of separate processing operations required to complete a part. The designer must possess a high degree of understanding of the processing functions involved in each method. Problems encountered in holding or positioning certain parts in production must also be considered. All design features of a part should be carefully examined to be sure that problems do not exist which may prevent the design of adequate tooling. Lugs and special locating holes on products may be added or contour changes of sections may be varied slightly in an effort to locate and hold the part more efficiently in a jig or fixture for a processing operation. Projecting pads or certain locating surfaces are often required in processing parts, regardless of quantity requirements.

4. *Specify finish and accuracy no greater than are actually needed.* Tolerance on finish and on dimensions plays an important part in achieving a practical production design. The specification of needlessly close tolerances

and an unreasonable degree of surface finish always results in excessive and, in some cases, prohibitive costs. As a consequence, low part production rate, extra operations, high tooling costs, high rejection rates, and scrap losses invariably result. Designers must be aware of the inherent variations present in any production process which are affected by economical dimensional tolerance limits and obtainable surface finish. If a particular process is found to be unreliable for producing the accuracy or surface finish required, another process must be chosen or the part must be redesigned.

Figure 1-1 illustrates relative cost relationships for producing parts to various degrees of accuracy. Cost savings may be made in cases where the designer exercises greater care in adapting his required part tolerance to suit the conditions of a particular manufacturing process. Figure 1-2 illustrates basic metal-removal processes, their overall and normal commercial range of surface roughness, and their accompanying range of practical dimensional tolerances.

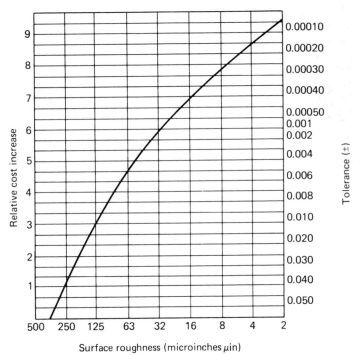

FIGURE 1-1 Cost relationship between dimensional tolerance and surface finish (cited from "Planning for Profit," *Tool and Manufacturing Engineer,* Jan. 1969. Used with permission)

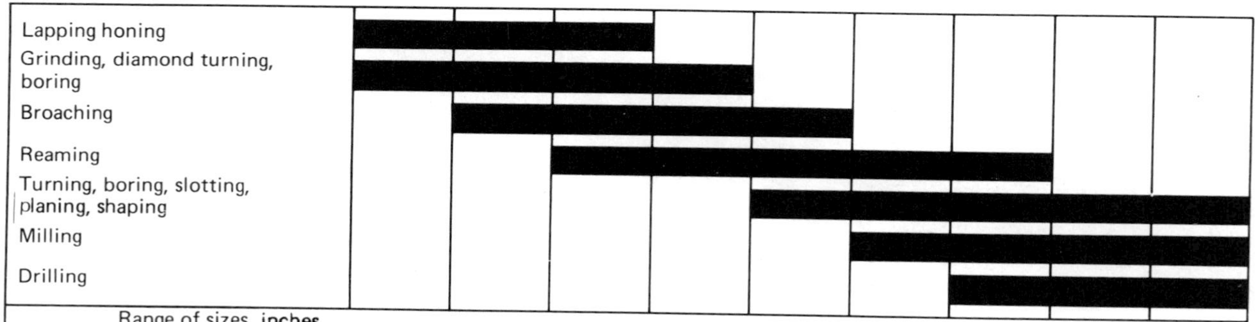

Machining processes	Range of sizes inches From	To and incl.	Tolerances (±)								
Lapping honing; Grinding, diamond turning, boring; Broaching; Reaming; Turning, boring, slotting, planing, shaping; Milling; Drilling	0.000	0.599	0.00015	0.0002	0.0003	0.0005	0.0008	0.0012	0.002	0.003	0.005
	0.600	0.999	0.00015	0.00025	0.0004	0.0006	0.001	0.0015	0.0025	0.004	0.006
	1.000	1.499	0.0002	0.0003	0.0005	0.0008	0.0012	0.002	0.003	0.005	0.008
	1.500	2.799	0.00025	0.0004	0.0006	0.001	0.0015	0.0025	0.004	0.006	0.010
	2.800	4.499	0.0003	0.0005	0.0008	0.0012	0.002	0.003	0.005	0.008	0.012
	4.500	7.799	0.0004	0.0006	0.001	0.0015	0.0025	0.004	0.006	0.010	0.015
	7.800	13.599	0.0005	0.0008	0.0012	0.002	0.003	0.005	0.008	0.012	0.020
	13.600	20.999	0.0006	0.001	0.0015	0.0025	0.004	0.006	0.010	0.015	0.025

FIGURE 1-2 Tolerance and size ranges of machining processes (cited from "Planning for Profit," *Tool and Manufacturing Engineer,* Jan. 1969. Used with permission)

8

PROCESSING CHECKLIST

Once a product has been conceived, the various methods by which the component parts can be made should be thoroughly evaluated by applying the items as listed in the processing checklist which follows. For every manufactured part there usually is an optimum cost manufacturing process. There are many acceptable methods of evaluating the relative processing capabilities. Most designers employ a method that begins by systematically examining the advantages and disadvantages of every possible process. The final step involves eliminating the obviously inappropriate processes and in narrowing down the search to a few choice alternatives which can then be researched in greater detail.

The following processing checklist consists of important points that the product designer should consider in evaluating specific process capabilities:

1. Suitability of materials
2. Properties of materials
   a. Strength (tensile, yield, shear, creep, compressive, fatigue, impact)
   b. Hardness
   c. Load deformation
   e. Resistance to corrosion
   f. Conductivity
   g. Strength-to-weight ratio
   h. Weldability
3. Dimensional accuracy
4. Typical size and weight ranges (maximum and minimum)
5. Lead time
6. Minimum and maximum production quantities
7. Production rate (i.e., pieces/hour)
8. Surface finish (as-fabricated or as-machined)
9. In-plant production capability
10. Heat treatment (if needed)
11. Inventory
12. Cost factors
    a. Influence of special desired features (complexity of part shape, inserts, coring, external surface detail)
    b. Materials (availability, scrap)
    c. Tooling (jigs and fixtures)
    d. Subsequent machining operations
    e. Subsequent finishing operations
    f. Handling equipment
    g. Gaging

9

PROBLEMS

1.1. Prepare a list of product specifications which you think are necessary before developing each of the following new products: (a) self-inking rubber stamp (b) manually operated nonelectric vacuum cleaner (c) gasoline-driven rock crusher (d) golf-ball retriever at a driving range (e) self-closing toothpaste tube (f) silencer for a power lawn mower (g) self-propelled water skis (h) adjustable child's tricycle (i) keyless lock (j) alarm device to warn of possible danger to a farm tractor on a steep incline (k) tirewasher device in an automatic car wash.

1.2. Using one or more of the products listed in problem 1.1, develop as long a list as possible of potential ideas which you feel might lead to an acceptable solution.

1.3. Compare the process of functional design to that of production design.

1.4. List five products which you feel are particularly noteworthy because of their design simplicity.

1.5. Referring to Figure 1–1, calculate the percentage decrease in cost for a part machined to a tolerance of ±0.001 (0.03 mm) instead of ±0.0004 in. (0.010 mm). Indicate the surface-roughness value for a tolerance of 0.001 in. (0.0254 mm) as compared to that for a tolerance of ±0.0004 in. (0.01016 mm).

1.6. Referring to Figure 1–2, what is the minimum tolerance for (a) a ½-in. (12.7-mm) drilled hole? (b) a 2¾-in. (69.85-mm)-wide broached slot? (c) a 5-in. (127-mm) diameter turned on a shaft? (d) a ¾-in. (19.05-mm) reamed hole?

BIBLIOGRAPHY

ARMAREGO, E. J. A., and R. H. BROWN, *The Machining of Metals,* Englewood Cliffs, N.J.: Prentice-Hall, Inc., 1969, pp. 157–173.

BELLOWS, G., *Machining—A Process Checklist,* Cincinnati, Ohio: Machinability Data Center, Metcut Research Associates Inc., 1976.

CHADDOCK, D. H., "Sparkling Ideas in the Design Process," *Engineering,* July 1974, pp. 550–552.

DE VRIES, M. F., S. M. HARVEY, and V. A. TIPNIS, *Group Technology, Cincinnati, Ohio Machinability Data Center, Metcut Research Associates Inc.,* 1976.

"Economics in Machining," *Machining Data Handbook,* 2nd edition, Dearborn, Mich.: Society of Manufacturing Engineers, 1972, pp. 859–882.

How to Design and Buy Investment Castings, Evanston, Ill.: Investment Casting Institute, 1960.

JACOBSON, R. A., "Design with Manufacturing in Mind," *Machine Design,* Nov. 14, 1974, pp. 144–149.

"Machinability," *Machining Data Handbook,* 2nd edition, Dearborn, Mich.: Society of Manufacturing Engineers, 1972, pp. 859–882.

"Machining Standards," *Machining Data Handbook,* 2nd edition, Dearborn, Mich.: Society of Manufacturing Engineers, 1972, pp. 957–966.

Modern Aspects of Manufacturing Management, Dearborn, Mich.: Society of Manufacturing Engineers, 1975, 340 pp.

NIEBEL, B. W., and A. B. DRAPER, *Product Design and Process Engineering,* New York: McGraw-Hill Book Company, 1974.

A Product Design Through the Engineering Analysis of Its End Use, G47, New York: American Society of Mechanical Engineers, 1967.

TRUCKS, H. E., *Designing for Economical Production,* Dearborn, Mich.: Society of Manufacturing Engineers, 1974.

* Abstracted with permission.

2

CAST OR MOLDED METAL PRODUCTS

Casting is the process of producing objects by filling a mold cavity by gravity or pressure with liquid metal and allowing it to cool and solidify. This historically significant process of forming metal shapes is known to have been used to make copper arrowheads as early as 4000 B.C. There is no other metalworking process which gives the designer such unlimited freedom or as many options in terms of required product complexity, size range, choice of available materials, desirable properties, range of surface texture, close dimensional tolerance or high-quantity production. Except for the -cost of the metal itself, the most significant cost is in making the molds and removing the flash, gates, and risers.

Several different important casting processes have been developed to fill specific manufacturing needs, each allowing potential advantages. Their similarity lies in the fact that each process employs a mold cavity which ultimately determines the desired shape and size of the casting. The mold may be filled either by gravity or by forcing the desired amount of liquid metal under pressure.

CASTING ECONOMIES

The economies of any particular casting process can be fully realized only when the part has been designed specifically for a particular method of manufacture. Frequently, a minor design change having a slight or no effect upon the appearance or function of the product can result in impressive savings. The ideal design for a particular casting should be determined in consultation with experienced casting specialists. Knowledgeable product designers have found that a successful casting design is usually the result of close cooperation with experts in the field, especially during initial design stages.

CASTING DESIGN

It is helpful in the design of any casting to try to visualize it inside a pair of die blocks or mold halves having cavities which give it its shape, often by using cores and movable slides and from which the casting must be removed without breakage or distortion. It is not generally necessary for the product designer to design the die or mold in which the casting will be produced. However, the design of the casting and its suitability for manufacture by a particular casting process is likely to be more logical if he can visualize what the major elements of the die or mold must be like. Further, a thorough knowledge of the techniques that are related to the particular casting process is of equal importance to the designer. As mentioned in Chapter 1, it is strongly recommended that the designer consult with experts in the appropriate field of materials processing before the design of the part has progressed so far that it becomes very costly to change. Product designers usually consult with foundry engineers, patternmakers, and foundrymen on matters pertaining to casting processes.

The design stage is considered to be the most important point in determining the ultimate performance requirements of a part. The design for a given part can vary greatly depending upon the material from which the part is to be made and the method of forming it into the final shape. What may be considered good design for one metal or manufacturing process is not necessarily good for another metal or process. It is not possible to produce a satisfactory part unless the design has been created specifically with the intent of unifying a metal and a manufacturing process to make that part. Attempts to adapt a design intended for one manufacturing process to some other process usually result in production problems and often with a part having a performance record far below expectations.

Casting design must be carefully adapted to the requirements of a particular process. It must also make the best use of the properties of the metal to be cast. It is only in this way that a cast part can yield the optimum in mechanical properties and serviceability. The basic requirements of good casting design must be followed. In cases where the design of the part violates a basic law of casting, the parts are rarely produced successfully or economically. Deciding upon the proper design *before* a part gets into production is important from the standpoint of both quality and economy.

Avoid Overdesign

Too often, certain design specifications may be unnecessarily severe in relation to the desired performance and practical life of the part. Excessive tolerances, parts designed to outlast the end product, the use of wasteful quantities of expensive metal in producing a part, and a surface finish finer than is actually required are considered expensive luxuries. Nonfunctional quality has no place in the competitive market.

Design Criteria

Designers agree there seems to be no single procedure that can be adopted for evaluating a design of a product as a basis for determining whether or not it should be cast. There are certain basic casting process selection criteria that should be considered. These are: product configuration, machinability, size and weight, material properties, dimensional tolerances, quantity, and, of course, cost.

Configuration and Quantity: Configuration and quantity are early factors to be reviewed in determining whether or not to produce a product by casting. Casting is often the only practical and economical method of producing many parts simply because of the demands imposed by the intricate exterior and interior shape requirements. A part that can be entirely machined from commercially available bar stock, for example, requiring only a minimum amount of stock removal and with a limited number of machining operations, is normally the most economical choice.

In other cases, the *material requirements* of a given part may dictate the selection of a particular casting process. For parts whose materials may be too difficult to machine or form by conventional methods, casting may be the only acceptable method of production.

Size and Weight: Size and weight of a given product are often determining factors in the decision to produce a product by casting. For products weighing less than 10 lb (4.54 kg), any of the casting processes to be discussed in the following chapters may be considered. The choice of a suitable casting process is somewhat restricted for products weighing over 10 lb (4.54 kg).

Physical Properties: Physical properties are considered to be important factors. Consultation with metallurgical engineers is recommended in the early stage of product design. It is important to be able to predict accurately the physical properties a particular alloy will have when cast by a particular process. Ferrous products are limited to sand casting, shell-mold casting, or investment casting. Recently developed techniques in die casting have

14

led to successful but limited production of products from ferrous materials. Commercial applications are predicted for the very near future. Any of the casting processes may be used for nonferrous products.

As will be explained in some detail, each of the different casting processes has definite *dimensional limitations* and must be carefully reviewed when selecting a particular casting process, especially when estimating the final cost. Certain "trade-offs" are usually necessary. For products with very precise as-cast tolerances, the savings in machining can sometimes be offset by the adoption of a more expensive casting process.

The required production quantity of a cast product generally influences the specific casting process that is ultimately selected. *Cost* is another important factor in selecting one casting process over another. The choice may normally be reduced to two or three processes. The selection of a final casting process will ultimately be dictated by the most economical method of producing the product in the desired quantity. Final determination usually follows an exhaustive survey of all the factors that contribute to the final cost of the product.

The following series of examples, Figures 2-1 to 2-12, illustrate some of the more common and important design factors, which apply to practically all the various casting processes.

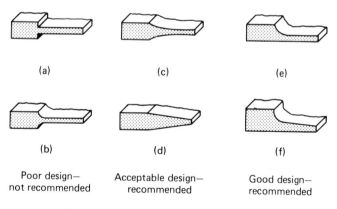

| (a) | (c) | (e) |
| (b) | (d) | (f) |

Poor design— not recommended   Acceptable design— recommended   Good design— recommended

Changes in section thickness should be gradual

FIGURE 2-1 (Courtesy, Steel Founders' Society of America)

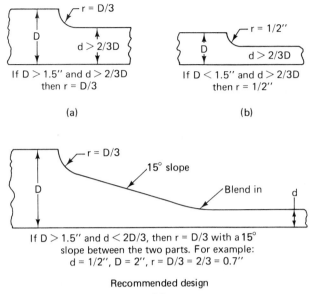

If D > 1.5" and d > 2/3D
then r = D/3

(a)

If D < 1.5" and d > 2/3D
then r = 1/2"

(b)

If D > 1.5" and d < 2D/3, then r = D/3 with a 15°
slope between the two parts. For example:
d = 1/2", D = 2", r = D/3 = 2/3 = 0.7"

Recommended design

(c)

FIGURE 2-2 (Courtesy, Steel Founders' Society of America)

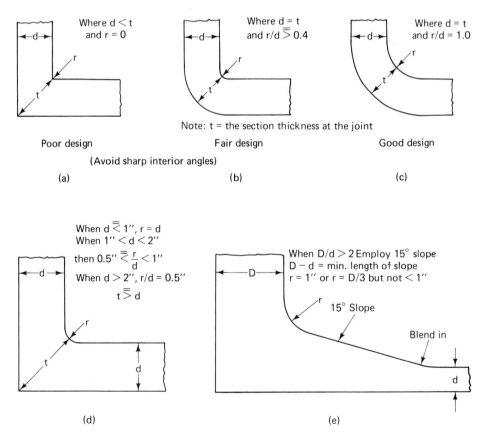

Where d < t
and r = 0

Where d = t
and r/d ≥ 0.4

Where d = t
and r/d = 1.0

Note: t = the section thickness at the joint

Poor design

Fair design

Good design

(Avoid sharp interior angles)

(a)

(b)

(c)

When d ≤ 1", r = d
When 1" < d < 2"
then 0.5" ≤ r/d < 1"
When d > 2", r/d = 0.5"
t ≥ d

When D/d > 2 Employ 15° slope
D − d = min. length of slope
r = 1" or r = D/3 but not < 1"

15° Slope

Blend in

(d)

(e)

FIGURE 2-3 L-junctions (Courtesy, Steel Founders' Society of America)

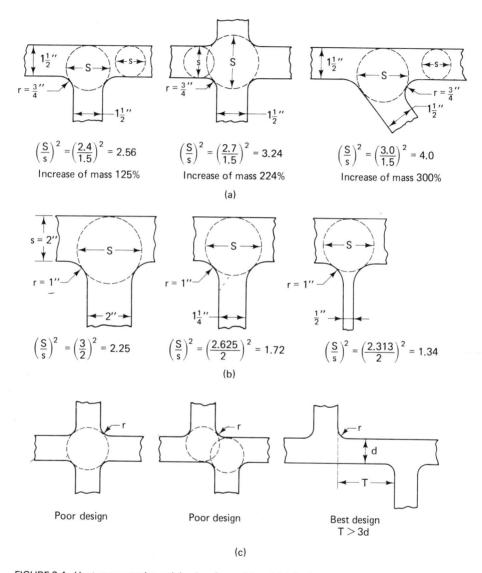

FIGURE 2-4. Heat concentration at joined surfaces: (a) and (b) the increased mass at a joint can be estimated by the inscribed method shown; (c) at 'X' junctions, offset the two parts of one section (Courtesy, Steel Founders' Society of America)

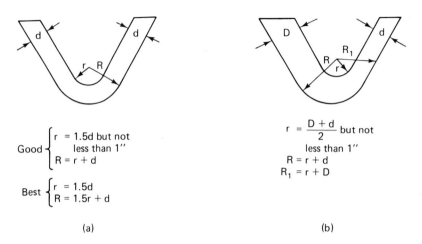

$$r = 1.5d \text{ but not}$$
$$\text{less than } 1''$$
Good $\left\{ \begin{array}{l} r = 1.5d \text{ but not} \\ \qquad \text{less than } 1'' \\ R = r + d \end{array} \right.$

Best $\left\{ \begin{array}{l} r = 1.5d \\ R = 1.5r + d \end{array} \right.$

(a)

$$r = \frac{D + d}{2} \text{ but not}$$
$$\text{less than } 1''$$
$$R = r + d$$
$$R_1 = r + D$$

(b)

FIGURE 2-5 V-junctions (Courtesy, Steel Founders' Society of America)

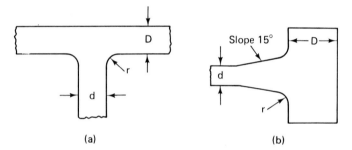

(a)   (b)

r = D, but never less than 1/2″ or greater than 1″
If D is < 1.5d; then r = D as shown in sketch (a)
If D is > 1.5d; then r = D as above with a 15°
slope to fit the radius as shown in sketch (b)

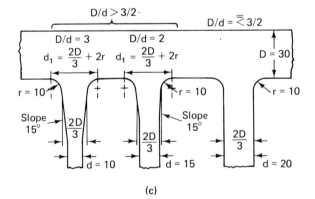

$D/d > 3/2 \cdot$ $D/d = \overline{\overline{<}} \, 3/2$

$D/d = 3$ $D/d = 2$

$d_1 = \frac{2D}{3} + 2r$ $d_1 = \frac{2D}{3} + 2r$ $D = 30$

$r = 10$ $r = 10$ $r = 10$

Slope 15° $\frac{2D}{3}$ Slope 15° $\frac{2D}{3}$ $\frac{2D}{3}$

$d = 10$ $d = 15$ $d = 20$

(c)

FIGURE 2-6 T-junctions (Courtesy, Steel Founders' Society of America)

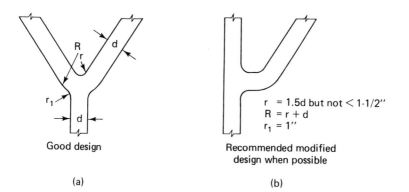

Good design

r = 1.5d but not $< 1\text{-}1/2''$
R = r + d
r_1 = 1''

Recommended modified
design when possible

(a)   (b)

FIGURE 2-7 Y-junctions (Courtesy, Steel Founders' Society of America)

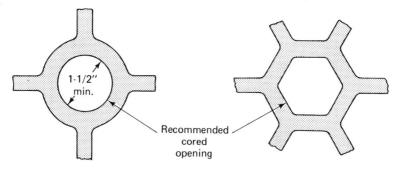

1-1/2''
min.

Recommended
cored
opening

(a)

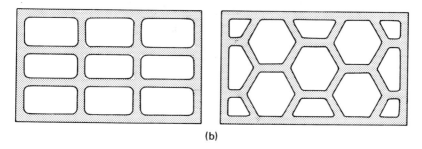

(b)

FIGURE 2-8 X-junctions (Courtesy, Steel Founders' Society of America)

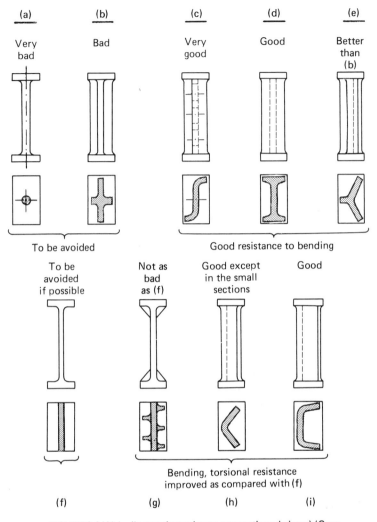

(a)	(b)	(c)	(d)	(e)
Very bad	Bad	Very good	Good	Better than (b)

To be avoided

Good resistance to bending

To be avoided if possible	Not as bad as (f)	Good except in the small sections	Good

Bending, torsional resistance improved as compared with (f)

(f)	(g)	(h)	(i)

FIGURE 2-9 Webs (lower views shown are sectioned views) (Courtesy, Steel Founders' Society of America)

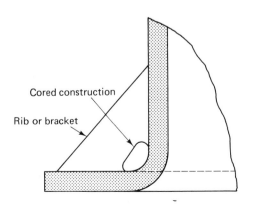

Cored construction

Rib or bracket

FIGURE 2-10 Eliminating hot spots. The sand at the junction of the rib, flange, and the body is subjected to heat from three sides. To eliminate this hot spot, a cored hole is used (Courtesy, Steel Founders' Society of America)

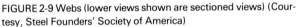

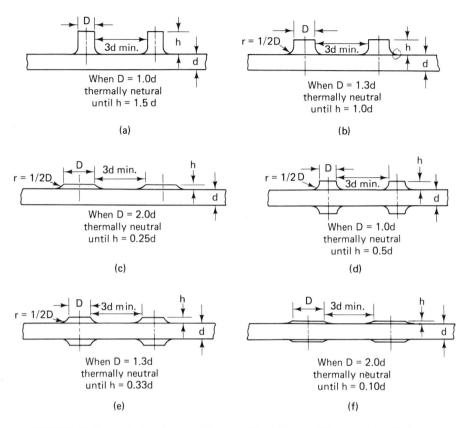

FIGURE 2-11 Proper design of bosses (Courtesy, Steel Founders' Society of America)

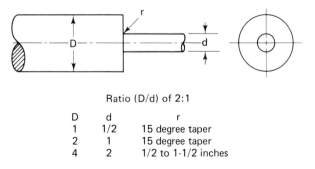

Ratio (D/d) of 2:1

D	d	r
1	1/2	15 degree taper
2	1	15 degree taper
4	2	1/2 to 1-1/2 inches

FIGURE 2-12 Joining cylinders (Courtesy, Steel Founders' Society of America)

CASTING PROCESSES

Important aspects of sand casting, shell-mold casting, investment mold casting, permanent mold casting, die casting, plaster mold casting, centrifugal casting, and slush mold casting are explained in Chapters 3 through 10. Chapter 11 contrasts the various selection criteria of each process.

PROBLEMS

2.1. Despite an early patent date (1846) the continuous casting process in its present high level state of development is considered a relatively new process. Write a paper describing the principles of the process using sketches where appropriate to illustrate important points. List the materials most commonly cast by this process and give the important advantages and limitations.

BIBLIOGRAPHY

BEELEY, P. R., *Foundry Technology,* Metals Park, Ohio: American Society for Metals, 1972.

"External Forces Mold Process Selection Decisions," *Metals Progress,* June 1974, p. 71.

KAUSERUD, H., "Guide to Vertical Gating for Flaskless Molding," *Foundry Management Technology,* Herlev, Denmark, Jan. 1975, pp. 94–96, 98.

SYLVIA, J. G., *Cast Metal Technology,* Reading, Mass.: Addison-Wesley Publishing Company, Inc., 1972.

3

SAND CASTING

Upon first examination, there would appear to be a seemingly disproportionate emphasis upon sand casting as discussed in this chapter compared to
the other casting processes found in later chapters. Sand casting is a major
casting process and, in terms of metal tonnage, it accounts for over 90% of
all metal poured. Another reason for the apparent disparity in emphasis is
that many of the terms and procedures used in sand casting apply as well
to other casting processes. The similarities and differences in molding techniques will become apparent as the various other units on casting processes
are studied. Each process offers its own potential and unique advantages.

DESCRIPTION OF THE PROCESS

Sand casting consists of making a cavity in sand with a pattern, removing
the pattern, and filling the cavity with molten metal. Following a period of
cooling, the metal solidifies and the sand is broken away. The casting is
then removed, trimmed, and cleaned. All ferrous and nonferrous metals
may be sand-cast. Sand molds are made in two sections. Figure 3-1 shows
the parts of a flask; the bottom section is called the *drag* and the top section
is called the *cope*. The joint that lies between the sections is called the *parting
line*. Molten metal is poured into a hole called a *sprue,* and connecting
runners conduct the metal to the casting cavity. Gravity causes the metal to
flow down the sprue and into the cavity. The riser ensures complete filling
of the cavity and gives additional metal to compensate for contraction on
cooling. It should be emphasized that all sand-casting molds are expendable
and must be destroyed as the solidified casting is removed from the mold
cavity. Expendable molds are sometimes referred to as "sacrificial" molds.

23

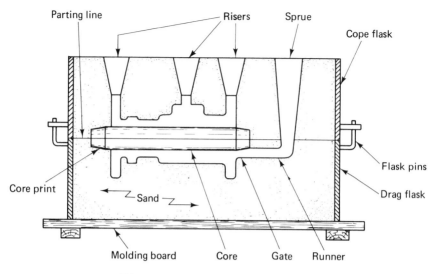

FIGURE 3-1 Sand casting mold terminology

Principal Methods

There are two principal methods of sand casting, each characterized by the type of sand that is used in the molding processes: green sand molds and dry sand molds.

Green Sand Molds: Green sand (called this because of the moisture content) *molding* is perhaps the most commonly employed method. Sand, moistened with a water binder, is used and is purposely prevented from allowing to dry out during the molding process. In fact, molten metal is poured just as soon as possible after formation of the desired mold cavity. Green sand molding is used to produce castings of intricate configurations, since green sand offers less resistance to normal mold contraction when the metal solidifies as it cools in the mold.

Dry Sand Molds: In *dry sand molds,* in which an organic binder is used, the moisture is completely removed by heating the molds in an oven. This results in a harder, stronger mold with less tendency for mold gases to form. Dry sand molds are used to produce castings which are more dimensionally accurate than those produced in green sand molds. Dry sand molds are used in the production of large, heavy castings. Used mold sand from both processes is generally reclaimed by cleaning and filtrating, and is reused.

Patterns

A basic requirement in the production of sand castings is a *pattern.* With some modifications, the pattern is a duplicate of the product to be produced. The pattern is crafted by skilled patternmakers to precise specifications. It is made slightly larger than the required final product, to compensate for the shrinkage of the metal as it solidifies and cools in the mold. Such a condition is called *shrinkage allowance.* Patterns are also modified by the application of a slight taper called *draft,* applied to those surfaces which are parallel to the direction of withdrawal of the pattern. The draft facilitates the removal of the pattern from the mold and lessens the tendency of the mold walls to break away.

Another modification to the pattern is the *machining allowance.* This is the excess stock which is added to the pattern so that the casting will be oversize at surfaces requiring finish by subsequent machining operations. Such surfaces are necessary in critical locations on products where mating surfaces contact one another or when the as-cast dimensional precision is insufficient.

Types of Patterns

There are a number of different kinds of patterns used in sand casting.

Removable Patterns: A *removable pattern* is used to make a mold by packing sand firmly around it and then removing it to form the mold cavity. Patterns such as this are used again and again. Patterns are commonly made of wood, especially when only a few castings are required. Larger numbers of castings are usually produced by using patterns made from cast iron, aluminum, magnesium, or, in some cases, rigid plastics such as phenolic or epoxy resins. Removable patterns may be made in a one-piece solid form. Patterns of this type are the easiest and least expensive to produce.

Removable patterns are also made of two pieces with split sections. *Split-type* patterns speed up the molding process because of the ease by which they may be removed from the mold. Figure 3-2 illustrates how the lower half of a split-type pattern is used to form the cavity in the drag and how the upper half forms the cope portion in the mold cavity. Tapered locating pins are often used to align the two halves of the split pattern.

Another type of removable pattern is the *match-plate pattern,* Figure 3-3. Patterns such as these are used for large-quantity production of castings. The cope and drag pattern halves are permanently mounted on opposite sides of a wood or metal plate. The match plate is inserted between the cope and drag sections of the flash and is held in position by pins or guide bars which mate with those on the flask. Sand is firmly rammed around the match-plate pattern beginning with the drag flask and then the cope. Next, the mold is disassembled and the match plate is removed. Finally, the mold is reassembled. It is common practice to mount more than one pattern on a single matchplate as a way of producing multiple castings.

Expendable Patterns: The *expendable* or *disposable pattern,* a patented process, employs a polystyrene or styrofoam pattern. This is also known as the full-mold process. The pattern is permitted to remain in the mold after ramming sand completely around it. When molten metal is poured into the mold, the pattern becomes vaporized and burns out, thus resulting in an exchange of the vaporized expendable pattern with the liquid metal. The gases escape through the permeable sand and through one or more vent holes which the molder provides. After cooling, the casting is removed by breaking away the sand. It is interesting to note that such patterns are actually one-unit, glued assemblies consisting of the pouring basin, sprue, and the runner system.

Cores and Core Prints: Cores are used in molds wherever it is necessary to produce a hole or undercut in a casting. The core consists of a firmly oven-baked rigid mixture of synthetic sand which has been bonded by a special core oil or proprietary binders. *Core prints* (see Figures 3-2 and 3-3) position and hold the core in the mold cavity. These extensions on the pattern form a recess at the outer edges of the mold cavity into which identical extensions of the core are fitted. The core is manually set or inserted into its proper position after the pattern has been withdrawn from the sand mold. It should be noted that after the casting has cooled and solidified, the cores are destructively removed from their positions in the casting.

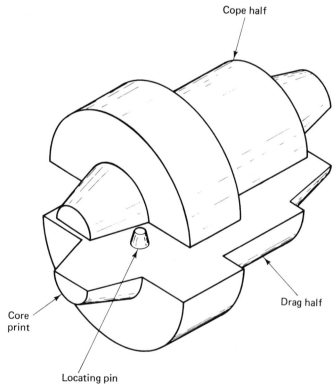

FIGURE 3-2 Split-type pattern

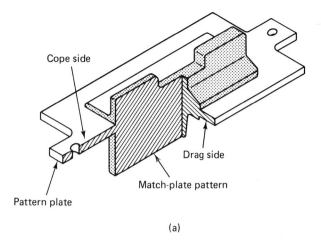

Cope side

Drag side

Match-plate pattern

Pattern plate

(a)

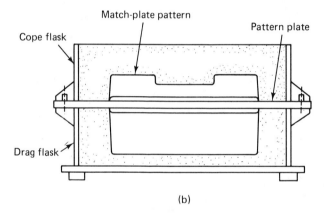

Match-plate pattern

Pattern plate

Cope flask

Drag flask

(b)

FIGURE 3-3 Match-plate pattern

PRODUCT APPLICATIONS

Figure 3-4 illustrates some product applications for sand casting. Examples of products cast by this process are not difficult to find. Sand castings are used as components for tanks and other military vehicles, railroad equipment, earth-moving equipment, aircraft, space vehicles, machine tools, low-horsepower vehicles (snow blowers, garden tractors, power lawn mowers, chain saws, etc.), automotive parts, heating and plumbing accessories, hand tools, and various hardware items.

PROCESS SELECTION FACTORS

In general, sand casting presents the designer with the option of obtaining relatively complicated shapes by taking advantage of intricate cores, special loose piece patterns, and other special molding techniques. A sand casting may eliminate the need for two or more parts which may have been previously

27

FIGURE 3-4 Typical sand castings (Courtesy, Burgess-Norton Manufacturing Co.)

joined by welding, threaded fasteners, or by other means. An important consideration which the designer must make is whether a given product might best be made by other, competitive processes, perhaps by forging or by welding. All aspects of each process must be carefully weighed to determine comparable costs for forging dies or for jigs and fixtures which are sometimes required for production weldments. All things considered, sand casting offers the least expensive method for producing general-purpose castings. Pattern equipment is relatively inexpensive and long-lasting. Pattern expense can be varied depending upon the quantities of castings required.

Physical Properties: Sand castings do not have directional properties. Since metal is not displaced after solidification, no laminated structures exist. Physical properties of strength, ductility, and toughness are equal in all directions, and designers take full advantage of this important factor. Cast iron products are used in applications where damaging vibrations exist, thus utilizing the unique damping characteristics of this material.

Materials: Any metal that can be melted can be cast in a sand mold. Metals most commonly sand-cast are cast iron, steel, brass, bronze, aluminum alloys, and magnesium alloys.

Waste: Discounting the material required for machining allowance, there is little waste of material. Metal left by the sprue and risers and the gating system are cut off and may be remelted and used again.

Size of Castings: There is almost no limit on the size and shape of products to be cast. Sand casting is an extremely versatile process. Castings weighing from a few ounces to 500 lb (225 kg) are common, while castings weighing as much as 6000 to 7000 lb (2700 to 3150 kg) have been successfully made.

Features: Parts may be cast with *openings* or *holes* as small as ³⁄₁₆ to ¼ in. (4.763 to 6.35 mm). It is possible to produce castings with *undercuts* and special inserts at additional cost.

Lead Time: In general, many simple patterns for sand casting may be constructed and made available within 2 or 3 days. In special cases, vital lead time may well be the determining factor in the designer's decision to use a casting for a particular product.

Heat Treatment: Castings may be heat-treated not only for stress relief but for grain refinement, thus increasing strength. Using certain molding techniques, a given surface may be made extra hard by employing a *chill,* shown in Figure 3-1.

Production Rates: Depending upon the sand-casting technique used and the particular molding equipment, production rates may vary widely from just a few parts per hour for large or complicated castings to 10, 50, or even several hundred or more per hour for small castings. Production techniques often include the use of multiple mold cavities, which increase the quantity output.

Some Limitations

Trial Runs: Sand casting may require one or two trial runs in order to test the castability of a given part configuration. Problems involving solidification, core shifting, sand pressure, mold growth, pouring rate, and casting shrinkage are not always predictable. Trial-and-error modifications may be required to compensate for unexpected situations. In all fairness to other casting processes, it should be noted that a certain degree of unpredictability is inherent in all casting processes.

Molding Skill: Sand castings are more subject to human control than parts made by most other casting processes. A sound casting is the direct result of an outstanding initial design coupled with the skill and expertise of the foundryman. Inadequate ramming of sand can result in sand inclusions or shrinkage cracks due to lack of collapsibility of the sand. A mold too tightly rammed may cause porosity as a result of gas entrapment. Overrapping a pattern as it is withdrawn from the mold can produce an oversize casting. Cores shifting in the mold cavity may increase the rejection rate.

Configuration: Some shapes of parts are not considered practical to make by sand casting. Long, thin projections extending from conventional shapes are difficult to produce as well as parts which, in themselves, have shapes that are long and thin. Parts with delicate thin sections are not suitable as sand castings.

Chilled-Skin Effect: Sand castings are characterized by a chilled skin, which usually results in making machining somewhat difficult. Tool life is noticeably decreased as a result of the wearing effects of the hardened skin. The sand inclusions present in most sand castings also cause faster tool-edge deterioration.

Molds: There are obvious disadvantages associated with moving and storing large and heavy sand molds for later use. Unfortunately it is impractical to stockpile previously prepared green sand molds, which, if not used almost immediately after preparation, dry out and lose their moisture content.

Casting Cleaning: For nonferrous castings, cleaning does not present much of a problem. Sand has less tendency to cling to the surfaces of such castings. After the gate and sprue have been cut off, the castings are generally machine-brushed or shot-blasted in preparation for the required machining operations. Iron and steel castings, however, are covered with a coating of sand and scale. Any one of several methods explained in Chapter 40 may be employed to clean ferrous castings. Depending upon their size, they may be rotated in a tumbling mill, sand-blasted, or hydraulically cleaned by a stream of water under high pressure. Some castings may require further refinement by chipping or grinding to remove certain excesses of unwanted metal. While most cleaning operations are accomplished within a short time, such operations add to the cost of producing sand castings.

PRODUCT DESIGN FACTORS

In general, regardless of the casting process being considered, there are two basic considerations: (1) Will the product do the job for which it is intended? (2) Can it be practically and economically cast?

Once it has been determined that a sand casting will do the job, considerable attention must be devoted to the special design requirements of the casting configuration. Since all casting processes are essentially a freezing process, the shape of the casting and the resulting mold cavity directly affects the progress of solidification. The designer must be concerned with the distribution of the molten metal. If the channels that feed the casting are uniform in section, the molten metal will freeze uniformly and quickly. It is helpful to visualize the metal in the cavity shrinking constantly as it freezes from the liquid to the solid state, cooling in layers from the outside inward. Thin sections will be frozen first while the thicker sections remain molten. If the transition between the thicker and thinner sections is abrupt, the contracting thin section draws metal from the still-molten thicker sections. Stress concentrations may tend to build up at locations where the sections intersect. If, however, the transition is gradual, undesirable stress concentrations are reduced or, in most cases, totally eliminated.

Metal that is well distributed in the mold cavity will solidify in a uniform manner, producing a consistent structure throughout the casting. A flow line must be carefully established as the metal is poured to provide a path of stress distribution for the product when it is in service.

An inherent advantage of a part produced by casting lies in the fact that designers have almost unlimited freedom of shape. They can use practically any configuration that will provide maximum load-carrying ability with minimum weight. There is usually no reason for having unnecessary metal in the original design. In-service, functional loads seldom produce uniformly distributed stresses. Therefore, both the casting shape and wall thickness can be varied accordingly to allow for applied stresses in locations where additional support is actually needed on the product.

Firm rules and formulas cannot be used to solve the many variables usually encountered in a typical casting design situation. However, most of the serious problems can be eliminated if the designer considers the layout of a casting as he/she might a hydraulic system. In this way, he/she can better anticipate and make corresponding allowances in the original design so that the flow of hot metal will occur smoothly and without interruption throughout all sections of the mold cavity. Casting engineering is a specialty. Decisions involving special types of patterns, areas affected by gating and feeding, dimensional limitations, machining allowances, pattern draft, surface projections, coring for both interior as well as exterior detail, fillet radii, section uniformity for stress distribution, and other considerations are often made in consultation with foundry engineers.

Regardless of the type of casting process being evaluated for a particular product application, there are some well-established, generally agreed upon, guidelines which apply to good casting design principles which we shall now describe.

Wall Thickness: Unrealistic specification of section thickness may lead to additional production costs. A thickness down to about $\frac{3}{16}$ in. (4.763 mm) is practical and not difficult to produce in aluminum-alloy sand castings. Walls measuring $\frac{3}{32}$ in. (2.3813 mm) can also be cast in cast iron and in copper, but special attention must be given to the mold preparation, venting, pouring techniques, and other factors. Designers often make certain "trade-offs" in special cases. For example, a part requiring a ⅛-in. (3.18 mm) wall may be more economical to cast oversize at 3⁄16 in. (4.763 mm) or more and later machined to the ⅛-in. (3.18 mm) dimension, despite the need for excess material and additional machining time.

Successful sand castings are designed with a minimum uniform wall thickness so as to eliminate possible distortion and internal stresses. Castings should have sections which are no thicker than necessary to meet the required stress imposed upon the size of the section thickness. It is important that the section thickness will not be so thin as to impede the flow of metal and allow it to solidify. Where variations in section thickness are absolutely necessary, the transition should consist of a gradual blend in an effort to reduce shrinkage stresses.

Bending Stresses: Once it has been decided to produce a product by casting, the designer should reevaluate carefully the functional requirements of the part to determine the effect of in-service bending stresses. Figure 3-5a shows how a compressive stress may occur at one face of a beam, and a tensile stress at the opposite face. A position approximately midpoint betwen them, called the *neutral axis,* is the least-stressed area of the part. Figure 3-5b shows how the designer may concentrate metal away from the neutral axis *toward* the edges of the part as in Figure 3-5c. In this manner, a favorable distribution of stresses is obtained.

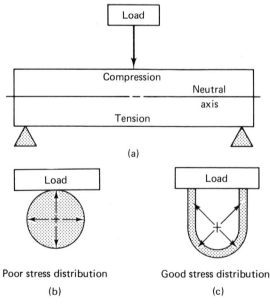

FIGURE 3-5 Bending stresses

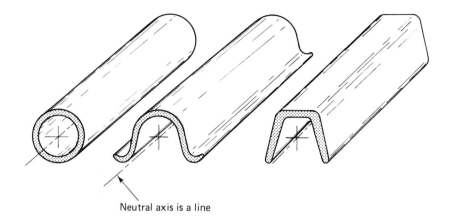

Neutral axis is a line

FIGURE 3-6 Torsional stresses

Torsional Stresses: For a product subject to in-service torsional stresses, the *neutral axis,* is the least-stressed area of the part. Figure 3-5b shows how the designer distributes metal *away* from the neutral axis and *toward* the perimeter of the part. The tubular cross section, shown in Figure 3-6, would offer the greatest resistance to twisting, but since it requires a core, a trade-off might be made in favor of one or the other more easily obtainable shapes.

Sharp Angles: Sharp angles are to be avoided. Fillets streamline the metal flow and distribute stress concentrations as shown in Figure 3-7. It is much more difficult to maintain a satisfactory flow of metal around a sharp or restricted corner than it is around a curved section. The recommended transition curve between two sections joining at a 90° angle is an arc with a radius at least equal to the thickness of the joining section. The outside radius is made proportionally larger to keep the section thickness uniform. Designers specify gentle contours on products wherever possible. To specify excessive fillet radii is considered just as bad as specifying radii that are too small.

Ribs, Beads, and Flanges: Stiffening members are used to reduce the weight of a casting and to lessen the tendency to warp. Such features increase the overall strength and rigidity of a part without increasing the section thickness. They are usually made to about 80% of the section thickness and are rounded and filleted. There is a tendency for some designers to add more ribs and stiffening members to a casting than is necessary. Such features should only be added after carefully evaluating the net effect upon the additional cost of patternmaking and improvement in the strength of the casting.

If the flat area on a product is subjected to in-service tensile stress, a rib should be avoided. Any rib involves a t-junction. Metal cast to this shape does not solidify as well as does a section of uniform thickness. Corrugation

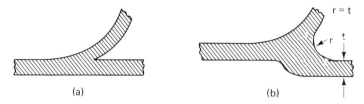

FIGURE 3-7 Use of fillets on sharp angles

keeps the section thickness uniform. When a product is subjected to in-service tension, the designer attempts to spread stresses over as wide an area as possible. Ribs interfere with such a distribution pattern. Stresses tend to concentrate in the narrow cross section of the rib, thus increasing the possibility of fatigue failure. A corrugation actually increases the area over which the tensile stresses can be distributed. Ribs should be used, however, when compressive loads are applied to flat areas on a casting. In this situation the compressive strength of the metal, which is greater than the tensile strength, carries much of the load.

Multiple Sections: When more than three walls join at one common location, the practice is to stagger the joint (see Figure 2-8). Designers may also specify a cored hole through the joint intersection for cooling purposes (see Figures 2-8 and 2-10). Features such as bosses, lugs, or pads are avoided where possible, since they also create hot spots.

Cores: Side cores on parts that cannot be supported on both ends should be avoided unless absolutely necessary. It may be more practical and economical to machine out cavities in spite of an increase in material and machining costs as compared to coring a particular part. When cores must be used, it is important to remember a satisfactory method for supporting the cores in the mold cavity must be found. Core prints (shown in Figures 3-1 and 3-2) may usually be adequately supported from opposite sides of the mold cavity. Other devices, called *chaplets,* Figure 3-8, may be used by the molder in special cases to support a core in the desired position. Adequate clean-out holes in cored areas should be provided. Whenever possible, such holes should be located away from the highly stressed areas of the part. Recommended minimum diameters for cored holes are given in Figure 3-9. If small holes are needed, coring should be omitted and the holes drilled by a later operation. Figure 3-10 gives the recommended lengths of cored openings.

Parting Lines: Considerable attention is devoted by the patternmaker to the position of the parting line. Insofar as possible, straight parting lines are used. Irregular parting lines require considerably more difficult molding techniques.

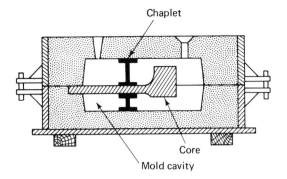

FIGURE 3-8 Use of chaplets

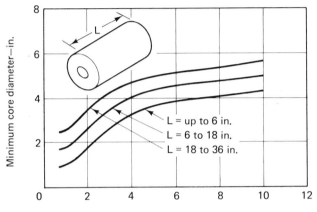

Section thickness surrounding core—in.

(a)

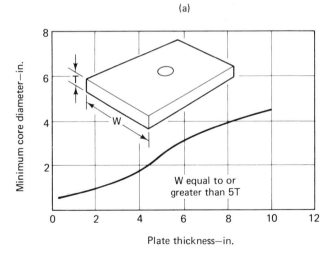

Plate thickness—in.

(b)

FIGURE 3-9 Minimum core diameters

35

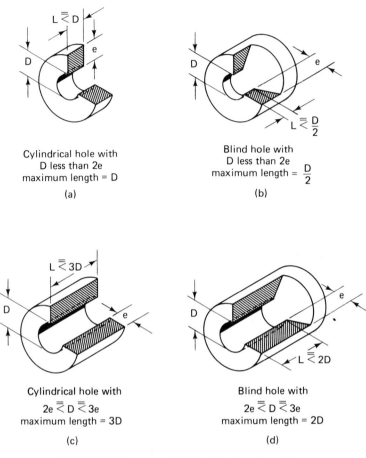

Cylindrical hole with
D less than 2e
maximum length = D

(a)

Blind hole with
D less than 2e
maximum length = $\frac{D}{2}$

(b)

Cylindrical hole with
$2e \stackrel{=}{<} D \stackrel{=}{<} 3e$
maximum length = 3D

(c)

Blind hole with
$2e \stackrel{=}{<} D \stackrel{=}{<} 3e$
maximum length = 2D

(d)

FIGURE 3-10 Recommended lengths of cored openings

Draft: Patternmakers generally provide a liberal amount of draft on patterns for sand casting. Draft usually averages about $\frac{1}{16}$ in./ft (5.207 mm/m). As a general case, draft allowance on internal surfaces exceeds that for external surfaces as a convenience in molding.

Machining Allowance: The amount of machining allowance varies with the metal used, the configuration of the casting and its size, mold characteristics, the tendency of the casting to warp, and the subsequent machining methods required. All such variables should be carefully evaluated.

Surface Texture: The physical characteristics of the molding sand, the quality of the pattern surfaces, and the molding techniques are the primary factors that govern the as-cast surface texture of sand castings. Improved surfaces [100 microinches (μ in.) for example] can be obtained with a layer of special facing sand which is rammed in intimate contact with the pattern surfaces, but which is backed with regular molding sand. Aluminum sand castings have better surface texture than sand castings of the heavier metals.

Since the requirements of surface texture on castings directly influence cost, the designer should carefully consider the necessity of specifying an excessively smooth surface. If such a condition is the principal objective in determining which of the casting processes to select for a given product, consideration should be given to a process other than sand casting where a superior as-cast surface texture may be obtained more economically.

Tolerances: One important measure of good casting design is how the designer selects and uses realistic tolerances. No matter how efficient the foundry, excessively close tolerances add needlessly to the cost. In sand casting, a tolerance of ± 0.030 in./in. (0.03 mm/mm) can be cast, but a close tolerance such as this usually results in a higher rejection rate, which, in turn, results in higher costs. If a close tolerance is necessary for one dimension, it does not logically follow that it must be similarly specified for all dimensions. With each additional tolerance dimension, the probability of a higher rejection rate rises dramatically.

In general, a tolerance of $\frac{1}{16}$ in. (1.588 mm) for the first inch is practical for small castings, but increased tolerances are required for larger castings. Tolerances of $\frac{1}{32}$ in. (0.794 mm) for the first inch are possible in special cases at additional cost. It should be noted that the closest tolerances can be held only for those dimensions which lie entirely in one part of the mold. A greater tolerance will be required for dimensions that bridge or extend from one part of the mold to another. The reason for this condition is that the dimension across the parting line is subject to variations caused by closing a mold. Other variations on the part may be due to cores that shift from their original position.

Dimensions of sand castings may also vary according to the degree of tightness with which the sand is packed into the mold, by the operation of withdrawing the pattern from the sand, and by both the temperature of the molten metal and the speed with which it is fed into the mold cavity.

REVIEW QUESTIONS

3.1. Explain the following terms as applied to sand casting: (a) drag (b) cope (c) parting line (d) sprue (e) runners (f) riser (g) pattern (h) draft (i) core (j) core print (k) chill (l) fillet.

3.2. Compare green sand molds to dry sand molds.

3.3. What is meant by shrinkage allowance?

3.4. What is meant by machining allowance?

3.5. Compare a removal-type pattern with a match-plate pattern.

3.6. Describe the main advantage of an expendable pattern.

3.7. Name some considerations in deciding whether to make a part by the sand-casting process.

3.8. Describe a procedure for producing multiple mold cavities.

3.9. Explain why it is usually difficult to produce long, thin projections extending from parts.

3.10. Explain why it is advisable to design sand castings with the least amount of wall thickness.

3.11. Explain the steps involved between the time a casting is poured and the time it is first machined.

3.12. Describe some of the considerations in designing a casting for minimum weight.

3.13. Make a brief outline describing how a typical mold is prepared. Make a sketch and label the principal parts.

PROBLEMS

3.1. Calculate the approximate cost of the castings illustrated in Figures 3-P1(a), 3-P1(b), 3-P1(c), and 3-P1(d) in accordance with the data given in the following table. Some irrelevant dimensions on the castings have been intentionally omitted.

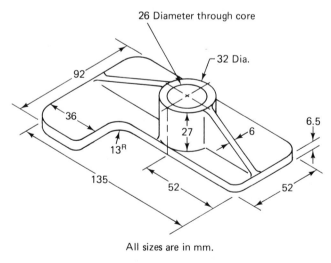

All sizes are in mm.

FIGURE 3-P1 (a)

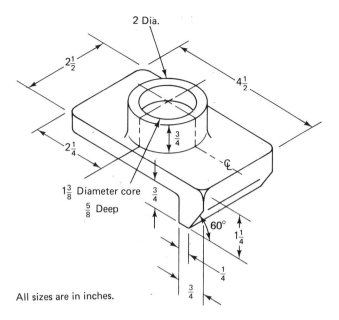

2 Dia.

$2\frac{1}{2}$

$4\frac{1}{2}$

$\frac{3}{4}$

$2\frac{1}{4}$

℄

$1\frac{3}{8}$ Diameter core $\frac{3}{4}$
$\frac{5}{8}$ Deep

60°

$1\frac{1}{4}$

$\frac{1}{4}$

All sizes are in inches.

$\frac{3}{4}$

FIGURE 3-P1 (b)

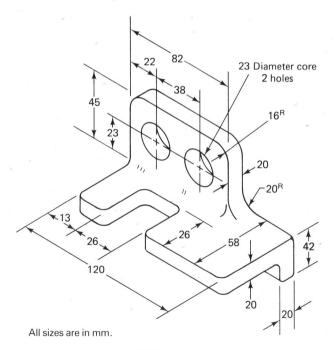

82

22

23 Diameter core
2 holes

38

45

16^R

23

20

20^R

13

26

26

58

120

42

20

20

All sizes are in mm.

FIGURE 3-P1 (c)

39

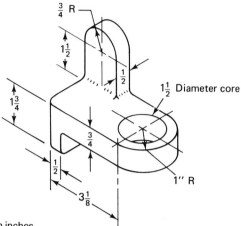

$\frac{3}{4}$ R

$1\frac{1}{2}$

$\frac{1}{2}$

$1\frac{3}{4}$

$1\frac{1}{2}$ Diameter core

$\frac{3}{4}$

$\frac{1}{2}$

1" R

$3\frac{1}{8}$

All sizes are in inches.

FIGURE 3-P1 (d)

Data for problem 3-1.	3-P1(a)	3-P1(b)	3-P1(c)	3-P1(d)
Size of fillets and rounds[1]	3 mm	1/8 in.	3 mm	1/8 in.
Quantity of castings required	5000	750	1500	500
Material[2,3]	Iron	Steel	Iron	Steel
Time Required[4]:				
to make pattern (5 required)	3 hr each	2½ hr each	2½ hr each	2 hr each
to make corebox and core	1 hr	1 hr	1 hr	1 hr
to make mold	5 min	5 min	7 min	5 min
to set core (s)	2 min	2 min	3 min	2 min
pouring	1 min	1 min	2 min	1 min
shakeout	3 min	3 min	3 min	3 min
cutting sprues and risers	3 min	3 min	3 min	3 min
additional handling: i.e. cleaning, etc.	5 min	5 min	6 min	5 min
cooling[5] (50 castings at a time)	2 hr	2 hr	2 hr	2 hr

[1]Weight should be increased by 2% for fillets.
[2]Cast iron weighs approximately 0.26 lb/in.[3] and costs $0.40/lb.
[3]Steel weighs approximately 0.283 lb/in.[3] and costs $0.60/lb.
[4]Labor and overhead charge is $20/hr.
[5]Cooling time is charged at $8/hr.

3.2. Make a tracing of the castings shown in Figure 3-P1(a) to 3-P1(d). Using a pencil with a red lead, indicate in each case the best location for the parting line. Use the symbol ℙ at the ends of the parting line.

3.3. Prepare a sketch of the sand casting shown in Figure 3-P1(b) to illustrate how the casting would appear in the flask with a core and split pattern if necessary. Label all of the important elements of the mold, i.e., parting line, sprue, etc.

3.4. Using a freehand sketch show how the various casting designs illustrated in Figure 3-P4 may be improved. Explain the reason for your choices.

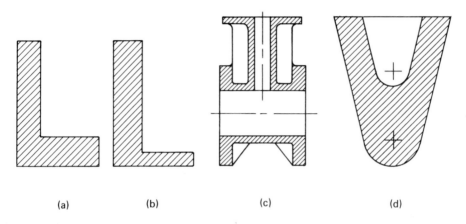

(a) (b) (c) (d)

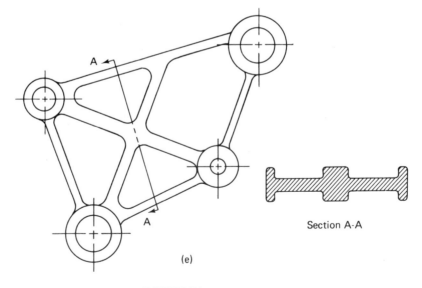

Section A-A

(e)

FIGURE 3-P4

3.5. Figure 3-P5 shows some possible designs for castings consisting of two end walls and a connecting center section. Indicate which designs are to be avoided and give your reasons.

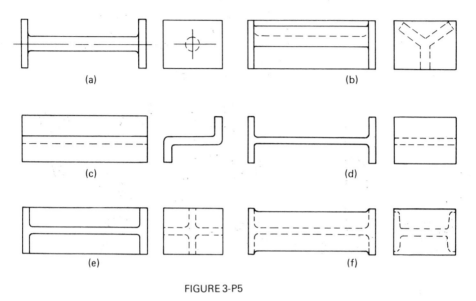

(a)   (b)

(c)   (d)

(e)   (f)

FIGURE 3-P5

BIBLIOGRAPHY

DOYLE, L. E., *Manufacturing Processes and Materials for Engineers,* Englewood Cliffs, N.J.: Prentice-Hall, Inc., 1961, pp. 126–173.

Foundry Sand Handbook, 7th edition, DesPlaines, Ill.: American Foundrymen's Society, 1963.

Gating and "Risering" Gray and Iron Castings, Report on Research Project, Des-Plaines, Ill.: American Foundrymen's Society, 1962.

HINES, C. R., *Machine Tools and Processes for Engineers,* New York: McGraw-Hill Book Company, 1971, pp. 49–72.

JOHNSON, H. V., *Manufacturing Processes,* Peoria, Ill.: Chas. A. Bennett Co., Inc., 1973, pp. 85–122.

"Sand Casting," *Metals Handbook,* Vol. 5, *Forging and Casting,* 8th edition, Metals Park, Ohio: American Society for Metals, pp. 155–181.

Sand Castings—Aluminum and Magnesium, Rockleigh, N.J.: Arwood Corporation.

Steel Casting Design—Engineering Data File, Rocky River, Ohio: Steel Founders' Society of America.

*Abstracted with permission.

4

PLASTER MOLD CASTING

The process of plaster mold casting requires that a new mold be prepared for each casting produced and, as such, the process is restricted to making one casting at a time. The mold material is basically composed of a loose slurry mixture consisting of 70 to 80% gypsum plaster, 20 to 30% fibrous strengthener, and water. In some cases, other additives are necessary, some that prevent the mold from cracking and others intended either to hasten or retard the setting time.

DESCRIPTION OF THE PROCESS

Patterns are made of highly polished yellow brass, plaster, molded rubber, phenolic resin, or wood with pores carefully sealed and highly finished surfaces. Single, loose patterns are used for experimental or prototype parts. For production quantities, patterns are mounted on bottom plates with two-piece match plates similar to those used in sand casting. One or more patterns may be assembled on the plates. Cores are used in a way similar to those which are used in the sand-casting process.

Plaster mold casting starts with the placement of the bottom plate in the flask. To facilitate removal of the patterns from the mold, a thin film of parting compound is first sprayed over the surfaces of the pattern. The plaster slurry mix is then poured into the flask directly over the pattern.

43

FIGURE 4-1 Separating the pattern from the mold (Courtesy, United States Gypsum Co.)

FIGURE 4-2 Pouring the molds (Courtesy, United States Gypsum Co.)

The mold is then vibrated or slightly jolted to aid in permitting the plaster slurry to fill all the small cavities in and around the pattern. After the plaster has set, which is usually only a matter of minutes, the pattern is carefully removed from the plaster mold, as in Figure 4-1. Both the cope and drag molds are made in this manner. A parting line is established as in sand casting. The molds are baked in a drying oven for a number of hours to remove all the moisture. Cores are next set in place if required. The castings are then formed by gravity-feeding molten metal into the mold cavities, as shown in Figure 4-2. After cooling and solidifying, the completed castings are removed by breaking open the mold.

44

FIGURE 4-3 Typical plaster mold castings (Courtesy, United States Gypsum Co.)

PRODUCT APPLICATIONS

Figure 4-3 illustrates some common applications of plaster mold castings. The process is particularly suitable as a method for the development and tryout of components that will later be produced by other precision casting methods. However, more and more important product applications are being found for this casting process. Its use is especially well adapted for producing products with hard-to-machine surfaces, such as irregularly shaped exterior surfaces and where a superior as-cast surface is important. Commonly produced plaster molded products include aluminum match plates, cores and coreboxes, miscellaneous parts for aircraft structures and engines, plumbing, automotive, household appliances, hand tools, toys, and ornaments.

PROCESS SELECTION FACTORS

Leading Advantages:

The process is particularly well suited to producing parts requiring high-dimensional accuracy, superior as-cast surface texture, thin walls, and intricate detail. In some cases, parts made by this process cannot be produced by any other casting process, particularly those parts which require complicated coring.

Quantity: When compared to aluminum die casting, plaster mold casting is considered economical if the total number of parts produced is less than 1000. A production run up to 2000 parts may be considered economical if the part configuration is especially complex. Plaster mold castings can also compete favorably with parts that have been entirely machined from stock, particularly if the parts require extensive profile milling in three planes or where extensive internal machining operations must be performed. Thin, plaster mold castings can also compete with sheet-metal assemblies where high precision is required or where distortion due to welding becomes a problem.

Wall Thickness: Because of the low thermal conductivity of plaster, the metal does not chill rapidly. Walls may be cast as thin as 0.020 in. (0.51 mm).

Structure: High-grade castings are obtained with fine-grained structures which are free from massive shrinkages, hot tearing, or blow holes.

Weight: Castings weighing less than 1 lb (0.45 kg) can be cast in plaster molds, although successful castings weighing as much as 3000 lb (1350 kg) have been produced in plaster molds.

Size: As a general rule, the process is restricted to castings measuring not more than 6 in. (152.4 mm) in any direction.

Inserts: Dissimilar metal inserts can be conveniently set and cast at the same time as the cores. This is the period in the casting cycle when the metal is in a semifluid state in the plaster mold just before setting takes place.

Tolerances: The process can compete with the accuracy of die-cast parts and is capable of remarkably close limits of accuracy. Tolerances of ± 0.005 in./in. (0.005 mm/mm) across the parting line are practical.

Surface Quality: Of all the metal-casting processes, plaster mold casting results in parts with the best as-cast surface texture, which may range between 30 to 50µ in. Plaster molded products often resemble parts produced by the die-casting process.

Some Limitations

The process is suitable only for casting nonferrous metals. The sulfur in the gypsum slurry reacts chemically with ferrous metals. Unsuitable casting surfaces result from mold deterioration under high temperatures. Aluminum is the most commonly employed metal for product applications. Yellow brass, zinc, magnesium, and some copper alloys can also be cast in plaster molds. Plaster molds are not reusable. When compared to metal molds, which may be used again and again, plaster-mold costs are generally higher.

PRODUCT DESIGN FACTORS

In general, casting design principles which apply to other casting processes also apply when designing plaster molded parts.

Machining Allowance: Plaster mold castings are often preferred over sand castings, because it may be possible either to completely eliminate or drastically reduce machining or finishing operations. When the required tolerances between related surfaces of a given part are more precise than possible to produce by casting, a machining allowance of $\frac{1}{64}$ in. (0.397 mm) to $\frac{1}{32}$ in. (0.794 mm) per surface is generally added.

Fillets and Rounds: Radii of 0.020 in. (0.51 mm) are sufficient for this process.

REVIEW QUESTIONS

4.1. Make a brief outline describing how a typical plaster mold is prepared. Make a sketch and label the principal parts.

4.2. Prepare a table contrasting sand casting to plaster mold casting from the standpoint of (a) minimum wall thickness (b) as-cast tolerance (c) surface quality (d) machining allowance.

4.3. Contrast the principal limitations of sand casting to those of plaster mold casting.

4.4. What are some of the advantages of plaster mold casting?

PROBLEMS

4.1. Pattern extraction in the plaster mold casting process may sometimes be difficult to achieve when compared to the relatively easy withdrawal of patterns from molding sand. Generally, a properly prepared and lubricated pattern may be easily released by using air pressure applied to the mold at the correct location. Give some other important methods which may also be employed to successfully extract a pattern.

4.2. In actual practice, an excess of water is added to the slurry mixture to promote the finest possible replication of detail from the pattern surfaces. A drying operation is necessary to drive off the free moisture in the mold before pouring the metal. The oven drying temperature will vary with the type of the oven used but generally it will be from 250° to 500°F (120° to 240°C). List some methods which could be used to determine when the plaster mold is sufficiently dried.

4.3. It is advisable to maintain casting temperatures of heavy castings to below 2000°F (approximately 1100°C) and to cast metal in plaster molds at as low temperatures as possible. The metal should be well skimmed and cast without delay. Molds should not be disturbed until the casting is completely solidified. Explain why each of these recommendations must be followed.

4.4. Explain why solidification time is much longer for a plaster mold casting than for a similar part made in a sand mold. Give a method to shorten the time required between pouring and shakeout for the plaster mold casting process.

4.5. Flatness and squareness on plaster mold castings can be held to a tolerance of ±0.002 in./in. (±0.05 mm/25.4 mm). Calculate the maximum amount of deviation in size which should be expected across the diagonal corners on the 12 × 28½ in. base surface on the casting shown in Figure 4-P5.

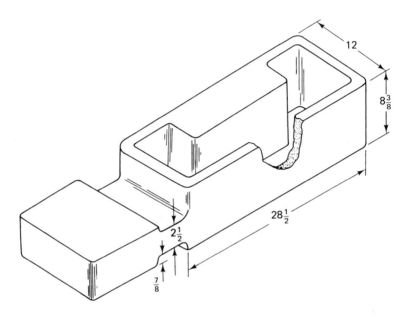

FIGURE 4-P5

4.6. Make a sketch of Figure 4-P5 and show a suitable location for a parting line. Also indicate where the draft should be applied to the pattern.

4.7. Show how the design of the channel section in Figure 4-P7 may be improved by reinforcing.

4.8. Prepare a sketch to show how the use of a core can reduce the accumulation of mass within the encircled region shown on Figure 4-P8.

FIGURE 4-P7

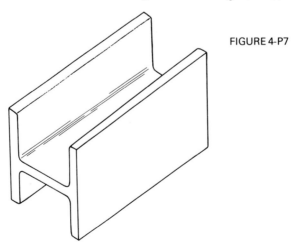

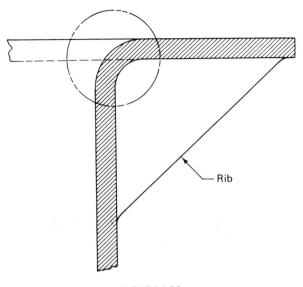

FIGURE 4-P8

BIBLIOGRAPHY

JOHNSON, H. V., *Manufacturing Processes,* Peoria, Ill.: Chas. A. Bennett Co., Inc., 1973, p. 141.

"Plastic Molding," *Metals Handbook,* Vol. 5, *Forging and Casting,* 8th edition, Metals Park, Ohio: American Society for Metals, pp. 222–236.

5

SHELL MOLDING

The shell mold casting process, as a production method, is relatively new in the United States, dating back to about 1949. It was developed in Germany by Johannes Croning about 1937. Shell molding is basically a modification of the sand molding process. Any metal that can be poured in a sand mold can be poured in shell molds.

DESCRIPTION OF THE PROCESS

Shell molding involves the use of metal match-plate patterns similar to the cope and drag patterns which are used for green sand casting. Castings are formed by pouring molten metal into an opening in an ovenbaked rigid shell which serves as the mold cavity. The shell wall thickness may vary from ⅛ in. (3.18 mm) to ⅜ in. (9.53 mm) thick. The mold cavity is formed by the space left between two matching shell halves which are clamped or glued together. As in the sand-casting process, the mold is expendable and must be broken away to facilitate removal of the solidified casting.

The process consists of the following sequence:

    1. *Prepare the metal match plate.* Figure 5-1 illustrates a typical match plate. Most patterns are comparatively expensive to produce because of special requirements regarding the dimensional accuracy and smooth surface of the

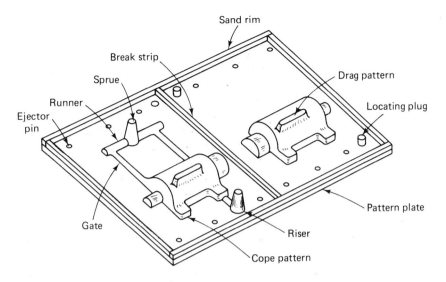

FIGURE 5-1 Typical match-plate pattern

product. Match plate patterns are made by metal patternmakers or by tool or die makers. Gates and runner systems are connected to one or sometimes several patterns mounted on the same plate to form a cluster of similar or different parts, which results in the simultaneous production of multiple castings. Shell-mold match plates are generally machined from iron or aluminum, bronze or steel to meet the exacting production requirements.

2. *Mix the resin and sand.* A dry fine sand, usually washed silica, is thoroughly mixed together with about 5% of a thermosetting phenolic resin. The resin serves as a binder.

3. *Heat the pattern.* The pattern plate is preheated to 450 to 550 °F (approximately 232 to 288 °C) and sprayed with a release agent, usually silicone.

4. *Invest the pattern.* The resin–sand mixture is placed into a specially designed dump box on which the heated metal matchplate is clamped. The pattern surface is placed toward the opening on the box. The box is then inverted, causing the sand–resin mixture to drop onto the heated pattern surface. The thickness of the adhering shell is dependent upon the length of time the mixture is allowed to remain in contact with the preheated pattern. The time generally does not exceed 30 to 45 seconds. A rollover cycle completes the investment step. The box is then returned to its original position, thus causing any excess resin–sand mix to fall off. The metal match plate now has a "soft" shell of the desired thickness which covers the entire pattern surface. It should be mentioned that investing the pattern may be accomplished by a shell blowing method. In this case, the heated pattern is invested by blowing the molding material onto the desired surface. One advantage of this method results from the fact that the mixture is deposited more uniformly because the desired quantity can be more carefully controlled. Also, the resin–sand mixture tends to pack more firmly into crevices and abrupt changes of pattern configurations.

51

5. *Cure the shell.* The pattern, with its undisturbed shell still attached, is next placed in an oven when the shell is hardened by curing. An oven temperature of 900 to 1400 °F (approximately 480 to 760 °C) over a 20- to 40-second interval is usually sufficient for this step.

6. *Remove the investment.* Figure 5-2 illustrates how the hardened shell or investment is removed or stripped from the pattern by means of special spring-activated ejector pins which have been designed as an integral part of the pattern plate. In some cases, special equipment operating independently of the pattern plate, by mechanical means, is used to remove the invested shell. Two considerations are important in this step: (1) the process must be rapid, and (2) precautions should be taken to prevent damage or warpage caused by unequal pressure of the pins as they press against the hard, but brittle shell.

7. *Insert cores.* Cores, if necessary, are set in place by hand. The core material is often the same as the resin–sand shell mixture.

8. *Repeat the process for the other half of the metal match plate.* Two mating shells must be produced which, when assembled, act as a complete closed mold.

9. *Assemble the shells.* The mating shells are assembled by clamping or bonding them together at the parting line with a quick-acting thermosetting resin adhesive such as phenol–formaldehyde, as shown in Figure 5-3.

FIGURE 5-2 Stripping the shell from the pattern (Courtesy, Cooper Alloy Corp.)

FIGURE 5-3 Applying the bond prior to shell assembly (Courtesy, Cooper Alloy Corp.)

10. *Pour the mold.* With the two halves securely locked together, the molds are positioned with the mating faces held either vertically or horizontally, as shown in Figure 5-4. To offset possible effects of pouring pressures or the possibility of tipping, which might be unsafe or damaging to the mold, the mold may be supported or embedded into various kinds of enclosures. A backup material of shot, plain sand, or gravel may be piled against the outer shell walls as reinforcement. Molten metal, which is gravity-fed into the pouring sprue, flows down into the multiple cavities, causing the simultaneous formation of multiple castings. Aluminum is often used for short-run production of shell-cast parts. Iron, bronze, and steel are used for higher production runs.

11. *Remove the casting.* After cooling and solidification, the shells are broken or shaken away from the castings.

12. *Clean and trim.* Sand that clings to the surfaces of the castings may be removed by shaking or by tumbling. Prior to any required machining operations, the removal of the sprue, the gating, and the runner system is accomplished in much the same manner as for sand castings. A variety of finishing methods may be used. These include acetylene cutting, high-speed sawing and friction wheel cutting, grinding, and sand and shot blasting. Care is taken to get as close to final tolerances and appearance as possible so that subsequent machining time can be held to a minimum.

53

FIGURE 5-4 Pouring the mold (Courtesy, Cooper Alloy Corp.)

PRODUCT APPLICATIONS

Figure 5-5 illustrates some typical product applications for the shell mold casting process. Shell molded castings can be used in practically every application that sand castings are used. Another application of shell molded products is for making cores, which are rapidly replacing expensive cores formerly used in the green sand and permanent mold casting processes. The use of sand-resin mixture cores instead of clay or sand is becoming more and more prevalent. Shell molded cores give a high precision and smooth finish on difficult-to-machine interior surfaces. They have excellent venting, collapsibility, and burnout properties.

PROCESS SELECTION FACTORS

Cost: Shell mold casting is not a cure-all for all foundry or design problems, nor is it an automatic solution for manufacturing parts that are impossible to cast by any other method. It may be, however, a means of reducing the cost of producing a given part which may formerly have been produced by a more costly method. If the maximum benefits of shell mold casting are to be obtained, the process must be selected only after considerable inquiry and evaluation. Case studies are on record which show that savings ranging from 20 to 80% for shell mold castings have been realized over other methods of production.

54

FIGURE 5-5 Typical product applications (Courtesy, Cooper Alloy Corp. and Shalco Systems)

Surface Finish: Shell molded castings have as-cast surfaces which are considerably better than parts made by sand casting. Surfaces are virtually free from surface defects such as voids and inclusions. In general, the quality of the surface texture is dependent upon such variables as casting size and configuration, grain size of the sand used, and certain techniques of shell making and metal pouring. Finish machining may in some cases be totally unnecessary on parts where a good surface may be required for outward appearance. The surface texture will also vary with the casting design and the alloy selected. Low-carbon steel alloys with cross sections under ¼ in. (6.35 mm) seldom have good as-cast finishes. It is believed the carbon reacts with certain elements in the resin–sand mixture, causing rough surfaces. Other ferrous alloys, including most types of steel alloys, are molded with surface-finish values ranging from 150 to 250 μ in. For the best as-cast finish, however, no ferrous alloy seems to be the equal of ductile iron, which consistently has a surface finish value of between 125 and 200 μ in. Shell molds are highly permeable. They vent readily. As compared to sand casting, blow holes, inclusions, and casting porosity are greatly reduced.

Materials: Any castable ferrous or nonferrous metal may be cast by this process.

Grain Structure: Photomicrographic inspections have shown that shell mold castings have what might be called "cooling-rate inversions." Unlike sand, permanent mold, or die castings, which have hardened skins and soft interiors, shell mold castings have soft skins and hard or dense uniform interior structures. As a result, shell mold castings have superior machining qualities. The pouring of shell mold castings is usually accomplished at temperatures ranging from 150 to 200 °F (approximately 65 to 90 °C) lower than sand castings. Reduced pouring temperatures are known to contribute to improved density of the casting structure.

Machining: Shell molding may replace sand castings for some applications where as-cast outer surfaces are suitable for in-service applications. It should be obvious that the greatest savings will be realized in cases where machining on castings can be eliminated altogether. Machining requirements on some shell molded castings may often be entirely eliminated because of the inherent ability of the process to yield parts with a superior surface texture and with acceptable detail in an as-cast condition. Substantial savings are possible when difficult-to-machine metals such as high-alloy and stainless steels can be used in service directly from the foundry.

Tooling Costs: For short-run production, shell mold casting, as compared to the die and permanent mold casting processes, often results in significant savings, owing to lower tooling costs. Only metal patterns can be used for shell mold casting. Aluminum match-plate patterns are used principally for short-run jobs. These may vary in cost from $200 to $1000. Iron patterns which have been stress-relieved are used for longer production runs. They typically cost from 50 to 75% more than aluminum patterns, but the increase in cost is generally offset by savings over periods of longer use.

Accuracy: Unlike parts made by the sand-casting process, products made by the shell mold process are less subject to human handling. As a result, the tendency of cores slipping in the mold or the possibility of a mold being enlarged by excessive rapping is practically eliminated. It is virtually assured that each casting will be a close reproduction of all other castings which are produced from a given match plate. The rejection rate of shell molded castings is extremely low.

Foundry Costs: Studies have been made that compare foundry costs of sand casting to those in shell molding. Most studies show that costs are generally less in shell molding. There are a number of good reasons. The initial investment is considered to be moderate in shell molding. There is much less sand to handle with easy-to-lift, lightweight molds. Shell molds range from one-eighth to one-tenth less in weight than sand-casting molds. Shell molds may be made in large quantities in advance and stored to await the pouring of the molds, as in Figure 5-6. Productivity of machines and equipment is considered higher than mold-making operations required in sand casting. Rejects are kept to a minimum since the process has a more satisfactory success rate. While certain basic labor skills are necessary in any molding process, shell molding can be described as a process that requires minimum skills with a resultant low labor cost. In large shell molding operations, it is practical to recover most of the silica sand from the used shells. The reclamation of the mix requires the resin to be burned out at a high temperature or by a process known as "dryscrubbing."

FIGURE 5-6 Assembled shell molds (Courtesy, Cooper Alloy Corp.)

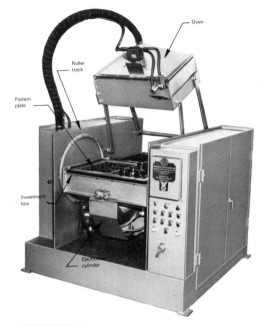

FIGURE 5-7 A shell mold machine (Courtesy, Shalco Systems)

Production Rate: Production-rate figures must, of necessity, be based upon actual casting products. Depending upon the pattern size, shell molding machinery is commercially available and, with many recent refinements, is now fully mechanized. One type of shell mold machine is shown in Figure 5-7.

Some Limitations

Shell mold casting is subject to many of the same limitations as other foundry processes, as well as to some of its own. Some complicated casting configurations cannot be adapted to the process, owing to the inability to arrange a suitable parting line or a gating system. Large, heavy castings with thick walls and sections are not adapted to shell molding. Castings weighing less than 30 lb (13.5 kg) are best suited for the process.

The resin used in shell molding costs more than the binder for sand-casting molds. Patterns are generally more costly than those used for sand casting. The polished surface finish and higher limits of accuracy for shell mold metal patterns tend to accelerate the costs of these patterns. The sprue and gating system must be provided as a part of the match-plate pattern, a fact that adds to the increased cost over sand casting. Rather large production runs are necessary to amortize the higher initial costs of shell molding equipment and many companies have established a minimum quantity of 50 parts as a practical limit.

PRODUCT DESIGN FACTORS

A thorough evaluation is necessary to determine if the advantages of shell molding are sufficient to offset any increased costs. A review of all aspects of each of the various competing casting processes is usually made so that the greatest combinations of advantages may be obtained. Molding techniques with regard to fillets, disposition of mass, undercuts, parting lines, and so on, are treated in much the same manner as for other casting processes.

Draft Angles: Draft angles as small as $\frac{1}{2}°$ are common, although angles of 1 to 2° are frequently used. Under favorable conditions, draws (or pattern removal) up to 2 in. (50.8 mm) can be accomplished without any draft. The standard rule for draw is to allow twice the diameter of draw depth for blind holes. This is not to be considered an absolute rule, since exceptions may be made under special production conditions. It should be noted that in spite of the fact that the cured shell is "stripped" from the match-plate pattern, the shell must be removed in a direction that is perpendicular to the pattern. The removal of the shell should not result in injury to the cavity walls.

Bosses, Undercuts, and Inserts: These types of features may usually be adapted to the original product design without difficulty.

Section and Wall Thickness: Shell mold castings may be made with a $\frac{1}{8}$-in. (3.18 mm) minimum wall for gray and ductile iron. Products cast from steel, aluminum, and magnesium alloys require at least a $\frac{3}{16}$-in. (4.763 mm) minimum wall. As a result, shell mold castings are affected far less from varying thicknesses of cross sections than are green sand molds. Feather edges can be cast in aluminum, gray iron and ductile iron. Such edges are generally brittle, however, because of the chill condition. Feather edges for steel and most nonferrous alloys are difficult to cast and are not recommended.

Special As-Cast Details: The general rule is to cast-in as many required detail features as possible. All things considered, it is usually more economical to cast a feature than to produce it by one or more subsequent machining operations. Shell mold casting yields parts with as-cast detail features which are superior to those produced by sand casting. Coarse helical threads may be readily cast. Fine threads, however, are impractical to cast. The process is particularly suited to producing detail features on castings such as gear and sprocket teeth, sharp corners, $\frac{1}{8}$-in. (3.18-mm)-wide keyways, slots as small as $\frac{1}{16}$ in. (1.588 mm), and in special cases, holes $\frac{1}{8}$ in. (3.18 mm) deep and fine recessed or raised lettering of $\frac{1}{16}$ in. (1.588 mm). In special cases, holes ⅛ in. (3.18 mm) in diameter may be economically cast, but, in general, it is more practical to machine holes less than ½ in. (12.7 mm) in diameter.

Machining Allowance: In shell mold casting, a machining allowance of ¹⁄₁₆ in. (1.588 mm) is adequate. A recent study conducted by one firm shows that cutting-tool life is in direct proportion to the amount of stock removal. The study also showed that the tool-life ratio was as high as 12:1 when machining shell mold castings as compared to green sand castings. For test purposes, the same amount of stock was removed from both types of castings. A minimum amount of stock left on the casting for machining results in a substantial savings of machining costs, as well as in a savings of materials, particularly in large-quantity production runs.

Tolerances: In general, standard tolerances for most nonferrous and steel-alloy shell molded castings are ±0.020 in./in. (0.020 mm/mm). Tolerances for iron castings are usually ±0.015 in./in. (0.015 mm/mm). In cases where the initial tooling may be offset by finish-machining savings, tolerances may be as close as ±0.005 to ±0.010 in./in. (0.005 mm/mm to 0.010 mm/mm). All tolerances for shell molding can be held closer than for sand castings. The process is more dimensionally stable, particularly for straightness and concentricity of cored holes.

Coring: In general, the shell molding process is not as dependent upon the use of cores as a means of producing intricate interior configurations as in sand casting. An inherent advantage of shell molding is an ability to produce deep draws, thus making it possible to form intricate interior surfaces of a casting as a part of the mold itself.

Gates and Runners: Match-plate patterns consist of an assembly of the sprue, the gate, and runners that connect one or more patterns, which are all mounted together on one plate. An optimum size must be used for the sprue, gates, and runners, since the flow rate of the molten metal directly affects the subsequent success of the resulting castings. Metal used for making the sprue, gates, and runners is usually the same as the pattern material.

Shrinkage Allowance: The inherent qualities of shell molded castings make the metal cooling rate more readily controllable, which eliminates difficult shrinkage problems. The amount of shrinkage allowance required for a given metal is much the same as for sand castings.

Size and Weight: The shell molding process is best suited for small castings weighing up to 30 lb (13.5 kg). Larger castings are not so well suited because of the cost of increased resin requirements and, except for special cases, equipment limitations. Shell molded castings weighing as much as 400 lb (180 kg) have been successfully produced, however.

REVIEW QUESTIONS

5.1. Make a brief outline describing how a typical shell mold is prepared.

5.2. Contrast the type of patterns used in shell molding to those used in sand casting.

5.3. Explain why the pattern must be preheated.

5.4. What is meant by "curing" the shell?

5.5. Indicate which of the following casting processes have expendable molds:
(a) sand casting (b) plaster mold casting (c) shell molding.

5.6. Contrast the as-cast surface quality of shell molded products to those of sand castings.

5.7. Explain why shell molded products are less porous than those of sand castings.

5.8. List the main advantages of shell molding.

5.9. Discuss the relative foundry costs of shell molding to sand casting.

5.10. Compare the maximum weight limit of shell mold casting to sand castings.

5.11. Compare the recommended machining allowance provided for shell mold castings to that for sand castings and plaster mold castings.

5.12. What is the recommended tolerance for an aluminum shell mold casting?

5.13. Why is shell molding not economical for small quantities of parts?

BIBLIOGRAPHY

JOHNSON, H. V., *Manufacturing Processes,* Peoria, Ill.: Chas. A. Bennett Co., Inc., 1973, p. 140.

"Shell Molding," *Metals Handbook,* Vol. 5, *Forging and Casting,* 8th edition, Metals Park, Ohio: American Society for Metals, pp. 181–202.

Shell Process Foundry Practice, 2nd edition, 1st revision, American Foundrymen's Society, DesPlaines, Ill.: 1973.

6

INVESTMENT CASTING

Investment casting has historical roots which extend well back in time. The basic principles of this process were used thousands of years ago by artisans for the casting of statues and other artistic forms in Japan, China, and Egypt. Benvenuto Cellini used a form of the investment casting process during the Renaissance in the sixteenth century in casting his bronze statue *Perseus*. Techniques of the investment casting process were used in the early 1930s in jewelry, dental, and medical applications. There was considerable industrial interest in the process during the early 1940s when requirements increased for producing parts of complex shape in alloys with very high melting points. In many cases, the configurations required were associated with metals which, because of their unique physical properties, were too difficult to machine or otherwise shape using conventional machine tools. Because of its uniqueness, the investment casting process provided a suitable means of producing these parts. Investment casting is also known as the ''lost-wax'' and as the ''precision casting'' process.

DESCRIPTION OF THE PROCESS

Basically, the investment casting process involves the use of expendable (heat-disposable) patterns surrounded with a shell of refractory material to form the casting mold. Castings are formed in the cavities created by melting out the pattern.

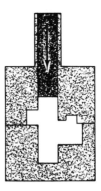

Wax or plastic is injected into a metal die to form a disposable pattern.

FIGURE 6-1 Preparing the wax pattern by injecting the die (Courtesy, Arwood Corp.)

The process starts with the preparation of wax or plastic patterns. These are made by pouring or injecting the pattern material into master die sets, shown in Figure 6-1, which have been accurately made by skilled die makers. A cluster or "tree" is formed by assembling several patterns together by wax-welding a gating system connected to a central sprue, as shown in Figure 6-2. The longest dimension of a pattern is normally placed in a vertical position.

Solid patterns are formed by placing a metal container-type flask over the cluster of patterns and then pouring a hard-setting molding material into the flask. The material completely invests the pattern cluster, as shown in Figure 6-3. Shell patterns are formed as shown in Figure 6-4. The clustered patterns are dipped in a ceramic slurry and the procedure is repeated until the required thickness of the mold or shell is achieved. Some companies use

FIGURE 6-2   Gating the wax patterns (Courtesy, Arwood Corp.)

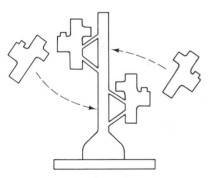

Patterns are gated to a sprue by wax "welding" to form a tree or cluster, and a base of the pattern material is attached to the tree.

(a)

(b)

63

FIGURE 6-3 Preparing a solid pattern (Courtesy, Arwood Corp.)

as a refractory material for this purpose a ceramic mixture of zircon flour and colloidal silicon. A coating thickness of from ⅛ to ⅝ in. (3.18 to 15.88 mm) is allowed to build up a series of as many as eight investments. The flask with its contents is then placed in an inverted position in an oven, where the wax patterns are allowed to melt and drain out of the mold. Molten metal is then gravity poured, as in Figure 6-5, into the cavity. Other pouring methods, including vacuum, centrifugal, and air pressure, are also used.

After cooling, the fragile mold material is broken away, freeing the castings. Cores are removed by a pressurized water blast or by leaching with a caustic soda. The final foundry operation consists of separating the castings from the gating system with band saws or abrasive wheels, as shown in Figure 6-6.

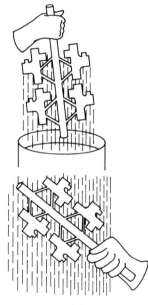

FIGURE 6-4 (a) Preparing a shell pattern; (b) dipping patterns into a ceramic slurry (Courtesy, Arwood Corp.)

(a)   (b)

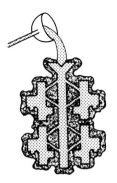

FIGURE 6-5 Pouring the mold
(Courtesy, Arwood Corp.)

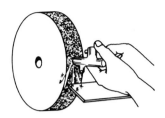

FIGURE 6-6 Separating castings
from the cluster (Courtesy,
Arwood Corp.)

PRODUCT APPLICATIONS

There are over 200 foundries in the United States that produce investment castings for the industrial markets. Figure 6-7 illustrates some typical investment cast parts. Investment castings are used on products for the aerospace industry; in aircraft engines, frames, fuel systems, and instruments; as parts for computers and data-processing equipment; food and beverage machinery; machine tools and accessories; scientific instruments; sewing machines; nozzles, buckets, vanes, and blades for gas turbines; radar waveguides; structural hardware; golf clubs; costume jewelry; movie cameras and projectors; and so on. Dentures and special metal implants for orthopedic surgery are also made by this process.

PROCESS SELECTION FACTORS

Materials: Virtually any metal can be investment-cast. Castings of high-melting-temperature metals are most prominently made by this process. Such materials are often difficult to machine or otherwise work to shape using conventional machining processes. Parts of stainless steel, Stellite, and tool and die steels are commonly investment-cast. Magnesium and copper base alloys are also cast by this method. The process is especially suited to casting metals which melt at temperatures too high for plaster or metal molds.

Surface Finish: The surface-finish values vary depending upon the pattern material and the type of metal being cast. Under normal production conditions, this ranges from 125 μin. in ferrous castings to between 63 and 125 μin. for nonferrous castings. A minimum surface finish of 30 μin. is obtainable using plastic patterns and other special process control techniques.

Parting Line: Because the mold consists of only one piece, there is no parting line. It should be understood, however, that the original pattern is prepared

(a)

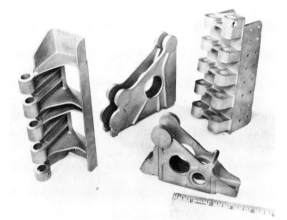

(b)

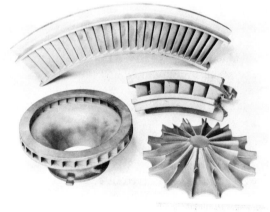

(c)

FIGURE 6-7 Typical investment castings (Courtesy, Arwood Corp.)

66

by injecting molten wax or plastic under pressure into the rubber or metal master die halves. After each individual pattern has been removed from the master die, it is necessary to remove any traces of the parting line formed by the die halves prior to investing the patterns to prevent this feature from appearing on the final casting.

Part Quantities: For small parts, quantities in the range 5000 to 10,000 are desirable for economy runs. For larger parts, quantities of only one to as few as 5 to 10 pieces are sometimes economical. Many factors must be considered in evaluating optimum part quantities. If the part is of such intricacy that it cannot be made by another casting method or it cannot be forged, if the cost of machining it from raw stock is too costly, or if the metal is too difficult to machine, then a production run of nearly any size is feasible. Investment casting foundries have reported production runs of up to 2,000,000 castings of a single design in 1 month. For foundries making a wide variety of castings, production of more than 300 different castings a month at a volume flow of 250,000 castings is not unusual.

Tolerances: Parts with tolerances that range from ± 0.003 to 0.005 in./in. (0.003 mm/mm to 0.005 mm/mm) can be obtained, with closer tolerances at additional cost. The general rule is to provide as large a tolerance as possible.

Casting Detail: Intricate cast features are possible because the rigid mold will not deform when the metal is poured into it. Slots as fine as $\frac{1}{16}$ in. (1.588 mm) wide and raised cast lettering are practical. Frozen mercury patterns are used in some cases in preference to wax or plastic patterns to obtain more intricate detail.

Post-Casting Operations: Castings may be heat-treated, welded, machined, and otherwise finished in accordance with the requirements and properties of the alloy used.

Some Limitations

Part Weight and Size: Investment castings are generally restricted to those weighing less than 5 to 6 lb (2.25 to 2.7 kg) and parts weighing less than 1 oz (28.3 g) are common. Successful production castings have been produced as large as 50 lb (22.5 kg). Prototype castings weighing as much as 100 lb (45 kg) are also possible. Parts may be produced with a size range extending to about 40 in. (1016 mm) in any direction.

Costs: Tooling and labor costs are generally more expensive than for sand, shell mold, or plaster casting. However, for some special product applications, the cost of the finished part may prove to be considerably less.

Quantities: Production runs are not competitive with permanent mold or die casting processes for low-melting-point alloys except in cases where small quantities are required or for particularly complex parts.

Maximum Section Thickness: The process is limited to castings having a maximum section thickness of 1 in. (25.4 mm).

Threads: It is usually more economical to machine both internal and external threads rather than to produce them by casting. Production of threaded holes may be facilitated by casting a hole corresponding in size to that produced by a tap drill and later tapping the hole. In special cases such as threads in hard-to-machine alloys or for unusual thread forms, investment cast threads could prove less costly than machining.

PRODUCT DESIGN FACTORS

Patterns: The design features and the methods of producing the master die sets are critical factors, because the dies are, in turn, used to produce the wax or plastic patterns. Superior quality patterns are required, since the resulting castings exactly duplicate the original patterns.

Cores: A wide variety of special-purpose cores are used to produce hollow interiors or other openings in investment castings. The use of cores increases the cost of producing a pattern and, in turn, a casting. Three types of cores are commonly used: *single-piece, collapsible,* and *preformed cores.* Single-piece cores generally consist of a steel pin or rod which is placed in the master die before injecting the wax. After the wax is poured, the core is withdrawn, the die opened, and the wax pattern removed. The core forms the desired internal configuration. Collapsible cores, also made of steel, consist of two or more movable segments which are retractable. These are used in cases where a one-piece core could not be removed without damage to the pattern. Preformed cores are made of an expendable wax-base material in a die that is built solely for this purpose. The preformed core is positioned in the master die in such a way that the wax injected to make the pattern will surround the functional area of the core. The core, when removed from the pattern, forms a hollow interior or opening. Cores are also made of ceramic materials, which are becoming popular.

Minimum Section Thickness

Limited or tapered areas:
   Low-melting-point alloys—0.020 in. (0.51 mm)
   High-melting point alloys—0.030 in. (0.76 mm)

Large areas:

Low-melting-point alloys—0.040 in. (1.02 mm)

High-melting-point alloys—0.060 in. (1.52 mm)

Tolerance Variables: The accuracy of investment-cast parts will vary according to the following factors: pattern die tooling, cross-sectional thickness of pattern, location of pattern-die parting line, wax-injection conditions, temperature control of stored expendable patterns, pattern assembly, and techniques used in the cleaning and finishing of castings. It is possible, however, to specify some general tolerances, which follow.

Machining Allowance: Standard practice is to allow 0.010 in. (0.25 mm) for a grinding finish with up to 0.032 in. (0.81 mm) for general machining finish. (Cutting tool life is extended when there is sufficient extra metal to let the tool get below the somewhat abrasive as-cast surface during its first cut.)

Radii: For best results and lowest cost, the size of radii on castings should be as generous as possible. The general tolerance is ±1/64 in. (0.397 mm). Sharp corners should be avoided.

Flatness and Straightness: The amount of tolerance needed for axial straightness and for flatness which is characterized by bowing, waviness, and twisting varies with the part length, thickness, and the general configuration. A practical as-cast tolerance range is from ±0.010 to 0.030 in. (0.25 to 0.76 mm). With mechanical straightening operations the as-cast tolerance can be reduced.

Concentricity: Concentricity between I.D.s and O.D.s on as-cast parts may be held within 0.003 in. (0.08 mm) T.I.R. (total indicator reading) per 1/2 in. (12.7 mm) of wall thickness.

Roundness: A general rule of ±0.005 in./in. (0.005 mm/mm) may be applied.

Angles: As-cast angles may range from ±1/2 to 2° closer tolerance than those obtainable by mechanical straightening operations.

Hole Locations: Hole locations may be as accurate as ±0.005 in./in. (0.005 mm/mm) from any single reference point.

Blind Holes: Blind holes may be cast if the length does not exceed the diameter. In nonferrous metals, blind holes less than 2 in. (50.8 mm) in diameter may be cast if the length does not exceed twice the diameter.

Serrations: A practical limit of 16 serrations to the inch (25.4 mm) in low-melting-point alloys is recommended with only eight serrations to the inch for high-melting-point alloys. Radii at the crests of serrations can be held to 0.005 in. (0.13 mm).

Fillets: A minimum size of $\frac{1}{32}$ in. (0.794 mm) is recommended, but fillet radii from $\frac{1}{16}$ to $\frac{1}{8}$ in. (1.588 to 3.18 mm) are preferable.

Draft: Most parts can be produced without any draft allowance, although a small amount is desirable for parts with long, extended surfaces to facilitate removal of the pattern from the die.

NEW DEVELOPMENTS

Considerable attention has been recently focused on methods of mechanization in an effort to reduce the costly labor requirements of the investment-casting process.

One new approach to the problem is the use of industrial robots (see also Chapter 21) in the mold-making process. It is now possible to obtain a more uniform shell thickness by use of robots which are programmed to mechanically dip and invest the patterns. One company has developed a patented method of coating and drying the molds entirely by automatic means which minimizes the skill and experience necessary to prepare a uniform shell. Still other foundries are experimenting with ceramic cores, particularly those for castings with complex internal passages. Great progress has been made leading to the quantity production of ceramic cores which are both strong and stable at room and at elevated temperatures. Considerable promise is reported in this 3000-year-old process by the development of new pattern materials and mold binders.

REVIEW QUESTIONS

6.1. Make a brief outline describing the entire process of investment casting. Prepare sketches where appropriate.

6.2. What materials are used for the molds?

6.3. In most other casting processes, parts with undercuts or reentrant sections are difficult or impractical to produce. Explain why features such as these normally present no unusual difficulty in investment casting.

6.4. Explain how it is possible to produce a casting without a parting line.

6.5. List the principal advantages of investment casting.

6.6. What materials are used for cores?

6.7. Prepare a table showing the as-cast tolerances for flatness and straightness, concentricity, roundness, angularity, and for center-distance locations for holes.

PROBLEMS

6.1. The following alloys are commonly selected for parts made by the investment casting process. Prepare a table listing each alloy showing the tensile strength, yield strength, percent of elongation, castability, and under "remarks" give a short description of other important characteristics such as weldability, wear resistance, corrosion resistance, machinability, suitability for heat treatment, etc.

ALLOYS: *Aluminum alloy* A 356; Cobalt 21; Manganese bronze; Yellow brass; *Magnesium alloy* AZ91C; *Alloy steel* AISI 1020; *Alloy steel* AISI 4140; *Austenitic stainless steel* AISI Type 302; *Nickel base alloy* *Hastelloy B; Beryllium copper (20 C).

6.2. The prime goal of good casting design is to avoid concentration of stress in one or a few sections of a casting. The strength of a casting may be improved by distributing the stress more evenly over the entire casting. Using a sketch, show how the design of each of the castings shown in Figure 6-P2 may be strengthened.

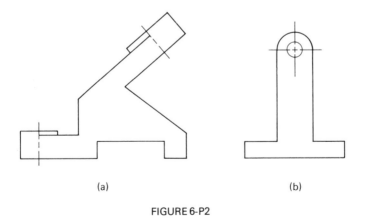

(a)   (b)

FIGURE 6-P2

6.3. Using sketches show how the undesirable effects of shrinkage on the heavy section of the casting shown in Figure 6-P3 can be avoided by a simple change in the design.

*Registered trademark of Union Carbide Corporation.

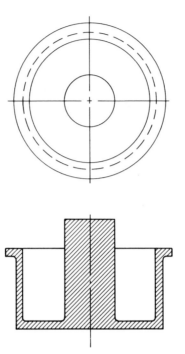

FIGURE 6-P3

6.4. Due to a poor original design, costly straightening operations may be necessary on some castings to correct distortion. Using a sketch, show how the design of each of the castings shown in Figure 6-P4 (a) and (b) may be improved to avoid straightening operations.

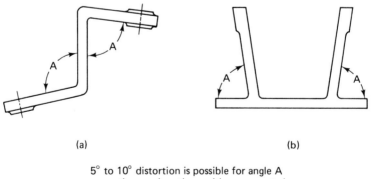

(a)   (b)

5° to 10° distortion is possible for angle A
on sections such as these with unsupported
members.

FIGURE 6-P4

6.5. The use of split cores are avoided whenever possible. Using a sketch, show how the casting in Figure 6-P5 could be redesigned so that a single core can be used thus eliminating the parting line on the bosses.

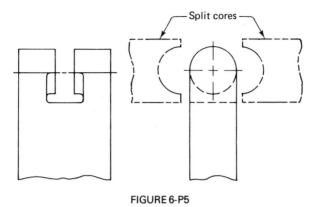

FIGURE 6-P5

6.6. The designer should make every effort to reduce the mass of metal at a junction wherever possible. Explain how the design of the junction on the casting in Figure 6-P6 could be changed thereby reducing the "hot spot" which is likely to occur within the encircled region.

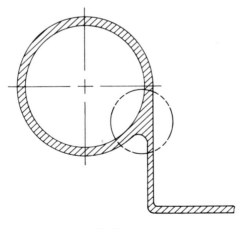

FIGURE 6-P6

6.7. Give a principal reason why the investment casting process was selected to make the part shown in Figure 6-P7. All sizes are as-cast. Only representative sizes are given on the drawing.

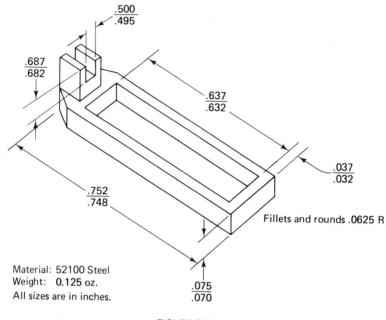

Material: 52100 Steel
Weight: 0.125 oz.
All sizes are in inches.

Fillets and rounds .0625 R

FIGURE 6-P7

6.8. The recommended size for correctly cored holes in investment castings are:
(a) for through holes the length should not be greater than from 4 to 5 times
the diameter; (b) for small blind holes, the maximum depth should not exceed
twice the diameter. Which of the cored holes shown in Figure 6-P8 correspond
correctly to these recommendations?

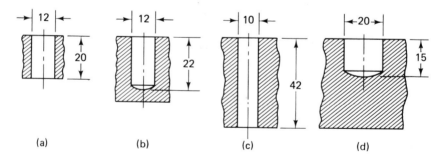

(a)   (b)   (c)   (d)

All sizes are in mm

FIGURE 6-P8

6.9. Explain why the casting design shown in Figure 6-P9 (b) is preferred over
that shown in Figure 6-P9 (a).

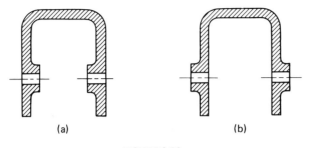

(a)   (b)

FIGURE 6-P9

BIBLIOGRAPHY

*HERRMANN, R. H., *How to Design and buy Investment C astings,* Evanston, Ill.: Investment Casting Institute.

"Investment Casting," *Machinery's Handbook,* 20th edition, New York: Industrial Press, Inc., 1975, pp. 2181–2183.

"Investment Casting," *Metals Handbook,* Vol. 5, *Forging and Casting,* 8th edition, Metals Park, Ohio: American Society for Metals, pp. 237–261.

JACKSON, H., *Precision Casting,* Flagstaff, Ariz.: Northland Press, 1972.

*Abstracted with permission.

7

PERMANENT MOLD CASTING

Permanent mold casting may be described simply as the process of pouring molten metal under gravity force into metal molds. Because no pressure is used in supplying metal to the mold cavity, the process is also known as "gravity die casting." Both ferrous and nonferrous metals may be cast. Production runs of aluminum parts have been reported of up to 250,000 castings from the same mold before requiring repair. The process is especially adaptable to high-volume production of small, simple castings which have reasonably uniform wall thickness, no undercuts, and limited coring requirements. It is possible to adapt this process to moderately complex castings in cases where high-volume production requirements must be met.

Both metal cores and sand cores are used to form the cavities in cast parts. The term "permanent mold casting" is used when metal cores are employed. "Semipermanent mold casting" is generally used to denote the use of sand cores.

DESCRIPTION OF THE PROCESS

Molds are made in two or more sections. The simplest form, shown in Figure 7-1, is constructed with hinges at one end and clamps at the other. Molds are hinged so that the solidified casting may be conveniently removed.

76

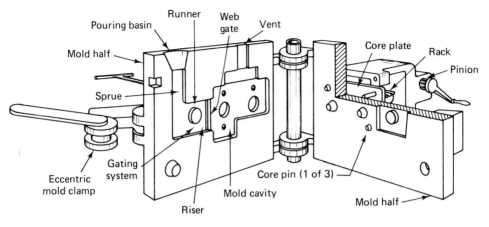

FIGURE 7-1 A hinged permanent mold

Permanent mold sections are securely clamped together before pouring to resist the fluid pressure of the molten metal and to eliminate the possibility of mold warpage. Simple hand-operated latches or screw or toggle clamps are used for small molds. Larger molds require heavier clamping devices, which are generally pneumatically operated. Gray iron and steel castings are generally made in molds of graphite or other suitable refractories. Castings of aluminum, magnesium, or copper-base alloys are made in molds of close-grained alloy cast iron such as Meehanite or die steels. Lead, tin, and zinc castings are usually made in bronze molds. Ceramic materials have also been successfully used as a mold material. Whenever practical, the mold cavity is formed to its approximate shape by a casting process and machined to final size. This results in a mold with a superior surface finish, close dimensional tolerance, and proper draft angles. Molds are integrally constructed with the necessary gating system required for feeding the mold cavity. Most molds are constructed with an ejector pin system for casting removal. Whenever possible, molds of multiple cavities are used.

Production Machines

Limited production of permanent mold castings may be done manually by using a simple "book" mold arrangement. Quantity production runs are accomplished by using mass production machines. One type of production machine, shown in figure 7-3, consists of a circular turntable with 12 or more timed-cycle stations. The machine automatically blows out hot molds, cleans and coats, places the dry sand cores (if required), locks the mold sections together, preheats the mold (if necessary), pours the molten metal, cools, unlocks, and finally ejects the castings. Two operators are necessary with an additional operator required if cores are to be placed. The

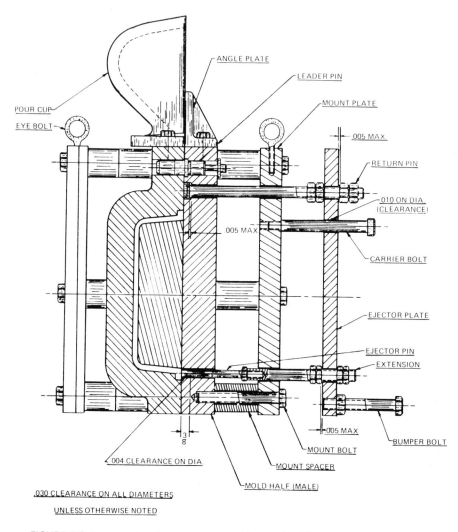

FIGURE 7-2 An example of a permanent mold assembly (Courtesy, Stahl Specialty Co.)

molds are heated at the start of a production run and kept at an elevated temperature. The rotational speed of the turntable may be adjusted to obtain an optimum casting rate and to avoid too rapid chilling of the metal as it solidifies. The mold cavities, together with the sprue, gate, and riser, are coated with a refractory wash which is applied by spraying or by brushing. The wash protects the mold cavity from premature erosion and checking and also assists in the release of the casting from the mold. For castings that require localized chilled surfaces, the wash is restricted to only a thin coat or it may be totally omitted in those areas. Acetylene soot is also used as a coating.

PRODUCTION APPLICATIONS

Castings made by the permanent mold process generally resemble parts made by sand casting in appearance as well as in many in-service applications (see Figure 7-3). Permanent mold castings are used when parts with high hardness, high strength, leak tightness, or corrosion resistance are needed. Such parts are commonly found on household appliances, typewriters, machine tools, and in many hardware applications. This important and competitive casting process is also used to produce pistons, connecting rods, fuel pump bodies for automobile engines, pipe fittings for marine uses, various aircraft fittings, flat-iron sole plates, cooking utensils, ornamental grilles, gear blanks and housings, special bolts, and for countless other commercial applications. Figure 7-4 illustrates some typical product applications.

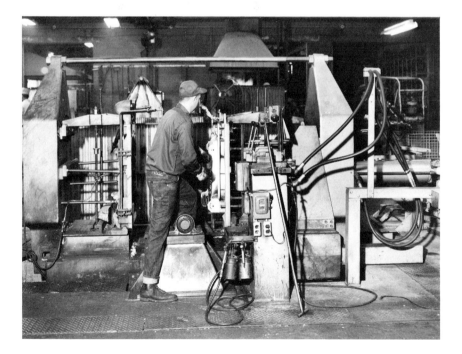

FIGURE 7-3 Operator is shown removing the finished casting from a modern permanent mold casting machine equipped with a 36 × 36 in. mold (Courtesy, Stahl Specialty Co.)

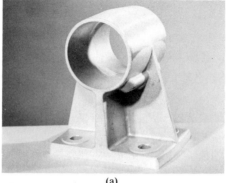

(a)

(b)

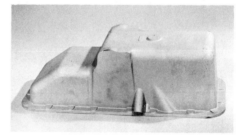

(c)

(d)

FIGURE 7-4 Examples of typical parts made by permanent mold casting; (a) a highway bridge rail part; (b) assembled components of a barbecue unit; (c) oil pan for a small engine; (d) intake manifold for a diesel engine (Courtesy, Stahl Specialty Co.)

PROCESS SELECTION FACTORS

Permanent mold castings may be selected as an alternative to sand castings, its closest competitor. Unless otherwise stated, the process selection factors which follow are based upon a comparison of the two methods.

Inserts and Cores: Necessary inserts, cores, and other features can be readily incorporated into the design of the molds and may be set by totally automatic or by manual means. Cores may be of gray iron, steel, sand, or plaster. It is generally cheaper to machine out holes less than ¼ in. (6.35 mm) in diameter than it is to cast such holes.

Cost: There is a low scrap loss. The casting gates are cut off and are reused. The molds are reusable and, though more costly initially than the expendable-type molds used in sand casting or shell molding, there is a substantial savings in mold preparation time. Depending upon the shape, size, and complexity of the required casting, permanent mold castings may well represent a substantial overall savings over other methods of casting. Another important cost-saving consideration is that less floor space is required for making permanent mold castings.

Size and Weight of Castings: Castings made by this process range in size from a few ounces (90 to 100 g) to 300 lb (135 kg). Castings weighing up to 500 lb (225 kg) have been made in special cases. The usual weight range for copper alloy castings is from 1 to 20 lb (0.45 to 9 kg). As a practical general rule, gray iron castings are limited in size to about 12 in. (304.8 mm) for the largest dimension and to about 30 lb (13.5 kg) weight. A practical weight limit for high production of aluminum alloy parts is also about 30 lb; however, parts in limited quantities have been successfully produced which weigh considerably more than 30 lb.

Surface Texture: The as-cast surface texture is superior to sand castings. It is possible to obtain an as-cast surface-finish value of 90 to 125 μin. The quality of the surface finish is dependent upon the condition of the mold with respect to the surface of the cavity, the mold coating, draft, gating design and size, venting, mold temperature, and certain design features of the casting itself.

Grain Structure: Owing to the rapid heat transfer from the molten metal to the mold, castings have a finer grain structure and better strength properties by as much as 20%. Permanent mold castings are less subject to gas porosity. A sound casting is produced with improved density because the metal enters the mold gently and without turbulence.

Wall Thickness: In general, thinner walls may be produced by this process, because of the more rapid heat transfer in cooling.

Shrinkage: Castings are less subject to shrinking during the cooling period in the mold.

Production Rate: The production rate may vary depending upon a number of factors, such as part size and complexity and the time required for the metal to solidify. A good production estimate for a single cavity die is about 20 to 50 castings per hour. Multiple cavities proportionately increase the yield per hour. The process is ordinarily classified as faster than sand casting but slower than die casting.

Materials: Gray iron, magnesium alloys, and aluminum alloys are of greatest importance to this process. Metals selected for product applications must have low pouring temperatures sufficiently fluid to fill all parts of the mold. Copper-base alloys and lead and zinc alloys, though sometimes cast, present problems because of their insufficient fluidity. Steel may be cast using special graphite dies.

Tolerances: For aluminum and magnesium alloy parts, tolerances of from ± 0.010 to 0.015 in. (0.25 to 0.38 mm) for the first inch (25.4 mm) with an additional ± 0.001 to 0.002 in. (0.03 to 0.05 mm) for each additional inch are recommended. In special cases, a tolerance of ± 0.005 to 0.010 in. (0.13 to 0.25 mm) total may be obtained, usually without an appreciable additional cost. Metal molds used in this process are redressed or overhauled periodically as they show signs of wear. They may be discarded when it becomes apparent that the castings no longer meet the designed tolerance. The dimensional accuracy is affected by cycle-to-cycle variations in the mold closure or by the changes in position of moving parts in the mold, foreign material on mold faces, distortion of the mold, and gradual and progressive mold distortion and wear.

Finish Allowance: Dimensional accuracy is better than sand casting and there is less variation from part to part. A generally accepted allowance for subsequent machining operations is $\frac{1}{32}$ in. (0.794 mm) for parts up to 4 in. (101.6 mm) and $\frac{1}{16}$ in. (1.588 mm) for parts greater in size than 4 in.

Some Limitations

High Initial Cost of Tooling: The quantity requirements of the product must be sufficiently large before this process can be considered. A permanent mold may cost more than three times that of a good metal pattern.

Warpage: Many castings and forgings have a tendency to warp when cooling and, depending upon the configuration, some more than others. Sometimes a straightening operation is necessary, which adds to the cost of producing the parts.

Size and Shape: There are certain practical restrictions governing the size and shape of a casting which may be produced in permanent molds. Since the shape of the metal mold cavity is produced by machining methods, the configuration is limited to relatively straightforward lines with limited coring. The complexity of the part configuration is also limited to some extent by the ability to eject the casting from the mold.

Materials: There is a limitation of castable materials. As previously mentioned, the process is generally unsuitable for casting high-melting-point metals such as steel unless special graphite die blocks are used. Trial runs are often necessary to determine the precise gating conditions needed for proper feeding of the mold cavity. Compared to the ease by which a change may be made to a pattern used in sand casting, it is difficult to alter steel die blocks in cases where alterations of the molds are necessary.

PRODUCT DESIGN FACTORS

Certain features which are incorporated into the design of the molds for permanent mold castings resemble many of the procedures used for some of the other casting processes.

Venting: Vent channels are cut on the parting surface of the die block which allows the gases in the poured metal or the trapped air to escape. Some natural venting takes place between the joint formed by the hinged die block sections along the parting line as well as through the gating system.

Mold Wall Thickness: Walls may vary from ¾ to 2 in. (19.05 to 50.8 mm) in thickness. The wall thickness is usually increased in areas that are opposite thicker sections of the casting, which helps to remove increased amounts of heat. Thicker walls help to provide a uniform temperature for mold cavity surfaces, thus promoting the desired chilling effect on certain required surfaces.

Bosses, Undercuts, and Inserts: Bosses on castings may usually be produced at a small additional cost. Undercuts add to the cost of producing a casting, because of the necessity for multipiece cores. It is generally more economical to machine undercuts than to cast them. Inserts such as bushings, bearings, and studs usually may be incorporated without difficulty into the casting design.

Draft: A draft of 3° minimum on each side of the mold cavity is recommended with as much as 5° in recesses.

Cores: Cores for permanent mold castings are made of steel, green sand, shell molded resin sand, or gypsum plaster, depending on the application. As might be expected, the surface finish of cored openings is governed by the quality of the finish of the core that is used. Metal cores must be removed as soon as practicable before the poured metal shrinks around the cores. Unless the cores are withdrawn before final casting solidification takes place, they will be very difficult to remove without cracking the casting. Some cores may be withdrawn by means of simple hand or power-operated mechanisms. Steel cores are not used for permanent mold casting of iron. Nonmetallic cores are used where the core cannot be withdrawn through the openings in the cavity and are removed by destroying them after the casting solidifies. As a general rule, cores should not be smaller than ⅜ in. (9.53 mm) in diameter, with a maximum length equal to six times the diameter.

Gates and Risers: The basic principles of gating for permanent mold casting are the same as for most other casting processes. In essence, the design of the mold should promote progressive solidification of the molten metal over a continuous sequence from the extreme end of the casting to the point of entry. The feeding system may be entirely cut into one die block face or it may be partially cut on both mold faces.

Checking: Checking is a condition characterized by one or more small cracks in the mold, extending from the mold cavity surface. Checks are caused by the alternate heating and cooling cycle of the molds. Close inspection of the mold halves must be made to ensure that the quality of the surfaces of the castings is not becoming affected by these cracks. Fortunately, metal exhibits little tendency to flow into these minute cracks. Serious mold cavity defects reduce the quality of castings. Some of the higher-pouring-temperature metals may eventually contribute to mold failure, particularly in locations of thin walls.

Heat Treatment: In some cases, as in some other casting processes, it is possible to obtain improved mechanical properties by heat-treating permanent mold castings.

Surface Texture: Defects in the original casting design may play an important part in the quality of the as-cast finish of permanent mold castings. Adverse effects upon the quality of the surface finish may result from severe changes in sectional thickness, complexity of the casting, abrupt changes in the direction of metal flow, and special requirements for large, flat surfaces.

REVIEW QUESTIONS

7.1. Describe the complete process of permanent mold casting.

7.2. How do the molds differ in this process from those in sand casting?

7.3. What materials are used for the molds?

7.4. Describe why the mold cavities are coated with a refractory wash before pouring.

7.5. Explain why the as-cast surface texture is superior to sand castings.

7.6. Why would you expect the structure of a permanent mold casting to be less porous and denser than that of a sand casting?

7.7. What factors contribute to the classification of permanent mold casting as "faster than sand casting" regarding comparative production rates?

7.8. Under what special conditions may steel be cast in permanent molds?

7.9. Compare the dimensional tolerance achieved on castings produced in permanent molds to castings from sand molds.

7.10. Compare the amount of finish allowance on permanent mold castings to that on sand castings.

7.11. Explain why it is generally more economical to machine undercuts on parts rather than to cast them.

7.12. What factors contribute to ultimate mold failure?

7.13. List the main advantages and some principal limitations of permanent mold casting.

BIBLIOGRAPHY

JOHNSON, H. V., *Manufacturing Processes,* Peoria, Ill.: Chas. A. Bennett Co., Inc., 1973, p. 124.

"Permanent Mold," *Metals Handbook,* Vol. 5, *Forging and Casting,* 8th edition, Metals Park, Ohio: American Society for Metals, pp. 265–284.

Permanent Molding of Iron and Steel Castings, Report of Research Project, Des-Plaines, Ill.: American Foundrymen's Society, 1968.

Standards for Aluminum Sand and Permanent Mold Castings, SPC-18, New York: The Aluminum Association, 1976.

8

CENTRIFUGAL CASTING

The basic principles of centrifugal casting were established over a century and a half ago. While it has been a routine production technique for over 60 years, its full design potential has evolved during the past 10 to 15 years. Greatly expanded possibilities for new product applications have been developed as a result of continuing research. Improvement in methods of melting and control of chemical composition, equipment and foundry techniques, heat treating, and in other important areas have led to cost economies and ever-increasing product versatility. As compared to sand castings, centrifugal castings provide a number of significant commercial and technical advantages. Perhaps the most outstanding advantage is that the mechanical properties of most metals are definitely improved when centrifugally cast.

DESCRIPTION OF THE PROCESS

There are three principal methods of producing centrifugal castings: *true centrifugal, semicentrifugal,* and *centrifuge.* In each method, molten metal is poured into a rotating mold, which causes it to be thrown to a mold wall, where it is held by centrifugal force until solidified. All of the commonly sand cast metals may be centrifugally cast. Castings produced by any of the three methods result in parts with a dense and fine-grained structure having uniform physical properties and high strength. It is important to understand the significant differences between the three casting methods.

True Centrifugal Casting

Hollow cylinders and, in some cases, pipes with tapered walls may be produced by this method. The outer shape may be round, square, hexagonal, or fluted. The process is best suited for parts having shapes that are symmetrical about a central axis to avoid unbalance. Two types of molds are used: sand-lined and permanent.

In the *sand-lined mold,* a mixture of dry sand or clay or loose graphite is packed around a pattern which is positioned in the center of a hollow, cylindrical flask made of steel or iron. After the pattern is withdrawn from the flask, a thin refractory coating is applied to the liner thus produced. The coating increases mold life, promotes faster solidification, and prevents the casting from fusing to the mold surface. After the mold is baked, the molten metal is poured into the preheated flask, which is caused to spin rapidly about an axis. Parts that require a circular center hole are always rotated about a horizontal axis. Figure 8-1 shows a cross section of a true centrifugal casting mold, while Figure 8-2 illustrates a mold in position on the casting machine. Rotating the mold about a vertical axis produces a part having a section of a parabola as its inner shape. Since the slope of the sides depends upon the speed of mold rotation, internal tapers may be cast in this way. Other, special internal shapes may be produced by positioning the flask so that the rotation may take place about an inclined axis.

Another method of forming a mold cavity, resulting in a *permanent mold,* is to insert a cylindrical metal liner into a cylindrical flask so that the two objects are concentric. The outside diameter of the liner is smaller than the inside diameter (hole) of the flask, thus establishing the desired wall thickness of the casting. Following the application of a thin coating of a refractory wash, the metal is poured into the spinning flask. Upon cooling, the casting is shaken out of the mold and cleaned. Castings may be heat-treated if required. There is no need for a sprue, risers, or even a gating system for this unique casting method.

The revolution of molten metal about a central axis in the centrifugal casting process causes considerable pressure against the mold walls. Pressures have been calculated to exceed a centrifugal force of 100g. Mold walls are thicker than for static casting processes, which are carried out at atmospheric pressure. Pressure casting results in parts with a dense grain structure and, often, a superior exterior surface texture. Impurities consisting of oxide and/or sand particles are collected at the inner walls of the hollow

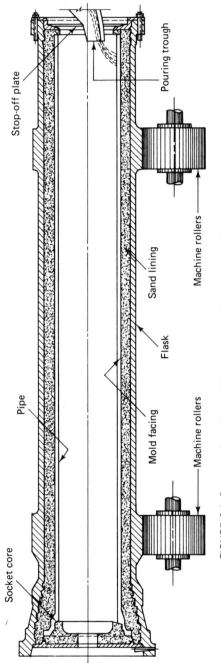

Stop-off plate

Pouring trough

Sand lining

Machine rollers

Flask

Pipe

Mold facing

Machine rollers

Socket core

FIGURE 8-1 Cross section of a mold for true centrifugal casting (Courtesy, Acipco Steel Products)

FIGURE 8-2 True centrifugal casting machine (Courtesy, Acipco Steel Products)

FIGURE 8-3 A view of a machine with a finished pipe being stripped from the mold. In the upper right hand corner, a machine ladle can be seen pouring a pipe in the adjacent machine (Courtesy, American Cast Iron Pipe Company)

89

castings and can be removed by subsequent machining operations. The wall thickness of the part can be controlled by the amount of molten metal poured into the mold cavity. The true centrifugal casting process is sometimes used to produce rings. Long tubular lengths are cast first and then slit or cut off into bands of the desired length. Figure 8-3 shows a view of a finished pipe being stripped from a centrifugal casting machine.

Semicentrifugal Casting

The main difference between this process and true centrifugal casting is that cores are frequently employed. Semicentrifugal casting is better adapted to the production of more complicated shapes with special central and offset cored features than the former process. Figure 8-4 shows head cores being carried on a conveyor to casting machines. Several identical molds may be stacked, one on the other, to increase production. A central gate is used which "feeds" two or more molds. Molds are commonly made of dry sand, green sand, or of metal. Special drag and cope-type flasks are used. The mold cavities are spun exclusively about a vertical axis in this process. While the rotational speed is lower than true centrifugal casting, the advantageous effects of the centrifugal force obtained by pouring molten metal into a spinning mold continue to be available; that is, outer portions of the casting result in a sound and dense grain structure. Semicentrifugal casting is often

FIGURE 8-4 A portion of the conveyor carrying head cores to the casting machines (Courtesy, American Cast Iron Pipe Company)

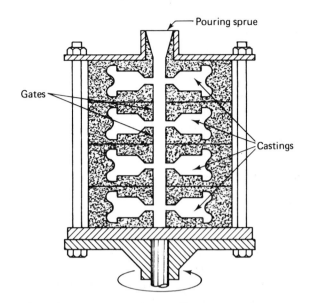

Pouring sprue

Gates

Castings

FIGURE 8-5 Parts cast by the centrifuge process. Centrifugal force is used to fill molds in cases where gravitational force would be insufficient

the preferred method when producing parts with hollow centers either circular or noncircular in shape.

Centrifuging

Centrifuging as compared to sand casting is a considerably faster production method. This process employs several molds arranged in wheel-like fashion around a circle, as shown in Figure 8-5. The pouring sprue is located at the hub of the "wheel," with radial connecting gates extending to each of the mold cavities. Molds, constructed of green or dry sand, plaster, steel, cast iron, or graphite, are prepared in special cope and drag flasks. The molds, rotated at relatively low speeds, are filled under pressure by the centrifugal force of molten metal supplied at the pouring sprue. The central axis of rotation may be either vertical or horizontal. This method, unlike castings produced by the true or semicentrifugal methods, is adaptable to an almost unlimited variety of irregular product shapes. While there is no symmetry requirement for parts made in this manner, the castings must be arranged in a mold so that the mold is balanced as it spins. In general, sand molds are used for single-piece or large parts. Permanent molds of graphite are used for small or medium production runs with a life up to 100 pours per mold. High production runs require metal molds whose life may average 500 or more castings. Cores may be used in the same way as for the semicentrifugal casting process.

91

Subsequent finishing operations on some parts are sometimes entirely unnecessary due to careful placement of gate locations. Cost reductions on some parts are often possible due to the adaptability of this process to quantity production. Special precautions must be taken in some cases to offset the possibility of severe centrifugal forces, which may cause the casting metal to penetrate into the mold walls. Molds must be extremely strong. As in sand casting, the skill of the foundryman plays an important part in the quality of the final castings.

PRODUCT APPLICATIONS

Cylindrical forms, produced by the true centrifugal process, are used for general purposes in industry, such as pump components, rolls, liners, hollow shafting, bearings, sleeves, hydraulic jack cylinders, and air cylinders. Figure 8-6 illustrates some typically massive parts. Centrifugal castings have been used, for many years, as components on surface and underwater ocean vessels. Such applications include large propeller shaft sleeves and stern tubes. The papermaking industry has long been a user of centrifugally cast cylinders for machine suction rolls, as in Figure 8-7. Textile machines use various roll bodies made by this process. Cylinders made of heat-resisting materials are found in industry as furnace rolls, glassmaking rolls, blowpipes,

(a)

FIGURE 8-6(a) Typical centrifugal castings: (a) chromium-copper mold liners (Courtesy, Ampco Metal Products Company)

FIGURE 8-6(b) A 900-lb trunion for a steel mill plating roll (Courtesy, Ampco Metal Products Company)

FIGURE 8-6(c) Manganese-bronze knuckle bushings for earth moving equipment. Each casting weighs 1300 lb (Courtesy, Ampco Metal Products Company)

93

FIGURE 8-7 Paper machine suction roll shell over 4 ft O.D. by 34 ft long.
It weighs 74,380 lb (Courtesy, Sandusky Foundry & Machine Company)

FIGURE 8-8 Nuclear test loop vessel completely assembled for a hydro-
static pressure test (Courtesy, Sandusky Foundry & Machine Company)

muffles, and retorts. Ammonia synthesis converter cartridges, chemical
reactor vessels, and heat-exchanger bodies are typical uses of centrifugally
cast products in the refinery and petrochemical fields. Designers of land
and sea nuclear power equipment are finding uses for centrifugal castings
as pressure-vessel bodies (Figure 8-8), reactor tubes, pressure piping, motor-
pump stator shells, and for other applications.

Disk-shaped parts such as wheels, rings, rollers, sheaves, pulleys,
flywheels, and gear blanks made by the semicentrifugal casting process are
important applications.

The shape of centrifuged castings, unlike those produced by the true or
semicentrifugal processes, may be irregular or nonsymmetrical. Such parts
compete favorably with other casting processes and forgings. Some common
examples are special hardware, bearing caps, machine parts, and even jewelry.

94

PROCESS SELECTION FACTORS

Shape: As a general rule, the choice of the particular method to be used is determined by the shape of the part to be cast. The true centrifugal casting process can only be used to produce hollow, thick-walled cylinders. Semi-centrifugal castings are restricted to modifications of disk shapes. Centrifuged castings may have a more complex shape and one that is not necessarily symmetrical about a central axis.

Properties: Compared to static casting, the mechanical properties of most metals are improved as much as 10 to 20% by centrifugal casting. Castings made in quick-chilled metal molds, as compared to those made in slow-chilled sand molds, have smaller grain size. Also, when liquid metal is poured into a spinning mold, it is accelerated outward toward the mold face, resulting in a part with a denser grain structure which is virtually free of porosity. Impurities consisting of slag and other nonmetallic inclusions are lighter than the poured metal and cling to the inside surface, where they may be removed by subsequent machining operations.

Size Range: Depending upon available equipment and methods, the maximum castable part size varies from foundry to foundry. One foundry reports a maximum weight limit for true centrifugal castings of 80,000 lb (approximately 36,000 kg) with an approximate range of 10 to 54 in. (or approximately 250 to 1370 mm) in outside diameter up to lengths of 34 ft (approximately 10 m), depending upon diameter and wall thickness. Cylinders with a diameter as small as 1½ in. (38.1 mm) have been successfully cast, although a practical minimum O.D. starts at 7 in. (177.8 mm). Wall thicknesses usually range from ½ to 5 in. (12.7 to 127 mm). As a general rule, the length should not exceed 15 times the diameter.

In semicentrifugal casting, the practical size limit for maximum diameter approximates 6 ft (approximately 1.82 m), with a recommended maximum length of about 4 ft (approximately 1.22 m).

The practical size range for centrifuged castings is largely dictated by equipment and mold capabilities. Castings have been successfully made with walls as thin as 0.015 in. (0.38 mm).

Materials: A wide choice of materials is available to the user of centrifugal castings. Carbon, low-alloy, or stainless steel may be cast as well as brass, bronze, copper alloys of nickel and aluminum, Monel, gray and ductile irons, and titanium.

Centrifugally Cast Laminates: A two-layer casting may be produced by first preparing a solidified thin-walled hollow cylinder in the usual way, adding a brazing flux and then pouring a dissimilar metal into the rotating mold.

Laminating in this way provides a corrosion-resistant surface to a cylinder, thus conserving a more expensive metal. Tubular products up to 9 in. (228.6 mm) O.D. and in lengths of 8 ft (approximately 2.43 m) have been processed in this novel way.

Dimensional Accuracy: Tolerances as large as $\pm \frac{1}{16}$ to $\frac{1}{8}$ in. (1.588 to 3.18 mm) in length and diameter are the general rule for true centrifugal castings. A similar tolerance applies to most of the larger categories of castings produced by the semicentrifugal process. Smaller centrifuged parts may be manufactured with tolerances comparable to sand castings, and even closer tolerances can be obtained under controlled conditions.

Some Limitations

It should be restated that true and semicentrifugal casting methods are restricted to the production of symmetrical circular-shaped parts. While a single external flange may be integrally cast at the center or at one end on pipes, tubing, shafts, and so on, it is not commercially practical to cast a flange at *both* ends. The physical properties of centrifuged castings are not equal to those obtained by true or semicentrifugal methods. Metal structure is inferior because of an inability to obtain uniform directional solidification. There is a tendency of alloy segregation in the case of some alloys due to separation by centrifugal force of the heavier elements from the base metal. As an example, copper-base alloys that contain more than 10% lead cannot be easily cast, and it is generally not practical to cast alloys that contain more than 5% tin and 10% antimony.

PRODUCT DESIGN FACTORS

Although centrifugal casting is one of the most specialized fields of the casting processes, most of the considerations normally associated with sand casting apply to a varying degree. Designers are encouraged to consult with casting engineers when considering the feasibility of manufacturing a given part by this process.

Fillets and Rounds: Fillets and rounds should be as large and generous as possible. Sharp corners on molds tend to reduce mold life.

Finish Allowance: The finish allowance is closely related to the size and shape, the metal, and the method of casting the part. The finish allowance for centrifugal castings may be as small as $\frac{1}{8}$ in. (3.18 mm) per surface on giant parts. The surface quality and accuracy limits on parts produced in graphite or metal molds is generally better than for parts made in sand molds, and the finish allowance, therefore, is correspondingly less. The process engineer must carefully evaluate the intended function of the part to determine which, if any, of the surfaces may be installed in service as-cast, or whether a light cleanup or a heavy cut is required.

Draft: Parts that are made in expendable centrifugal casting sand molds generally do not require any draft allowance. Permanent molds, on the other hand, are made with draft ranging from ⅛ to ¼ in. (3.18 to 6.35 mm) per foot. Because of the influence of many variables, it is not possible to specify a precise amount of draft allowance for all casting conditions. Some important variables are the techniques used in casting, the type of mold, and, of course, the size and shape of the part. Adequate draft allowance generally simplifies removal of the pattern from the mold.

Special Features: Procedures in the design of ribs, webs, spokes, rims, and other elements of a casting closely follow those for other casting processes. Successful castings require careful attention to important techniques of metal feeding. Premature solidification can be caused by feeding metal into partially restricted areas. Warpage and shrinkage due to problems associated with cooling can often lead to excessive casting rejections.

Surface Quality: Centrifugal sand mold castings have surface qualities that closely approximate conventional sand castings. Improved surface quality with an approximate value of 125 μin. is sometimes possible on parts made in metal or graphite molds.

REVIEW QUESTIONS

8.1. Compare the significant advantages of centrifugal casting to those of sand casting.

8.2. List the important differences between the three principal methods of producing centrifugal castings.

8.3. Explain why parts that require circular center holes are rotated about a horizontal axis in the true centrifugal casting method.

8.4. Explain how a square external shape may be produced in the true centrifugal casting method.

8.5. Which of the two methods, true centrifugal casting and semicentrifugal casting, are best suited to the production of complicated shapes? Why?

8.6. Explain why centrifuging is considered to be a faster production method than sand casting.

8.7. Contrast the type of molds and materials used in mold making for each of the three principal methods of producing centrifugal castings.

8.8. Insofar as the range of configurations that can reasonably be produced by centrifugal casting, which of the three principal methods would favorably compete with other casting methods and with forging? Explain the reason for your selection.

8.9. Why would you expect the mechanical properties of most metals to be improved by as much as 10 to 20% by centrifugal casting?

8.10. Contrast the materials commonly sand-cast to those in centrifugal casting.

8.11. Describe how centrifugally cast "laminates" are produced.

8.12. Explain the advantage of a laminated product.

8.13. Discuss the principal limitations of centrifugal casting.

8.14. Contrast centrifugal casting to sand casting in terms of dimensional accuracy, finish allowance, draft, and as-cast surface quality.

BIBLIOGRAPHY

*BASTIN, S. B., and D. R. BELL, "Centrifugal Casting and Its Applications," *Engineer's Digest,* Vol. 33, No. 3, Mar. 1972.

*BRYSON, F. E., "What's Happening in Centrifugal Casting," *Machine Design,* Oct. 19, 1972.

JOHNSON, H. V., *Manufacturing Processes,* Peoria, Ill.: Chas. A. Bennett Co., Inc., 1973, pp. 132–133.

*MISKA, K. H., "Centrifugal Casting Cuts Costs of Large Symmetrical Parts," *Materials Engineering,* June 1971.

*Abstracted with permission.

9

DIE CASTING

This popular and economical method of manufacturing has aptly been termed "the shortest distance between the raw material and the finished part." Of all the casting processes, it is considered the fastest. In die casting, molten metal is forced into the cavities of metal dies under high pressure. The molten metal is then held for a short time under pressure in the dies as the metal solidifies. The die blocks are then opened and the casting, with its assembly of sprue, runners, and gates, is ejected by pins. The die is closed again and the same cycle is repeated. Figure 9-1 illustrates a typical die casting production line. While die casting employs permanent molds or dies, it should not be confused with the permanent mold casting process, whereby the mold cavities are fed entirely by gravity force.

DESCRIPTION OF THE PROCESS

Dies

The first step in the die casting process is to construct an alloy steel die usually consisting of two sections which meet at a vertical parting line so they may be opened for the removal of the casting. Figure 9-2 shows a tool and die maker working on a die section. These sections are called the cover

FIGURE 9-1 A typical die casting production line (Courtesy, Koering Company)

FIGURE 9-2 A tool and die maker working on a die section for the die casting process (Courtesy, The Newton-New Haven Company)

FIGURE 9-3 An illustration showing a cover die half with four cavities (Courtesy, Zinc Institute, Inc.)

half and the ejector half. In Figure 9-3 there are four cavities machined into the two mating faces of the die block. The die block is held in the die casting machine and is locked securely by massive clamps or toggle linkages.

The heart of the die casting process is, of course, the die metal itself. The die must be stronger than those used in other casting processes because of the high pressures under which the molten metal is fed into the cavities. The metal dies must be able to withstand the repeated effects of the high temperatures of the molten metal; hence heat-resistant chromium or tungsten steels are generally used. A well-designed die, while costing up to and sometimes in excess of $5000, can produce over 500,000 castings. To hasten cooling, and to avoid the formation of air pockets in the casting, the die is often elaborately water- or air-cooled and vented. Cooling preserves the life of the dies. Outer surfaces on die castings tend to be harder than the interior as a result of the chill effect of the metal dies.

Types of Machines

While casting machines may differ in their method of forcing the molten metal into the die, they all work on the same principle: holding the die in place, opening and closing it, using the necessary pressure to keep it closed while casting, and finally, forcing the metal into the die.

101

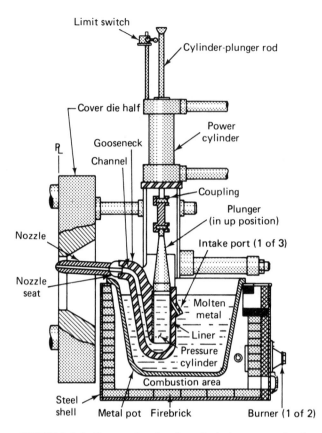

Limit switch

Cylinder-plunger rod

Cover die half

Power
cylinder

Gooseneck

Channel

P_L

Coupling

Plunger
(in up position)

Nozzle

Intake port (1 of 3)

Nozzle
seat

Molten
metal

Liner

Pressure
cylinder

Combustion area

Steel
shell

Metal pot   Firebrick

Burner (1 of 2)

FIGURE 9-4 A diagram showing the principal components of a
hot chamber die casting machine

There are two types of die casting machines in use. Figure 9-4 illustrates
a *gooseneck-type* machine, commonly called a hot-chamber machine, which
is used for low-melting-point materials such as zinc- and tin-base alloys.
Molten metal enters through an open port when the piston is raised to fill
the gooseneck. The molten metal is forced out of the gooseneck and into
the die by the plunger, which is actuated by the air cylinder. Hot-chamber
machines are fast in operation. One such machine can operate at 150
"shots" per minute. The *cold chamber-type* machine, Figure 9-5, employs
a plunger which takes the metal fed into a cold chamber and forces it into
the die. A shot of metal, sufficient in volume for one casting, is fed individually into the cold chamber. Die castings of aluminum, magnesium,
brass, and bronze are made on cold chamber machines.

Die casting machines may be either fully automatic or semiautomatic.
They range greatly in size. Some of the larger machines stand over 9 feet
(2.75 m) high and are 25 ft (7.6 m) long. There are a number of companies
whose entire production is devoted to making die castings of maximum size

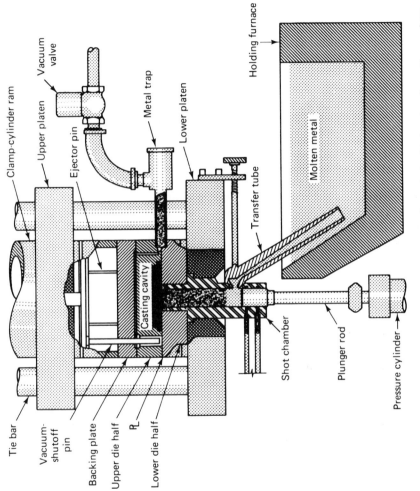

FIGURE 9-5 A diagram showing the principal components of a vertical cold chamber die casting machine. The die parting line is in a horizontal plane

103

and weight. Other companies specialize in producing small parts weighing just a few ounces. Some casting machines are capable of delivering the molten metal into the die at pressures up to 20,000 psi (1406 kg/cm^2) and are designed with provisions for ejecting the finished product from the mold cavity. While aluminum is usually cast at 5 psi (0.352 kg/cm^2), casting at higher pressures increases the density and reduces the porosity of the castings.

Secondary Operations

Recent developments in die casting technology now enable some secondary operations to be integrally performed in a single casting cycle. With this technique, milling, drilling, threading, counterboring, countersinking, and even some assembly operations (explained later in this chapter), can be performed directly on the casting before it is ejected from the die block.

PRODUCT APPLICATIONS

Die castings are characterized by their economy, durable finish, strength, lasting qualities, and mass-production techniques.

The average automobile uses up to 150 lb (68 kg) of die-cast parts. Die castings are also widely used in the aircraft industry, for household appliances, business machines, portable power tools, outboard motors, and for a number of other parts used in small engines, power lawn mowers, jewelry, parking meters, traffic signals, and so on. Figure 9-6 illustrates some typical product applications. The following list shows the tremendous versatility of product applications for die castings.

Household Equipment: Widely used in labor-saving home appliances as major structural parts as well as decorative and mechanical parts for mixers, fans, dishwashers, meat slicers, can openers, polishers, washer and dryer components, stoves and refrigerators, and so on.

Business Equipment: Components for printing register meters, time recorders, package sealers, offset duplicating machines, dictating machines, pencil sharpeners, staplers, marking machines, typewriters, adding machines, data-processing machines, and so on.

Hardware: The hardware industry has long been one of the principal users of die castings. Die-cast hardware includes a wide variety of cabinet handles, knobs, catches, latches, casters, pulleys, bathroom fixtures, electrical fixtures, and so on.

Industrial Equipment: Die castings are used for motor housings, components for vending machines, wheels, parts for hoisting equipment, motors, switches, weighing scales, and so on.

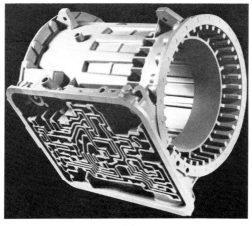

(a)

FIGURE 9-6 Typical product applications for die castings: (a) a die cast aluminum transmission case which weighs about 70 lb (Courtesy, Doehler-Jarvis, Division of National Lead Company)

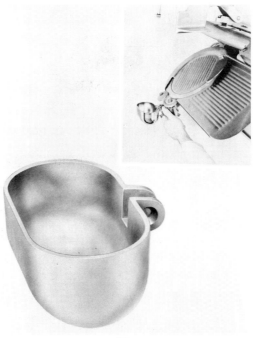

(b)

FIGURE 9-6(b) Wheel cover for a food machine. Casting is 6 in. long by 3 in. wide (Courtesy, The Newton—New Haven Company)

105

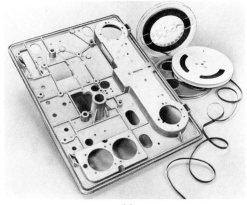

FIGURE 9-6(c) A die cast tape with 73 accurately positioned core holes (Courtesy, American Die Casting Institute)

(c)

FIGURE 9-6(d) A handle for a fire extinguisher. Previously the handle had been a four piece assembly (Courtesy, The Newton—New Haven Company)

(d)

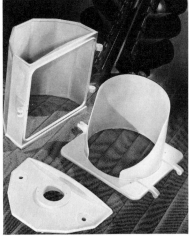

FIGURE 9-6(e) Die cast traffic signal hoods (Courtesy, Aluminum Company of America)

(e)

106

FIGURE 9-6(f) Small zinc alloy die cast gears in almost every conceivable configuration (Courtesy, Gries Reproducer Company)

(f)

Automotive: Windshield frames, window channels, fuel pumps, carburetors, decorative grilles, instrument panel housings, body trim, emblems, handles, heat and air-conditioning components, rear-view-mirror parts, switch housings, steering wheel hubs, and so on.

Tools: Unusual and complex die-cast shapes which are both tough and durable are used in the manufacture of modern electric and hand tools, such as electric planes, sanders, drills, saws, routers, mandrels, boring jigs, cutter-grinder attachments, tubing cutters, knife handles, air compresser parts, steel rule housings, die stocks, transit-level components, lawn sprinklers, grass shears, components for snow blowers, lawn mowers, sweepers, and so on.

Music and Communications: Components for microphones, fire alarm systems, telephones, television sets, speakers, drums, clarinets, record players, intercommunication units, and so on.

Toys: Pistols, electric trains, model aircraft, automobiles, game equipment, and so on.

Photographic: Components for film holders, slide projectors, X-ray meters cameras, movie projectors, and so on.

PROCESS SELECTION FACTORS

Materials: There are six principal families of die casting alloys: zinc, aluminum, magnesium, copper, lead, and tin. Of the six, zinc accounts for the greatest tonnage and is by far the most widely cast metal. The leading reasons for the popularity of zinc as a die casting metal include its ability to be cast in thin sections with relatively sharp corners and the fact that zinc is more rapidly castable than other die casting alloys. In short, zinc die castings can be produced faster and cheaper than other metals.

107

Size and Weight: From the standpoint of size, die castings are extremely versatile. There is a case on record of a die cast watch part so tiny that it takes 250,000 of them to weigh 1 lb (0.4536 kg). At the other extreme, a die-cast automotive dash panel has been cast weighing 33 lb (approximately 15 kg) and measuring 60 in. long, 11 ¾ in. deep, and 12½ in. high (or 1524 mm by 298.45 mm by 317.5 mm). Aluminum die castings weighing over 100 lb (45 kg) are not unusual.

Inserts: Both metallic and nonmetallic inserts may be used. Special features such as threaded or knurled studs, steel shafts, stiffening devices, sleeves or bearings, screens, special fittings, or wear pads of harder metals may be permanently anchored into the die casting. It is possible to combine the assembly of two parallel strips of fabric with die casting zipper components in one operation.

Production Rate: As previously stated, die casting is considered the fastest of all the metal casting processes. The production rate of the higher-melting-point alloys (magnesium, aluminum, and copper) using multiple cavity dies may reach 200 to 300 castings per hour with a progressive increase in production rate for the lower-melting-point alloys. In general, zinc die castings may be produced at an average rate of up to 500 parts per hour.

Scrap Loss: There is very little waste. Excess flash, and the sprue, runners, and gates once removed, may be remelted and reused.

Tolerances: One of the outstanding advantages of die casting is the precision with which parts can be made. As an example, the as-cast linear dimensional tolerance for a zinc part is ±0.003 in. (0.08 mm) up to 1 in. (25.4 mm) and an additional tolerance of ±0.001 in. (0.03 mm) for each additional inch can be maintained. Since die castings are produced in accurately machined dies, most parts may be used in the as-cast condition, with no machining necessary.

Complexity of Parts: The use of pressure in die casting permits the designer to consider die casting for the production of extremely complex parts. Cavities fed by gravity (sand, permanent mold, etc.) will not produce the detail found in most die casting applications. Special die casting techniques can be incorporated so that two or more small parts may be integrally cast into pivoting or movable assemblies. Products cast in this manner include notebook binder rings, small link chains, latches, and so on.

Surface Texture: The as-cast surface-finish value of die castings, sometimes called "hardware finish," may range from 40 to 100 μin. A variety of interesting and decorative textured surfaces are also available. Such surfaces add to the appearance of the part, and actually improve the subsurface quality of the part. Diecast surfaces later may be finished with a variety of mechanical, plated, chemical, or organic finishes.

Some Limitations

Minimum Production: The process is not suitable for short-run jobs and, compared to other casting processes, tool and die costs are higher.

Materials: The high-pressure die casting process is limited to nonferrous metals. Care should be exercised in selecting appropriate metals for specific product applications. Because of the high melting point, brass alloys are damaging to the dies. While zinc alloys have a lower melting point and die life is at a maximum, zinc castings exhibit low creep strength, become increasingly brittle at low temperatures, and corrode in some environments. Magnesium alloys are severely attacked by seawater and are affected in a humid tropical environment.

Flash Removal: Practically all die castings have a thin projecting fin or flash of excess metal at the parting line. While this condition is not necessarily restricted to this particular process, the processing sequence must include a trimming operation following the removal of the castings from the dies.

Secondary Finishing: In general, except for buffing, die-cast parts may be used in as-cast condition. Excessive finishing may break through the dense skin and expose underlying porosity.

PRODUCT DESIGN FACTORS

Requirements of Die Casting Metals: To be castable, the metal must have sufficient fluidity to permit filling out fine detail and thin sections as well as in the production of a good surface texture. Metals with low melting points conserve fuel and minimize the heat shock on the die surfaces. To be castable, an alloy must fulfill a considerable number of specific requirements. These may be divided into two groups: (1) those which affect the casting operation, and (2) those which affect the properties, both mechanical and physical, of the casting produced.

In general, the choice of an alloy for a particular die casting is based upon the following considerations: cost (usually figured on a per casting basis rather than per pound); ease of casting; physical, mechanical, and chemical properties; ease of machining or finishing; cost of dies (including upkeep); overall cost in comparison with other manufacturing processes, such as permanent mold casting, powder metallurgy, stampings, and forgings.

Dies: The shape of the die cavity must closely correspond to the dimensions of the required part. Allowances must be made to compensate for draft and shrinkage. Dies may consist of only one or two mold cavities or they may contain up to 10 or 12 cavities. Single-cavity dies are generally used for large or complex castings. When the quantity of parts required is large and the parts are relatively small, a multiple cavity is used. Combination dies may also be used to produce two or more different parts gated from a common sprue. To prolong die life, most dies are made with cooling passages through which water or air is allowed to flow. Ejector pins are installed into the ejector half to facilitate the removal of the castings.

Cores: While cores add to the cost of the dies, their use may usually be justified in terms of cost reductions in subsequent machining operations. In some cases, cores may be the only practical way of producing functional internal configurations. Coring may also be employed to reduce the weight of the castings as well as to help keep the sections more uniform in size.

Two types of cores are used, fixed and movable. *Fixed cores* consist of stationary steel pins integrally built into the die. To operate, the axis of the fixed core must be parallel to the die motion so that it slides on or within the dies. These are the least expensive to produce and generally give the most accurate results. It is not necessary to remove fixed cores before ejecting from the casting. *Knockout pins* (also called *ejector pins*) force the casting from the fixed cores as the casting is pushed out of the die cavity. Cores with an axis not parallel to the die movement, shown in Figure 9-7, are called *movable cores.* Considerable attention must be given to the direction from which the cores are moved or retracted from the mold. Some movable cores are withdrawn by a separate mechanism before the casting is ejected from the die. Other types of movable cores consist of loose pieces which are inserted into the die at the beginning of each casting cycle and, of necessity, removed from the casting after it is ejected from the die.

Cores require draft which varies in amount with the casting metal, the size of the hole diameter, and whether the core type is fixed or movable.

The Gating System: The gating system ensures that a stable flow of liquid metal is supplied to all parts of the casting during the feeding period. Overflow wells are normally machined in the die faces, which prevents the entrapment of undesirable oxides, lubricants, and other impurities within the die cavity. These recesses are connected to the die cavity by a runner. Properly designed, the gating system controls metal turbulence and provides a sufficient amount of feeding material to reduce shrinkage.

Venting the Mold: Entrapped air may be intentionally permitted to escape between the surfaces of the die halves and around the core walls. Some dies are made with small vent holes for this purpose. Excess metal that flows into vent holes or overflow wells is trimmed off after the casting solidifies.

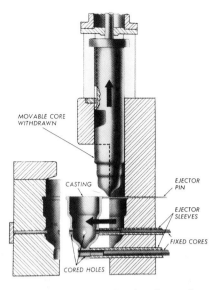

MOVABLE CORE
WITHDRAWN

CASTING

EJECTOR
PIN

EJECTOR
SLEEVES

FIXED CORES

CORED HOLES

FIGURE 9-7 A set-up showing the application of a movable core (Courtesy, Aluminum Company of America)

Section Thickness: The minimum section thickness depends on the metal to be cast and, generally, on the size of the casting. Recommended minimum thicknesses for medium-size castings are as follows: 0.015 in. (0.38 mm) for zinc; 0.030 in. (0.76 mm) for aluminum, tin, and lead; and 0.050 in. (1.27 mm) for magnesium, brass, bronze, and copper. The recommended maximum section thickness is ½ inch (12.7 mm). Deviations from recommended maximum and minimum section thickness result in an increased part-rejection rate.

Draft: The draft allowance for die castings can be considerably less than for sand casting or for shell molding. The amount of draft varies with the alloy used and the depth of the wall. As a general rule, most zinc-base alloys require about 0.005 in./in. (0.005 mm/mm) of draft. The recommended draft allowance for magnesium and aluminum alloys is 0.010 in./in. (0.010 mm/mm) and 0.015 in./in. (0.015 mm/mm) for brass alloys.

Machining Allowance: Most die castings are used in the as-cast condition, thus eliminating the need for machining operations. Excess flash projections around the edges of the castings must be trimmed, and in some cases, functional holes must be drilled or tapped. When subsequent machining operations are necessary, a recommended allowance is from 0.010 to 0.020 in. (0.25 to 0.51 mm) per surface. In general, the need for machining is kept to a minimum to avoid uncovering subsurface porosity and to reduce the cost.

The following series of examples, Figures 9-8 through 9-13, illustrate some important design factors which the designer must consider when designing parts for the die casting process.

111

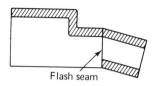
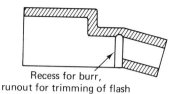

Flash seam

Recess for burr,
runout for trimming of flash

(a)

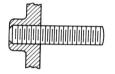
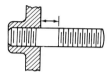

(b)

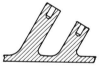

(c)

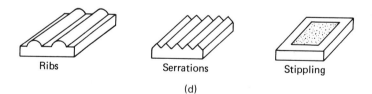

Ribs   Serrations   Stippling

(d)

FIGURE 9-8 Design factors in die casting: (a) simplify flash and burr removal on inside surfaces; (b) to prevent flash from forming in threads of inserts, leave space between surface of casting and the threads; (c) sloping bosses should, when possible, be designed for straight die parting to avoid expensive die construction; (d) use these methods to break large, plain areas and improve appearance (Courtesy, Zinc Institute, Inc.)

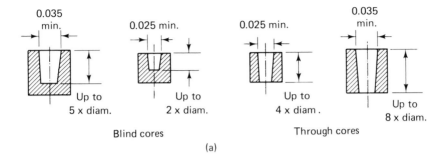

Blind cores   Through cores

(a)

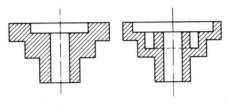

(b)

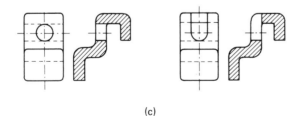

(c)

FIGURE 9-9 Design factors in die casting: (a) cored holes must be tapered. Where straight holes are required, use a secondary reaming operation. The recommended taper is 0.016 inch per inch on diameter; minimum taper on core diameter is 0.005 inch per inch; (b) add ribs for strength where a heavy section has been cored to gain a uniform wall; (c) side cores can often be avoided by a modification such as this (Courtesy, Zinc Institute, Inc.)

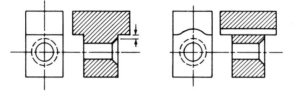

(a)

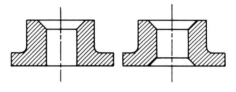

(b)

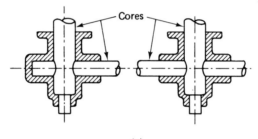

(c)

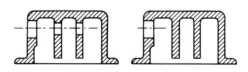

(d)

FIGURE 9-10 Design factors in die casting: (a) avoid thin walls in die around cores; (b) cored through holes that are to be tapped should have countersink on both sides; (c) interlocking cores are rarely justified. Preferred design is two small cores bearing against a larger core; (d) avoid moving cores that must fit accurately in both die halves. Holes on inner elements can be produced by drilling after casting (Courtesy, Zinc Institute, Inc.)

(a)

(b)

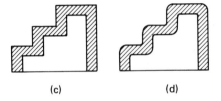

(c)   (d)

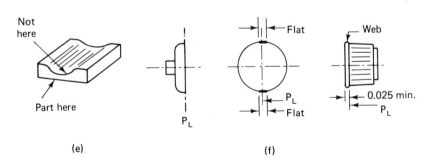

(e)   (f)

FIGURE 9-11 Design factors in die casting: (a) fillets between beads avoid knife edges on die; (b) provide maximum radius on all corners; (c) keep parting lines in one plane. Avoid irregular parting lines; (d) parting along an edge usually results in lower tooling costs, and facilitates production; (e) flats allow for closer diametral tolerance when parting on a diameter; (f) when casting knurls, gears, or serrations, an added web simplifies the trimming operation (Courtesy, Zinc Institute, Inc.)

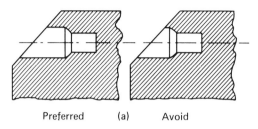

Preferred   (a)   Avoid

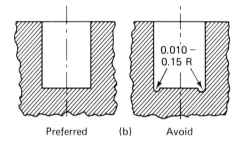

Preferred   (b)   Avoid

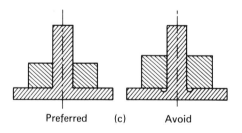

Preferred   (c)   Avoid

FIGURE 9-12 Design factors in die casting: (a) thin cast section is difficult to fill; (b) cast sharp interior corners are difficult to maintain, but radius improves design; (c) where corners must fit together on two castings, allow a relief (Courtesy, Zinc Institute, Inc.)

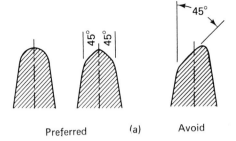

Preferred   (a)   Avoid

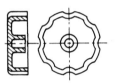 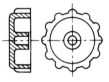

Preferred   Avoid

(b)

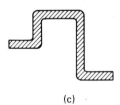

(c)

FIGURE 9-13 Design factors in die casting: (a) some contours are easier than others to mill into die; (b) convex finger grips are easy to mill; concave notches are difficult and expensive; (c) wall sections should be uniform where possible (Courtesy, Zinc Institute, Inc.)

OTHER DIE CASTING TECHNIQUES

Assembly by Die Casting

Assembly by die casting is a technique that can be employed to join two or more components by casting metal around small sections of the parts. In Figure 9-14, for example, a gear is die-cast directly on a steel shaft. The shaft is held in the die in the required position by means of a specially designed fixture. The metal is injected into the die cavity. Solidification occurs almost instantaneously. The gear is formed in the die as the metal contracts and interlocks with the knurled section machined on the shaft. Parts are rigidly held together by the combined effects of radial and axial shrinkage of the cooled metal. The process does not rely on adhesion or bonding.

This innovative technique of die casting is used to assemble a wide range of parts, including hubs cast into fan blades, armature and shafts for small fan motors, small gears and shafts for meters and clocks, governor-case assemblies for telephones, magnet assemblies for speedometers and synchronous motors, cable ends for small-vehicle brakes, mandrels assembled with abrasive wheels, and film gates for motion picture cameras. Figure 9-8(b) illustrates a typical example of components integrally assembled by die casting.

INJECTED
ZINC ALLOY

FIGURE 9-14 A method of assembling a drive gear to a steel shaft. During the torque test, the shaft sheared at 26 ft lb. Injected metal hub showed no sign of failure (Courtesy, Fisher Gauge Ltd.)

There are a number of significant advantages of this technique. Substantial cost reductions are obtained as a result of the high assembly speed. In the case of a simple two-piece assembly, for example, as many as 750 assemblies per hour can be obtained. Integrally cast parts can be accurately assembled because fixtures are used to position each part independently of every other part. Good size uniformity from assembly to assembly is possible. The process requires fewer finishing operations. Prior to being ejected from the holding fixture, the assemblies are rotated within the die to shear the flash and sprues from the assembled components.

Zinc alloys and lead alloys are the principal metals used in this process. The capacity of fixture dies currently available is restricted to about 0.3 in.^{3} (4.917 cm^3) maximum of molten metal. Because the volume of metal ejected for each assembly is relatively small, little heat is dissipated through the fixture and the parts being assembled. Thus, the parts are not damaged by the heat and can be handled immediately after assembly. Various components made of materials such as steel, abrasives, brass, plastic, paper, and glass may often by assembled by this method.

Limitations of the assembly-by-die-casting technique include the size of the parts that can be assembled, their shape, and the stress they can withstand.

Parts assembled by this method should have surfaces onto which the metal can shrink or lock. For this purpose, various configurations of dovetails, knurls, or slots are commonly used. There is no need for special cleaning and surface preparation.

Low-Pressure Die Casting

Low-pressure die casting is a process that is adaptable for producing parts in quantities which exceed those made in permanent molds but are not large enough to warrant conventional high-pressure die casting techniques. While a comparatively new process, it has developed to the point where it is now regarded as a major foundry technique.

The basic principle of this novel technique is explained as follows. A metal die is installed above a sealed furnace. The lower die half (fixed die) is connected to a metal feed tube. The casting cycle begins with preheating the dies. When air is introduced into the furnace under relatively low pressure (usually between 5 and 10 psi (0.352 to 0.704 kg/cm^2) the molten metal rises up the central feed tube to enter and fill the die cavity. Gas and air are vented from the die in advance of the metal. The metal is allowed to solidify and the air pressure is then removed, exhausting the furnace to atmosphere. The die is opened, the casting is removed, and the whole procedure is repeated for subsequent castings. There is no need for risers or conventional gravity runners. Electric and pneumatic controls are provided so that castings can be produced on a fully automatic basis. The only manual operation is the unloading of the casting from the removal arm, which, in some cases, is mechanized.

Low-pressure die casting was one of the best and longest kept secrets of a few foundries. As a result of this secrecy and the scarcity of available technical knowledge, the few foundries that had the courage to try the method had to develop their own techniques and equipment, and often encountered serious difficulties. Considerable facilities and finance were needed to mount a constructive development program.

Dies for this process are normally made of Meehanite cast iron. Special steel inserts are often positioned in the die at appropriate cooling locations. Dies may be made on a single- or multiple-impression basis. The die cavities are coated with a refractory wash to facilitate casting removal. The die material used must enable the die to operate efficiently, taking into account thermal balance and shocks, coating adhesion, cooling media, erosion, wear resistance, and maintenance. Die design and construction is kept as compact as possible, as the heavier the die, the more heat that is required for preheating. Also, an unnecessarily large die tends to retain its heat, which slows the rate of heat dispersion during casting.

While the process of nonturbulent filling of die cavities avoids oxides and gas inclusion, filling must still be rapid enough to ensure good formation of the casting. When compared to the permanent mold casting process, low-pressure die casting is reported to yield sounder castings because casting pressures can be adjusted to suit particular casting conditions. In addition, the process is adaptable to the casting of parts with irregular wall thicknesses. The combination of precise control, both of the speed of the cavity filling and of the pressure applied, makes it possible to eliminate flow marks completely, to feed thick sections through thin ones, and to reduce minimum wall thickness by as much as 20%. Some shapes that cannot be produced by permanent mold casting are possible by die casting.

The major application of the low-pressure process at present is in the casting of aluminum alloys. However, several foundries already cast magnesium very successfully and economically, and the extension of the process to other metals is inevitable in the future.

Ferrous Die Casting

Ferrous die casting is a process for producing most shapes in ferrous materials that would easily be die-cast in aluminum. It is considered to be a lower-cost alternative to some types of forging and investment casting. While it is an outgrowth of the traditional die casting process, there are some important differences. The two processes substantially differ in die design concepts as related to gates, runners, and injection systems; in the materials used to make the dies; and in certain other casting procedures. It should be noted that an impressive amount of new technology has been specifically developed for ferrous die casting which has led to nearly 100 patents on the various technical aspects.

Early developments in ferrous die casting began over a decade ago in the Refractory Metals Department of General Electric Company. Soon afterward, the Federal Die Casting Company of Chicago became interested in this technique. Finally, the Ferrodyne Corporation was formed by GE to investigate the commercial possibilities of ferrous die casting.

According to recently published technical reports, any ferrous metal or alloy that has been cast successfully by any other process can also be die cast. Stainless steels are reported to be the easiest metal to cast, followed by the high-alloy steels. Other ferrous die cast metals that have been cast include gray iron, malleable irons, low-carbon steels, nickel alloy steels, and certain nickel and cobalt alloys.

One of the leading problems associated with the process is the manufacture of a die so that the pouring of alloys of high melting points (around 2600°F) does not damage the molding surfaces. Refractory metals such as tungsten and molybdenum show promise of overcoming this limitation. Molybdenum continues to be the leading die material because it can be machined and because there is less chance of heat checking due to thermal stresses.

Even at this comparatively early stage of development, parts made by ferrous die casting are regarded as competitive, under some conditions, with similar parts made by forging, shell molding, and sand and investment casting. It is possible to cast ferrous parts with a better surface finish and with physical properties at least equal to, and in some cases superior to, those produced by other casting processes. One of the strong points which favor ferrous die casting is that the parts are relatively sound and free from porosity. When compared to forgings, die-cast parts can be produced to closer tolerances and, in many respects, have equal or superior physical properties. The structure of a heat-treated ferrous die casting is often finer than that of a forging. The rapid chill of surface metal against the die produces a fine-grained structure that is good for resisting fatigue.

The size limitations of parts produced by this process are governed by the size of the equipment. Current production has led to parts with a maximum dimension of 12 in. (304.8 mm), with wall thicknesses ranging from ⅛ in. (3.18 mm) to over ½ in. (12.7 mm). Parts weighing up to 8 lb (3.6 kg) have been successfully produced. Parts with surface finishes up to 50 μin. can be die-cast. Holes can be formed perpendicular to the parting line.

There are two important limitations of ferrous die casting: (1) compared to sand casting or shell molding, the complexity of shape and coring remains restricted; and (2) the comparative cost of tooling is said to be 25 to 50% higher than that of conventional nonferrous die casting.

The best potential, according to available reports, is in builders' and marine hardware and for other small and relatively intricate components, such as latches, hand tools, and valve parts. The process shows considerable promise for both simple and sophisticated parts. It is predicted that ferrous die casting will soon be adopted as a major production process.

REVIEW QUESTIONS

9.1. Make a brief outline describing the process of die casting. Prepare a sketch where appropriate.

9.2. Explain the essential differences between the hot-chamber die casting process and the cold-chamber die casting process.

9.3. List the principal advantages of die casting.

9.4. List materials that are commonly die cast.

9.5. Describe the principal advantages obtained by using metallic and nonmetallic inserts in a die-cast product.

9.6. What are the important factors in the die casting process which result in a production rate superior to all other casting processes?

9.7. Explain why die casting is not a suitable process for short-run jobs.

9.8. Explain why excessive secondary finishing is not recommended on die castings.

9.9. List the factors that must be considered when selecting a particular alloy material for a die casting.

9.10. Describe the difference between fixed and movable cores.

9.11. Explain how the use of cores in die casting usually provides greater versatility in product design.

9.12. Prepare a table showing the relative recommendations for minimum section thickness, draft, machining allowance and tolerances for die casting, investment casting, and permanent mold casting.

9.13. Explain the significant advantages of the technique known as "assembly by die casting."

9.14. List the important limitations that should be considered when evaluating the technique of integrally cast parts.

9.15. Describe the similarities and differences between conventional high-pressure die casting and low-pressure die casting.

9.16. List the principal limitations of ferrous die casting.

PROBLEMS

9.1. Figures 9-P1 (a-b) illustrates small dies. Each is shown with a casting in position in the die cavity. Due to the tendency of the castings to cling to the cavity, ejector pins are used to push the casting out of the ejector portion and off the core. There is also a tendency of a casting to shrink on a core as it cools. Ejector pins should be made to bear at position on the casting where the impression they leave will not be particularly noticeable or objectionable. In addition, they should be placed so that the casting is not permanently distorted by the pressure they exert. Using a sketch show where you would place the ejector pins on the dies shown in Figure 9-P1 (a-b).

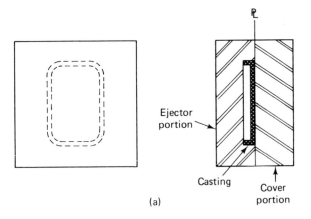

(a)

(Sprue hole, runner and gate not shown)

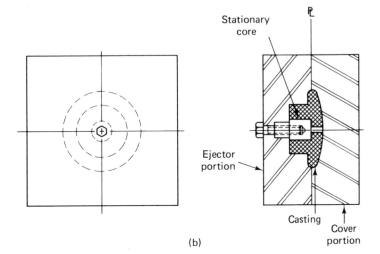

(b)

(Sprue hole, runner and gate not shown)

FIGURE 9-P1

9.2. Figure 9-P2 shows a die for producing a casting with a hollow circular shape. Two cores are illustrated in their final casting position. Which of the cores can be an integral part of the die (stationary) and which core should be movable? State the reasons for your recommendations.

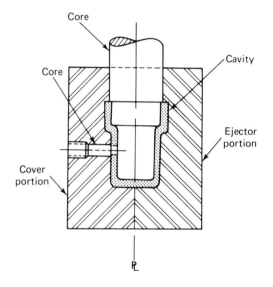

(Sprue hole, runner and gate not shown)

FIGURE 9-P2

9.3. Figure 9-P3 (a) illustrates the design originally adopted for a die casting. Figure 9-P3 (b) shows how the design was altered resulting in a better die parting location. Explain how the alterations improved the final appearance of the part from the standpoint of residual blemishes from flash removal.

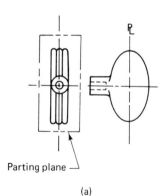

(a)

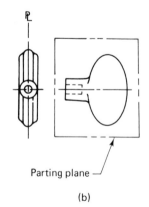

(b)

FIGURE 9-P3

9.4. When casting thin parts from ductile alloys (zinc-based alloys, for example) it is often advantageous to cast them flat and then bend them to the desired arc shape as shown in Figure 9-P4. Explain the main advantages of this technique. Also, list a factor which might influence the die designer against a secondary bending operation.

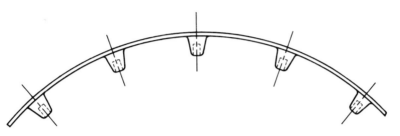

FIGURE 9-P4

9.5. From the point of view of flash removal and final product appearance, explain why the casting condition shown in Figure 9-P5 (a) is considered superior to that shown in Figure 9-P5 (b).

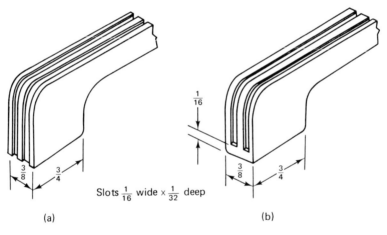

Slots $\frac{1}{16}$ wide × $\frac{1}{32}$ deep

(a)   (b)

FIGURE 9-P5

9.6. Explain why the design of the casting shown in Figure 9P6(a) would be more difficult and costly to produce than the one shown in Figure 9-P6 (b).

FIGURE 9-P6

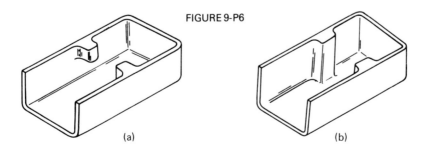

(a)   (b)

9.7. Explain why the junctions shown in Figure 9-P7 are not considered good casting design.

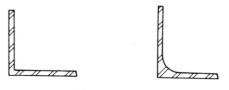

FIGURE 9-P7

9.8. Explain why a casting with a sharp irregular edge as shown in Figure 9-P8 results in production difficulties.

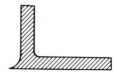

FIGURE 9-P8

9.9. Explain why the depression shown under the junction (a) and the cylindrical boss (b) in Figure 9-P9 is recommended.

FIGURE 9-P9

9.10. The design for the part shown in Figure 9-P10 (a) is recommended. Explain why the design shown in **Figure 9-P10** (b) is *not* recommended.

FIGURE 9-P10

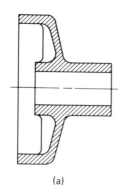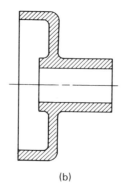

(a)   (b)

9.11. Why are the rib designs shown in Figure 9-P11 not recommended?

FIGURE 9-P11

9.12. In the die casting shown in Figure 9-P12, the designer has specified a drill spot in preference to coring a through hole. Explain why this technique may be preferred in some cases.

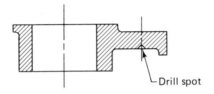

Drill spot

FIGURE 9-P12

9.13. Figure 9-P13 illustrates a case where two die castings must fit together. How should the design of one or both castings be altered to ensure a good contact along face A?

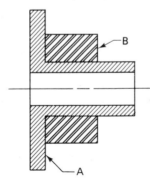

B

A

FIGURE 9-P13

9.14. Figure 9-P14 shows a steel stud which is integrally assembled during casting with the part shown. Suggest a design change so that flash will not accumulate in the threads on the extended portion of the insert.

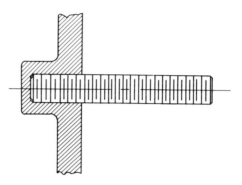

FIGURE 9-P14

BIBLIOGRAPHY

"Die Casting," *Machinery's Handbok,* 20th edition, New York: Industrial Press, Inc., 1975, pp. 2178–2181.

"Die Casting," *Metals Handbook,* Vol. 5, *Forging and Casting,* 8th edition, Metals Park, Ohio: American Society for Metals, pp. 285–334.

*Die Casting by Arwood, Rockleigh, N.J.: Arwood Corporation.

*Die Casting with Zinc, New York: Zinc Institute, Inc., 1965.

HURD, D. T., "Status Report on Ferrous Die Casting," *Metal Progress,* Feb. 1974, pp. 73–74.

*JAY, F. H., "Joining Parts by Die Casting," *Machine Design,* Apr. 15, 1971.

JOHNSON, H. V., *Manufacturing Processes,* Peoria, Ill.: Chas A. Bennett Co., Inc., 1973, pp. 125–131.

*KOHL, R. "What's with Ferrous Die Casting," *Machine Design,* Nov. 2, 1972.

"Machining of Zinc Alloy Die Castings," *Metals Handbook,* Vol. 3, *Machining,* 8th edition, Metals Park, Ohio: American Society for Metals, pp. 507–509.

*WHARTEY, G. G., "Low Pressure Casting," Paper 7372, *7th SDCE International Die Casting Conference,* Oct. 16-19, 1972.

Why Zinc?, BZ-5, New York: Zinc Institute, Inc., 1975.

Zinc Die Casting—Molten Metal to Finished Part—Direct, BZ-1, New York: Zinc Institute, Inc., 1975.

    *Abstracted with permission.

10

SLUSH CASTING

DESCRIPTION OF THE PROCESS

Slush casting is a method by which hollow, thin-walled products are made. The process consists of pouring molten metal into a closed permanent mold, generally of bronze. The mold is preheated to a suitable temperature before pouring. Molds may consist of two, three, or four sections which are held together with C-clamps during casting operations. Bronze is preferred for molds because it can be easily cast and machined into the desired design at moderate expense as compared to costly steel dies. The poured metal immediately reacts by chilling and solidifying against the inner walls of the mold. The molds are then quickly inverted to permit the remaining liquid metal to run out. In some cases, the molds are slung or otherwise agitated to enhance the ability of the metal to flow into finely detailed crevices in the mold. Castings are removed by separating the mold halves. A coating of lamp black is often applied to the mold walls to prevent sticking of the casting. Cores are not used in this process.

The approximate wall thickness of the casting is regulated by the timing of the pouring and inversion operations. The process is not suitable for producing products with uniform wall thicknesses. The process results in the formation of a product having a thin shell whose outer surfaces precisely correspond to the mold cavity configuration.

129

PRODUCT APPLICATIONS

In general, products are restricted to art, decorative, and ornamental uses, although industrial product applications are increasingly being developed. Spouts for kettles and teapots, toys, bookends, lamp bases, and ornamental objects such as souvenir statues, various kinds of trophies, and other novelties are produced by the slush casting process. Certain types of thin-walled plumbing fixtures are also produced by this process.

PROCESS SELECTION FACTORS

Surface Quality: The outer surfaces of slush castings have a bright and superior surface texture that is well suited for plating. Castings made by this process are often painted or otherwise finished to resemble bronze, silver, or other, more expensive materials.

Materials: The materials commonly used to produce slush molded castings are lead, zinc, and tin alloys. Presently, only nonferrous metals may be cast by this process.

Production Rate: The process is considered economical for low-quantity production runs. On the average, 60 to 80 units per hour may be cast.

Dies: Die costs are relatively low. With reasonable care, a bronze mold can be expected to deliver as many as 50,000 castings without any upkeep expense.

Some Limitations: The interior surface of the castings is rough. When this consideration is undesirable for a specific product application, this may become an important disadvantage. Because of the thin wall, the products are weaker than parts produced by die casting.

Also, a casting made by this process does not have a dense surface skin like a die casting to give it increased strength.

REVIEW QUESTIONS

10.1. Make a brief outline describing the process of slush casting. Prepare a sketch where appropriate.

10.2. Explain the major reason why bronze is the preferred mold material.

10.3. Explain how the mold wall thickness of a slush casting is regulated.

10.4. List some commonly used metals for slush casting.

10.5. Give the principal limitations of this process.

BIBLIOGRAPHY

CHOATE, S., *Creative Casting,* New York: Crown Publishers, Inc., 1966, pp. 77–89.

11

SUMMARY OF CASTING PRINCIPLES

Casting process	Leading advantages	Leading disadvantages
Green sand castings	Least expensive method of producing a mold; flasks are ready for use in a minimum amount of time; dimensional accuracy good across the parting line; less danger of hot tearing of castings than in other casting processes; almost no size or shape limit.	Sand control more critical than dry sand method; erosion of the mold more common for large castings; as weight of casting increases, the surface finish deteriorates and the dimensional accuracy decreases; some machining is necessary on castings.
Dry sand castings	Stronger and less susceptible to handling damage than green	More susceptible to hot tears than green sand molds; more flask equipment is needed to

Casting process	Leading advantages	Leading disadvantages
	sand molds; overall dimensional accuracy is better; surface finish is better due to the wash coating.	produce the same number of castings because the process cycle is longer; the production rate is slower; limited to production of smaller size parts.
Shell mold castings	Greater dimensional accuracy than sand casting; smoother surfaces, resulting in minimizing machining operations; higher production rate.	Maximum size and weight of casting is limited; high cost of metal patterns; high cost of resin binder; relative inflexibility in gating and risering which must be incorporated into the shell mold pattern; more equipment and control facilities are needed, ovens for heating metal patterns, etc.; there is a restriction on metals which may be cast.
Investment castings	Can produce castings with extremely fine detail, excellent surface finish; greater accuracy possible than by other methods, no parting-line tolerance; can mass-produce complex shapes which are difficult or impossible to produce by other methods; almost any metal can be cast; castings up to 50 lb (approx. 23 kg) or more sometimes feasible; little or no finishing (except gate removal) with no flash; close control of grain size, orientation,	There is a practical size and weight limit due to physical and economic considerations; in general, castings not practical over 10 lb (4.54 kg); initial tooling costs are high; labor costs are high due to many steps which are required to produce casting from start to finish.

Casting process	Leading advantages	Leading disadvantages
	and directional solidification, which results in close control of mechanical properties.	
Permanent mold castings	Castings may be produced with more uniform grain structures than by sand casting; with closer dimensional tolerances; superior surface finish; higher production rate; minimum of scrap; mold life may exceed 25,000 pours; permanent molded castings are inherently stronger than die castings.	Process most suitable for small castings with reasonably simple shapes; some shapes not possible to produce due to location of parting line or difficulty in removing casting from the mold; not all alloys are suitable due to inability to cast high melting point metals such as steel; considered expensive for low quantity production due to high mold costs; relatively slow production rate; difficult to control metal temperature from one pour to the next.
Die castings	More complex shapes are possible than by permanent mold casting; castings with thinner walls, greater length-to-thickness ratio, and dimensional accuracy are possible; due to multiple cavities in nonexpendable molds operating within a limited timing cycle, production rates are highest of all casting processes; metal cost is lowest due to the	There is a practical size and weight limit generally less than 10 lb (4.5 kg), but may be as much as 50 lb (23 kg) in special cases; porosity may be a problem due to air entrapment in the dies; specialized facilities, machines, and equipment costs are high; die preparation costs are high; process is limited to producing nonferrous metals whose melting points do not exceed that of copper base alloys; steel castings have been produced in special cases.

Casting process	Leading advantages	Leading disadvantages
	ability to cast thinner sections; products can be plated with a minimum of surface preparation; under special conditions, some aluminum castings can develop higher strength than those produced as sand castings.	
Plaster mold castings	Process yields the best possible as-cast surface; parts have high dimensional accuracy; low porosity; very intricate configurations possible.	Only nonferrous metals may be cast; molds are expendable; the process is restricted to relatively small parts; because of the intricate nature of some parts, mold-making time may be comparatively long.
Centrifugal castings	An ideal method for producing extremely large, cylindrical parts; high production rates are possible; products are characterized by sound internal structure.	For true or semicentrifugal casting methods, only limited shapes may be produced; specialized rotational equipment is expensive.
Slush mold castings	A comparatively rapid method of producing hollow nonsymmetrical shapes with thin shells; outer surfaces result in bright, smooth texture; products may be plated usually without machining; die costs relatively low.	Process not suitable for products requiring uniform wall thickness; principally restricted to nonindustrial applications; inner surfaces are rough; castings lack strength; only nonferrous metals may be cast.

Casting process	Principal materials processed	Minimum production quantities
Sand castings	Iron, low-melting-point steel alloys, copper-base alloys, aluminum, magnesium, and nickel alloys.	Considered a high production process but 5 to 10 units considered economical; 10 to 50 units per hour practical.
Shell mold castings	Iron, aluminum, copper-base alloys.	30 to 200 units per hour with usual runs of 1000 to 100,000 units; rate per hour can be improved in special cases.
Investment castings	High-melting-point steel alloys, aluminum, nickel, copper-base alloys, magnesium.	5 to 30 units or more per minute; a minimum production run of 500 units is generally preferred.
Permanent mold castings	Iron, aluminum, magnesium, copper-base alloys.	20 to 50 or more units per hour; minimum production run of 50 to 100 units is preferred; considered to be a low–medium production process.
Die castings	Only nonferrous metals—zinc, magnesium, nickel, copper-base alloys; steel under special conditions.	150 to 200 shots per hour but up to 500 shots per hour for some pieces is possible; minimum of 2000 parts preferred.
Plaster mold castings	Only nonferrous metals—aluminum alloys, zinc, copper-base alloys.	150 to 450 units per pattern per week is usual; minimum production run of 1 to 10 parts can be justified.
Centrifugal castings	All metals.	30 to 50 units per hour; for centrifuge casting, higher production is possible with multiple cavities.

Casting process	Principal materials processed	Minimum production quantities
Slush mold castings	Only nonferrous metals—lead, zinc, and tin alloys.	5 to 10 units is considered a practical minimum.

Casting process	Practical weight limits	
	Minimum	Maximum
Sand castings	A few ounces (approximately 75 to 100 g).	Green sand—almost no weight limit. Dry sand—5000 to 6000 lb (approximately 2300 to 2700 kg).
Shell mold castings	A few ounces (approximately 75 to 100 g).	Usually not more than 30 lb (13.5 kg), but 100 to 200 lb (45 to 90 kg) is possible.
Investment castings	1 oz (28.3 g).	Usually not more than 5 to 6 lb (2.3 to 2.7 kg), but 100 lb (45 kg) is possible.
Permanent mold castings	Several ounces (approximately 100 g).	Usually not more than 20 lb (9 kg), but 300 lb (135 kg) is possible for nonferrous metals.
Die castings	Less than 1 oz (28.3 g).	10 lb (4.5 kg) for magnesium; 40 lb (18 kg) for zinc; 100 lb (45 kg) for aluminum.
Plaster mold castings	1 oz (28.3 g).	Usually not more than 25 lb (11 kg), but 100 lb (45 kg) or more is possible.
Centrifugal castings	Semi- and true centrifugal—a few lb (approximately 1.35 kg);	Semi- and true centrifugal— over 10 tons (approximately 9000 kg); centrifuge—5 to 10 lb

Casting process	Practical weight limits	
	Minimum	Maximum
	centrifuge—a few oz (approximately 100g).	(approximately 2.3 to 4.5 kg) and heavier in special cases.
Slush mold castings	1 oz (28.3 g).	Usually not more than 5 to 10 lb (approximately 2.3 to 4.5 kg).

Casting process	Type of mold	As-cast surface quality (μ in.)	Draft allowance	Machining allowance
Sand castings	Expendable (sand)	250 to 650; varies with alloy and casting size and configurations; below 250 is possible with special facing sands.	$\frac{1}{16}$ in./ft (5.233 mm/m).	$\frac{1}{8}$ in. (3.18 mm) for small castings; $\frac{1}{4}$ in. (6.35 mm) for larger castings.
Shell mold castings	Expendable (resin-sand)	150 to 250 for ferrous alloys; 125 to 200 for non-ferrous alloys; 100 to 150 for ductile iron.	$\frac{1}{2}$° common; 1 to 2° more common.	$\frac{1}{16}$ in. (1.588 mm).
Investment castings	Expendable (plaster slurry)	40 to 125; a better surface is obtainable under special conditions.		Generally, none is required.
Permanent mold castings	Nonexpendable (metal)	90 to 125; but size and casting configuration may affect the as-cast surface quality.	2° min. on each side; 5° in recesses.	$\frac{1}{32}$ in. (0.794 mm) for castings up to 4 in. (101.6 mm); $\frac{1}{16}$ in. (1.588 mm) for castings greater than 4 in.

Casting process	Type of mold	As-cast surface quality (μin.)	Draft allowance	Machining allowance
Die castings	Nonexpendable (metal)	40 to 100; a better surface is obtainable under special conditions.	0.005 in./in. (0.005 mm/mm) for zinc; 0.010 in./in. (0.010 mm/mm) for aluminum and magnesium.	0.010 to 0.020 in. (0.25 to 0.51 mm); but generally, none is required.
Plaster mold castings	Expendable (plaster)	30 to 50; yields the best as-cast surface finish of all casting processes; a better surface is obtainable under special conditions.	$\frac{1}{2}$ to $1°$	$\frac{1}{64}$ to $\frac{1}{32}$ in. (0.397 to 0.794 mm); generally, no machining is required.
Centrifugal castings	Either sand, metal, or graphite	For semi- and true centrifugal casting —100 to 500; for centrifuge—down to 60 to 70 in special cases.	$\frac{1}{8}$ in./ft. (10.46 mm/m)	See sand casting.
Slush mold castings	Nonexpendable (metal)	40 to 100	Generally, none is required.	Generally, none is required.

Casting process	Section thickness	
	Minimum	Maximum
Sand castings	Al, $\frac{3}{16}$ in. (4.763 mm); Cu, $\frac{3}{32}$ in. (2.381 mm); Fe, $\frac{3}{32}$ in. (2.381 mm); Mg, $\frac{5}{32}$ in. (3.969 mm);	No limit in floor and pit molds; sections 4 ft (1.22 m) thick have been cast in steel.

Casting process	Section thickness	
	Minimum	*Maximum*
	steel, ¼ to ½ in. (6.35 to 12.7 mm)	
Shell mold castings	Gray iron, ⅛ in. (3.18 mm); steel, aluminum, magnesium ³⁄₁₆ in. (4.7625 mm)	¼ in. (6.35 mm) or less is standard.
Investment castings	0.010 to 0.050 in. (0.25 to 1.27 mm)	1 to 3 in. (25.4 to 76.2 mm).
Permanent mold castings	Fe, ³⁄₁₆ in. (4.763 mm) Al, ³⁄₃₂ to ⅛ in. (2.381 to 3.18 mm) Mg, ⁵⁄₃₂ in. (3.969 mm) Cu, ³⁄₃₂ to ⁵⁄₁₆ in. (2.381 to 7.938 mm)	2 in. (50.8 mm).
Die castings	Cu, Mg, 0.050 to 0.080 in. (1.27 to 2.03 mm); Al, 0.030 to 0.080 in. (0.76 to 2.03 mm); Zn, 0.015 to 0.050 in. (0.38 to 1.27 mm); Pb, Sn, 0.030 to 0.060 in. (0.76 to 1.52 mm).	⁵⁄₁₆ in. (7.938 mm) preferable but usually less than 0.50 in. (12.7 mm); in special cases, if voids are acceptable, 1 to 1½ in. (25.4 to 38.1 mm).
Plaster mold castings	0.020 in. (0.51 mm) for areas less than 2 in.^{2} (approximately 13 cm^2).	0.052 in. (1.32 mm) for areas less than 4 to 6 in.^{2} (approximately 25 to 37 cm^2); 0.093 in. (2.36 mm) for areas up to 30 in.^{2} (approximately 190 cm^2).
Centrifugal castings	0.060 to 0.250 in. (1.52 to 6.35 mm).	4 in. (101.6 mm).

Casting process	Section thickness	
	Minimum	*Maximum*
Slush mold castings	$\frac{1}{16}$ in. (1.588 mm) in most cases.	Usually, $\frac{1}{8}$ in. (3.18 mm) is sufficient.

Casting process	Lead time required	
	Samples	*Production*
Sand castings	2 to 6 weeks.	2 to 4 weeks.
Shell mold castings	6 to 8 weeks.	2 to 4 weeks.
Investment castings	4 to 6 weeks.	2 to 4 weeks.
Permanent mold castings	6 to 14 weeks.	2 to 4 weeks.
Die castings	8 to 16 weeks.	4 to 8 weeks.
Plaster mold castings	4 to 6 weeks.	2 to 4 weeks.
Centrifugal castings	Varies.	Varies.
Slush mold castings	Varies.	Varies.

Casting process	Relative cost factors	As-cast tolerances
Sand castings	Lowest tooling costs when compared to other casting processes.	⅟₁₆ in. (1.588 mm) to ¹¹⁄₆₄ in. (4.37 mm) depending upon the kind of metal being cast and size of casting.
Shell mold castings	Aluminum match plates cost from $200 to $1000. Iron match plates may cost from 50 to 75% more.	For nonferrous and steel alloys ±0.020 in./in. (0.020 mm/mm); iron ±0.015 in./ (0.015 mm/mm); ±0.010 in./in. (0.010 mm/mm) across the parting line.
Investment castings	High tooling cost. The costs are lowest when the tolerance does not exceed ±0.010 in./in. (0.010 mm/mm).	±0.003 to 0.005 in. (0.08 to 0.13 mm) for first inch; over 1 in. add ±0.003 in. (0.08 mm) for ferrous metals.
Permanent mold castings	Medium tooling costs when compared to other casting processes.	±0.015 in. (0.38 mm) for first inch; over 1 in. add ±0.002 in./in. (0.002 mm/mm); ± 0.030 in./in. (0.030 mm/mm) across the parting line.
Die castings	High tooling costs— generally comparable to investment casting costs.	±0.003 in. (0.08 mm) for first inch; over 1 in. add ±0.0015 in./in. (0.0015 mm/mm); Zn, ±0.001 in./in. (0.001 mm/mm); Al, Mg, ±0.0015 in./in. (0.0015 mm/mm); Cu-base alloys, ±0.003 in./in. (0.003 mm/mm).

Casting process	Relative cost factors	As-cast tolerances
Plaster mold castings	Costs generally higher than for metal molds.	±0.005 in./in. (0.005 mm/mm); ±0.010 in./in. (0.010 mm/mm) across the parting line.
Centrifugal castings	Medium tooling costs when compared to other casting processes.	±0.010 in. (0.254 mm) for first inch in metal molds; pipe O.D.—3 to 12 in. (76.2 to 304.8 mm), ±0.060 in. (1.52 mm); 14 to 24 in. (355.6 to 609.6 mm), ±0.080 in. (2.03 mm); 30 to 36 in. (762 to 914 mm), ±0.100 in. (2.54 mm); 42 to 48 in. (1066.8 to 1219.2 mm), ±0.120 in. (3.05 mm); I.D. tolerance approximately 50% greater.
Slush mold castings	Die costs relatively low.	Considered a relatively imprecise casting process.

Casting process	Special casting capabilities			
	Bosses	Undercuts	Inserts	Minimum diameter of cored hole
Sand castings	Yes, small added cost.	Yes, small added cost.	Yes, small added cost.	³⁄₁₆ to ¼ in. (4.763 to 6.35 mm), but ½ in. (12.7 mm) is a more practical minimum.
Shell mold castings	Yes.	Yes.	Yes.	⅛ to ¼ in. (3.18 to 6.35 mm).

| Casting process | Special casting capabilities | | | |
	Bosses	Undercuts	Inserts	Minimum diameter of cored hole
Investment castings	Yes, some difficulty.	Yes, moderate cost.	No.	0.020 to 0.050 in. (0.51 to 1.27 mm).
Permanent mold castings	Yes, small added cost.	Yes, large added cost with reduced production rate.	Yes.	$\frac{3}{16}$ to $\frac{1}{4}$ in. (4.763 to 6.35 mm).
Die castings	Yes, small added cost.	Yes, large added cost with reduced production rate.	Yes, with a somewhat reduced production rate.	Cu, $\frac{1}{8}$ in. (3.18 mm); Al, Mg, $\frac{3}{32}$ in. (2.381 mm); Zn, Sn, Pb, $\frac{1}{32}$ in. (0.794 mm).
Plaster mold castings	Yes, moderate added cost.	Yes.	Yes.	$\frac{1}{2}$ in. (12.7 mm).
Centrifugal castings	Possible.	No.	Yes.	—
Slush mold castings	Yes.	Yes.	Yes.	—

SURFACE GENERATION METHODS

Figure 12-2 shows two different methods of generating surfaces by peripheral milling operations: *up* (or conventional) *milling* and *down* (or climb) *milling.*

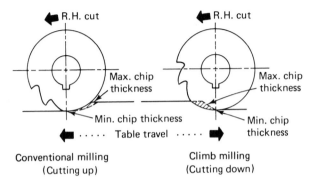

Conventional milling
(Cutting up)

Climb milling
(Cutting down)

FIGURE 12-2 Surfaces may be generated by up-milling or down-milling

Up Milling

Up milling is the condition when the work is fed against the direction of the rotating milling cutter. The chip is very thin where the tooth first contacts the work and increases in thickness to a maximum where the tooth breaks out of the work. The initial tooth contact usually occurs in clean metal and ends by lifting or peeling off the rough surface scale. Because of this effect, this is the preferred method for machining sand castings, forgings, or metals that have a rough or hard abrasive surface scale. Workpieces must be rigidly held to offset the tendency of the cutting forces to lift or pull the work out of the vise or fixture.

Down Milling

Down milling differs from up milling in that the work is fed in the same direction as the rotating cutter. The cutter tooth starts into the work with a maximum cut thickness and ends with a thin chip, resulting in less cutter wear. The tendency of the cutter is to hold the work down and, in fact, to pull the workpiece *under* the cutter. Down milling must be accomplished only on machines that have been specifically designed for this method of cutting. Machines that are made with backlash in the leadscrew or feeding mechanism are unsuitable, since any looseness will allow the cutter to draw

147

the workpiece ahead and take bites that are too large. Down milling usually produces a better surface finish on harder steels than does up milling. Chips are disposed of more readily and are less likely to be carried along by the teeth. It is the method that is usually selected for milling operations on slender and intricate parts. Fixture design is simplified. Less power is required. Increased cutting speeds and feeds are practical.

MILLING CUTTERS

The selection of the precise type of cutter for milling operations is normally dictated by the requirements of the general part design or the particular surface configuration on the part to be machined, the surface quality, workpiece material, and available production equipment. Milling cutters vary widely in size, material, and type.

Cutter Materials

General-purpose solid milling cutters are commonly made of high-speed steels. Some cutters use tungsten-carbide teeth which may be brazed on the tips of the teeth or individually inserted and held in the body of the cutter by some mechanical means. Carbide-tipped cutters are especially adapted to heavy cuts and increased cutting speeds.

Types of Cutters

Milling cutters are classified in several different ways. Their classification may be based upon construction characteristics (solid, braze-tipped, or inserted-tooth); relief of teeth (profile-relieved, form-relieved); purpose or use (t-slot, keyseat, gear milling); or method of mounting (arbor type, shank type, facing type).

Arbor-Type Cutters: Arbor-type cutters have an accurately ground center hole for mounting on a machine arbor. Milling cutters, commonly grouped within this classification, are shown in Figure 12-3.

Shank-Type Cutters: Shank-type cutters have straight or tapered extensions for the purpose of mounting or driving. Cutters with tapered shanks may be inserted directly into the spindle on the milling machine while straight-shank cutters are held in a chuck. Figure 12-4 illustrates the various types of cutters commonly grouped within this classification.

Facing-Type Cutters: Facing-type cutters, designed with provisions for mounting directly on the milling machine spindle nose, are attached to a stub arbor or used with an adapter. Figure 12-5 illustrates a roughing cutter with carbide-tipped blades used for heavy-duty milling on cast iron.

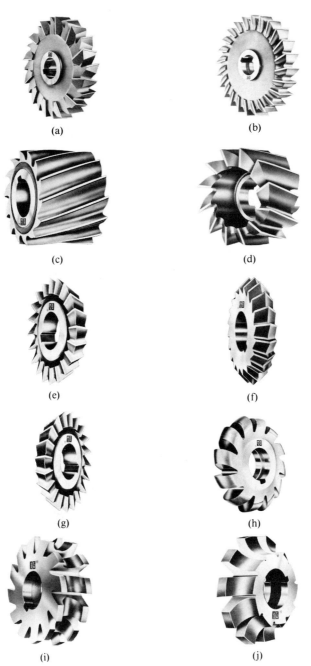

FIGURE 12-3 Arbor-type milling cutters: (a) staggered tooth cutter; (b) side milling cutter; (c) plain milling cutter; (d) shell mill; (e) single angle (RH) milling cutter; (f) double angle milling cutter; (g) single angle (LH) milling cutter; (h) convex milling cutter; (i) concave milling cutter; (j) corner rounding milling cutter (Courtesy, Illinois/Eclipse, a Division of Illinois Tool Works, Inc.)

FIGURE 12-4 Shank-type milling cutters (Courtesy, The Ingersoll Milling Machine Company)

FIGURE 12-5 A roughing cutter with carbide tipped blades used for heavy duty milling on cast iron workpieces (Courtesy, The Ingersoll Milling Machine Company)

MILLING MACHINES

There are a wide range of milling machine types and sizes. Some machines are capable not only of performing the usual surface milling operations but can be used to cut threads and gear teeth, to drill, and to ream, bore, and counterbore holes. Milling machines, depending upon their use, can be classified into three general groups: low to medium production, high production, and special types.

Low- to Medium-production Milling Machines

Low-to-medium-production milling machines consist of the various models of column and knee types. A typical horizontal spindle, column, and knee milling machine is shown in Figure 12-6.

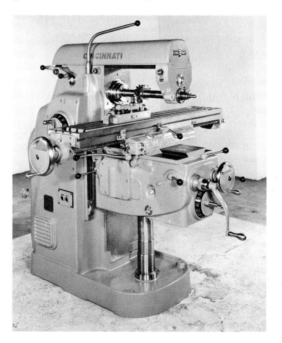

FIGURE 12-6 A column and knee milling machine (Courtesy, Cincinnati Milacron Company)

High-production Milling Machines

High-production milling machines include the fixed-bed types, shown in Figure 12-7. *Special* milling machines are designed and built to do specific jobs, particularly when production quantities are sufficient to offset the initial cost.

151

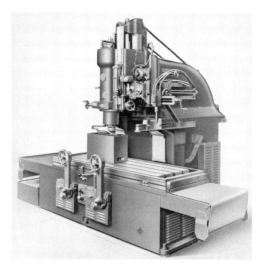

FIGURE 12-7 A fixed-bed type milling machine
(Courtesy, Cincinnati Milacron Company)

Special Types

Figure 12-8 illustrates a duplicating milling machine equipped with a tracing probe which follows a three-dimensional master. Machines of this type, sometimes called *die sinking machines,* are used to make molds and dies. Figure 12-9 illustrates a planer-type milling machine. The machine is used to cut large amounts of metal from massive workpieces.

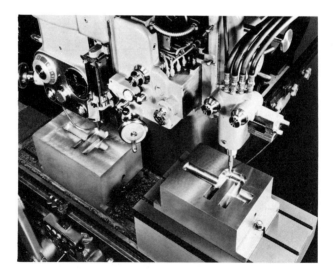

FIGURE 12-8 A hydro-tel duplicating milling machine. The cutter on the left is duplicating the 3-dimensional master on the right (Courtesy, Cincinnati Milacron Company)

FIGURE 12-9 A planer-type milling machine (Courtesy, National Machine Tool Builders Association)

Milling Machine Accessories

There are a number of special accessories that can be used with milling machines to enlarge the scope and variety of machining operations. Many machines have an adjustable tilting swivel vise which facilitates work holding. Special rotary tables or dividing heads, shown in Figure 12-10, (a) and (b), are available for spacing various cuts around a workpiece. There is a universal

FIGURE 12-10(a) A rotary table is used to position a workpiece with a radial slot.

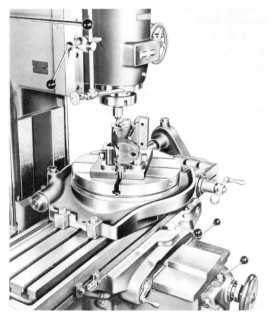

(a)

(b)

FIGURE 12-10(b) Teeth on spur gears may be properly spaced by using a dividing head (Courtesy, Cincinnati Milacron Company)

milling attachment available for use on horizontal milling machines which permits milling to be done at any angle. A vertical milling attachment can be obtained to convert a horizontal milling machine so that it can do the work of a vertical milling machine. A slotting attachment can also be applied to a horizontal milling machine to permit vertical shaping operations such as cutting a keyway to be performed. Another special milling machine attachment which is available from some manufacturers is a special overhead spindle that has as many as eight speeds.

PROCESS SELECTION FACTORS

Milling is particularly well suited for both low-volume and high-volume production. Parts can be machined at a high rate and with a very satisfactory surface finish. Milling machine cutters rotate at speeds ranging from 20 to 3800 rpm. Feed rates vary from 0.001 to 0.022 in. (0.03 to 0.56 mm) per minute per tooth. Specific cutting speeds and recommended feed rates are listed in most manufacturers' catalogs and in various handbooks. The selection of proper speeds and feeds is based upon various factors, such as workpiece and cutter materials, type of cutter, type of finish required, use of coolant, available horsepower, and other important items. One cut may be sufficient for some work, but a combination of roughing and finishing cuts are generally required to produce the best combination of desired surface finish and tolerance requirements.

There is generally a considerable latitude in the precise way a specific surface may be milled. Any surface that is accessible can usually be milled. A plane surface, for example, can be produced by slab milling, face milling, or by side milling. The methods by which any surface may be produced is governed by the type of milling machine used, the cutter, the special requirements of the surface on the workpiece, and the relative position of the surface on the machine to the cutter.

PRODUCT DESIGN FACTORS

Part Configuration: Parts should be designed wherever possible to incorporate the advantages of gang and/or straddle milling, shown in Figure 12-11, that is, to permit the maximum number of surfaces to be milled in one pass of the cutter. Considerable savings in setup time can be realized when more than one operation can be performed with the part in a single position. Errors may develop when a part is relocated, particularly in the dimensional relationship of one surface to another. Reclamping may lead to marring or otherwise damaging the workpiece. Good design requires that considerable attention be devoted to simplifying the configuration of the part. A simple part outline generally results in a less complicated work-holding fixture. Part design should be carefully scrutinized so that complex or difficult to machine surfaces are eliminated. Parts with surfaces to be finished by milling should have structures sufficiently strong so that reasonably heavy cuts, preferably accomplished in one pass, can be made at normal cutting speeds and feeds. A light feed may have to be chosen for a fragile workpiece.

FIGURE 12-11 A straddle milling operation (Courtesy, Cincinnati Milacron Company)

FIGURE 12-12 An end milling operation in a vertical spindle milling machine (Courtesy, Bridgeport Machines)

End mills, which are usually mounted in vertical spindle milling machines as in Figure 12-12, have a low cutting efficiency because the end of the cutter is not supported and the length-to-diameter ratio is usually high. Heavy cuts on machined surfaces with end mills are usually avoided in favor of arbor-mounted peripheral cutters.

Slot Cutting: Designers usually avoid specifying slot depths greater than three times the cutter width. Problems involving increased power requirements, the need for especially rigid work-holding fixtures, and difficulties in holding tolerance requirements are often encountered. Shell end mills may be used as an alternative to peripheral slot cutters.

Milling Fixtures: Parts whose surfaces require finishing by milling operations are generally subject to considerable cutting pressures. While the standard milling vise is often used to hold certain types of workpieces, it cannot be used without modifications for production purposes, since means for accurately locating the work are not incorporated into the original design. Special vise jaws are often used in a standard vise modified to particular workpiece requirements. While workholding fixtures may be designed in special cases for low production quantities, they are almost always required for high-quantity repetitive production. A fixture should contain the following features:

1. It should permit accurate positioning and location of parts.
2. It should allow parts to be rapidly and easily loaded and unloaded.
3. It should be rigid and durable.

156

4. It should have secure clamping features which are quick acting. Clamps must not cause deflection or distortion of parts. There should be provisions for backing up milling pressures by a solid component of the fixture.

5. Whenever possible, the cutter action should work in the direction of a clamp.

6. The design should be such that the part and the cutters are as close to the table surface as possible.

7. It should have facilities for rapid chip release. There should be no chip pockets which might restrict ease in cleaning or reloading parts. Coolant traps should be avoided.

8. It should not interfere with cutter action or necessitate a requirement for cutters of unusually large diameter.

9. It should be multipurpose and permit, if possible, the machining of several different surfaces with one location of the part.

10. The design should strive for the best concept with the least number of components and be kept as simple as possible. The use of commercially available fixture components is highly recommended.

11. It should be designed with provisions for attaching to the milling machine table.

In general, the quantity of parts required usually determines the extent of the fixturing.

Tolerances: Under normal milling conditions, it is reasonable to work to tolerances within ±0.001 to 0.002 in. (0.03 to 0.05 mm) for limited-quantity production. Tolerances of ±0.002 to 0.005 in. (0.05 to 0.13 mm) are more practical, however, for production quantity milling operations.

Surface Finish: A surface finish of 50 to 250 μin. may be obtained with high-speed cutters under normal milling conditions with improved surfaces of 20 to 40 μin. possible using carbide-tipped cutters.

REVIEW QUESTIONS

12.1. Defend the statement: "The milling machine is regarded by most authorities as the most versatile of all machine tools."

12.2. Explain why it may be necessary to design and build a special fixture to position and hold a workpiece in order to perform the required milling operations.

12.3. Describe the difference between peripheral milling and face milling.

12.4. Explain why it is necessary to use down milling only on machines specially designed for this method of cutting.

12.5. What are some of the considerations involved in selecting the appropriate milling cutter for a specific operation?

12.6. How may milling cutters be classified?

12.7. Under what conditions would a combination of roughing and finishing cuts be required?

12.8. List some important factors to consider when selecting proper milling speeds and feeds.

12.9. Describe the operation known as gang or straddle milling and list advantages of the operation.

12.10. Explain why it is considered good shop practice to accomplish the maximum number of machining operations on a given part while the part is held in the same position.

12.11. Give an example of four milling machine accessories.

12.12. Which of the following sizes (expressed in inches) could be satisfactorily obtained by production milling operations? (a) 0.7505/0.7501 (b) 1.375/1.370 (c) 2.015/2.017 (d) 3.6175/3.6170 (e) 9.312/9.314.

PROBLEMS

(See appendix for useful formulas.)

12.1. Prepare a sketch and label each of the following terms associated with the teeth on a plain milling cutter: land, tooth face, radial rake angle, clearance angle, cutting edge, gullet, tooth angle, back or flank of the tooth, and chip space.

12.2. (a) Using an 8-tooth 5½ in. diameter slab milling cutter, how long will it take to mill a 2-in. wide cut, 0.100 in. deep, 14 in. long at 160 rpm with a feed of 0.004 in./tooth? (b) Using the same rpm and feed, how long will it take to mill the same workpiece using a 2¼ in. diameter plain milling cutter?

12.3. Calculate the feed when the feed per tooth is 0.005 in., the number of teeth in the cutter is 6, and the rpm is 350.

12.4. Using a 1¼ in. diameter four-lip end mill at 100 (sfpm) surface feet per minute with a feed of 0.007 in. per tooth, what is the feed?

12.5. A 3 in. diameter milling cutter with 12 teeth is performing the operation

FIGURE 12-P5

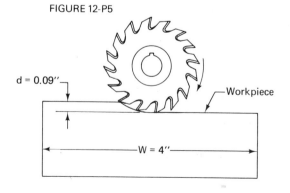

shown in Figure 12-P5. It is cutting at the rate of 0.009 in. per tooth and operating at 190 sfpm. Calculate the amount of stock removal.

12.6. Assume a milling operation with a removal rate of 3 cu in./mm. Cast iron requires a 1.6-unit horsepower (HP_u). Calculate the horsepower required at the cutter.

12.7. Using the data given in problem 12.6, calculate the horsepower required at the motor. Assume the idle horsepower to be 0.62 and the mechanical efficiency to be 0.8.

12.8. Calculate the rate of stock removal for a feedrate of 8.2 in. per min. with a width of cut of 1½ in. and a ½ in. depth of cut.

12.9. A 6 in. diameter high-speed cutter is to be used to machine a cast iron part. What spindle speed would be necessary to maintain a cutting speed of 80 sfpm?

12.10. Find the equation for calculating the cutting ratio and determine the cutting ratio when the length of the chip is 0.890 in. and the length of the path of the cutting tool while forming the chip is 1.937 in.

12.11. Explain the relationship of the cutting ratio to the quality of surface finish obtained and the type of cutting fluid used.

12.12. Cutting fluids may be grouped as emulsions of water and other elements, cutting oils, and air. List as many desirable characteristics as possible which a cutting fluid should have.

12.13. Define a condition occasionally encountered in machining operations known as *chatter*.

12.14. List some factors which may contribute to tool chatter.

12.15. Give some ways to eliminate chatter.

12.16. Explain how heat is generated by metal cutting operations.

12.17. List some important factors which are affected when the maximum temperature is attained in cutting operations.

12.18. Determine the rate of stock removal for the following data:

Feed per tooth 0.009 in.

Number of teeth 6

Revolutions per minute 275

Depth of cut 0.090 in.

Width of cut 4 in.

12.19. Considering the relative size of *w* as expressed in the previous problem, what types of milling cutters would be appropriate?

12.20. List the various operations necessary to mill all surfaces on the rough gray iron casting shown in Figure 12-P20. The machining allowance is ⅛ in. and the tolerance is ±0.003 in. In each case, specify the recommended operation, the type of machine, and the type of cutter. Also indicate how the workpiece is to be held for each change of position.

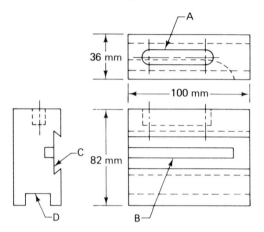

FIGURE 12-P20

BIBLIOGRAPHY

Armarego, E. J. A. and R. H. Brown, *The Machining of Metals,* Englewood Cliffs, N.J.: Prentice-Hall, Inc., 1969, pp. 174–188.

Carbide Blanks and Cutting Tools, B94.5-1966 (N18), New York: American Society of Mechanical Engineers, 1966.

*Gough, P. J. C., "Forging," *Engineer's Digest,* Vol. 34, No. 12, Dec. 1973.

Hines, C. R., *Machine Tools and Processes for Engineers,* New York: McGraw-Hill Book Company, 1971, pp. 317–363.

*"Hot Shaped Parts—Forging," *Machine Design, Reference Issue,* Feb. 14, 1974.

Inserted Blade Milling Cutter Bodies, B94.8-1967 (K44), New York: American Society of Mechanical Engineers, 1967.

"Milling," *Machining Data Handbook,* 2nd edition, Dearborn, Mich.: Society of Manufacturing Engineers, 1972, pp. 91–230.

"Milling," *Metals Handbook,* Vol. 3, *Machining,* 8th edition, Metals Park, Ohio: American Society for Metals, pp. 169–193.

*Abstracted with permission.

13

SHAPING

The shaper is a machine that produces flat surfaces in horizontal, vertical, and angular planes. Two advantages of shapers are: (1) they use relatively inexpensive tools, and (2) for most types of work, they require only a short time for setup. Except for those machines equipped with special attachments, most work is confined to small-quantity production lots.

DESCRIPTION OF THE PROCESS

In shaping, a reciprocating single-point tool bit is used which is accurately guided as it moves back and forth in a straight line across a rigidly held workpiece. A ram, which moves back and forth, carries the tool holder and the cutting tool. The machined surface on the workpiece is generated by the cutting action of the tool bit as it peels off a chip on its forward stroke. As the ram returns, the table that holds the workpiece feeds crosswise a preset incremental distance equal to the feed desired. This distance rarely exceeds $\frac{1}{16}$ in. (1.588 mm). The length of the stroke is adjustable from its full stroke to less than $\frac{1}{4}$ in. (6.35 mm). Since the return stroke of the ram represents nonproductive or lost time, the ram travels up to two times faster on the return stroke than on the forward or cutting stroke. The ram ranges from 10 to 200 strokes per minute on most machines. The ram may be driven by a crank motion or by means of a hydraulic system.

161

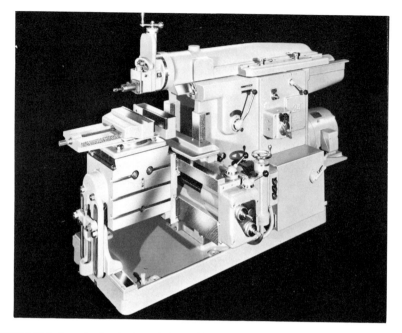

FIGURE 13-1 A production horizontal shaper (Courtesy, Rockford Machine Tool Company)

FIGURE 13-2 A hydraulic ram shaper in operation (Courtesy, Rockford Machine Tool Company)

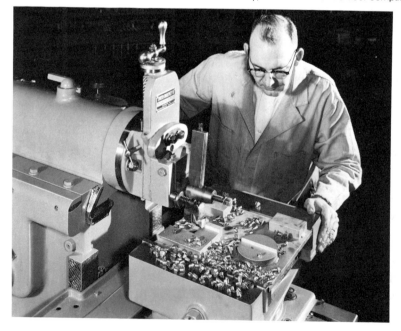

TYPES OF SHAPERS AND PRODUCT APPLICATIONS

Shapers are classified into three distinctly different types: horizontal, vertical, and special-purpose. Most of these types are designed and designated specifically for light-duty, medium- or standard-duty, or heavy-duty work.

Horizontal Shapers

Horizontal shapers are by far the most common of the three types. They are available with either a push-cut or a pull-cut stroke. The push-cut stroke shaper greatly outnumbers the pull- or draw-cut motion. Figure 13-1 illustrates a standard plain horizontal shaper. Figure 13-2 shows a hydraulic ram shaper in operation. It operates on the push-cut principle and is equipped with a universal table which may be swiveled at any desired angle for making cuts. Pull- or draw-cut shapers are built with a long stroke and are designed to take heavy cuts. Such machines are generally found in railroad shops and in plants where heavy-duty machines are produced. Horizontal machines may be equipped with special tools, attachments, and devices for holding and rotating the work. In some cases, work may be directly fastened to the table instead of securing it in a vise.

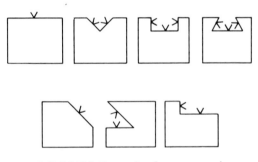

FIGURE 13-3 Types of surfaces commonly produced by shaping

Figure 13-3 illustrates the types of surfaces on workpieces most commonly produced by horizontal shapers. Curved or irregular surfaces and other miscellaneous shapes can be machined by carefully handfeeding the tool bit along a scribed line. Elaborate shapes can also be automatically produced by using machines equipped with hydraulically operated attachments for automatically contouring and duplicating, shown in Figure 13-4. In this case, a tracer or follower travels over a master metal template which is mounted on the shaper. The tool creates the corresponding surface on the workpiece as it raises and lowers in unison with the template.

163

FIGURE 13-4 Shaper with tracer attachment producing four identical propeller blades simultaneously (Courtesy, National Machine Tool Builders Association)

Vertical Shapers

Vertical shapers, shown in Figure 13-5, are also known as *slotters.* They have a vertical ram and a hand- or power-operated rotary table. On some machines, the ram may be inclined as much as 10° to either side of the vertical position when cutting inclined surfaces. Also included in this machine classification is a *keyseater,* which is a shaper principally designed for cutting keyways on the inside of gears, pulleys, cams, and wheel hubs. A typical production operation performed on a vertical shaper is shown in Figure 13-6.

Special-Purpose Shapers

Special-purpose shapers are principally confined to gear-cutting operations. Figures 13-7 and 13-8 show a gear shaper used to cut both internal and external spur and helical gears. The tool bit, or cutter, is shaped like

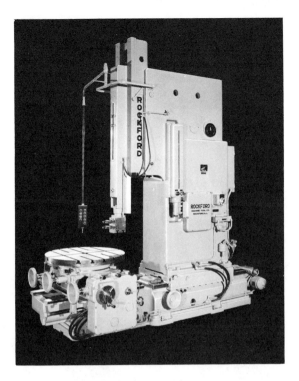

FIGURE 13-5 A vertical shaper or slotter
(Courtesy, Rockford Machine Tool Company)

FIGURE 13-6 An inside slotting operation
(Courtesy, Rockford Machine Tool Company)

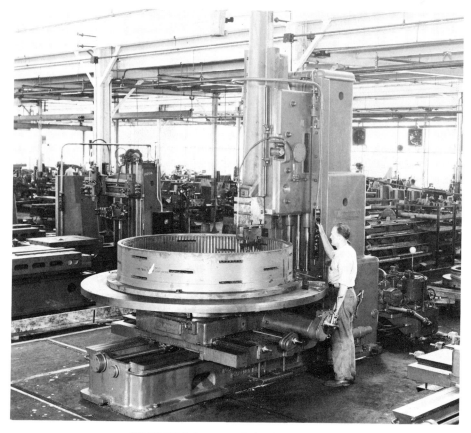

FIGURE 13-7 A gear tooth shaping operation
(Courtesy, Rockford Machine Tool Company)

FIGURE 13-8 A massive gear shaper
(Courtesy, Fellows Corporation)

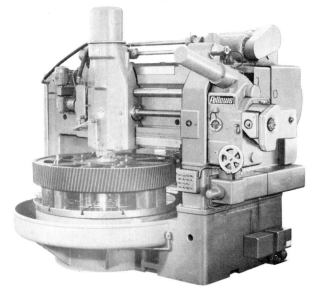

FIGURE 13-9 This high-speed gear shaper can handle involute and non-involute shapes up to 10 in. diameter and 4 in. face width (Courtesy, Fellows Corporation)

the gear it produces, except that each tooth is ground or relieved in such a way as to produce a cutting edge. The cutter is clamped on a ram and is gradually stroked into contact with one or more slowly rotating gear blanks. The up-and-down motion of the gear cutter generates or shapes the desired gear teeth. Figure 13-9 shows a high-production gear shaper.

Shaper Size

The size of shapers is classified according to the maximum length of stroke. Push-cut shapers can accept work sizes from 4 to 36 in. (102 to 915 mm). Pull-cut shapers are made for work requirements up to 72 in. (1.82 m). The maximum cross-feed distance is generally equivalent to the maximum ram-stroke distance. Therefore, a shaper with a 16-in. (406 mm) maximum stroke, for example, is capable of machining a part with a plane surface that measures at least 16×16 in. (406 mm $\times$ 406 mm) square.

PROCESS SELECTION FACTORS

Shapers are commonly found in production and maintenance shops, tool and die shops, and in school shops. Cutting tools used on the shaper are essentially the same as those used on the lathe except for slight differences in the tool angles. Tool bits may be made of cast alloy and cemented carbides, but cutting tools made of high-speed steel are the most common. The main advantage of the single-point cutting tool bit used on the shaper as compared to multiple-point cutting tools used on broaching or milling machines, for example, is that they may be made and sharpened more easily. While milling cutter sharpening requires special grinding equipment, a shaper tool bit may be conveniently sharpened on an ordinary bench grinder. Production costs can sometimes be reduced by initially rough-shaping required finished surfaces on sand castings or on workpieces made from hot-rolled steel stock. After the hardened skin or outer layer has been removed by shaping, the parts may be milled to final size. In this way, undue wear on expensive milling cutters may be held to a minimum.

While shaping operations are often associated with the production of only one or a limited number of identical pieces, an impressive number of high-production machine parts may be also processed by this method. A shaper is an extremely versatile machine tool. It can be put to many uses and can easily be changed and set up from one job to another. Most work can be conveniently positioned using the standard shaper vise or simply bolted directly to the table.

The leading disadvantage of a shaper is that its rate of metal removal is slow in comparison with a milling machine, because of the lost cutting time of the return stroke. Also, milling machines remove metal more rapidly because of the rapid cutting action of the multiple-tooth cutters.

PRODUCT DESIGN FACTORS

There are few limitations which restrict the design of parts that are to be finished by shaping. As was pointed out previously, shapers can generate surfaces that are as long as 72 in. (1.82 m). It may be possible for the designer to anticipate and simplify certain features of parts so that multiple parts may be setup, clamped, and simultaneously machined. In some cases, parts may be designed with special mounting or locating pads or clamping brackets which are specifically intended to simplify setup and clamping problems. This technique is often applied to odd-shaped or otherwise-difficult-to-clamp parts. Such special features are removed after the desired surfaces have been finished.

The accuracy of the work is largely dependent upon the skill of the machine operator. It is not practical to work to tolerances closer than ± 0.001

to 0.002 in (0.03 to 0.05 mm), since the feed dials on shapers are generally graduated to divisions no finer than 0.001 in. (0.03 mm). Closer tolerances may be obtained in special cases, particularly on small parts, when errors due to looseness in clamping or when table deflections are held to an absolute minimum.

The finer the feed, the better the surface finish. On most work, a surface finish from 75 to 500 μin. is readily obtainable. On cast iron, a 60-μin. surface finish is possible with normal operation.

REVIEW QUESTIONS

13.1. Describe the cutting action of a shaper.

13.2. Compare the types of surfaces most commonly produced on workpieces by horizontal shapers to those produced by vertical shapers.

13.3. Explain why it is difficult to produce curved surfaces by shaping.

13.4. What is the advantage of automatic contour and duplicating attachments which may be applied to a shaper?

13.5. How is the size of a shaper classified?

13.6. What is the principal advantage of the single-point cutting tool used on the shaper as compared to the multiple-point cutting tool used for milling?

13.7. What is the advantage of shaping a surface on a sand casting prior to milling the same surface?

13.8. Give one principal advantage of shaping.

13.9. Give a practical tolerance for shaping.

13.10. What are two disadvantages of using a shaper?

BIBLIOGRAPHY

ARMAREGO, E. J. A. and R. H. BROWN, *The Machining of Metals,* Englewood Cliffs, N.J.: Prentice-Hall, Inc., 1969, pp. 200–206.

HINES, C. R., *Machine Tools and Processes for Engineers,* New York: McGraw-Hill Book Company, 1971, pp. 287–301.

"Shaping," *Machinery's Handbook,* 20th edition, New York: Industrial Press, Inc., 1975, pp. 1791–1792.

"Shaping," *Metals Handbook,* Vol. 3, *Machining,* 8th edition, Metals Park, Ohio: American Society for Metals, pp. 52–57.

14

PLANING

The planer is a machine which, like the shaper, produces flat surfaces in horizontal, vertical, or angular planes. There are two major differences between the shaper and the planer. The first difference is that most planers operate with an action opposite to that of the shaper; that is, the workpiece reciprocates past one or more stationary single-point cutting tools. A second difference is that planers are designed to accommodate workpieces which are far greater in size than those machined on the shaper.

DESCRIPTION OF THE PROCESS

The work is clamped onto a large table which is constructed with T-slots and special holes that provide a means for positioning and securely holding the work. A wide assortment of clamps, jacks, stop pins, wedges, and special work-holding attachments are available to attach parts or fixtures to the table. The table rides in V-grooves on the bed of the machine and, accordingly, is accurately guided as it travels back and forth. Cutting tools are held in toolheads that can travel from side to side. The toolheads are mounted on a horizontal cross rail that can be moved up and down. Productivity is increased by using multiple toolheads, which permit several simultaneous cuts to be made with each relatively slow stroke of the table. Unlike the shaper, cutting may occur during *both* directions of the table travel. Toolheads may be installed with two or more cutter bits which are arranged in such a way that one or two cut in one direction of table travel and the remaining cut while the table travels in the opposite direction.

170

TYPES OF PLANERS

Planers are classified into four distinctly different types: double-housing planers, open-side planers, edge or plate planers, and pit-type planers.

Double-Housing Planer

The *double-housing planer,* shown in Figure 14-1, is considered to be the most common or conventional type. The block diagram in Figure 14-2 illustrates the various features of this type of planer. Essentially, it consists of a massive bed on which a table slides back and forth. The length of the bed must be slightly more than twice the length of the table. Twin vertical housings are mounted near the center of the bed. These support the horizontal cross rail. Both the cross rail and each of the housings are equipped with ways so that sliding motion may be obtained in both the vertical and

FIGURE 14-1 A 48-in. × 20-ft double-housing milling planer (Courtesy, G.A. Gray Company)

171

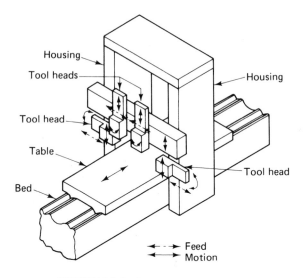

FIGURE 14-2 Basic components of a planer

horizontal directions. In most cases, two toolheads are mounted on the horizontal cross rail and one on each of the vertical housings. Toolheads may be swiveled so that angular cuts can be made. The tools can be fed into the work manually or operated with power. As in the case of shaping, if the cutting is to be done during one direction of table motion only, the return stroke speed is faster than the speed during the cutting stroke. The work capacity of double-housing planers is governed by the horizontal distance between the vertical housings, the vertical distance from the table to the cross rail in its uppermost position, and the maximum length of table travel. Double-housing planers have been built to accept work up to a maximum size of 10 ft (approximately 3 m) square by 60 ft (approximately 18 m) long, although conventional planers with stroke lengths of less than 20 ft (or approximately 6 m) are more common.

Open-Side Planer

The *open-side planer,* shown in Figure 14-3, has only one vertical housing, to which the cross rail is mounted. Such planers are used primarily for machining parts which exceed the normal width limitations of double-housing planers. In most cases, open-sided planers offer the added advantage of simplifying work setup as well as permitting wide workpieces to overhang beyond the edge of the table. Only three toolheads are used on this type of planer. The work capacity of open-side planers is governed by the maximum planing width, which is the distance from the table side of the housing to the maximum outward position of the tool in the outer tool holder.

FIGURE 14-3 A 48-in. × 34-ft openside planer being used to machine a giant casting (Courtesy, National Machine Tool Builders Association)

Edge or Plate Planer

Edge or plate planers are special-purpose machine tools designed specifically for squaring or leveling the edges of heavy steel plates. The plate is clamped to a bed and the side-mounted carriage supporting the cutting tool is moved back and forth along the edge. The operator rides along the work on the carriage. Cutting can take place during both directions of the carriage travel. In some cases, milling cutters are used instead of conventional planer cutting tools. The work capacity of the edge or plate planer is governed by the maximum length of a plate that can be machined.

Pit-Type Planer

A *pit-type planer* differs from the other types of planers in that the table remains stationary and the tool reciprocates. This type of planer is used for extremely heavy work when the weight of the workpiece makes reciprocation difficult. Massive rails are mounted on both sides of the table. Toolheads, usually restricted to two in number, are mounted on the crossrail for two-way planing. The worker rides on a specially built platform at the side of the carriage as it moves back and forth over the work.

173

PRODUCT APPLICATIONS

Planers are machine tools specifically designed for machining flat surfaces on very large parts. Planers are considered to be the largest machine tools manufactured. Planing operations may be performed on forgings, castings, billets, and heavy workpieces such as machine tool beds and ways, locomotive components, large military ordinance, earth-moving equipment, welded fabrications, and ship parts.

PROCESS SELECTION FACTORS

Planers are expensive machine tools. In most cases, they represent high overhead rates and, accordingly, must be operated at the highest possible efficiency and over as continuous a time period as possible. One method for obtaining maximum use of planers is to set up multiple small or medium-size parts for simultaneous machining. (This process is sometimes known as *string planing*.) In some companies, machining output is improved by using duplex-table planers. The process consists of setting up individual or multiple parts on one table while the other table on the same machine is at work, thus eliminating idle or lost planer time. Another technique used for cost reduction is in the economical use of as many toolheads as possible. Cutting efficiency can often be improved in cases when gang tools are used. In this case, three or four tools are set in a toolholder in such a way that each tool takes its proportional share of the total feed.

A fundamental requirement for economical planer operations is that all cuts should be as heavy as possible. Shaper and planer tools are similar in shape, but planer tools are much larger and heavier. Common practice is to use a heavy tool shank of plain carbon steel with a high-speed steel, cast-alloy, or carbide tip clamped or brazed to the shank. High-speed steel or cast-alloy tools are usually used for heavy roughing cuts and carbides for finishing cuts. Figure 14-4 illustrates some typical planer tool shapes. Some planers using carbide cutting tools can operate as fast as 500 sfpm. On heavy planers it is possible to take a roughing cut as deep as ⅛ to ¾ in. (15.88 to

FIGURE 14-4 Planer tools

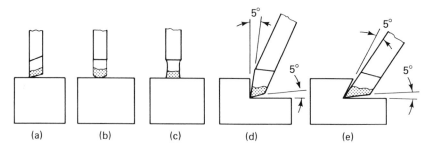

(a)   (b)   (c)   (d)   (e)

19.05 mm) with a feed of ⅟₁₆ to ⅛ in. (1.588 to 3.18 mm) per cut. As in shaping, two or three cuts are considered normal practice for producing an acceptable finished surface. Approximately 0.005 to 0.010 in. (0.13 to 0.25 mm) of stock is left on the workpiece for the final cut.

PRODUCT DESIGN FACTORS

Shaping and planing operations may usually be performed to approximately the same limits of accuracy. Feed dials on both machines are graduated with 0.001-in. (0.030-mm) divisions. It should be noted, however, that tolerances for typically heavy and large parts normally machined on planers usually have correspondingly larger ranges.

REVIEW QUESTIONS

14.1. Contrast the important similarities and differences between shaping and planing.

14.2. Explain how the planer table is guided as it travels back and forth.

14.3. How is work held on a planer?

14.4. How may productivity be increased in planing?

14.5. Briefly describe the four main types of planers and list their principal uses.

14.6. Describe the operation known as "string planing" and give a leading advantage of this technique.

14.7. For what main classification of product applications are planers intended?

14.8. What is the advantage of using gang tooling?

14.9. Contrast the tools used in planing to those used in shaping.

14.10. How many cuts are considered normal practice for producing an accurate surface on a workpiece?

BIBLIOGRAPHY

Armarego, E. J. A. and R. H. Brown, *The Machining of Metals,* Englewood Cliffs, N.J.: Prentice-Hall, Inc., 1960, pp. 200–206.

"Planing," *Machining Data Handbook,* 2nd edition, Dearborn, Mich.: Society of Manufacturing Engineers, 1972, pp. 465–478.

"Planing," *Metals Handbook,* Vol. 3, *Machining,* 8th edition, Metals Park, Ohio: American Society for Metals, pp. 45–49.

15

BROACHING

Broaching is a modern method of stock removal that has almost unlimited applications in metal working. It can also be used as a method to work wood, hard vulcanized fiber and other composites, hard rubber, graphite, and some plastics. Metals and alloys are the most commonly broached materials, however.

DESCRIPTION OF THE PROCESS

The machine used may be hand-operated (an arbor press) or electromechanically or hydraulically operated (a hydraulic press). Both the cutting tool (broach) and the workpiece are held in rigid fixtures. The principal function of the machine is to provide the speed necessary for cutting to take place. Cutting speed or travel of a broach across the work is usually 15 to 25 times faster than in milling. Speeds of 200 sfpm or more are common, although most broaching operations are done at lower speeds. The feed (or material removal rate) is regulated by the broach, which consists of a hardened steel bar with a series of cutting teeth. Each tooth is made progressively higher, to remove successively larger amounts of materials. Thus, the cut grows deeper as the operation progresses. Cutting fluid is generously applied along the cut. As a general rule, the required surface is frequently transformed to final size after one or more continuous passes or strokes of the broach. Depending upon the machine capacity and other important factors, material removal in a single pass can be as much as ¼ in. (6.35 mm).

The process of surface finishing by broaching may be compared roughly to planing, except that in broaching a multitoothed tool is used. In both cases, the cutting tool may be either passed across a fixed workpiece or the tool may be held stationary and the work moved in a continuous stroke. In some machines, the tool is pushed along the surface to be broached while other machines pull the broaches through or over the work.

Broaching Tools

Except for variations in tooth sizes, the teeth on a broaching tool resemble those on a wood rasp. The design of a broaching tool is based upon a concept unique to the process, in which rough, semifinish, and finish cutting teeth are combined in one tool. As shown in Figure 15-1, the first roughing tooth is, proportionately, the smallest tooth on the tool. The subsequent teeth progressively increase in size up to and including the first finishing tooth. The difference in height between each tooth, or tooth *rise,* usually is greater along the roughing section and less along the finishing section. All finishing teeth are the same size.

Many factors govern the design of a broach other than the basic size and general configuration of the hole or surface desired. The length of cut (thickness of the part to be broached), type of material, hardness of the part, and required surface finish all have a bearing on the overall length of the broach. The shank length, tooth pitch, and relief and rake angles at the cutting edges of the teeth, as well as the number of teeth, are all important design factors. Each broach, whether standard or special, must be designed for specific minimum and maximum lengths of cut. When the maximum specified length of cut of a broach is exceeded, it is quite easy to visualize that the chips will crowd into the tooth gullets until they are packed full. Breakage of the tool or damage to the workpiece is almost certain to result.

As each broaching tooth enters the workpiece, it cuts a fixed thickness of metal. The fixed chip length and thickness produced by broaching create a *chip load,* which is determined by the design of the broach tool. The chip load cannot be altered by the machine operator as it can in most other machining operations. The entire chip produced by each tooth in one complete pass of the broach tool must be freely contained within the preceding tooth gullet as in Figure 15-2. Standard nomenclature of a typical surface-broaching tool is shown in Figure 15-3. The size of the tooth gullet (which determines tooth spacing) is a function of the chip load and the types of chips produced. It should be understood that the form that each chip takes largely depends upon the workpiece material, but also upon the tooth design and depth of cut. Brittle materials produce flakes, whereas ductile or malleable materials form spiral chips. Long cuts in ductile material or interrupted cuts producing two or more chips would soon fill a circular gullet with chips.

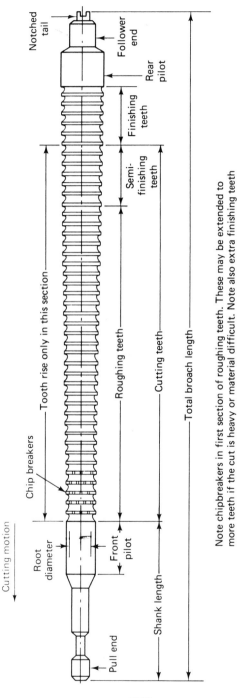

Cutting motion

Notched tail

Follower end

Rear pilot

Finishing teeth

Semi-finishing teeth

Tooth rise only in this section

Roughing teeth

Cutting teeth

Total broach length

Chip breakers

Root diameter

Front pilot

Shank length

Pull end

Note chipbreakers in first section of roughing teeth. These may be extended to more teeth if the cut is heavy or material difficult. Note also extra finishing teeth

FIGURE 15-1 Basic shape of a round hole, pull-type broaching tool

178

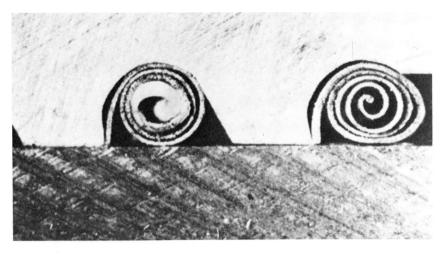

FIGURE 15-2 Single pass steel chip formation (Courtesy, National Broach and Machine Company)

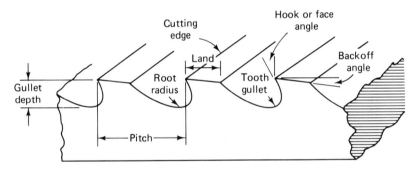

FIGURE 15-3 Standard nonenclature for teeth on a typical surface-broaching tool. Gullets here are average size (Courtesy, National Broach and Machine Company)

Figure 15-4 illustrates how the design of a broach may be altered by use of a flat-bottomed gullet with extra wide spacing. The extended space provides room for two or more spiral chips or for a large quantity of chip flakes. Broach designers may also specify notches, called *chipbreakers,* also shown in Figure 15-1, which eliminate packing and facilitate chip removal.

FIGURE 15-4 Flat-bottom gullet is an extended space between teeth for more chips (Courtesy, National Broach and Machine Company)

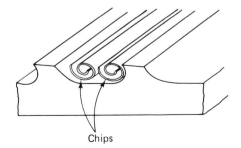

FIGURE 15-5 Internal broaching (Courtesy, National Broach and Machine Company)

Kinds of Broaching Tools

All broaching tools are specially designed and made for a specific use and for a specific machining operation. However, certain types of broaches may be singled out and identified for the type of work they do.

Internal Broaching Tools: Internal broaching tools are used to enlarge and finish a wide variety of holes previously rough-formed by casting, forging, punching, drilling, boring, and so on. In Figure 15-5, a solid internal broaching tool is used to simultaneously form 55 internal teeth in a coupling. Figure

FIGURE 15-6 Holes of practically any shape can be broached

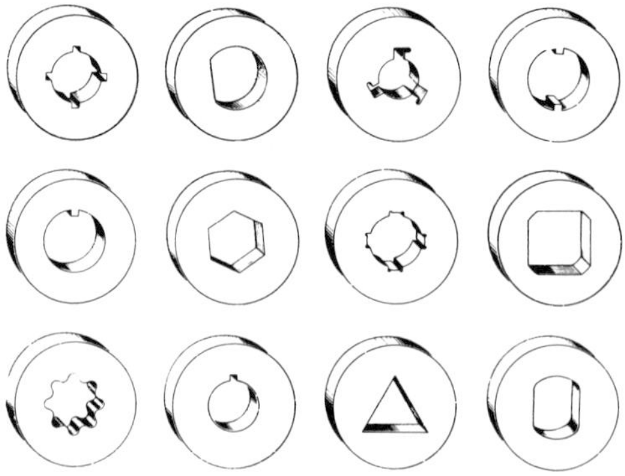

15-6 illustrates a variety of internal shapes which may be produced by broaching. Most internal broaching operations are done with pull broaches, since they can take longer cuts and consequently can remove more stock than push broaches.

While most internal broaching tools are of one-piece solid construction, shell broaches, shown in Figure 15-7, may also be used. These consist of a broach body, an arbor section over which the shell fits, and the removal shell (that is, finishing section). Shell broaches are used in cases where the broach may be subject to rapid wear and are particularly useful because of the convenience in replacing worn or broken sections.

Burnishing broaches are used on parts when surface finish and accuracy requirements are critical. Burnishers are generally designed as push broaches. They are designed to polish (by cold working) rather than to cut a hole. The change in size is normally not more than 0.0005 to 0.001 in. (0.013 to 0.03 mm). Special sizing broaches are pulled or pushed through a semifinished hole to remove the remaining amount of stock faster and more efficiently than possible with a fine-feed broaching tool. Burnishing is generally restricted to soft, ductile materials.

Helical internal splines, rifling in gun barrels or internal gear teeth may be produced by rotating the broach as it is pulled through the workpiece.

FIGURE 15-7 A shell broach used for broaching ring gears
(Courtesy, Detroit Broach and Machine Company)

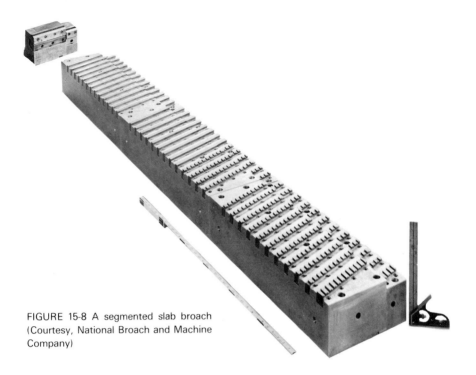

FIGURE 15-8 A segmented slab broach
(Courtesy, National Broach and Machine
Company)

External Broaching Tools: External broaching tools may be of one-piece
construction when used for small, short surfacing applications or they may
consist of a series of segmented sections, as in Figure 15-8, which are assembled end to end on a moving slide. The entire length of such broaches
can be supported. Sectional construction results in a broach which is easier
and cheaper to construct and sharpen. Another advantage is that broken
sections can be readily replaced without discarding other sections. The
broach may be made with either straight or angular teeth. Figure 15-9
illustrates a variety of external shapes that may be produced by broaching.

FIGURE 15-9 Typical external shapes produced by broaching

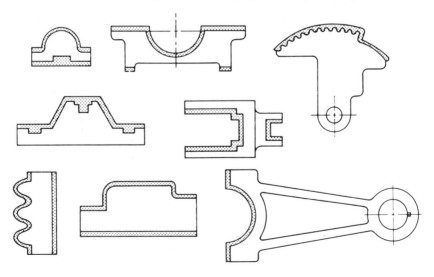

Broaching Tool Materials

High-speed tool steel is by far the most widely used material for broaches. Brazed carbide or disposable inserts are sometimes used for the cutting edges when machining cast iron parts which require close tolerances and high production rates. Carbide tools are also used to an advantage on steel castings to offset the damaging effect of local hard spots.

PRODUCT APPLICATIONS

It is not difficult to find examples of parts whose surfaces have been processed by broaching. Broaching is often appropriate where there is a need for straight or contoured external surfaces or special shapes on the inside surfaces of holes. One of the largest users of this mass-production process is the automotive industry, where finished surfaces on crankcases, cylinder heads, connecting rods and caps, power steering pistons, steering worms, and transmission parts are machined in this way. Other large users of this process are rifle and gun manufacturers, who produce a wide variety of components by this versatile process. Parts for aircraft engines and structures are commonly processed by broaching. Broaching is widely used in the manufacture of special gears, bushings, and sleeves; diesel engine parts; assorted machine parts such as crank pins and push rods; compressor wheels; rotors; chain sprocket teeth; turbine blades; propeller hubs; and countless other product applications. Figure 15-10 illustrates a variety of broaching tools and typical externally and internally broached products.

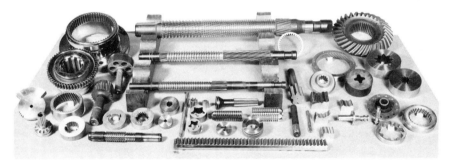

FIGURE 15-10 Typical broaching tools and products
(Courtesy, National Broach and Machine Company)

BROACHING MACHINES

In general, there are two types of broaching machines, horizontal and vertical, based upon the direction of broach travel. Most broaching machines are hydraulically powered, but some are driven electromechanically.

Horizontal Machines

Horizontal machines are general-purpose machines and may be used for high or low production quantities. (Internal operations account for the great majority of work performed on horizontal machines.) Machines of this type may be as long as 40 to 50 ft (12 to 15 m). There is convenient access to any part of the machine. Figure 15-11 illustrates a typical horizontal broaching machine.

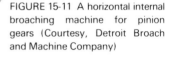

FIGURE 15-11 A horizontal internal broaching machine for pinion gears (Courtesy, Detroit Broach and Machine Company)

FIGURE 15-12 A single station vertical pull-down broaching machine (Courtesy, Detroit Broach and Machine Company)

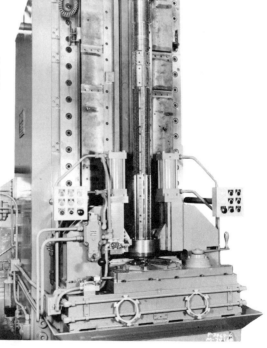

Vertical Machines

Vertical machines are available as push-down, pull-down, or pull-up types. Vertical broaching machines are adapted to both external and internal surfacing operations. Figure 15-12 illustrates one type of vertical broaching machine.

PROCESS SELECTION FACTORS

The decision to broach or not to broach a specific part is based upon many factors. A surface cannot be broached if it has an obstruction that interferes with a path of the tool, such as a blind hole. A broach always moves forward in a straight line. Except in special cases, blind holes or pockets are not processed by broaching. Almost any irregular cross section can be broached as long as all surfaces of the section remain parallel to the direction of broach travel. Soft materials may present problems in broaching because of their tendency to deform, rather than cut, when subjected to broaching tools. Broaching may be uneconomical on some hard materials as a result of rapid tool wear.

Broaching is a method that is usually employed for high-quantity production. The cost of designing and manufacturing the required broaching tools and fixtures must be justified. There are, however, some notable exceptions to this rule, particularly in cases of special part configurations which cannot be economically produced by any other method. Such operations include special hole configurations, keyways, and sizing operations. Short runs are usually justified in cases where standard broaching tools can be used.

Parts of less than 1 oz. to workpieces weighing several hundred pounds (or approximately 140 kg) may be broached. There are practically no size restrictions. Broaching tools can be as small as 0.050 in. (1.27 mm) or as large as 20 in. (508 mm) in diameter.

Most of the external surface configurations produced by broaching can be produced by other machining processes. The principal advantage of broaching is the speed of the machining operation.

BROACHING FIXTURES

There is a close relationship between the machine, the broaching tool, and tooling. When parts are to be mass-produced by broaching, the selection of the machine and the design of the broaching tool and fixture are usually done concurrently.

Fixtures are required to hold each part in correct alignment so that the production of accurate parts is assured. Because of the high forces that are imposed by the action of so many cutting teeth against the workpiece at any one time, the fixture must be rugged and strong. Tremendous cost savings can be obtained by using a fixture that reduces the work-handling time and labor. Whenever possible, fixtures are designed to permit stacking or nesting multiple parts. The type and quality of the fixture is usually determined by cost considerations and the required quantity of parts.

The characteristics of a good broaching fixture are as follows:

1. It can be loaded and unloaded quickly and easily.

2. It does not obstruct free chip removal.

3. It does not distort the workpiece.

4. It holds the workpiece solidly against the locating pads and in correct alignment with the broach under maximum cutting pressures when so required.

5. It allows complete freedom of motion of the workpiece in one plane when floating.

6. It allows free rotation of the workpiece when designed for helical cutting with a nonrotating broach.

7. It has no hardened surfaces that might come in contact with the broach in normal operation.

8. It is easy to mount on the broaching machine and, when mounted, holds correct alignment.

9. Its clamping and locking mechanisms are positive and not affected by vibration or maximum clamping pressures.

10. Elements subject to wear are adjustable for wear and may easily be replaced when necessary.

11. It does not obstruct free flow of coolant.

12. It is sturdy enough to support the work without deflection under broaching pressure.

13. It is easy to clean and keep clean; there are no blind corners or pockets for chips to lodge.

14. Indexing mechanism is positive and has no backlash.

15. When broach guides are incorporated, they are sturdy enough to prevent broach chatter that would occur otherwise.

16. Its operation requires minimum skill or judgment on the operator's part and a minimum of physical energy.

17. Clamping pressures are equalized mechanically.

18. Does not in any way hamper the normal operation of the machine or broaching tool.

19. Has minimum clearance between broach bar and wear plates consistent with free movement of the latter.

20. Its mechanism is as simple as circumstances permit.

PRODUCT DESIGN FACTORS

Process Selection: A close examination of a wide sampling of random parts reveals that most parts may be satisfactorily produced by more than one process. In actual practice, final selection of a particular processing operation for a given part is usually based upon available equipment, expected production rates and total quantity of parts required, and special workpiece conditions such as configuration, dimensional accuracy, surface finish, and material. Process engineers are guided in the final selection of the most efficient process by thorough comparisons of the technical and economic feasibility of one process to another.

Part Configuration: Parts to be broached should have uniform wall sections. Parts with frail or thin wall sections will fail because of the damaging tool forces and the possibility of damaging effects caused by the fixture.

Stock Allowance: The amount of stock allowance varies with the workpiece material and the primary method of forming: stamping, casting, forging, or whatever. Broached surfaces on stampings require only a minimum amount of stock allowance. When broaching holes in parts that have previously been formed by punching, drilling, or boring, the standard practice is to allow $\frac{1}{32}$ in. (0.794 mm) for small hole sizes, increasing to $\frac{1}{16}$ in. (1.588 mm) for larger sizes. Cast or forged parts require the greatest amount of stock allowance because of the need for a reasonably deep cut which should penetrate below the abrasive surface scale, hard spots, and inclusions. Cored holes have draft angles and size variations from part to part which may affect the amount of stock allowance needed. Careful size control must be exercised on castings or forgings to offset the possibilities of premature tool wear and breakage.

Surface Finish: Finishes can be obtained which are equal to and, in many cases, better than those achieved by milling. Surfaces on steel parts with a uniform microstructure have been repetitively produced in substantial quantities with $25\,\mu$in. A practical surface finish range for broaching operations is from 30 to $250\,\mu$in.

Tolerances: Studies have proved that in most cases broaching is capable of providing and maintaining closer tolerances over a long production run than is possible by reaming and in most milling operations. Broaching tolerances can satisfy the critical requirements of interchangeable manufacture and can be conveniently maintained. The distance between two or more gang- or straddle-broached surfaces can be held to within ±0.0005 in. (0.013 mm).

Round, square, and most odd-shaped holes can be produced to size within ±0.001 in. (0.03 mm) from locating surfaces. A similar tolerance can be held on spline or slot widths which are broached on circular parts. As in all other machining operations, even closer tolerances can be obtained, but always at additional cost.

REVIEW QUESTIONS

15.1. Describe how a finished surface is produced by use of a broach.

15.2. How does the action of a broach differ from that of a saw or file?

15.3. What is the principal function of a broaching machine?

15.4. Compare the relative cutting speed of a broach to that of a milling cutter.

15.5. Briefly describe the important elements of a broaching tool.

15.6. How is the "chip load" determined?

15.7. Explain the function of "chipbreakers" used on a broaching tool.

15.8. Contrast a one-piece broach to a shell-type broach.

15.9. Describe the function of a burnishing broach.

15.10. List two advantages of sectional external tools over one-piece construction.

15.11. Compare the main features of the two types of broaching machines.

15.12. What are some of the limitations of broaching?

15.13. What is the main advantage of broaching?

15.14. Explain the importance of work-holding fixtures to broaching operations.

15.15. When evaluating broaching as a potential machining process, what are some of the important considerations to be made with respect to (a) part configuration (b) stock allowance (c) surface finish?

15.16. How does the accuracy achieved in broaching compare to that in milling?

PROBLEMS

15.1. Continuous surface broaching machines are generally capable of higher hourly production rates than conventional surface broaching machines. Write a paper describing the principles of operation of a continuous broaching machine. Describe the importance of work-holding fixtures and give some product applications of this production process.

15.2. Extremely soft steels are difficult to broach because of their tendency toward metal pick-up on the sides of the spline broach teeth and on the lands, resulting in tears, galling, and a poor surface finish. Suggest ways to improve this condition.

15.3. Broach "pull" is defined as the amount of force necessary to operate a broach. Calculate the amount of pull necessary to broach the surface of a cast iron workpiece 2 in. wide by 10 in. long when the pitch of the broach teeth is $^{11}\!/_{16}$ in. and the rise per tooth is 0.003 in. The teeth are set at a 15° angle.

15.4. Determine the force required to broach a 3¼ dia. round hole in a 4 in. long copper workpiece. The broach has a pitch of 0.75 in. and a rise per tooth of 0.0025 in.

15.5. Discuss the effects chills or hard spots have upon broaching teeth.

15.6. Describe how the cutting conditions are affected by a worn and dull broach.

15.7. Explain some factors leading to an undesirable broaching condition known as *drifting*.

15.8. Prepare a table showing relative broaching speeds (fpm) obtainable when cutting the following materials with a feed of 0.003 ipt: Low alloy steel (free cutting), martensitic stainless steel, and gray iron (class 35 and 40, 150–190 Brinell hardness).

BIBLIOGRAPHY

"Broaching," *Machinery's Handbook,* 20th edition, New York: Industrial Press, Inc., 1975, pp. 1932–1939.

"Broaching," *Machining Data Handbook,* 2nd edition, Dearborn, Mich.: Society of Manufacturing Engineers, 1972, pp. 479–496.

"Broaching," *Metals Handbook,* Vol. 3, *Machining,* 8th edition, Metals Park, Ohio: American Society for Metals, pp. 58–74.

HINES, C. R., *Machine Tools and Processes for Engineers,* New York: McGraw-Hill Book Company, 1971, pp. 364–389.

*National Broach and Machine Company, *Broaching Practice,* Detroit, Mich: Lear Sigler, Inc., 1953.

   *Abstracted with permission.

16

TURNING

Turning operations are most commonly associated with a machine tool known as a *lathe*. Evidence of the basic beginning of a lathe has been documented in Egypt reaching as far back in history as the third century B.C. Perhaps the single most important improvement in recent years was the principle of screw cutting, which was developed in France in 1740. Steady improvements to this highly versatile machine tool have occurred ever since.

Turning may be defined as a machining process by which cylindrical, conical, or irregularly shaped external or internal surfaces are produced on a rotating workpiece. The cutting action is generated by one or more stationary single-pointed cutting tools which are held at an angle to the axis of the workpiece rotation.

DESCRIPTION OF THE PROCESS

Figure 16-1 illustrates how the shape of the desired workpiece surface is determined by the shape and size of the stationary cutting tool, which is being fed inwardly but without longitudinal feed.

FIGURE 16-1 Typical workpiece surfaces produced by in-feeding on a lathe

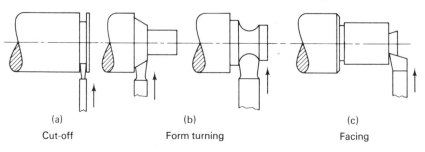

(a)   (b)   (c)

Cut-off   Form turning   Facing

Speed and Feed

Speed and feed are important cutting factors. *Speed* is a term that describes the cutting speed or velocity of the rotating workpiece as it moves past the cutting edge of the tool. The measurement of velocity is in units of surface feet per minute or (sfpm). Spindle speed, on the other hand, is measured in units of revolutions per minute (rpm) and may range from 10 to 2000 rpm. Turning *feed* is the rate of advance of the cutting tool per revolution of the spindle. Feeds may range from 0.001 to 0.075 in. (0.03 to 1.91 mm) per revolution. Considerable study has been devoted to the determination of appropriate turning speeds and feeds, with the result that there now are tables available in manufacturer's catalogs and in various machining handbooks which list recommended speeds and feeds for practically any classification of work. Such tables usually relate such variables as workpiece hardness, use of coolant, type of operation, and tool-bit material. In some plants selection of speed and feed is often heavily influenced by experience with similar parts or from previous production runs.

Effect of Speed and Feed on Cost: Speeds and feeds that are too low consume excessive time, which usually results in an increase in workpiece costs. However, optimum speeds and feeds are not necessarily the maximum that the workpiece and the machine will tolerate. Excessively high speeds and feeds result in shorter tool life and therefore in increased tool cost.

Depth of Cut

Depth of cut is a term that denotes the distance, in thousands of an inch, or in some convenient metric distance, to which the tool bit enters the work. The thickness of the metal chip thus removed is equivalent to the depth of cut. The depth of cut may vary depending upon such factors as rigidity of the particular type of lathe, the design of the tool bit, available horsepower, and the configuration and material of the workpiece. An in-feed of 0.005 in. (0.13 mm), for example, results in a size reduction on a workpiece diameter of *twice* this amount, or 0.010 in. (0.25 mm).

Tool Bits

Most metal lathe-turning operations employ single-point cutting tool bits which are ground to cut in only one direction. For proper cutting action to take place, the cutting edge must contact the workpiece *before* any other parts of the tool bit do. Figure 16-2a illustrates common shapes of single-point cutting tools which are produced by grinding one end of a solid bar of tungsten or molybdenum high-speed steel. Figure 16-2b illustrates examples of standard shapes of carbide tips or inserts, which may be attached to a

191

FIGURE 16-2(a) Common shapes of single point cutting tools (Courtesy, South Bend Lathe)

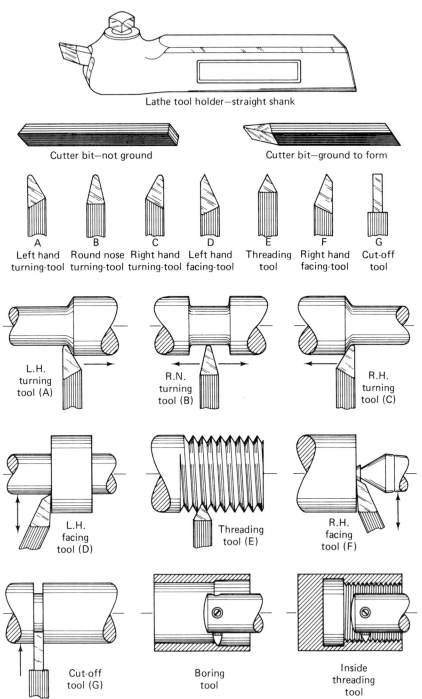

Lathe tool holder—straight shank

Cutter bit—not ground

Cutter bit—ground to form

A
Left hand turning-tool

B
Round nose turning-tool

C
Right hand turning-tool

D
Left hand facing-tool

E
Threading tool

F
Right hand facing-tool

G
Cut-off tool

L.H. turning tool (A)

R.N. turning tool (B)

R.H. turning tool (C)

L.H. facing tool (D)

Threading tool (E)

R.H. facing tool (F)

Cut-off tool (G)

Boring tool

Inside threading tool

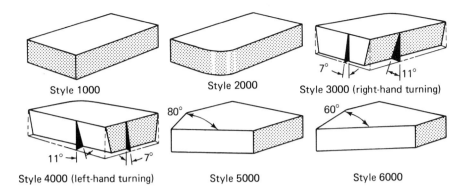

Style 1000
Style 2000
Style 3000 (right-hand turning)
$7°$ $11°$

Style 4000 (left-hand turning)
$11°$ $7°$
Style 5000
$80°$
Style 6000
$60°$

FIGURE 16-2(b) Standard shapes of carbide tool tips (Courtesy, South Bend Lathe)

less costly bar of steel by brazing, soldering, or welding. Clamp-on-type carbide or ceramic inserts are usually discarded when the cutting edge becomes dull. The brazed-on inserts are resharpened after dulling.

Lathe Operations

Lathes are the principal machine tools used for the vast majority of turning operations. Many other machining operations are possible on lathes. These include facing, parting (or cutoff), necking, drilling, boring, undercutting, reaming, chamfering, knurling, threading, tapping, and, by the use of special attachments, additional operations, such as taper turning and boring, grinding, and milling, can be performed. Figure 16-3 illustrates various lathe operations on work held in a chuck. It should be noted that two or more similar or dissimilar lathe operations may, in special production cases, be simultaneously performed. For example, the outside diameter of a workpiece may be turned while a hole is being drilled or reamed in its center. In addition, by use of a turning head, two or more surfaces may be simultaneously faced or turned with multiple tool bits.

FIGURE 16-3 Lathe operations on work held in a chuck

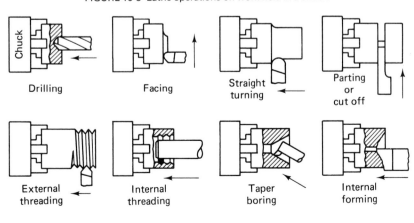

Drilling
Facing
Straight turning
Parting or cut off

External threading
Internal threading
Taper boring
Internal forming

TYPES OF TURNING MACHINES

There are many kinds of workpiece-rotating machine tools in use today. Close examination reveals a remarkable similarity in the manner by which all lathes operate. The type of lathe selected for a particular work application is a function of the size of the workpiece, the nature of the machining operation, and the quantity requirements. Turning machines may be conveniently grouped according to the manner of operation: manual, semi-automatic, and fully automatic. While it is reasonable to expect some overlapping, turning machines may also be roughly classified for evaluation as follows:

Limited or low-production types

Engine lathe, bench lathe, toolroom lathe, speed lathe

Medium-production types

Duplicating (or tracer) lathe, turret lathe

High-production types

Automatic screw machine, Swiss-type automatic screw machine

Limited or Low-Production Types

Engine Lathe: The *engine lathe* derives its name from its forerunner, which was powered by steam engines. The outstanding feature of an engine lathe is that most parts may be easily set up. Work-holding fixtures are not usually required for low production of parts. Because this type of lathe is limited to a single-point cutting tool and the fact that it can generate only one surface at a time, it is mainly used in school shops, toolrooms, model and home workshops, and for producing prototypes and low-quantity job-shop work. Engine lathes are available with several very useful attachments, which further expand its usefulness over a wide range of process applications.

The basic structure of the engine lathe may be conveniently divided into five main sections: the headstock, tailstock, bed, carriage, and the quick-change gearbox. These sections are shown in Figure 16-4 and will be discussed together with certain other related lathe accessories.

The *headstock* (see Figure 16-4) is mounted in a fixed position and is located to the operator's left. It houses and supports the driving mechanism which drives the spindle. As many as 8 to 27 different speeds are possible, depending upon the size and model of the particular machine. The headstock spindle is hollow to permit the loading of tubular or shaftlike workpieces whose length may be too long to hold by the usual machining methods. Work-holding attachments such as a driving plate, faceplate, or various types of chucks may be mounted on the threaded spindle nose. Some types of work may also be held in a collet shown in Figure 16-5, which is inserted into the hollow headstock spindle.

194

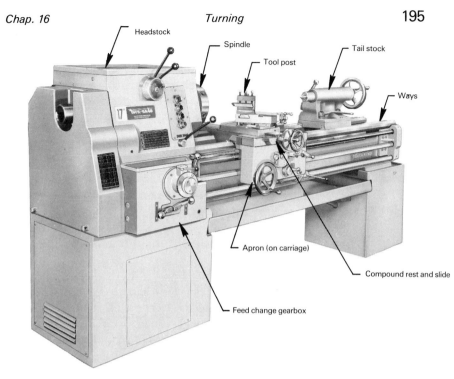

Headstock

Spindle

Tool post

Tail stock

Ways

Apron (on carriage)

Compound rest and slide

Feed change gearbox

FIGURE 16-4 A modern engine lathe (Courtesy, South Bend Lathe)

FIGURE 16-5 Turning a workpiece while held in a collet

The *tailstock* (see Figure 16-4) is located to the operator's right. It can be adjusted along the bed of the lathe and clamped in any desired position to accommodate different lengths of stock. A hardened steel part known as a "dead center," or, more popularly, a rotating part known as a "live center" is inserted into the tapered hole in the hollow tailstock quill. Alignment is made with a corresponding point on another center, which is similarly held in the headstock spindle. Work held in this manner is said to be in position "between centers." Tools such as drills and reamers fitted to the tailstock are controlled by rotating the tailstock handwheel.

The *bed* (see Figure 16-4) is the central foundation upon which the lathe is built. This rigid structure serves as a mounting frame for the headstock, the tailstock, the carriage, and the gearbox. Two sets of accurately machined parallel ways on the top of the bed guide the carriage and align the headstock and tailstock.

The *carriage* (see Figure 16-4) supports and controls the action of the cutting tool, which is firmly held in a tool post. The carriage can be moved longitudinally by manual or by power feed. A cross slide, mounted on top of the carriage, provides cutting-tool motion which is at right angles to the axis of the workpiece. Adjustments as small as 0.001 in. (0.03 mm) can be made in positioning the tool bit. The cross-feed movement may also be controlled by manual or by power feed. The compound rest which is mounted above the cross slide can be swiveled in a horizontal plane. In this manner, the cutting tool can be positioned and set at any desired angle to the axis of the workpiece. In and out feed adjustments may be made by manually turning the compound feed-rest knob, which has a graduated dial.

The *quick-change gearbox* (see Figure 16-4) is located just below the headstock. By arranging the position of the various control levers, the operator may select the amount of desired feed for the longitudinal or cross-cutting tool motion. The levers may also be set to obtain various pitches of screw threads.

Engine lathes are available with work capacities of 9 to 50 in. (228.6 to 1270 mm) in diameter (called "swing") and with bed lengths from 3 to 16 ft (approximately 1 to 5 m). In addition to diameter and length work-capacity considerations, other factors to consider when selecting an appropriate lathe size include the hole diameter in the headstock spindle, horsepower ratings, change gear equipment, and type of accessories.

Bench Lathe: A *bench lathe* is usually mounted on a benchtop. Bench lathes are designed for light work and, except for a size restriction of 12 to 13 in. (304 to 330 mm) in maximum workpiece diameter, most are capable of doing work similar to that of an engine lathe.

Toolroom Lathe: A *toolroom lathe* is built to more precise tolerances than an engine lathe and thus is used principally for highly accurate, low-production work on small parts. It may be either a pedestal or a bench type. Toolroom lathes are generally equipped with additional accessories for greater work versatility and often have a variable-speed mechanism and added facilities for obtaining a wider range of feeds. Such lathes are more expensive than comparable sizes of engine lathes.

Speed Lathe: Speed lathes consist simply of a headstock, a tailstock, and a toolpost, which is mounted on a light-frame bed. It may be driven by a step-cone pulley or it may be equipped with a variable-speed motor. Owing to the nature of the work performed on such a lathe, high spindle speeds are required. Speed lathes are used for making centerholes in workpieces prior to mounting work between centers for machining on other types of lathes. Speed lathes are also used for metal spinning (see Chapter 36), polishing (see Chapter 39), and for wood turning.

Medium-Production Lathes

Duplicating Lathe: Duplicating (or *tracer*) *lathes* are designed to produce duplicate parts by semiautomatic means. Except for tape-controlled lathes (discussed in Chapter 21), the operator is required to load and unload the workpieces and to start the automatic operation for each cutting cycle. One single-point cutting tool is used. Straight, tapered, and contoured turned or faced surfaces and bored holes are repetitively produced by the motion of a stylus as it follows the outline of a low-cost template. The template is generally mounted at the rear of the machine. The movements of the stylus are actuated by mechanical, air, hydraulic, or electrical methods. Work may be held between centers or chucked. Duplicating or tracer attachments are generally factory-installed on a standard engine lathe, although special machines are also available.

Turret Lathe: Turret lathes are semiautomatic machine tools which can produce parts in greater quantities, to closer tolerances, faster, and as a result more economically than is possible by conventional engine lathes. Unlike the engine lathe, the turret lathe is not restricted to a single cutting tool. A leading advantage is that several operations may be performed on a workpiece at a given time. It should be noted, however, that the range and types of turret lathe operations are essentially the same as those associated with the engine lathe.

Figure 16-6 illustrates examples of turret lathe production parts. To offset the need for frequent tool changes, a six-sided turret is used in place of a tailstock. The turret may easily be indexed so that any desired cutting tool may contact the workpiece. Additional cutting tools may be placed in square turrets on the cross slide and may be readily indexed to machine other surfaces. Unlike a cutting tool used on an engine lathe, once turret

FIGURE 16-6 Turret lathe production parts
(Courtesy, Jones & Lamson)

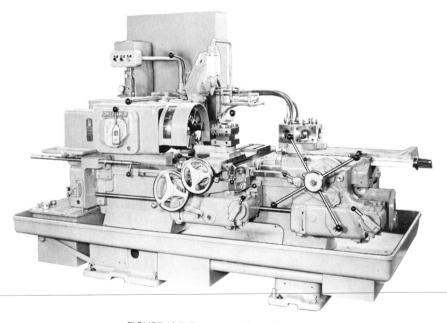

FIGURE 16-7 Ram-type universal turret lathe
(Courtesy, Jones & Lamson)

lathe tools have been properly inserted in one of the various tooling stations, they need not be removed until they require sharpening or after the production run has been completed. Finally, there is also available a single tool holder on a turret lathe which is mounted with a cutting tool on the rear of the cross slide.

Full advantage of the versatility and adaptability of the engine lathe can be obtained only in cases where the services of a "Highly" skilled machinist are available. Skilled labor is expensive. Parts produced by engine lathe operations are usually more costly than similar parts produced on a turret lathe. Turret lathe production costs are often minimized by using skilled setup men whose job it is to set and adjust the tools, leaving the simple and repetitive operations for the lower-pay-scale machine operators.

Turret lathes may be classified as *horizontal* and *vertical.* The two leading types of horizontal turret lathes are the *ram type,* shown in Figure 16-7, and the *saddle type.* Both types are suited to bar (cylindrical turning) and chucking work. The ram-type turret lathe is best suited for light barwork and small chucking jobs, while the saddle-type turret lathe is primarily used for longer barwork and for heavier chucking workpieces. Figure 16-8 shows a numerically controlled four-axis turret lathe. Ideal for bar and chucking work, they can also handle between-centers shaft work.

FIGURE 16-8 A four-axis numerically controlled turret lathe
(Courtesy, Giddings & Lewis-Bickford Machine Company)

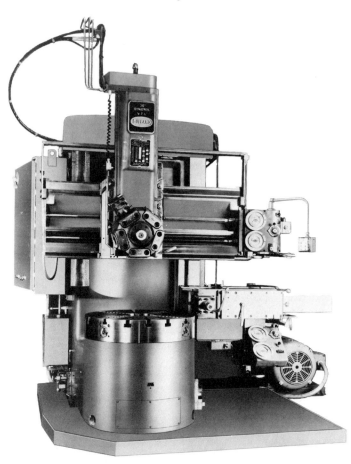

FIGURE 16-9 A vertical turret lathe with a 36 in. diameter table
(Courtesy, The Bullard Company)

Vertical turret lathes are designed for considerably larger and heavier work than is commonly associated with either type of horizontal turret lathes. Vertical machines are utilized solely for complex chucking work, particularly for boring operations, and are not adapted to bar work. An example of a vertical turret lathe is shown in Figure 16-9.

High-Production Lathes

It is difficult to precisely classify the various types of turning machines strictly according to their production output. Machines previously discussed are all reasonably similar, in that each type requires operator attention to an extent that varies from ''considerable'' to ''only occasional.'' The types of high-production lathes selected for the following discussion are those that run almost continuously. Some are fully automatic and require only occasional operator attention.

Automatic Screw Machines: Automatic screw machines (or *automatic bar machines*) were originally designed for the high production of screws and various other threaded fasteners. The basic beginnings of this machine date back to the 1880s, when the first machine was developed. The machine is essentially an advanced form of a turret lathe which fashions a wide range of parts in large quantities from bar stock. High output of interchangeable parts is possible because of intricate mechanisms designed to automatically feed single and continuous lengths of stock, index the turret for the desired sequence of the proper cutting tools, and retract the tool after cutting. Contrary to the designation "screw machine," production is not limited to threaded parts. The process is adaptable to the economical production of turned and formed parts of almost unlimited configuration, as shown in Figure 16-10.

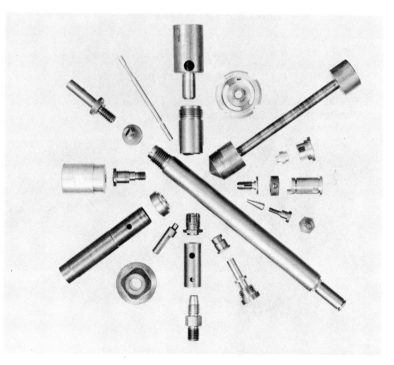

FIGURE 16-10 Typical parts produced on a screw machine (Courtesy, Brown & Sharpe Manufacturing Co.)

FIGURE 16-11 A single spindle screw machine (Courtesy, Brown & Sharpe Manufacturing Co.)

FIGURE 16-12 A close-up view of cutting tools in a six-station turret (Courtesy, Brown & Sharpe Manufacturing Co.)

Automatic screw machines may be classified as single-spindle automatics, multiple-spindle automatics, and automatic chucking machines. An example of a *single-spindle bar automatic screw machine* is shown in Figure 16-11. Bar stock (ranging from round, square, hexagonal, or other cross-sectional shapes) is fed through a revolving hollow spindle at the beginning of each cycle of operation. The stock is stopped at a predetermined distance and held during the cutting operations in a collet. Cutting tools are mounted around a six-station turret shown in Figure 16-12, which rotates in a vertical plane in a Ferris wheel motion. The turret is fixed to a slide which gives it longitudinal movement. There is a cross slide on which additional tools may be mounted in positions on the front as well as to the rear. The various cutting tool movements are obtained by means of cams mounted on shafts which are geared together on three sides of the machine. Mounted "dogs" engage various trip levers to control the sequence of operations on the machine.

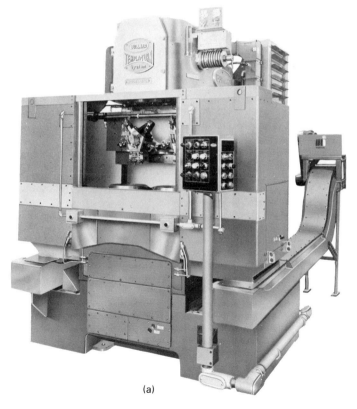

(a)

FIGURE 16-13(a) A vertical chucking machine
(Courtesy, The Bullard Co.)

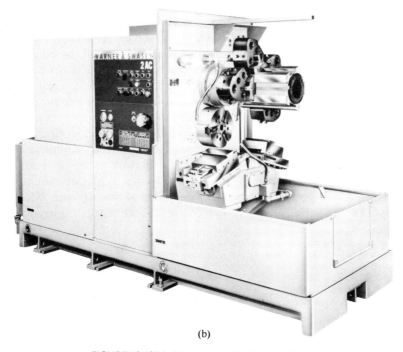

(b)

FIGURE 16-13(b) A horizontal chucking machine
(Courtesy, The Warner & Swasey Co.)

The principles that apply to the tooling arrangement of single-spindle-bar automatic screw machines apply also to *multiple-spindle-bar automatic screw machines*. Instead of a single spindle, however, multispindle machines may have four, five, six, or eight hollow spindles. Bars of stock are loaded into each continuously revolving work spindle. The spindle carrier indexes and automatically moves the stock from one cutting station to another. The required cutting operations are performed on the stock in progressive stages as the workpiece proceeds from one station to another, ending with the completed part as it is cut off from its original bar. Figure 16-13a and b illustrates two types of chucking machines designed for machining castings, forgings, pressed parts, and other shapes that cannot be machined from bar stock. *Automatic chucking machines* are capable of performing most of the operations normally associated with the multiple-spindle-bar automatic machine. A numerically controlled two-axis twin-turret bar chucker is shown in Figure 16-14. It is common practice for an operator to handle or service two or more automatic machines once they have been properly tooled. Figure 16-15 illustrates some typical parts machined on a vertical chucking machine.

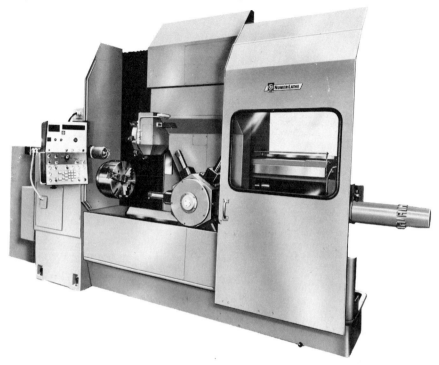

FIGURE 16-14 Two-axis twin bar chuckers are well suited for either bar or chucking work. This machine has a 14 tool capacity that inherently reduces tool-changing time. (Courtesy, Giddings & Lewis-Bickford Machine Co.)

FIGURE 16-15 Typical parts machined on a vertical chucking machine (Courtesy, The Warner & Swasey Co.)

FIGURE 16-16 A Swiss-automatic screw machine (Courtesy, American Bechler Corporation)

FIGURE 16-17 A close-up view showing the tooling and the slides on a Swiss automatic screw machine (Courtesy, American Bechler Corporation)

Swiss-Type Automatic Screw Machine: An example of a *Swiss-type automatic screw machine* is shown in Figure 16-16, with a closeup view showing the tooling and slides in Figure 16-17. It differs widely in design and function from other types of screw machines. A revolving piece of stock is fed through a carbide-lined guide bushing into the path of five radially mounted tools which are individually cam-controlled. The cutting edges of the single-point cutting tools which are used in this process are set to contact the stock as close as a few thousands of an inch but never more than $\frac{1}{32}$ in. (0.794 mm) away from the end of the guide bushing. Accurate parts may be produced in this manner because the firmly held work is prevented from springing away from the cutting tool. Special configurations such as tapers and multiple diameters are possible by combining the forward or dwell movements of the headstock and the in-and-out tool movements.

Swiss-type automatic screw machines are capable of an extremely high production output of minute parts. As an example, it is possible to produce parts down to 0.005 in. (0.13 mm) in diameter. Turn lengths can be produced on parts as short as $\frac{1}{32}$ in. (0.794 mm) up to 9 in. (228.6 mm). Parts for watches are commonly made on Swiss-type automatic screw machines. Examples of typical Swiss-type automatic machine production parts are shown in Figure 16-18.

FIGURE 16-18 Examples of Swiss automatic machine production parts (Courtesy, American Bechler Corporation)

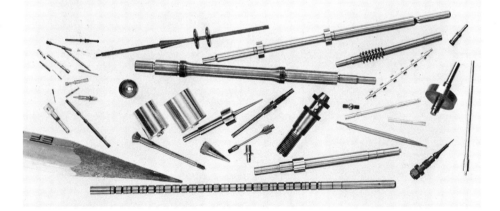

PRODUCT APPLICATIONS

Limited or Low-Quantity Turning Operations

Low-quantity operations are associated with work performed on engine lathes, bench lathes, and toolroom lathes. Considerable skill is required to properly set up and operate these types of lathes. Because it is not feasible for one man to operate more than one low-production-type lathe at a time, a disproportionate amount of an operator's time is spent waiting while the machine is cutting. Workpiece applications are limited to runs of a few to ordinarily not more than 10 pieces of a kind. Work is generally restricted to parts for prototype or experimental applications, or for producing one or two similar parts as required for maintenance, or for special tooling such as jigs and fixtures. Turning operations are required to finish giant shafts (see Figure 16-19) up to 80 ft (approximately 24 m) long and propellers

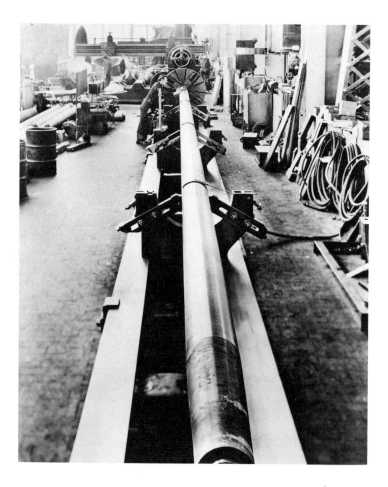

FIGURE 16-19 Lathe operations on a giant shaft (Courtesy, National Machine Tool Builders' Association)

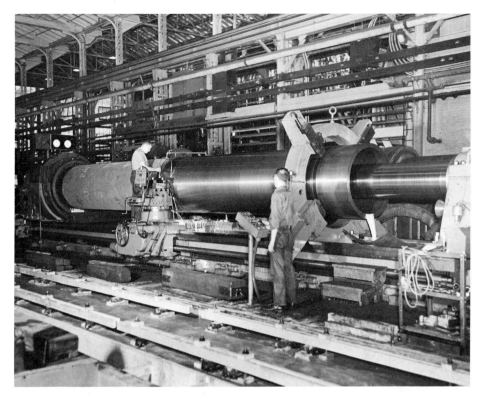

FIGURE 16-20 Lathe operations on a suction roll shell for a paper machine. Shell measures nearly 4 ft in diameter by over 32 ft long (Courtesy, Sandusky Foundry & Machine Co.)

for ships, and for turning long-range gun barrels; large steel or paper-mill rolls, as shown in Figure 16-20; missile parts; and parts for oil-drilling equipment. Heavy-duty special engine lathes are made for these special purposes. Speed lathes are used for long-run buffing and polishing operations on a wide variety of metal products, and, except in unusual cases, speed lathes are rarely used for normal turning operations.

Medium-Quantity Operations

Medium-quantity production requirements are usually associated with work performed on duplicating and turret lathes. The workpiece configuration generally regulates the type of machine in this category that is selected to perform the required operations. Duplicating lathes are used to produce parts whose required finished surfaces are to be irregular in shape or shapes which consist of complicated forms, tapers, and other shapes not readily machined by other methods. Such parts include valve plugs and bodies, nozzles, orifices, cluster gear blanks, automotive and aircraft pistons, stepped shafts, and contoured rolls. There are fewer restrictions on parts selected for machining on duplicating machines than on those machined on turret

209

lathes. The principal requirement is that the configuration be such that it can be adapted to holding between centers or supported in chucks.

Turret lathe operations are frequently performed on parts with complicated designs or on heavy forgings or castings whose sizes exceed the capacity of screw machines. Parts commonly processed on turret lathes include valve pistons, flywheels, rods and special bolts, gears, and a wide variety of machine parts, such as collars, liners, and spacers.

High-Quantity Operations

High-quantity turning requirements are reserved for automatic screw machines or Swiss-type automatic machines. Production runs of over 100,000 parts are made on such machines.

PRODUCT DESIGN FACTORS

Part Configuration: Turning and boring operations can be performed on an almost unlimited variety of cast, forged, or stamped parts. Further, turning operations can be even more readily performed on bar stock, tubing, and extrusions of various cross-sectional shapes. Basically, the process of reaching a decision as to whether to use one turning method in favor of another is often based upon an evaluation of practical methods of holding the work while the cutting operations are being performed. Special fixtures may be designed when the part cannot be conventionally held between centers, in a chuck, or in a spindle collet.

Part Sizes: As might be expected, there are definite size restrictions inherent in each type of turning machine. Turning operations are possible on "mini-parts" only a few thousands of an inch in diameter (0.08 mm, for example) and less than 1 in (25.4 mm) in length to giant parts that weigh several hundred pounds (150 kg, for example) and measure over 6 ft (1.82 m) in diameter. Size capacities for standard types of turning machines are given below:

TURRET LATHES

Ram type: Maximum swing to 22 in. (558.8 mm); maximum travel carriage 24 in. (609.6 mm); maximum travel cross slide 10 in. (254 mm); bar capacity ⅝ to 4½ in. (15.88 to 114.3 mm); chucking capacity up to 15 in. (381 mm) in diameter.

Saddle type: Swing range from 16 to 40 in. (406.4 to 1016 mm); maximum travel side carriage 93 in. (2.36 m); maximum travel cross slide 16 in. (406.4 mm); bar capacity from 2½ to 6 in. (63.5 to 152.4 mm); chucking capacity up to 32 in. (812.8 mm) in diameter.

Vertical type: Maximum swing to 78 in. (1.98 m); chucking capacity up to 73 in. (1.85 m) in length.

Single-spindle (bar): Minimum stock diameter capacity ³⁄₁₆ in. (4.763 mm); chucking capacity ranges from 2 to 8 in. (50.8 to 203.2 mm) for solid bar stock and 2 to 9½ in. (50.8 to 241.3 mm) for tubing; minimum length of turn with one movement ¾ in. (19.05 mm); maximum length of turn with one movement 9 in. (228.6 mm).

Multiple-spindle (bar): Minimum stock diameter capacity ⁹⁄₁₆ in. (14.29 mm); maximum stock diameter capacity 7 in. (177.8 mm); maximum length of turn 5 in. (127 mm); maximum stock length 20 in. (508 mm).

Chucker: Minimum capacity 8 spindle with 6 in. (152.4 mm) swing.

SWISS-TYPE AUTOMATIC MACHINES

Chuck capacities range from ³⁄₃₂ to ½ in. (2.381 to 12.7 mm); with a ⁵⁄₃₂-in. (3.969-mm) chuck, a maximum workpiece length of 1⁹⁄₁₆ in. (39.69 mm) is possible and with a ½-in. (12.7-mm) chuck, a maximum workpiece length of 2¾ in. (69.85 mm) is possible; a maximum workpiece length of 4 in. (101.6 mm) is possible by substituting a special headstock-feed bell plate cam.

Selection of Materials: The following materials and related cutting speeds are recommended for the various turning processes. (Of the materials listed below, AISI B1113 is rated highest in machinability of all the steels and is generally preferred, where applicable, because of its ability to yield the best finish at the highest cutting speeds.)

TURRET LATHES

Material	Cutting speed (sfpm)	Tool materials
Commercial brass, aluminum, and bronze	Up to 1000	Carbides
Commercial brass only	Up to 300	High-speed steel
Aluminum	Up to 400	High-speed steel
Bronze	Up to 150	High-speed steel
Hard, semi-, or medium-carbon steel	As low as 40 to 120 but up to 600	High-speed steel Carbides

AUTOMATIC SCREW MACHINES: As compared to steel, for example, brass workpieces may be produced at a high production rate. Aluminum and magnesium alloys may be worked at high cutting speeds, especially where the design of the part is reasonably simple. The stock tolerance should be +0.000 to −0.005 in. (0.13 mm) O.D.

211

SWISS-TYPE AUTOMATIC MACHINES: Carbon drill rod as well as various high-machinability steels are frequently used. Other materials in wide use are stainless steels and nonferrous metals such as brass, bronze, and aluminum. Centerless ground stock should be specified for Swiss-type-automatic operations, with such tolerances as: maximum out-of-roundness, 0.0003 in. (0.008 mm); uniformity of size over total length of bar, ±0.0003 in.; and uniformity of bar-to-bar diameters, ± 0.0005 in. (0.013 mm).

Production Rates: Several factors exert a marked influence upon the production rate of a given part. Careful planning is required to obtain the many cost benefits of multiple tooling. Production rates of most parts are usually estimated on a time-study basis, with the objective to obtain the lowest cost per unit over the shortest time period. Most designers start by closely evaluating each feature of the part configuration, in a sincere attempt to keep the part as simple as possible. Simple, straightforward shapes generally require minimum work-handling time and rarely require special setups or tooling. Standardized work holders and tools are used whenever possible. Substantial cost economies can usually be had by reducing cutting time and by machining workpieces at recommended speeds and feeds. Savings can also be had when multiple tooling can be incorporated, thus permitting the simultaneous machining of two or more surfaces of a given workpiece. Other cost reductions can often be obtained by using air, hydraulic, or electrically operated chucks, which reduce setup time.

TURRET LATHES: Depending upon the workpiece configuration and the required tooling, hand-operated turret lathes may be economical for producing as few as 25 to 50 pieces. As a general rule, however, automatic turret lathes are used for production runs of greater than 50 pieces.

AUTOMATIC SCREW MACHINES: While setup time may require as many as 15 to 20 hours, a part may be completed in just a few seconds. One company has developed a tooling sequence consisting of nine automatic operations which produced a completed carburetor adjusting screw every 3 seconds. In general, production runs of at least 4000 to 5000 parts are necessary to justify tooling costs on a multiple-spindle automatic screw machine.

SWISS-TYPE AUTOMATIC MACHINES: For most parts, the total cycle time averages about 1 minute with a reasonable cycle time range of 5 seconds to 4 or 5 minutes per part. Swiss-type automatic machines are considered to be the most efficient and economical method for producing small, slender, and intricate parts in quantities over 100,000.

Tolerances: Workpiece accuracy is largely dependent upon the workpiece and cutting-tool materials, as well as the speed and feed settings. Tolerances should be applied with considerable discretion. Workpiece costs always increase when the specified tolerance is more precise than is actually required. As an example, a reamed hole may be routinely produced on a part in a turret lathe within a tolerance of ±0.001 in./in. (0.001 mm/mm). A tolerance of ±0.0005 in. (0.013 mm) on the hole size could require one or more additional operations, resulting in elevated costs. A surface may be produced to accuracy limits of ±0.005 in. (0.13 mm) in one cut, but a tolerance of 0.0025 in. (0.06 mm) could require *three* cuts. Liberal tolerances result in maximum cost savings.

TURRET LATHES: *Tolerances* on workpieces of less than ±0.002 in. (0.05 mm) are difficult to obtain in high-production turret lathe operations, because of the increased cost imposed by the requirements of additional finishing cuts and fine feed settings.

Diameters: ±0.001 to 0.0005 in. (0.03 to 0.013 mm) is possible in special cases, but, in general, a tolerance of ±0.003 in. (0.08 mm) is less costly to produce.

Concentricity: It is possible to control the concentricity of two or more diameters on a part to a tolerance of 0.001 to 0.003 in. (0.03 to 0.08 mm) total indicator reading (T.I.R.) if the position of the part is not changed.

Runout: Using special precautions in tooling and setup, a total of 0.0005 to 0.001 (0.013 to 0.025 mm) can be obtained.

Lengths: A tolerance of ±0.001 in./in. (0.001 mm/mm) under usual conditions is possible, although a tolerance of ±0.005 in./in. (0.005 mm/mm) results in lower production costs.

Drilled-Hole Diameters: Tolerances vary according to size from 0.001 to 0.002 in. (0.03 to 0.05 mm) for No. 80 drills up to 0.010 in. (0.25 mm) for drills 1 to 2 in. (25.4 to 50.8 mm) in diameter.

Drilled-Hole Depths: Same tolerances as *lengths.*

Reamed-Hole Diameters: Holes initially formed by drilling: ±0.001 in./in. (0.001 mm/mm). Holes initially formed by boring: ±0.0005 in./in. (0.0005 mm/mm).

SCREW MACHINES: Closer tolerances are possible when doing large work on single-spindle screw machines than for similar size work on multispindle screw machines. Standard practice for average size work is to specify a tolerance of ±0.003 in./in. (0.003 mm/mm) on important dimensions and ±0.010 in./in. (0.010 mm/mm) on noncritical dimensions.

Diameters: For work sizes up to 1 in. (25.4 mm), ±0.003 in. (0.08 mm); sizes 1 to 2 in. (25.4 to 50.8 mm), ±0.004 in. (0.10 mm); and work sizes over 2 in. (50.8 mm) ±0.005 in. (0.13 mm). Tolerances less than ±0.0005 in. (0.013 mm) are not economically feasible.

Lengths: Tolerances of ±0.001 to 0.005 in./in. (0.001 to 0.005 mm/mm) are practical.

Concentricity: A tolerance of 0.005 in./in. (0.005 mm/mm) T.I.R. is practical if all operations are accomplished during one setup.

Squareness: Usual practice for small parts is to specify 0.0005 in./in. (0.0005 mm/mm) T.I.R. for faces that are normal to bores or turned surfaces.

Angles: Tolerances on taper angles are usually given as ± 1°, with closer limits obtainable by finishing with a special shaving or taper reaming tool.

SWISS AUTOMATICS: Workpiece accuracy is affected by the accuracy of the raw stock and how well it fits into the guide bushing on the machine.

Diameters: In cases of high production runs on small parts, tolerances may be specified as close as ± 0.0002 in./in. (0.0002 mm/mm) by using two or more light cuts and a slow feed. Usual practice is to specify a tolerance of ± 0.0005 in./in. (0.0005 mm/mm).

Lengths: If necessary, the tolerance on turned lengths to shoulders may be specified as close as ± 0.0005 in./in. (0.0005 mm/mm). A more reasonable tolerance of ± 0.002 in./in. (0.002 mm/mm) may be obtained without increased production cost.

Concentricity: It is possible to specify 0.0005 in./in. (0.0005 mm/mm) T.I.R. for special cases, but a tolerance as large as 0.005 in./in. (0.005 mm/mm) T.I.R. is considerably more economical.

Surface Finish: Surface quality is dependent to a great extent upon the type of material being machined, the type of tooling and depth of cut, and upon the speeds and feed selected to form the finished surface. Production time is always increased when an unusually fine surface finish is specified. High-quality finishes require fine feed settings over several different cuts, which *always* increase the cost of the part.

A 60-μin. finish can be obtained under normal conditions on an engine lathe or turret lathe. A better finish is obtainable with special handling. Screw machines are capable of producing parts with surface finishes of 32 μin. and with a finish of 16 μin. in special cases. Runs can be had on workpieces with surfaces as fine as 5 up to 50 μin. on Swiss-type automatic machines.

REVIEW QUESTIONS

16.1. Define the terms "speed" and "feed" and show how they are related to lathe work.

16.2. Prepare a sketch showing how an infeed (or depth of cut) of 0.010 in. (0.25 mm) results in a size reduction on the workpiece diameter of twice this amount.

16.3. Define the following lathe terms: (a) plain or straight turning (b) facing (c) parting (d) chamfering (e) knurling (f) swing (g) headstock (h) tailstock (i) carriage (j) bed.

16.4.   Contrast the major differences among a bench lathe, a toolroom lathe, and a speed lathe.

16.5.   Describe the movements of a lathe tool and tell how they are obtained.

16.6.   Discuss the important features of a duplicating lathe and describe the nature of the work performed.

16.7.   Discuss the important features of a horizontal turret lathe and describe the nature of the work performed.

16.8.   What are some of the principal advantages of an automatic screw machine?

16.9.   Discuss the essential differences between an automatic screw machine (or automatic bar machine) and the Swiss-type automatic screw machine.

16.10. On what types of lathes would the majority of limited or low-quantity turning operations be performed? What types of lathes would be suitable for medium-quantity work? For high-quantity work?

16.11. Explain why a decision to use one turning method over another often depends upon the availability of suitable workholding devices.

16.12. Discuss factors that may affect cost savings in estimating production rates.

16.13. Explain why liberal tolerances normally result in maximum cost savings.

16.14. Which of the following external diameter dimensions (expressed in inches) would be reasonable to specify on a drawing of a workpiece to be processed on a turret lathe? (a)   1.6255/1.6253 (b) 0.752/0.750 (c) 3.9375/3.9360 (d) 0.5055/0.5015.

16.15. Which of the following diameters (expressed in inches) is not practical to produce on a screw machine? (a) 1.755/1.750 (b) 2.3125/2.3120 (c) 0.501/0.500 (d) 0.6875/0.6835.

16.16. What important factors affect the quality of surface finish for turned work?

PROBLEMS
(See appendix for useful formulas.)

16.1.   Determine the angle of taper in inches per ft of the part shown in Figure 16-P1. Calculate the angle of taper.

FIGURE 16-P1

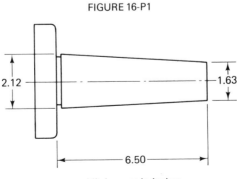

All sizes are in inches.

16.2. Determine the angle of taper and the taper per inch of the following tapers: (a) #4 Morse taper, (b) #8 Brown & Sharpe taper.

16.3. What is the minimum diameter at the small end of the following tapers? (a) #2 Morse taper, (b) #10 Brown & Sharpe taper?

16.4. Determine the horsepower that is required to turn the surface of a medium carbon steel shaft at 200 sfpm when the depth of cut is 0.06 in. and the feed is 0.0259 ipr. The feed correction factor is 1.05 and the unit horsepower is 1.4.

16.5. Calculate the cutting time required to turn the surface of a 12 ⅞ in. long ANSI 1010 steel workpiece when the feed is 0.007 ipr and the cutting speed is 120 rpm.

16.6. Solve for T to face the casting shown in **Figure 16-P6** when the feed is 0.005 ipr and the lathe is running at 175 rpm.

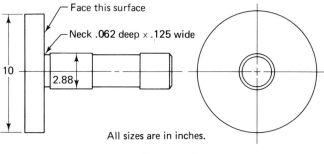

FIGURE 16-P6

16.7. Find the rpm for a 1 in. diameter low carbon steel workpiece with a cutting speed of 90 sfpm.

16.8. Find the rpm for a 4½ in. diameter medium carbon steel shaft with a cutting speed of 600 sfpm.

16.9. A 3 in. diameter × 12 in. long bar of medium steel is being turned down to size. The depth of cut is 0.187 in. and the feed is 0.011 ipr with a cutting speed of 450 rpm. How many cubic inches of stock are being removed per minute?

16.10. Suggest some points to check if the faceplate or the chuck does not run true.

FIGURE 16-P11

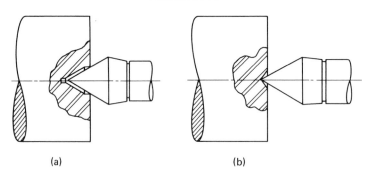

(a)   (b)

16.11. Figure 16-P11 illustrates two improperly drilled centerholes on stock to be turned on a lathe. Explain why these conditions are unsatisfactory. Prepare a sketch showing a correctly drilled and countersunk hole.

16.12. Before turning, the lathe centers should be checked for proper alignment. Other than a visual check by running the point of the tailstock center against the point of the headstock center, what procedure could be followed after taking the first roughing cut on a shaft?

16.13. Discuss the important difference between an independent chuck and a universal chuck.

16.14. Make a sketch of a machine part and indicate some typical processing operations that would make it a logical choice for chucking on a lathe.

16.15. Explain how a draw-in collect chuck is used on a lathe.

16.16. Explain how a mandrel for machining a workpiece on a lathe is applied to the workpiece. How does the use of a mandrel facilitate certain lathe operations?

16.17. A hole may be enlarged or an internal taper may be produced on a lathe by boring. Explain how the work may be held and describe the cutting tool used.

16.18. Prepare a sketch of a machine part you would consider as typical for manufacture on a duplicating lathe. Assume a production run of 5000 pieces.

16.19. Explain the causes of tool cratering.

BIBLIOGRAPHY

Aluminum for Automatic Screw Machined and Cold Headed Parts, AAS-12, New York: The Aluminum Association, 1976.

Hines, C. R., *Machine Tools and Processes for Engineers,* New York: McGraw-Hill Book Company, 1971, pp. 541–568.

"Turning," *Machining Data Handbook,* 2nd edition, Dearborn, Mich.: Society of Manufacturing Engineers, 1972, pp. 3–90.

*"Turning," *Metals Handbook,* Vol. 3, *Machining,* 8th edition, Metals Park, Ohio: American Society for Metals, pp. 1–19.

*Abstracted with permission.

17

POWER SAWING

Most machine tools can perform cutting operations to a limited extent. When compared to special-purpose power sawing machines, however, they are generally regarded as slow, cumbersome, and often wasteful. The cutting action in the power sawing process is accomplished by the action of a continuous series of single-point cutting tools as they pass over the workpiece. Tool life on the various cutters or blades (straight, circular, or band) is extended because of the sharing of the wear on the individual teeth.

SAWING MACHINE TYPES

Power-driven saws are classified according to the kind of motion used in the cutting action. Three general types are in common use in industrial shops: reciprocating saws, circular saws, and band saws. There are several categories of machines within the three general types.

Reciprocating Sawing Machines

Reciprocating sawing machines, shown in Figure 17-1, are perhaps the most common examples of power saws to be found in most production, toolroom, and maintenance departments. A power hacksaw is considered the simplest of all power sawing machines. Its function is principally restricted to cutoff operations. It is usually possible to position the stock so that multiple pieces may be simultaneously cut. Designs vary from light-duty crank-driven machines to hydraulically driven heavy-duty machines. The sawing

218

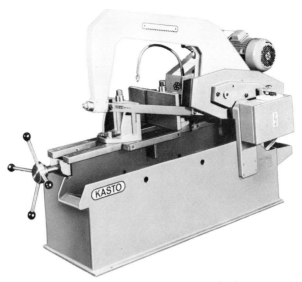

FIGURE 17-1 A reciprocating power hack saw
(Courtesy, Kasto Racine, Inc.)

action takes place in only one direction (on the forward or pull stroke), with the straight saw blade lifted slightly off the stock on the return stroke. Such action saves premature wear on the saw teeth.

Essentially, the power hacksaw consists of a bed or frame, a vise for rigidly clamping the stock, a straight saw blade, a holding frame, a power mechanism for reciprocating the saw frame, and a feeding mechanism. Most of the lightweight reciprocating sawing machines use a gravity feed in which the saw blade is lowered onto the work by the weight of the saw frame. The depth of the feed is usually regulated by spring loading or by weights clamped to the frame. Power saw manufacturers refer to this type of feed as *definite pressure,* which is the condition when the feed pressure of the saw teeth is uniform at all times. A *positive pressure* feed occurs when a machine has provisions for setting an exact depth of cut for each stroke. In this case, the pressure of the blade varies directly with the number of teeth in actual contact with the stock.

Power hacksaws with positive pressure feeds are usually operated by hydraulic mechanisms controlled by the operator. Such devices are found on large, heavy-duty machines. They also have facilities that may be set to automatically feed in or to discharge stock in predetermined lengths. Most power hacksaws have built-in coolant circulation systems as well as automatic mechanisms that shut off the machine after the cut has been completed. Power hacksaw blades are available in standard lengths from 12 to 36 in. (304.8 to 914.4 mm) with from 4 to 14 teeth per inch (or over a distance of 25.4 mm). Blade thickness ranges from 0.060 to 0.125 in. (1.52 to 3.18 mm). Most power hacksaw blades are made of high-speed tungsten or molybdenum steel.

219

Circular Sawing Machines

Circular sawing machines (or *cold sawing machines*) are used exclusively for cutoff operations. There are two major types of toothed saw blades which serve as the cutter. One type, used in cold sawing, is a toothed circular disk, shown in Figure 17-2. It strongly resembles a slitting saw used on milling machines, or the cutting blade used on a woodworking table saw.

(a)

(b)

FIGURE 17-2 (a) Cold sawing a heavy guage box channel (Courtesy, DeWalt-A Division of the Black & Decker Mfg. Co.); (b) cold sawing a billet for a rocket nozzle forging (Courtesy, Wyman-Godon Company)

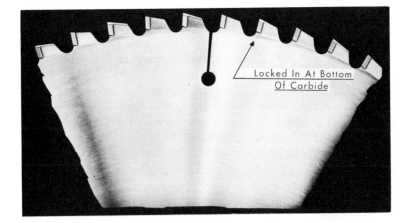

FIGURE 17-3 A carbide-tipped saw blade. This type is available in
diameters of 6 to 24 inches (Courtesy, Heinemann Saw Corporation)

Such blades are generally available in diameters of from 8 to 16 in. (203.2 to
406.4 mm). They are used principally for relatively minor cutoff operations.
A second type, shown in Figure 17-3, may have either individually inserted
teeth or detachable segments that may contain several teeth. One obvious
advantage of the segmented circular saw is that the disk can be made of a
less expensive steel, with more expensive high-speed or tungsten-carbide
steel reserved for the actual teeth. A cutting fluid is recommended when
cutting all metals except cast iron.

Friction sawing consists of a high-speed rotating disk which, except for
small indentions or projections spaced along the periphery, has no teeth at
all. Examples of two typical friction saw blades are shown in Figure 17-4.

FIGURE 17-4 (a) A *v*-tooth friction saw blade; (b) a *u*-tooth friction saw blade (Courtesy,
Heinemann Saw Corporation)

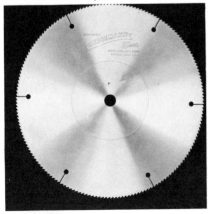 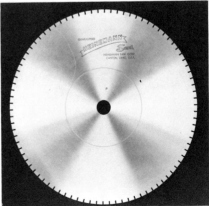

The cutting action is produced by friction of the metal disk, which heats the material to the softening point. The heat of friction creates a dramatic rise in temperature of the work stock, which actually melts the metal, with a resulting loss of strength. A path or cut is formed by the wiping action of the rotating disk. Friction sawing requires a coolant to prevent excessive heat buildup of the rotating disk, which is operated at speeds of up to 25,000 surface feet per minute (sfpm). While there is no reduced effect of metal workpiece hardness on cutting efficiency, the melting point and the structural characteristics of a given metal are critically important. Friction cutting disks range in size from 24 to 72 in. (609.6 to 1828.8 mm) in diameter.

Abrasive disk cutting, although not strictly a sawing technique, is another method of cutting off stock. An abrasive-disk cutoff operation may be selected in preference to power hack sawing or friction sawing when a better surface finish and closer limits of accuracy are required. Figure 17-5 illustrates an abrasive cutting machine. Most machines are readily adaptable to both wet and dry cutting operations. Such machines are principally intended for cutting hard materials that may be especially difficult (or impossible) to machine by other methods. Cutting soft materials with an abrasive disk

FIGURE 17-5 A dry abrasive cut-off machine (Courtesy, American Chain & Cable Company, Inc.)

FIGURE 17-6 Reinforced cut-off wheels are highly resistant to breakage combined with fast cutting action and low rate of wear. The unusual strength is due to the built-in fabric reinforcement. Wheels are available in both resinoid and rubber bonds

presents no particular problems. Essentially, an abrasive disk cutting machine consists of a powerful drive motor which drives the wheel head by means of one or more belts, a method of engaging the wheel against the work by means of a swing frame, and a work-holding vise or frame. Examples of reinforced cutoff wheels are shown in Figure 17-6. The wheels are characterized by their relative thinness, often measuring only 0.005 in. (0.13 mm) thick.

The wet-cutting process employs a rubber-bonded wheel that rotates with a surface speed of about 8000 sfpm. Any increase in rim speed above this point results in overheating the disk wheel, largely owing to an inability to supply a sufficient amount of coolant.

In dry cutting, resinoid-bonded wheels are used and operate at surface speeds up to 16,000 sfpm. Increased cutting efficiency is obtained at higher wheel speeds. This is due to the rapid heating of the metal being cut, with the result that the metal is softened and thus is easily removed. Abrasive disk-cutting wheels are made in sizes up to 36 in. (914.4 mm) in diameter. On most machines the cutting is generally accomplished by manually swinging the rotating wheel down onto the work. Some manufacturers produce a machine with a wheel that can automatically travel back and forth across the work.

Band Sawing Machines

Band sawing machines are considered more versatile than any sawing machine previously discussed. Band saw machining includes simple cutoff operations. It also encompasses continuous cutting along straight and curved lines. Band saws employ the continuous cutting action of an endless high-speed toothed blade. Most band saws may be classified into one of three general groups: horizontal (principally restricted to cutoff sawing), vertical machines (for straight and contour sawing), and vertical machines (used exclusively for friction sawing).

Horizontal band sawing machines provide a convenient method for holding long lengths of stock for cutting-off operations. Except for the continuous cutting action of the endless band saw blade, horizontal sawing machines incorporate most of the features of the power hack saw. Horizontal band saws, illustrated in Figure 17-7, have a manually or hydraulically operated swing frame which pivots the blade down onto the work. The cutting position of the blade is nearly horizontal. The feed pressure on light-duty types is gravity-fed. The work is manually clamped in a vise. Heavy-duty machines have hydraulically fed mechanisms which lower the blade down onto the work. Most heavy-duty models are equipped with a hydraulically operated vise as well as having facilities for varying the blade speed, calibrated work stops, automatic bar feeding, and so on.

Vertical band sawing machines resemble those used in woodworking. One type, classified as a standard-duty universal band saw shown in Figure 17-8, is used for practically all kinds of straight, angular, and contour cutting. The process of contour cutting of odd-shaped parts is often expedited by first compressing and welding together sheets of metal. Multiple parts are thus produced by cutting along the outline scribed on the top piece and later separating the individual pieces. Standard metal cutting band saws are extremely versatile. There are a number of useful accessories, such as attachments for automatic table feed, work-holding jaws for both straight and circle cutting, a ripping fence, a magnifying lens, a coolant system, and a flash butt resistance welding attachment for saw band assembly. Internal configurations may be produced by severing the blade, inserting it into a specially produced hole in the workpiece, and then welding it back together again to form a continuous blade. After cutting out the desired stock, the blade is once again broken to allow the part to be separated from the saw blade. Vertical band saw machines may also be used for filing and polishing operations when equipped with appropriate bands.

Friction sawing machines may also be used to produce straight, angular, and contour cuts. The process is not usually recommended for parts measuring greater than 1 in. in thickness. In contrast to conventional band saw machining, such machines operate at greatly increased speeds and cut through materials several times faster. The process is virtually chipless. A special low-cost "dull" blade contacts the work at super-high speeds (up to 16,200 (sfpm) and produces sufficient friction to heat the material to the softening point. The strength of the metal is lowered and the blade pulls or wipes the molten metal away from the sides and immediately ahead of the blade. No coolant is used, although a fine jet of compressed air may be directed upon the blade to promote the cutting action. Depending upon certain conditions,

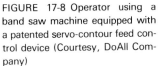

FIGURE 17-7 A horizontal band cut-off machine shown with discharge tray and stock stand (Courtesy, KTS Industries-Kalamazoo Saw Div.)

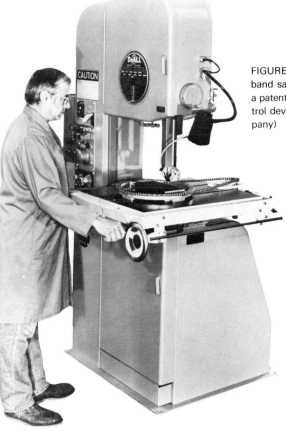

FIGURE 17-8 Operator using a band saw machine equipped with a patented servo-contour feed control device (Courtesy, DoAll Company)

the blades usually have a limited number of notches or teeth, although in some cases, blades with no notches are used. Friction band sawing is especially adaptable to the cutting of certain difficult-to-machine metals (armor plate, stainless and tool steels, etc.). It has been found that increased hardness of materials actually enhances the cutting efficiency of this process. Metals with hard tempers have a high coefficient of friction and are heated more readily.

PRODUCT APPLICATIONS

One or another method of power sawing generally precedes most machining operations. The power sawing cutoff operation is considered to be an important operation in the sequence of processing steps in the manufacture of many products. Power hacksawing and circular sawing machines are the principal methods by which desired lengths of different rolled and extruded shapes, tubing, flat strip, or plate stock are cut. Band sawing machines are

FIGURE 17-9 Applications of power sawing cut-off operations
(Courtesy, American Chain & Cable Co., Inc.)

commonly employed to remove unwanted flash from pressed sheet-metal parts or excess flash, gates, runners, risers, and sprues from castings. Typical shapes cut by power sawing are shown in Figure 17-9.

Contour sawing is often the method employed to produce simplified forging prototypes, which serve as samples in process design evaluation. Substantial savings can often be made in low-volume production of reasonably complicated three-dimensional parts by contour sawing. Contour sawing is often the method used to produce cam profiles, parts for dies, jigs, fixtures, and, in some cases, products in quantities of up to 1000.

PROCESS SELECTION FACTORS

One of the main advantages of power sawing is that material waste is held to a minimum. Because of the narrow width of the kerf or cut, only a small amount of material is lost as chips. Power sawing is faster than any other machining operation that performs a similar function. For example, an abrasive disk takes less than 10 seconds to cut through ordinary 2-in. (50.8-mm)-diameter steel pipe. In friction cutting, a 24-in. (609 mm) I-beam may be cut entirely through in less than 1 minute.

Friction sawing is restricted solely to ferrous metals. Nonferrous metals tend to smear and adhere to the friction disks. Power sawing machines require less electrical power to operate as compared to conventional machine tools. Straight, angular, and contour cutting can be performed with very basic fixtures and workholding devices. Except for abrasive disk cutting, power sawing operations produce a relatively rough as-cut surface of about 250 μin. Most plastics and ceramics require no special cutting techniques, although the resultant surface finish may be inferior to that obtained on ferrous metals. Products that require improved surfaces or greater accuracy are usually cut oversize and machined to final size by other methods.

Most power sawing operations result in a burr that projects on the underside of the metal being cut. This is particularly noticeable for softer metals. Certain alloys and high-carbon-steel parts that have been cut by the friction sawing method may require further machining, depending on the functional requirements or the required surface finish. In this case, tool or cutter life may be prolonged by annealing the parts after sawing to offset the effects of the heat of friction. Local hardening has been found to penetrate as much as 0.005 in. (0.13 mm) in depth in some cases.

Despite various feed mechanisms, stop rods, or cutoff gages found on most power sawing machines, it is impractical to saw multiple pieces of stock to consistently accurate lengths. Careful operators, however, are able to consistently maintain contour cuts on parts to within 0.007 to 0.010 in. (0.18 to 0.25 mm) of scribed lines, with closer tolerances obtainable in most cases by using the magnifying lens.

REVIEW QUESTIONS

17.1. In what ways do reciprocating sawing machines and band sawing machines differ?

17.2. What is the essential difference between "definite-pressure" feed and "positive-pressure" feed?

17.3. Of what materials are power hacksaw blades commonly made?

17.4. What is the main use of circular sawing machines?

17.5. Briefly explain the important aspects of friction sawing.

17.6. Contrast the process of abrasive disk cutting to friction sawing.

17.7. List the principal advantages of horizontal band sawing machines.

17.8. List the principal advantages of vertical band sawing machines.

17.9. Explain how internal cuts such as a hole in sheet-metal stock may be made using vertical band sawing machines.

17.10. In what ways is friction band sawing different from conventional band sawing?

17.11. List some power sawing operations commonly used prior to the actual machining operations in the manufacture of a product.

17.12. Explain the method of contour sawing. Give some examples of product applications.

17.13. Give some important advantages of power sawing.

17.14. Give a practical tolerance for power sawing operations.

17.15. Explain why power sawing is considered to be a relatively economical process.

PROBLEMS

17.1. Prepare a sketch and illustrate the following terms as applied to band saw blade teeth: set, tooth face, tooth back, tooth spacing, gullet, gullet depth, tooth rake angle, and tooth back clearance angle.

17.2. A 16 in. diameter circular saw for a cold sawing machine revolves at 75 sfpm and is fed at the rate of 0.050 ipt (inch per tooth). The pitch of the teeth is 0.18 in. and the kerf is $\frac{3}{32}$ in. Calculate the time required to cut through a 4 in. low carbon steel square bar.

17.3. It has been decided to cut the excess flash from 1500 low carbon steel castings illustrated in Figure 17-P3 by power sawing. Prepare a sketch showing a design for a fixture that will hold and position the part during the flash removal operation. It would be appropiate to saw some of the flash without using the fixture.

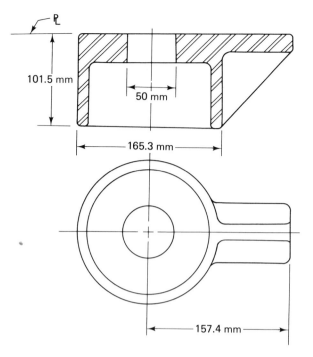

FIGURE 17-P3

17.4. Two hundred parts shown in Figure 17-P4 are required. Design a fixture that will hold and position the part while the slots are being cut with an abrasive disc. The tolerance is ±0.002 in. (0.05 mm).

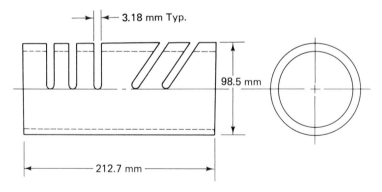

FIGURE 17-P4

17.5. Replacement costs may be reduced by regrinding the teeth on a segmented circular cold sawing blade. The cost of a 32 in. diameter segmented blade from one manufacturer is in excess of $500. This blade may be sharpened until a total of 1 in. from the radius is removed. Each regrinding operation removes approximately 0.030 in. How many regrinds per blade are possible?

229

17.6. The maximum allowable speed of abrasive cutoff wheels and circular saw blades is measured in units of surface feet per minute (sfpm). Calculate the maximum allowable spindle rpm of a 14 in. diameter abrasive wheel with a maximum speed of 7200 sfpm. Calculate the rpm of a 12 in. diameter toothed circular saw blade with a maximum speed of 6000 sfpm.

17.7. Give some reasons why an abrasive cutoff wheel should be operated at the maximum allowable speed.

17.8. An unusually high incidence of blade breakage on a reciprocating sawing machine has been reported in your plant. Investigate the possible causes which may contribute to this problem and list as many as possible.

17.9. Write a paper describing the principle of operation of "sawing with a wire," e.g., the use of a diamond-encrusted wire. Give some applications of this principle.

17.10. Manufacturers of circular saw blades often introduce controlled amounts of stress in blades (called *tensioning*) by hammering. What is the primary reason for this procedure?

BIBLIOGRAPHY

"Abrasive Cutoff," *Machining Data Handbook,* 2nd edition, Dearborn, Mich.: Society of Manufacturing Engineers, 1972, pp. 555–566.

"Circular Sawing," *Machining Data Handbook,* 2nd edition, Dearborn, Mich.: Society of Manufacturing Engineers, 1972, pp. 537–551.

"Contour Band Sawing," *Metals Handbook,* Vol. 3, *Machining,* 8th edition, Metals Park, Ohio: American Society for Metals, pp. 218–223.

"Cutoff Band Sawing," *Metals Handbook,* Vol. 3, *Machining,* 8th edition, Metals Park, Ohio: American Society for Metals, pp. 224–261.

Wilkie Brothers Foundation, *Fundamentals of Band Machining,* Albany, N.Y.: Delmar Publishers, Inc., 1964.

WILSON, R. A., "Low Speed Abrasive Sawing," *Iron Age,* July 26, 1973, pp. 212–252.

18

GRINDING

Precision grinding and rough grinding each consist of forming surfaces by the use of a rotating abrasive wheel composed of many small and hard bonded abrasive grains. Each individual and irregularly shaped grain acts as a cutting tool. *Rough grinding* is a commonly used method for removing excess material from castings, forgings, and weldments, or as a method of removing or snagging thin fins, sharp corners, burrs, or other unwanted projections from various shapes of workpieces. Small parts are often hand-held and moved into contact with a rotating abrasive wheel. Surfaces on larger work are ground by manually moving a portable abrasive tool over the workpiece surfaces. This chapter will deal only with precision grinding processes.

DESCRIPTION OF THE PROCESS

Precision grinding is the principal production method of cutting materials that are too hard to cut by other conventional tools or for producing surfaces on parts to tolerance or finish requirements more exacting than can be achieved by other manufacturing methods. Surfaces on workpieces are simultaneously produced by grinding to an accurate size and with a superior surface finish. The process of precision grinding is principally associated with the removal of small amounts of material to close tolerances and should not be confused with the process of *abrasive machining,* which is explained in Chapter 19. Abrasive machining relates to the rapid removal of relatively larger amounts of stock, with the added capability of special workpiece shape formation. As compared with precision grinding, abrasive machining requires heavier machines with increased power capacities.

Abrasives

There is no one abrasive or bond that is best for all grinding applications. Research and development are constantly taking place to guide manufacturing engineers in the task of selecting the best wheel for a particular job. As labor and overhead costs continue to rise, the grinding wheel price is becoming a decreasing and often insignificant proportion of the total cost. Progressive firms are finding that higher-cost wheels are generally more economical because they grind faster, last longer, and reduce downtime for dressing and changing. In many cases, the wheels also improve product quality by permitting closer tolerances to be maintained, minimizing distortion, and preserving the surface integrity of parts ground.

Abrasives are substances that are both hard and tough. Efficient cutting action can only take place when the abrasive is sufficiently hard to penetrate and scratch the workpiece material. Toughness is a property that prevents premature fracturing and breaking away of the abrasive grains at the beginning of the cut. Maximum cutting efficiency of a grinding wheel is obtained when the dulled abrasive grains are continually fractured, thus exposing a succession of fresh new cutting edges.

There are two types of abrasives, *natural* and *synthetic.* The *natural* abrasives include sandstone or solid quartz, emery, corundum, garnet, and diamonds. There are few on-line production applications of natural abrasives in precision grinding processes and, therefore, our attention will be concentrated upon the synthetic types of abrasives.

Synthetic abrasives include silicon carbide, aluminum oxide, diamonds, and boron nitride. *Silicon* carbide crystals are formed by crushing a mixture of raw materials consisting of silica sand, petroleum coke, sawdust, and salt which has been held at a high temperature in an electric furnace. Silicon carbide grains are very sharp and are extremely hard. Because of their extreme brittleness, silicon carbide abrasives are not recommended for grinding hard materials such as carbide tools, but are particularly well adapted to grinding operations on soft, nonferrous metals such as brass, copper, bronze, magnesium, and aluminum or on cast iron or chilled iron and nonmetallics. Silicon carbide abrasives are sold commercially as *Carborundum* or *Crystolon.* Aluminum oxide abrasive is produced by refining a mixture consisting of the mineral bauxite, iron filings, and coke. A huge "pig" is formed in an electric arc furnace by the fusion of the mixture and, after removal, is rolled and crushed into abrasive grains and graded according to the desired size. Aluminum oxide is softer than silicon carbide. Because of improved toughness properties, this abrasive is recommended for grinding high- and medium-carbon steels, nonferrous cast alloys, or annealed malleable and ductile iron. Aluminum oxide abrasive is the single most used material and is sold commercially as *Alundum* and *Aloxite.*

Diamonds are the hardest and most expensive of all abrasive materials. They may be jointly classified as natural and synthetic. In their natural form, the abrasive grains called "borts" are produced by crushing unsuitable gemstones to the required particle size. Grinding applications with natural diamond abrasives have traditionally been limited in scope because of the tendency of dulling and glazing of the cutting faces. Unlike silicon carbide, natural diamond grains do not readily fracture and break down to continually reexpose new cutting faces. Synthetic diamonds were first produced in 1955 on a commercial scale by the General Electric Company. This important discovery was soon followed by the development of the first diamond grinding wheel by the Carborundum Company. Synthetic or man-made diamonds, although more costly to produce than natural, exhibit improved breakdown characteristics, which result in a cooler and more efficient cutting action. Until lately, grinding wheels consisting of natural and man-made diamonds were principally restricted in use to truing and dressing other grinding wheels, for sharpening carbide tools, and for processing glass or ceramic parts.

A recent development, however, has led to the successful grinding of steel and other ductile metals. For many years, diamond grinding of these materials was not feasible, owing to the tendency to load the wheel. The DeBeers Company has introduced a new type of resin-bonded synthetic diamond wheel. Grit for this abrasive is called DXDA-MC. Tiny and well-formed chips are produced. A cool, free-cutting action takes place, thus minimizing damage to the workpiece. The DXDA-MC wheel has proved to be an efficient and economical replacement for aluminum oxide and silicon carbide for cutting stainless steels, cast iron, and tool and die steels as well.

The Carborundum Company has developed a new line of diamond wheels for carbide wet-grinding applications called the B_1 wheels. These feature a special high-tenacity resinoid-bond formulation that locks the diamonds tightly for longer life. High grinding ratios (defined as the volume of material removed per unit volume of wheel wear) and smooth surfaces are obtained without undue wear and change to the size and profile of the wheels. *Boron nitride* grinding wheels are being increasingly used on workpieces of tool and die steels as well as many types of hard high-alloy steels. General Electric Company's Specialty Materials Department manufactures a material known commercially as *Borazon,* which consists of cubic boron nitride abrasive crystals. Significant advantages are reported, such as superior workpiece surface integrity, ability to maintain close tolerances, and increased wheel life. Norton Company makes a boron carbide material under the resistered trademark *Norbide,* which is described as, except for diamonds, the hardest material ever made commercially by man.

Grain Size

After manufacture, the abrasive material is reduced to small particles by the action of roll or jaw crushers. Standardized sizing of grits or grains is accomplished by sorting or grading the material as it passes through screens. A 60-grit, for example, is such that it will just pass through a 60-mesh screen or one that has 60 openings per linear inch. Screen mesh sizes range from 4 to as small as 1000 grit or even finer. Grinding wheels are bonded with abrasive grits of nearly identical size, to accomplish a uniform cutting action.

Structure or density of a grinding wheel refers to the relationship of the pores or voids between the grains and the bond that unites the grains. While the bonding material thoroughly coats all the surfaces of the grains, it ordinarily does not fill in all the space between the grains.

The *cutting action* of a grinding wheel is similar to that of other cutting tools having multiple teeth. The pores on grinding wheels and the tooth spaces on milling cutters or broaching tools, for example, provide convenient spaces for the chips to escape and assist in the passage of cutting fluid to the cutting area. Wheels with open structures have less "teeth" than those with dense structures and, as a result, operate with a cooler cutting action. Coarse wheels with open structures are used principally for roughing cuts on soft materials, while dense or fine wheel structures are used for hard materials that require a maximum of cutting edges. Dense structures are stronger and help to prolong the wheel life for the specially formed profiles on the wheel peripheries.

Bonding Materials

A grinding wheel consists of abrasive grains that are supported or held together by a *bond*. There are six principal types of bonds: vitrified, silicate, resinoid, metal, rubber, and shellac. Each of these bonds has a specific advantage, depending upon the intended wheel application. All bonds must be sufficiently strong to withstand the stresses of the high-speed rotating grinding wheel. They must be capable of holding the abrasive grains firmly and yet must not be so dense as to impede the cutting action.

Vitrified bonded wheels are by far the most commonly used. Abrasive grains are baked in a kiln with a clay-like material similar to the process of pottery making and are shaped in metal molds by hydraulic pressure. They are subsequently ground or otherwise processed to finished dimensions. Vitrified wheels are strong, porous, and rigid. They are not affected by water, oils, and acids and are designed to operate at speeds up to 6500 sfpm.

Silicate bonded wheels consist of a mixture of sodium silicate and abrasive grains which is tamped or compressed in a metal mold and baked over a long period of time at 500°F (260°C). These wheels wear faster than vitrified wheels, because the abrasive grains are released more readily. An

234

important use of silicate bonded wheels is in sharpening cutting tools and in other applications where heat must be kept at a minimum.

Grinding wheels with a *resinoid* binder are made by mixing abrasives with a synthetic material such as Bakelite or Resinox. Various shapes of grinding wheels are obtained by molding and baking the mixture at relatively low temperatures. Such wheels are extremely strong and tough. Resinoid wheels can operate at extremely high speeds.

Metal bonds are used on diamond wheels and for electrolytic grinding.

Rubber bonded wheels consist of a combination of pure rubber, sulfur, and abrasive grains which are mixed, heated, and then rolled out into sheets of the desired thickness. Wheels are then die-cut and vulcanized. It is possible to produce wheels as thin as 0.005 in. (0.13 mm) by this process. Rubber-bonded wheels are used for high-speed grinding and can operate up to speeds of 16,000 sfpm. Because of their superior properties, they may be used for plastics, glass, porcelain, tile, and practically all metals as cutoff wheels as well as for snagging operations on castings and forgings. Rubber wheels are used to produce high-quality surface finishes.

Shellac bonded wheels are also used to produce a superior finish for some job applications. Shellac and abrasive grains are mixed and then rolled and pressed into steel molds to the desired wheel shape. Hardened wheels are produced by baking at relatively low temperature over a period of several hours. It is possible to produce extremely thin shellac-bonded wheels, which are both tough and relatively elastic.

Grinding Wheel Markings

Wheel manufacturers have adopted a standard marking system for identifying important grinding wheel information. These data are printed on each grinding wheel as an aid in selecting the best wheel for a particular

FIGURE 18-1 Standardized method for wheel markings (Courtesy, Norton Company)

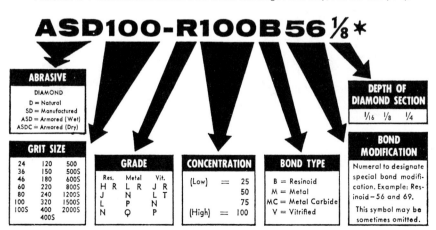

NOTE: No grade is shown for Hand Hones.
©1944 and 1969 by Norton Company

* Manufacturer's Identification Symbol

job application. It should be understood that, in spite of standardization, grinding wheels may differ, even though they are of the same general type. Wheels are processed differently from one manufacturer to another and, as a consequence, there may be differences in the precise significance of the symbols used to denote the grain size, grade, and structure of grinding wheels. Figure 18-1 illustrates the standardized method for wheel markings.

Grinding Wheel Shapes

Grinding wheel shapes have also been standardized by manufacturers. Some of the commonly used types are shown in Figure 18-2a. A variety of standard face contours for straight grinding wheels are shown in Figure 18-2b.

FIGURE 18-2(a) Standard grinding wheel shapes

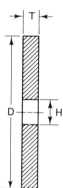

1. Straight

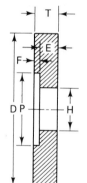

2. Recessed one side

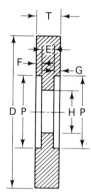

3. Recessed two sides

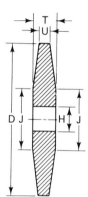

4. Tapered

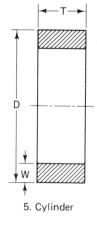

5. Cylinder

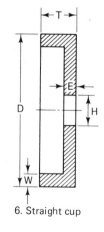

6. Straight cup

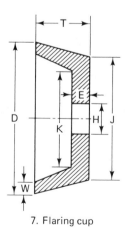

7. Flaring cup

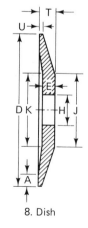

8. Dish

(a)

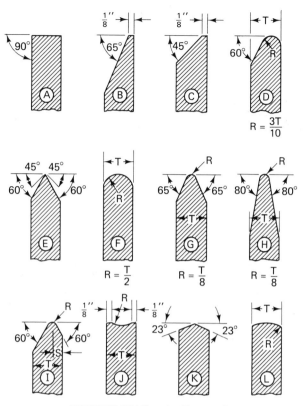

FIGURE 18-2(b) Standard face contours

Grinding Fluids

Grinding fluids are used when grinding metals, as well as for most non-metallic materials. The function of grinding fluids is twofold: to cool the workpiece and to maintain the wheel face. The amount of heat generated is dependent upon the sharpness of the abrasive grains. Attritious gradual wear without fracture is undesirable because the wheel face becomes dull and glazed and heat builds up as a result of the tendency of the wheel to gum or load. Grinding fluids carry the chips away and improve the surface finish of the workpiece. In conventional dry grinding, the air acts as a grinding fluid. The oxygen in the air oxidizes the workpiece surface. The action is sufficient to prevent the tendency of the chips to weld back onto the ground surface. Water-base fluids consisting of soluble-oil emulsions or synthetic compounds are also used as well as grinding oils, which are mineral oils containing fatty materials. While grinding oils effectively limit wheel wear and heat generation, they are messy to use and constitute a fire hazard. Grinding fluids are applied in ample volume directly to the grinding area.

On cylindrical pieces

On flat pieces

On hollow pieces

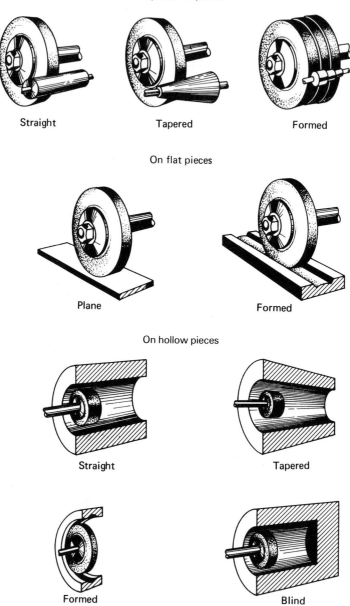

FIGURE 18-3 Typical applications of various grinding operations (Courtesy, The Carborundum Co.)

(a)

(b)

FIGURE 18-4(a) Close-up showing grinding operation on parts held in a magnetic chuck; (b) a horizontal-type surface grinder (Courtesy, Norton Company)

Grinding Ratio

Grinding ratio is defined as the volume of material removed from the work per unit volume of wheel wear. Manufacturing engineers consider this term to be a very useful unit of measure to describe the ease of grinding or "grindability" of a given material. The higher its grinding ratio, the easier a work material is to grind. The grinding ratio is found experimentally by using a specific set of governing conditions; thus, the ratio will vary for a given material according to the type of operation, nature of the grinding wheel, type of grinding fluid, speeds and feeds, and so on.

Grinding Machines

Broadly considered, there are three main categories of precision grinding machines for production applications: *surface grinders, external cylindrical grinders,* and *internal cylindrical grinders.* Typical applications of various grinding operations are shown in Figure 18-3.

Surface Grinders: It should be understood that grinding is a method that can be used to obtain a wide variety of surfaces by means of a rotating abrasive wheel. The term *surface grinding* is understood to relate specifically to the production of *flat* or *plain* surfaces. The two main types of surface grinders, *horizontal* and *vertical,* are classified according to the position of their wheel spindle. The horizontal-type surface grinder shown in Figure 18-4 uses a plain wheel and, depending upon the individual work requirements, may have any one of the face contours shown in Figure 18-2b. Cup and cylinder wheels, illustrated in Figure 18-2a, or segmental grinding wheels, shown in Figure 18-5, may be used for vertical spindle grinding machines, shown in Figure 18-6. Each type of machine is available with a rotary or reciprocating work table.

FIGURE 18-5 A chuck with a set of eight segments takes the place of the solid cylinder wheel and its mounting ring (Courtesy, Norton Company)

FIGURE 18-6 A vertical spindle precision grinding machine with dual wheel heads (Courtesy, The Bullard Co.)

External Cylindrical Grinders: External cylindrical grinders consist of two basic types: *center type* and *centerless.* The chucking-type external cylindrical grinder (not illustrated) is used for grinding relatively small parts with shapes that are adaptable for holding in a chuck or a collet. Chucking grinders are particularly useful for parts with shapes that cannot be readily held between centers. An external cylindrical grinder is shown in Figure 18-7 with a close-up view of the workpiece being held between centers.

(a)

(b)

FIGURE 18-7(a) Grinding the plunger shaft for a hydraulic arm. Diameter is ground to 1.7490 in. + 0.0001 − 0.0000 in. (Courtesy, Landis Tool Co. − Div. of Litton Industries); (b) workpieces ground on this cylindrical grinding machine range in length from ½ to 72 in. with diameters from $\frac{7}{16}$ to 14 in.

FIGURE 18-8 The O.D. of this jet engine part is crush ground in 7 minutes 55 seconds. Tolerances are held to ±0.002 in. on diameter and ±0.001 on lateral dimensions (Courtesy, Bendix—A & M division)

When a sufficient production quantity is required, a process known as *form, crush,* or *contour grinding* is used. This is a method of using cylindrical grinders to produce complex shapes on rough workpieces by combining many grinding operations into one (see Figure 18-8). One or more wheels may be used, depending upon the desired workpiece configuration. The face of the grinding wheel is formed by crush or diamond truing or by template dressing. Crush truing is accomplished by forcing a high-speed steel or carbide crushing roll against the periphery of a slowly revolving grinding wheel as shown in Figure 18-9. The face of the steel roll has the same contour of the part to be ground. Diamond truing consists of dressing the wheel to the desired shape by feeding a motor-driven cemented diamond particle roll into the face of the rotating wheel. The abrasive grains are fractured and become dislodged from the wheel face, thus exposing fresh, new sharp edges. Template dressing utilizes a single diamond cutter which contacts the rotating wheel. The traversing action on the grinding wheel surface generates a form duplicating that of the template.

FIGURE 18-9 Crushing roll for forming wheel used in producing sprockets (Courtesy, Grinding Wheel Institute)

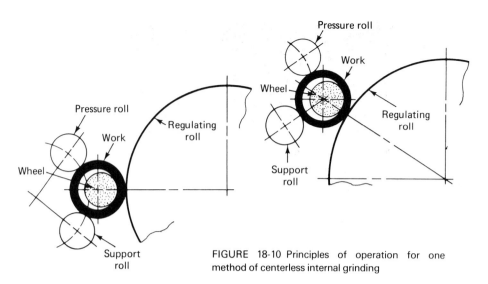

FIGURE 18-10 Principles of operation for one method of centerless internal grinding

Internal Cylindrical Grinders: Internal cylindrical grinders are designed to produce internally ground surfaces or holes on parts. There are three types of internal grinding machines: one that holds the workpiece in a rotating chuck, one in which the workpiece is rotated by the outside diameter between rolls shown schematically in Figure 18-10, and the planetary type, for heavy work, in which the workpiece is held stationary.

Other Grinding Machines

In addition to the three main categories of production grinding machines previously discussed, there are a number of other standard and special types, each intended to accomplish a variety of grinding functions. An example of a *tool* grinder is shown in Figure 18-11, and of a *multiple wheel camshaft*

FIGURE 18-11 Grinding the flutes of a tap on a tool grinder (Courtesy, Norton Company)

FIGURE 18-12 A multiple wheel grinder simultaneously grinds bearings of camshafts in a completely automatic operation (Courtesy, Landis Tool Co. — Div. of Litton Industries)

grinder, in Figure 18-12. *Jig* grinders operate at extremely high grinding speeds and are capable of working to production tolerances of ±0.0001 in. (0.003 mm). A special form of flat surface grinder, the *abrasive belt* machine, is also commonly found in production shops, as well as *disk* grinders, *cutter* grinders, and *gear* grinders.

PRODUCT APPLICATIONS

Chapter 19 will explain how grinding may, in some cases, supplant planing, shaping, turning, or milling as a method for removing as much as 0.025 in. (0.64 mm) of stock in one pass of the grinding wheel. Precision grinding is a term that is restricted to the method of removing *small* amounts of material, typically 0.001 in. (0.03 mm) per pass, for improving dimensional accuracy and surface smoothness. Grinding is not generally considered to be a primary forming process in the manufacture of parts. Precision grinding operations are performed on workpieces that have been previously rough-shaped by some other primary forming process such as forging or casting. Grinding operations usually follow one or more sizing operations which have removed the bulk of the stock from the rough workpieces. Precision grinding is not considered to be a product-making process, and therefore no product applications can be listed. The main types of surfaces that are commonly precision-ground are shown in Figure 18-3.

PROCESS SELECTION FACTORS

Materials: Any material that can be machined by any other method can also be ground. All hardened or soft ferrous metals, as well as all the nonferrous metals, can be ground, but some are more difficult to grind than others. Nonmetallics that are difficult or impossible to machine by conventional methods can be ground. These include ceramics, tile, marble, glass, rubber, plastics, and a wide variety of composite materials. There are practically no limitations to the kinds of materials that can be ground.

Tolerances: The dimensional accuracy of surfaces that are sized and finished by precision grinding can vary widely depending upon the type of operation, grinding wheel used, use of a grinding fluid, and the grinding procedure (regarding the amount of stock removal). As a general rule, tolerances as small as ±0.0001 in. (0.003 mm) may be regularly obtained by production grinding methods. Special size-control devices on grinding machines are available that will allow grinding to tolerances as small as ±0.000025 in. (0.0006 mm). As in all other machining processes, close tolerances increase the cost of grinding.

Surface Finish: Surface finishes as low as 5 to 15 μin. may be regularly obtained by general production grinding. The cost of obtaining the required surface finish varies with the type of grinding wheel and equipment, the number of dressings required, workpiece hardness, and other important factors, including feed rate and in-feed per pass.

PRODUCT DESIGN FACTORS

Part Configuration: Production economy is best accomplished when parts are designed for manufacture on standard equipment. For this reason, designers constantly strive to keep the shape of production parts as simple as possible. Owing to the versatility of available grinding machines, the majority of parts may usually be conveniently positioned without the need for special workholding attachments. As the design of a given production part progresses, careful attention should be devoted to possible grinding problems which may be expected to occur as a result of part distortion, wheel interference, or hard-to-reach or unnecessarily complex surfaces. Simple accessible flat surfaces and straightforward cylindrical shapes are always the most economical to produce. Grinding operations on some circular parts, because of their unique shape, may be adapted *only* to center-less grinding. In this case, designers control the end shape of the workpiece as in Figure 18-13, which facilitates the entry of the workpieces into the space between the grinding and the regulating wheels.

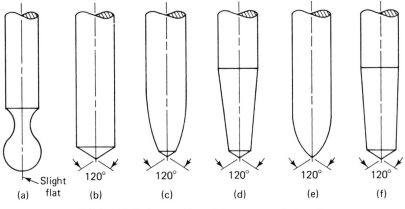

Slight flat
(a)   120°   120°   120°   120°   120°
       (b)   (c)   (d)   (e)   (f)

FIGURE 18-13 Acceptable end shapes on workpieces for centerless grinding operations

Grinding Reliefs: Grinding reliefs are recommended when grinding a diameter and an adjoining square shoulder or when grinding adjoining diameters. Repeated wheel dressings are required to maintain a consistent size radius on the corner of a wheel. Except in cases of critical stress concentrations, or for blending adjacent surfaces, a relief should be used instead of a fillet. Standard practice is to use a relief with a minimum width of ⅛ in. (3.18 mm) and to a sufficient depth so that the wheel will not bottom in the relief before cutting to the required workpiece diameter.

Fillets: Fillets should be made as large as possible when they are essential to the function or strength of the part. Fillet radii of less than 0.010 in. (0.25 mm) should be avoided because of the difficulty in maintaining the radius dimension on the wheel. Whenever possible, all fillet sizes on a given workpiece should be the same.

Centerholes: Designers strive to incorporate centerholes in the overall configuration of a part that will be cylindrically ground between centers. The centerholes should not be removed from the completed part unless absolutely necessary, since they may be needed for workpiece maintenance at a later time as a means of reinstalling on a machine for regrinding. Centerholes should be made as large and as accurate as possible and in close alignment.

Holes: It is impractical to specify production grinding of holes less than $\frac{1}{16}$ in. (1.588 mm) in diameter on standard internal grinders. Limited production of holes of this size and smaller may be produced on a special machine known as a jig grinder under carefully controlled conditions. Designers also avoid specifying workpiece operations that require the grinding of blind holes to full depth.

Stock Allowance: Unless the full effects of all the variables are known, it is not possible to recommend an amount of stock allowance that will satisfy all production grinding requirements. Among the variables that must be considered are size variations from part to part, possible warpage effects from hardening, stock eccentricity, twist or warpage, tolerance range of previous turning operations, and accuracy and alignment of centerholes. Standard industrial practice is to critically evaluate each set of job conditions, on an individual basis, before determining the amount of stock allowance required for a given part. As a general rule for work of average size, a minimum stock allowance of 0.005 to 0.007 in. (0.13 to 0.18 mm) per side is **recommended for external (between-centers) cylindrical grinding and for** centerless grinding operations. Surface grinding operations may require as much as 0.010 to 0.015 in. (0.25 to 0.38 mm) for stock allowance.

REVIEW QUESTIONS

18.1. List the materials that are classified as natural abrasives.

18.2. List the materials that are classified as synthetic abrasives.

18.3. Explain why only the synthetic types of abrasives have specific application to production grinding operations.

18.4. Which of the synthetic materials is the "single most used material"? Why?

18.5. Describe how a vitrified bonded wheel is produced.

18.6. What is the purpose of grinding fluid?

18.7. Define "grinding ratio." How is it usually determined?

18.8. List three main categories of precision grinding machines for production applications.

18.9. On what type of a grinding machine are flat or plain surfaces on workpieces usually produced?

18.10. Explain the process of "crush truing." Why is this operation necessary?

18.11. Define the term "precision grinding."

18.12. Give a practical tolerance for production grinding operations.

18.13. Compare the surface finish range achievable for grinding to that of milling.

18.14. List the variables that should be considered when determining the stock allowance for grinding.

PROBLEMS

18.1. In Figure 18-P1, the grinding wheel leaves an undesirable radius on the shoulder. Show how a square shoulder can be produced.

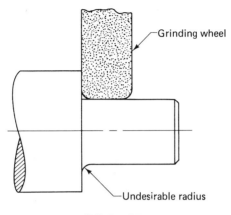

Grinding wheel

Undesirable radius

FIGURE 18-P1

18.2. Of aluminum oxide and silicon carbide types of grinding wheels, which type would be recommended for grinding operations on hard and brittle materials? Why?

18.3. Assume a surface grinding operation on tool steel stock. The surfaces have to be ground flat and to a commercial finish. Which type of wheel is recommended? Why?

18.4. What two basic ingredients make up a grinding wheel?

18.5. Define *grit size.*

18.6. Define *structure* of a grinding wheel.

18.7. Sizes of abrasive grit are expressed by numbers. When compared to fine grains, are coarse grains expressed in higher or lower numbers? Briefly explain the meaning of the numbering system.

18.8. What is the normal range of grit sizes?

18.9. Explain the following grinding wheel marking: 32A-46-H-8-V52.

18.10. On what surface (A or B) is grinding performed on the straight cup wheel shown in Figure 18-P10?

FIGURE 18-P10

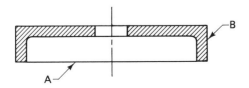

B

A

18.11. In center type cylindrical grinding, the work speed is determined by the diameter and rpm of the cylindrical workpiece being ground measured in terms of sfpm. What is the normal recommended range of workpiece speeds?

18.12. Define *plunge grinding.*

18.13. Explain why plunge grinding techniques would most often be used in mass production operations with automatic grinding setups.

18.14. The surface of a grinding wheel is observed to be "glazed." What does this condition indicate?

18.15. Assume you have been consulted for the purpose of correcting grinding problems. What would be the probable cause for the following problems affecting the workpiece surface? (a) chatter marks; (b) "burned" surface; (c) random scratches?

18.16. Describe a method of checking for imperfect wheels.

18.17. Explain the difference between "dressing" and "truing" a grinding wheel.

18.18. On a benchstand or a floorstand grinder, why should the work rest be kept adjusted close to the wheel? What is the recommended distance that should be maintained between the work rest and the wheel?

BIBLIOGRAPHY

*"Abrasives: Where They Stand Today," *Manufacturing Engineering and Management,* Oct. 1972.

External Cylindrical Grinding Machines—Centerless, B5.37-1970 (J28), New York: American Society of Mechanical Engineers, 1970.

External Cylindrical Grinding Machines—Universal, B5.42-1970 (M40), New York: American Society of Mechanical Engineers, 1970.

"Grinding," *Machining Data Handbook,* 2nd edition, Dearborn, Mich.: Society of Manufacturing Engineers, pp. 627–690.

"Grinding," *Metals Handbook,* Vol. 3, *Machining,* 8th edition, Metals Park, Ohio: American Society for Metals, pp. 257–277.

   *Abstracted with permission.

19

ABRASIVE MACHINING

Abrasive machining, unlike conventional grinding, is considered a primary metal-removal process. It is competitive with cutting-tool machining processes which generate forms by milling, turning, planing, broaching, and so on. The function of conventional grinding is to machine a part to the desired size and finish. While this is also a function of abrasive machining, there are some significant differences in the two processes.

DESCRIPTION OF THE PROCESS

It is important to establish the fact that abrasive machining is a *high-production process.* It is a method of simultaneously generating precise part geometry, with a superior surface finish, by the rapid removal of large amounts of stock, usually with one pass of the grinding wheel. As a process, it is classified as an intermediate step between snagging (or rough grinding) and finish grinding.

The process of abrasive machining is the result of a cooperative development program with machine builders and abrasive manufacturers. Early problems centered on ways to offset the tendency of wheel dulling and burning of the workpiece surfaces, coupled with methods of preventing wheel stalling as high-volume feed rates were attempted. It was found that as the feed rate is increased, the grinding wheel reaches a "self-sharpening" or "pivot point," where, instead of dulling at a faster rate, the cutting points of the abrasive grains are automatically replaced faster than they dull. At

251

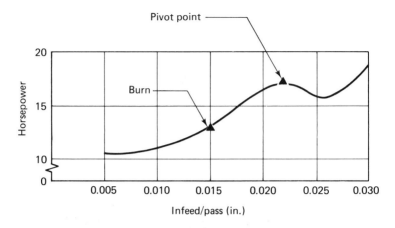

FIGURE 19-1 The pivot point as related to abrasive machining (Courtesy, Roy A. Lindberg, Materials and Manufacturing Technology, Allyn and Bacon, pp 506)

this point, shown in Figure 19-1, the wheel face sharpness is maintained and, in fact, the vast reserve of fresh, sharp cutting points is fully utilized.

Abrasive machining requires rugged high-powered machines, some of which operate over 100 hp. Wheel speeds for vitrified grinding wheels range from 12,000 to 16,000 sfpm, with surface speeds up to 17,000 sfpm often used for resinoid wheels. Table 19-1 lists various types of abrasive wheels commonly recommended for various workpiece materials.

TABLE 19-1 Materials and types of grinding wheels

Type of abrasive wheel	Materials machined or cut off
Aluminum oxide	Aluminum, nickel, most steels, refractory metals
Silicon carbide	Brass, bronze, cemented carbide, iron, titanium, clay products, marble, plastics, most stone
Diamond	Alumina and other ceramics, cemented carbide, glass, germanium, granite, silicon, vanadium steel, molybdenum steel

PRODUCT APPLICATIONS

Figure 19-2 illustrates some typical parts that have been abrasive-machined. There are a wide range of applications. As in conventional grinding, the type of grinding machine employed is determined by the shape requirements

(a)

(b)

FIGURE 19-2 (a) Automobile wheel spindles are abrasively machined from the rough casting in 42 seconds to finished tolerance of 0.0007 in. on the bearing diameters (Courtesy, Bendix—A & M Division); (b) ten teeth in these clippercombs are ground to size in 13 seconds to a depth of 0.859 in. The tooth width is 0.050 in. with a tolerance of ±0.002 in. (Courtesy, Bendix—A & M Division); (c) parts that are ideally suited to abrasive machining operations (Courtesy, Landis Tool Co.—A Division of Litton Industries)

(c)

FIGURE 19-3 Abrasive machining of a massive weldment
(Courtesy, Grinding Wheel Institute)

of the workpiece. Figure 19-3 illustrates a large vertical-spindle grinder used for abrasive machining of a turbine casing. Specially built or retrofitted horizontal-spindle and vertical-spindle surface grinders may be used for abrasive machining operations on brackets and clutch disks, or for gear housings, transmission covers, motor and machine bases, and other similar types of flat workpieces. Heavy-duty centerless and center-type cylindrical grinders are used to produce cylindrical parts with shapes resembling those produced on conventional turning machines. Typical parts in this category include pinions, steering knuckles, and various forms of spindles or shafts. Abrasive cutoff operations (discussed in Chapter 17) are generally considered to be a form of abrasive machining.

PROCESS SELECTION FACTORS

Speed and Economy: As compared to lathe operations, abrasive machining is considered faster and more economical, particularly when machining circular parts whose length-times-diameter product is 4 or greater and in cases where ¼ in. (6.35 mm) or less of metal is to be removed from the diameter. Figure 19-4 is a cost comparison for a part machined by conventional methods and the same part by abrasive machining.

FIGURE 19-4 Case history of an abrasively machined part showing reduced cost and increased productivity (Courtesy, Norton Company)

Material: 1015 Seamless steel
tubing—piston rod

The Job: Present

Rough turn 1-3/8 tube to
1.268/1.263

Proposed

Abrasive machining 1-5/16
tube to 1.268/1.263

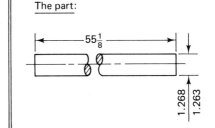

The part:

Job Data	Present Method	Proposed Method
Machine	Engine lathe	#2 Norton centerless
Tool	3/8″ Square HSS	A46M5VG
Horsepower	15	15
How loaded	1 pc. at a time	1 pc. at a time
Speeds	377 r.p.m.	42.4 in/min
Feeds	0.021″/Rev.	0.20″/pass
Stock removal per piece	12.6 in³	5.3 in³
Machining time per piece	7.29 min.	3.90 min.
Rate of stock removal	0.67 in³/min.	1.36 in³/min.
Floor to floor time per piece	10.79 min.	4.10 min.
Total cost per piece	$1.69	$0.78
Setting up costs	$2.87	$4.80

Remarks: Abrasive machining reduced costs 50% and
increased productivity 160%

Materials: Practically any material can be abrasive-machined. Figure 19-2 lists materials that are commonly machined.

Size of Parts: The capacity of the machine available is the principal basis for determining the adaptability of the size of the parts. In general, the size ranges listed in Chapter 18 pertain to machines used for this process.

Metal Removal Rates: For flat surfaces this may be as much as 40 in.^{3} (656 cm^3) per minute for iron and about 15 in.^{3} (246 cm^3) per minute for steel. Rates as high as 80 in.^{3} (1312 cm^3) per minute have been reported for machining aluminum.

Surface Finish: A range from 19 to 90 μ in. is practical, depending upon the abrasive grit size used.

Variety of Forms: A range of almost unlimited contoured and flat surfaces may be produced by abrasive machining. These include blends and combinations of both curved and flat surfaces of internal and external shapes, corners, tapers, and shoulders.

Some Limitations

As described in Chapter 18, there are various ways to form a desired wheel contour, which, in turn, produces the corresponding workpiece shape. Certain, complex shapes cannot be abrasive-machined on parts, because of limitations in forming the abrasive wheel contour. Certain part geometries are limited: for example, depth of form is held to ¾ in. (19.05 mm) and the radii at the bottom of a cut cannot be held closer than ±0.005 in. (0.13 mm). Another important limitation is that many of the existing grinding machines lack the rigidity and power necessary for abrasive machining operations.

PRODUCT DESIGN FACTORS

Finish Allowance: An allowance of 0.015 in. (0.38 mm) is generally considered adequate. If the part contains an unusually heavy coating of scale or if the part is warped or severely out-of-round, an additional allowance is generally necessary. Unlike cutting tools, which lose their edge on hard or abrasive surface scale, abrasive wheels penetrate the scale with little more effort than through the solid metal beneath.

Reduction in Areas: Reduction in the area to be abrasive-machined results in obvious cost economies. Figure 19-5 illustrates a part that has been redesigned to reduce weight and machining time.

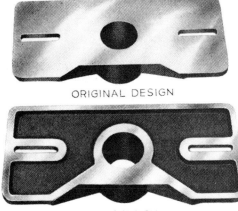

ORIGINAL DESIGN

NEW DESIGN

FIGURE 19-5 A new design resulted in a time-savings, over 80% in addition to material saving (Courtesy, Grinding Wheel Institute)

Parallel Planes: Designers try to keep surfaces in parallel planes whenever possible so that they may be conveniently abrasive-machined in one setup. Figure 19-6 illustrates this principle.

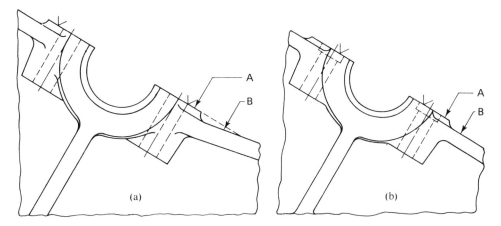

FIGURE 19-6(a) The face grinder would hit surface B when grinding surface A; (b) review of original design showed that surface B could be redesigned parallel to surface A (Courtesy, Grinding Wheel Institute)

Design Freedom: Some parts can be economically abrasive-machined to final form from solid, raw stock. Typical examples are shown in Figure 19-7. The material may be hardened prior to machining without adverse effects upon the grinding operations. Hardening usually results in a more accurately finished product, free of distortion.

257

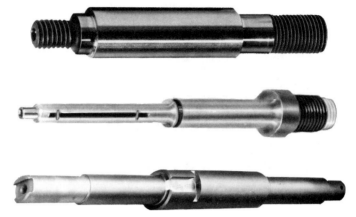

FIGURE 19-7 Diameters of these parts are produced in one grinding operation (Courtesy, Landis Tool Co. — A Division of Litton Industries)

REVIEW QUESTIONS

19.1. In what important ways do abrasive machining and precision grinding differ?

19.2. Explain why abrasive machining is regarded as a high-production process.

19.3. Explain what is meant by the "pivot point."

19.4. Explain why rugged, high-power machinery is recommended for abrasive machining.

19.5. Give some advantages of abrasive machining over corresponding lathe or milling operations.

19.6. Give the important limitations of the process.

19.7. Compare the surface finish and the recommended finish allowance of this process to milling.

BIBLIOGRAPHY

*Abrasive Machining, Bulletin 2, Case History 11, Norton Company, Worcester, Mass.

*MUELLER, J. A., "How to Design for Abrasive Machining," Machine Design, June 1, 1972.

    *Abstracted with permission.

20

ABRASIVE JET MACHINING

Abrasive jet machining (AJM) is a process that removes material by directing a high-velocity stream of abrasive particles onto a workpiece.

The process differs from sand blasting (discussed in Chapter 40) in three basic ways. One, the main purpose of sand blasting is to clean workpiece surfaces. Abrasive jet machining is used to *cut* materials. Two, the abrasive particle size used in sand blasting is larger than that of the abrasive powder used in AJM. Finally, closer control of the high-speed stream of particles is possible with AJM.

DESCRIPTION OF THE PROCESS

Figure 20-1 shows a model of an abrasive jet machining unit weighing only 40 lb (18 kg). This tabletop unit measures $14\frac{1}{2} \times 14\frac{3}{4} \times 13\frac{3}{4}$ (or approximately $370 \times 375 \times 350$ mm). Abrasive jet machining is used chiefly on materials that are sensitive to heat damage and for cutting thin sections of hard, brittle materials. The carrier gas or propellant used may be air, nitrogen, or carbon dioxide. While pressurized abrasive powders are usually propelled at about 75 psi (5.27 kg/cm^2), gas pressures may range from a low of 30 psi (2.1 kg/cm^2) to a high of 125 psi (8.79 kg/cm^2).

Abrasive materials generally range in size from 10 to 50 microns with larger sizes available for special job applications. Aluminum oxide and silicon carbide powders are used for the majority of cutting operations.

FIGURE 20-1 An abrasive jet unit. The handpiece with the attached nozzle tip is stored to the left and the footswitch with cord is in the foreground (Courtesy, S.S. White Industrial Products, Div. of Pennwalt Corp.)

Other materials, such as dolomite (calcium magnesium carbonate), sodium bicarbonate, and glass beads, are used for cleaning, etching, polishing, or deburring.

The abrasive stream is directed to the workpiece locations by a nozzle, which may be manually positioned or mounted in a specially designed fixture for automatic operation. Cam drives, pantographs, or tracer mechanisms are generally used to automatically position either the nozzle or the workpiece. Nozzles are made from highly abrasion-resistant materials such as tungsten carbide or synthetic sapphire. Various circular, rectangular, or square nozzle openings are available, depending upon the type of abrasive used and the special workpiece requirements. Circular openings range in size from 0.005 to 0.075 in. (0.13 to 1.91 mm) in diameter. The minimum rectangular opening is 0.003 by 0.020 in. (0.08 by 0.51 mm), ranging to a maximum size of 0.26 by 0.02 in. (6.61 by 0.51 mm). For average conditions of material removal, tungsten carbide nozzles have a useful life of up to 30 hours, while sapphire nozzles may last up to 300 hours.

PRODUCT APPLICATIONS

Abrasive jet machining is particularly adaptable to the machining of materials considered to be too hard or brittle to process successfully by other methods. As a result, most product applications of this relatively new process apply to somewhat specialized secondary operations on this class of materials.

Among the interesting current metal product applications are drilling and slicing thin sections or wafers of hardened metal; cutting or etching trade names and number markings on parts; removing broken tools from holes; producing shallow inclined crevices or otherwise making small adjustments on production parts or on molds; producing a matte finish or generally cleaning mold cavities; deburring (as in Figure 20-2) and cleaning contaminants; exposing an area on a part for electrical contact; and removing coatings from electronic or missile components or from medical equipment. In addition, the process has some highly useful applications on nonmetallic materials.

FIGURE 20-2 Burrs are removed from hypodermic needles without altering the size of the hole or creating a radius. Heavy burrs and metal slivers are shown at A. The clean micro-deburred needle is shown at B. The magnification is 8X (Courtesy, S.S. White Industrial Products, Div. of Pennwalt Corp.)

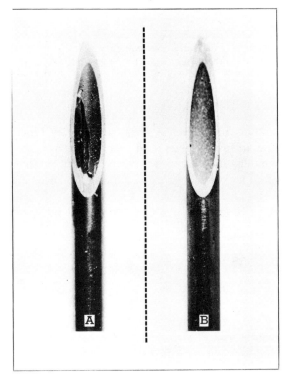
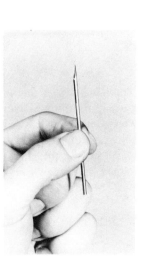

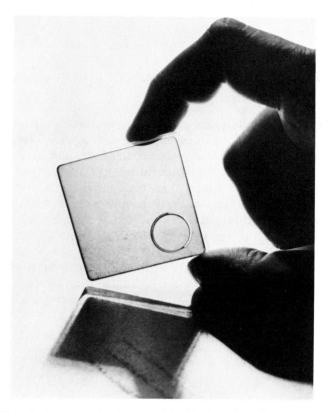

FIGURE 20-3 A circular groove has been machined halfway through the thickness of this 2 × 2 in. filter glass slide. The dark area of the cut is the laminating layer and the light area to the right is the bottom (uncut) layer of the filter glass. The process cuts evenly with no chipping or delaminating. (Courtesy, S.S. White Industrial Products, Div. of Pennwalt Corp.)

These include cutting or inscribing fine reference marks on glass level indicators; abrading or frosting the insides of glass tubes; etching quartz, sapphire, mica, or glass parts; trimming thin silicon wafers and various machining operations on other semiconductors such as germanium; or for removing flash and parting lines from molded plastic parts. Figure 20-3 shows an AJM application for scribing a circular groove to a required depth on a piece of filter glass.

PROCESS SELECTION FACTORS

Surface Finish: The quality of the surface finish depends upon the type of abrasive used and upon the particle size. A practical production finish ranges from 20 to 50 μin. A 6-μin. surface finish has been accomplished on glass using a 10 micron aluminum oxide abrasive. A wide variety of surface textures may be produced ranging from a highly polished appearance to a frosted or matte surface.

Heat Effects: Because temperatures associated with AJM do not exceed room temperature, the workpiece is not damaged by heat effects.

Mechanical Effects: There is no mechanical contact between the tool and the workpiece.

Workpiece Hardness: The process is adaptable to machining materials of any hardness.

Material Removal Rate: Compared to conventional chip-removal processes, AJM is considered slow. A practical metal removal rate is 0.001 in.^{3} (0.0164 cm^3) per minute, which, in more easily understood terms, is roughly the equivalent of machining a slot 0.020 in. (0.51 mm) wide, 0.010 in. (0.25 mm) deep, and 5 in. (127 mm) long. The metal-removal rate is governed by the type of abrasive, particle size, nozzle pressure, and the distance between the nozzle tip and the workpiece.

Material Removal Control: The abrasive particles from any nozzle follow parallel paths for a short distance, and then the abrasive jet flares outward like a narrow cone. Masks of copper, glass, or rubber may be used to concentrate the jet stream of abrasive particles to confined locations on workpieces. Intricate and precise shapes may be produced by using masks with corresponding cutouts.

Minimum Slot Widths: A practical minimum is 0.005 in. (0.13 mm).

Tolerances: Normal production operations can be held to ±0.005 in./in. (0.005 mm/mm) with a tolerance of ±0.002 in./in. (0.002 mm/mm) possible under carefully controlled conditions.

Some Limitations

Expendable Abrasives: Unlike abrasive particles used in sand blasting, the abrasive is not generally reused. Cutting edges on abrasive powders, once used, tend to lose their sharpness. Also, there is a tendency of the abrasives to mix with workpiece waste particles, which may clog the equipment.

Workpiece Materials: AJM is best suited to hard materials, because the abrasives cling to and become embedded into the surfaces of soft materials. The process is not suitable for materials such as rubber, lead, copper, and some plastics.

Versatility: The process is not a cure-all for all secondary operations. Only a small amount of material can be removed. Also, the nature of the types of operations that may be reasonably produced are largely restricted to workpieces whose configurations permit free access of the nozzle.

PRODUCT DESIGN FACTORS

At the present stage of development, AJM offers designers and process engineers an extremely unique and useful extension of secondary operation capabilities. For its intended application, that of machining intricate shapes in hard, brittle materials, it may well have no equal.

REVIEW QUESTIONS

20.1. What types of workpiece materials are particularly appropriate for abrasive jet machining operations?

20.2. Explain how the blast of abrasive powders may be accurately directed on specific locations on the workpiece.

20.3. List the various types of propellants commonly used for cutting and those used for other purposes.

20.4. What types of abrasive powder materials are used in this process?

20.5. Compare (by percentage) the material removal rate of this process to abrasive machining.

20.6. What important factors control the rate of material removal?

20.7. Give some important limitations of this process.

20.8. Give the significant advantages of this process.

BIBLIOGRAPHY

*"Abrasive Jet Machining," *Metals Handbook,* Vol. 3, *Machining,* 8th edition, Metals Park, Ohio: American Society for Metals, p. 252.

LaCourte, N. J., "How to Jet Slice Glass," *Glass Industry,* Oct. 1973, pp. 12–15.

*Lavoie, F. J., "Abrasive Jet Machining," *Machine Design,* Sept./Oct. 1973.

Non-traditional Machining Processes, Dearborn, Mich.: American Society of Manufacturing Engineers, 1975.

Springborn, R. K., Editor, *Non-traditional Machining Processes,* Detroit, Mich.: ASTME, 1967, pp. 15–24.

*Abstracted with permission.

21

NUMERICAL CONTROL AND AUTOMATED PROCESSES

Numerical control (N/C) encompasses many areas. It consists of a broad number of combinations of elements. The Bendix Industrial Controls *NC Handbook* defines the term N/C as "running a machine tool by tape to make it produce more for less." The Electronics Industries Association (EIA) defines N/C as "a system in which actions are controlled by the direct insertion of numerical data at some point with at least some portion of the data automatically interpreted." Finally, the ASTME defines N/C more simply, perhaps, as "a technique for automatically controlling machine tools, equipment or processes." While the precise terms of the definitions of N/C may vary, most authorities agree that N/C provides one of the major technical innovations of our age. In fact, some predict that N/C is the start of the second industrial revolution. It seems inevitable that N/C will ultimately invade all repetitive manufacturing processes, large and small.

EARLY DEVELOPMENTS

An early forerunner of the principle of N/C began with the development of steel cards with punched holes which were installed on weaving looms. This invention by Joseph M. Jacquard, a Frenchman, enabled designs on woven patterns to be changed more readily. A patent was granted in 1801.

N/C, as applied to machine tools, had its beginning in 1949, when the United States Air Force combined with the Parsons Corporation in ways to produce contoured surfaces from instructions in the form of punched tape. N/C equipment was substituted for the tracer controls on a three-axis Cincinnati Hydrotel vertical planer mill. A successful feasibility demonstration of continuous contour milling was held in March 1952, followed by a final report published in May 1953. Soon after, both the Air Force and private industry began further work to extend the commercial possibilities of the basic N/C technology developed at MIT.

In most cases, early pioneering efforts in adapting N/C concepts to the production of parts involved extensive retrofitting of existing machine tools. At the time, this method proved to be the quickest and most practical way to obtain the benefits of the increased production output of the machines. In the late 1950s, however, the practice of equipping standard machine tools with N/C equipment was largely abandoned in favor of the specialized stand-alone tape-controlled N/C machines.

DESCRIPTION OF THE PROCESS

First-Generation N/C Units

The earliest versions of N/C units worked on a vacuum-tube principle and were primarily concerned with *rate* and *motion* control only. Basically, these *first-generation* units consisted of servo systems which controlled the machine's axes. There was little data processing performed by the control device, and no control of other machine functions. Tape or data input preparation was tedious and required a very high level of skill. The principal users of vacuum-unit N/C systems were the aircraft industries (to machine small quantities of very complicated parts).

Second-Generation N/C Units

The development of the transistor is regarded as the *second generation* of N/C. Introduced in 1960, transistorized N/C units took over more control functions of the machine tool, such as tool changers and spindle speed. These more capable controls also provided for more of the data-processing functions, including circular interpolation and compensation for cutter radius, tool offset, and tool length.

Automatically Programmed Tools

Transistorized N/C controls led to the development and implementation of a programming language called Automatically Programmed Tools (APT). The APT Language (or software package) consists of a program of over

30,000 simplified, English-like words which are preprogrammed and stored in a computer. APT is recognized as the universal standard for the metal-working industry. It can be used to control as many as 50 different machine tools and can be applied to five axes of machine motion. APT provides arithmetic and equation-solving capabilities to eliminate much of the need for manual calculations. It makes repetitive programming easier via the use of macros (stored program statements that can be called up in whole or in part by simple entry of a macro code), as well as index and copy statements. It also embodies essential curve-fitting routines. APT enables the programmer to instantly shift the full coordinates of the workpiece to a different machine tool-cutting location, eliminating a tremendous amount of statement writing.

The main disadvantage of APT is that it requires a big computer, resulting in heavy expenses in computer time. Consequently, only very large manufacturing companies can afford to use this system. A condensed-parts programming version of APT called ADAPT (Adaptation of APT) has been developed which can be used with a small computer. Thus, the system is more readily adaptable to smaller companies. Finally, another method of processing APT programs, called UNIAPT, is now available which makes possible the use of a small desk-top-type computer, such as Digital Equipment Corporation's PDP-11.

Third-Generation N/C Units

The first integrated-circuit (I.C.)-oriented control units were introduced in 1967. These constitute the *third generation* of N/C development. The lower cost and miniaturization of solid-state I.C. technology resulted in controls with greatly improved capabilities in data processing. As a result of advanced developments in control-unit capabilities, it became practical to prepare data input for some applications by simpler means than the sophisticated "hard-wired" APT procedures. N/C units with I.C. have virtually replaced and made obsolete all vacuum-tube or transistorized equipment. It was the advent of the I.C.-oriented control that brought many new N/C builders into the market, particularly users outside of the aerospace industries.

Current-Generation N/C Units

The current generation of N/C development involves the use of the small digital computers, often referred to as minicomputers. Capabilities possible in computer-aided manufacturing will be discussed in later paragraphs, entitled Direct Numerical Control (DNC), and Computer-Aided Manufacturing (CAM).

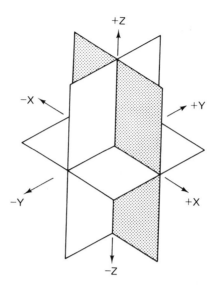

FIGURE 21-1 The conventional cartesian coordinate system is the basis for numerically controlled axis movement

FIGURE 21-2 A commonly employed six-axis nomenclature system for a machining center

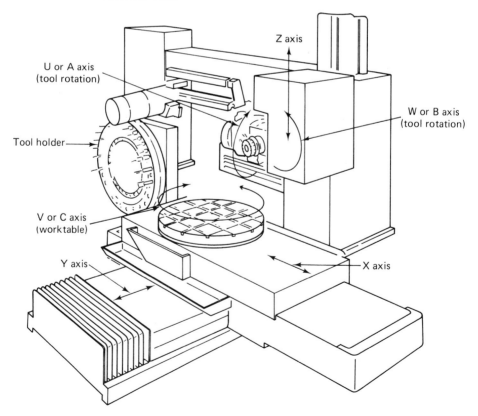

Machine Axes

The capability of a numerically controlled machine tool depends upon the number of axes in which it can operate. In specifying motions, the conventional Cartesian coordinate system is used with linear motion along and about each of the three axes X, Y, and Z, as shown in Figure 21-1. While some N/C machines are restricted to only an X and Y capability, other more versatile machines are equipped with X, Y, and Z travel or motion. Considerably more sophisticated six-axis control is available with the addition of rotary axes U, W, and V, as shown in Figure 21-2. While the X, Y, and Z axes nomenclature is standard throughout the industry, letters U, W, and V are not standardized. Instead, many firms use the letter A in place of U, letter B in place of W, and letter C instead of V.

Types of Numerical Control

There are two basic types of numerically controlled machine tools: *point-to-point* and *continuous-path* (also called *contouring*). Point-to-point machines use unsynchronized motors, with the result that the position of the machining head can be assured only upon completion of a movement, or while only one motor is running. Machines of this type are principally used for straight-line cuts or for drilling or boring. Figure 21-3 illustrates a typical sequence of a point-to-point movement from *xyz* coordinates 2,0,0 (point A) to 0,1,3 (point D). In this example, all three servomotors would begin operating and then each would shut off as it reached the proper station for its axis. The action would start at coordinate 2,0,0 (point A). The three motors operating together would carry the machining head to the vicinity of 1,1,1 (point B), where the y motor would stop. The head would then continue to 0,1,2 (point C), where the x motor would stop.

FIGURE 21-3 Point-to-point motion

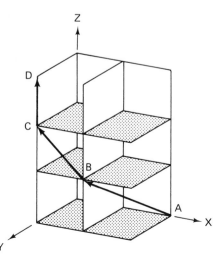

Finally, the head would complete its movement to 0,1,3 (point *D*) under the action of motor *z*. Machine tools with point-to-point system controls are the simplest and least expensive.

A comparison of continuous-path motion to point-to-point motion is illustrated in Figure 21-4. In this example, the motors would run continuously at proportional speeds. A straight line would be generated from *A* to *D*. Machine tools equipped with continuous-path capabilities are normally operated by computers.

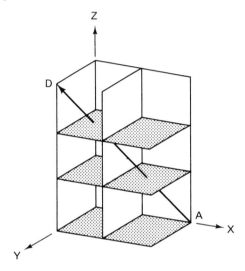

FIGURE 21-4 Continuous path motion

CONVENTIONAL NUMERICAL CONTROL

The N/C system consists of the following components: data input, the tape reader with the control unit, feedback devices, and the metal-cutting machine tool or other type of N/C equipment.

Data input, also called "man-to-control link," may be provided to the machine tool manually, or entirely by automatic means. *Manual methods* when used as the sole source of input data are restricted to a relatively small number of inputs. Examples of manually operated devices are keyboard dials, pushbuttons, switches, or thumbwheel selectors. These are located on a console near the machine. Dials are analog devices usually connected to a synchro-type resolver or potentiometer. In most cases, pushbuttons, switches, and other similar types of selectors are digital input devices. Manual input requires that the operator set the controls for each operation. It is a slow and tedious process and is seldom justified except in elementary machining applications or in special cases.

In practically all cases, information is *automatically* supplied to the control unit and the machine tool by cards, punched tape, or by magnetic tape. Eight-channel punched paper tape is the most commonly used form of data input for conventional N/C systems. The coded instructions on the tape consist of sections of punched holes called *blocks*. Each block represents a machine function, a machining operation, or a combination of the two. The entire N/C program on a tape is made up of an accumulation of these successive data blocks. Programs resulting in long tapes are wound on reels like motion-picture film. Programs on relatively short tapes may be continuously repeated by joining the two ends of the tape to form a loop. Once installed, the tape is used again and again without further handling. In this case, the operator simply loads and unloads the parts. Punched tapes are prepared on typewriters with special tape-punching attachments or in tape punching units connected directly to a computer system. Tape production is rarely error-free. Errors may be initially caused by the part programmer, in card punching or compilation, or as a result of physical damage to the tape during handling, etc. Several trial runs are often necessary to remove all errors and produce an acceptable working tape.

While the data on the tape is fed automatically, the actual programming steps are done manually. Before the coded tape may be prepared, the programmer, often working with a planner or a process engineer, must select the appropriate N/C machine tool, determine the kind of material to be machined, calculate the speeds and feeds, and decide upon the type of tooling needed. The dimensions on the part print are closely examined to determine a suitable zero reference point from which to start the program. Dimensions on drawings prepared for N/C processing are given as point-to-point or arranged in a coordinate system as shown in Figure 21-5. A program manuscript is then written which gives coded numerical instructions describing the sequence of operations that the machine tool is required to follow to cut the part to the drawing specifications.

The *control unit* receives and stores all coded data until a complete block of information has been accumulated. It then interprets the coded instruction and directs the machine tool through the required motions.

The function of the control unit may be better understood by comparing it to the action of a dial telephone, where, as each digit is dialed, it is stored. When the entire number has been dialed, the equipment becomes activated and the call is completed.

Silicon photo diodes, located in the *tape reader head* on the control unit, detect light as it passes through the holes in the moving tape. The light beams are converted to electrical energy, which is amplified to further strengthen the signal. The signals are then sent to registers in the control unit, where actuation signals are relayed to the machine tool drives.

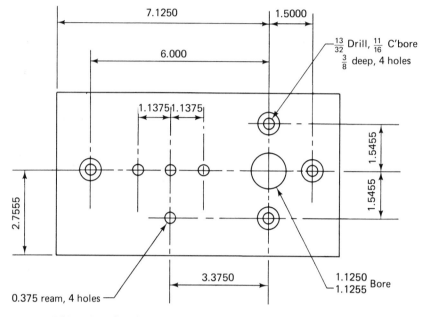

FIGURE 21-5 A typical part print prepared for numerical control machining

Some photoelectric devices are capable of reading at rates up to 1000 characters per second. High reading rates are necessary to maintain continuous machine-tool motion; otherwise, dwell marks may be generated by the cutter on the part during contouring operations. The reading device must be capable of reading data blocks at a rate faster than the control system can process the data.

A feedback device is a safeguard used on some N/C installations to constantly compensate for errors between the commanded position and the actual location of the moving slides of the machine tool. An N/C machine equipped with this kind of a direct feedback checking device has what is known as a *closed-loop* system. Positioning control is accomplished by a sensor which, during the actual operation, records the position of the slides and relays this information back to the control unit. Signals thus received are compared to input signals on the tape, and any discrepancy between them is automatically rectified.

In an alternative system, called an *open-loop* system, the machine is positioned solely by stepping motor drives in response to commands by a controller. The degree of work precision depends almost entirely upon the accuracy of the lead screw and the rigidity of the machine structure. With this system, there is no self-correcting action or feedback of information to the control unit. In the event of an unexpected malfunction, the control unit continues to put out pulses of electrical current. If, for example, the table on a N/C milling machine were suddenly to become overloaded, no response would be sent back to the controller. Because stepping motors are not sensitive to load variations, many N/C systems are designed to permit the motors to stall when the resisting torque exceeds the motor torque. Other systems are in use, however, which, in spite of the possibility of damage to the machine structure or to the mechanical system, are designed with special high-torque stepping motors. In this case, the motors have sufficient capacity to "overpower" the system in the event of almost any contingency.

The original N/C used the closed-loop system. Of the two systems, closed and open loop, closed loop is more accurate and, as a consequence, is generally more expensive. Initially, open-loop systems were used almost entirely for light-duty applications because of inherent power limitations previously associated with conventional electric stepping motors. Recent advances in the development of electrohydraulic stepping motors have led to increasingly heavier machine load applications.

APPLICATIONS

A current estimate reveals that there are over 40,000 numerically controlled devices presently in use in the United States. This total does not include any N/C machines and equipment installed in plants outside the industries defined as metalworking. The most widely used applications for N/C operations continue to be found in the *metal-cutting* industry, in the major production categories of turning, boring, drilling, milling, cutter grinding, and multifunction (machining centers). The principal N/C applications in the *metal-forming* industry are largely applied to conventional punching and shearing operations. Figure 21-6 illustrates how multiple parts are flame-cut under numerical control.

There are many other N/C applications on equipment which, in some cases, are totally unrelated to the metal-cutting or metalforming industries. Some well-established examples are numerically controlled drafting and plotting machines, assembly positioning machines, and machines used for wire wrapping, shot peening, laser beam applications, EDM operations (such as cutting with a wire electrode or hole cutting), riveting, flame cutting, and tube bending. In addition, there are other types of special machines and equipment used for garment cutting and in tire-mold and bottle-mold plants; food and tobacco packing plants; and in aircraft maintenance shops. In fact, almost any application that involves movement or positioning in a prescribed manner between the tool and the workpiece is a possible candidate for numerical control.

FIGURE 21-6 A numerically controlled double-beam flame cutting machine cutting parts from large plate stock (Courtesy, C-R-O Engineering Co., Inc.)

Numerically controlled machine tools are especially economical when used to manufacture parts in small quantities. There are many instances when spare parts must be made for equipment that is no longer in production. The standard program written for a particular part can be preserved for future use without any reprogramming expense or danger that the machine part will not meet specifications.

Perhaps the most outstanding disadvantage of existing (conventional) N/C machine tools is that they have no facilities for checking or inspecting their own production.

DIRECT NUMERICAL CONTROL

The newest generation of N/C development (DNC) is now upon us. In this system, punched tape is no longer used. Instead, all information flow is provided by a computer that interfaces with the machine control unit. In operation, the DNC computer receives the equivalent of punched tape information from a compiling facility and directs storage of that information

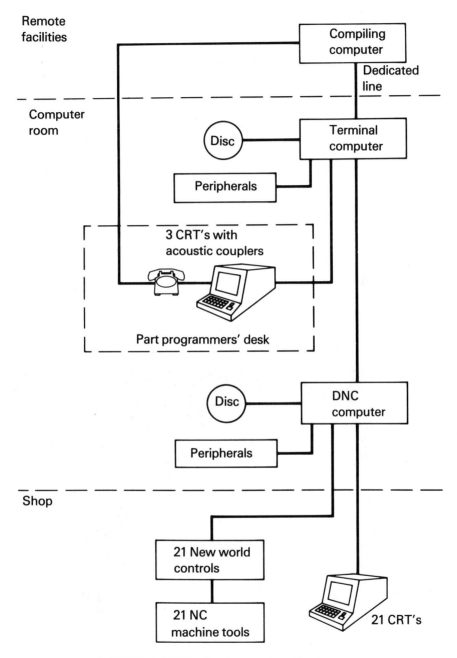

FIGURE 21-7 A DNC system in present use (Courtesy,
Westinghouse Electric Corporation)

in a disk or magnetic-tape data storage unit. The computer may be a mini-computer, several minicomputers linked together, a minicomputer connected to a large computer or one large computer. In effect, all control originates from a remotely located but central data-processing room. There is a small console at the machine, which operates as an input–output device, which is connected to the control unit in the data-processing room. Figure 21-7 illustrates a typical DNC system presently in use. Of considerable advantage is the fact that a portable cathode-ray tube (CRT) editor, equipped with a keyboard, can be installed at the machine tool site, as illustrated in Figure 21-8. The CRT is used to display and edit or optimize a program during the time period of part prove-out and/or machine operation. The CRT, a television tube especially interfaced to an electronic driver, displays data in alphanumerical form. The system can operate in a "conversational" mode in direct communication between the machine and the computer. Other systems are available in which a special on-site typewriter can be used to establish "live" communication with the computer. A programmer or pro-duction supervisor working at the machine tool is able to make on-the-spot

FIGURE 21-8 A DNC system using a CRT display and keyboard allows a programmer to change or optimize a part program at the machine (Courtesy, Sunstrand Machine Tool)

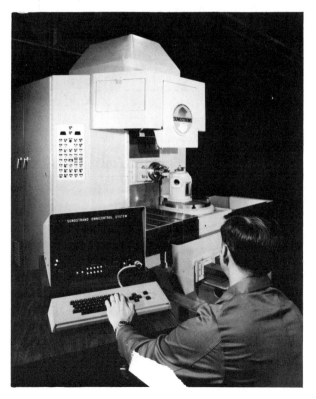

program changes which are instantly relayed and permanently stored in the computer's library. This new concept of machine control permits program tryout, debugging, and modifications in a fraction of the time that is required with a tape system. Furthermore, it eliminates any practical need to operate with less than an optimum program. CRT terminals can be effectively used to increase or decrease feeds and speeds in compensating for actual versus perceived conditions during the processing of a workpiece. It is also possible to give information on a CRT such as *turn coolant on* or *off, unload the workpiece, end of lot,* and so on. CRTs may be located at each individual machine, or a portable CRT may be used to serve a battery of machines to provide editing capability at a much lower cost.

The maximum number of machine tools that can be controlled by a single DNC computer is governed primarily by the types of machines under control and the nature of the parts being produced. As a minimum, a range of 5 to 10 machine tools is usually required to make such a system economically justifiable. The minimum quantity depends upon the extent of work required on different parts being produced and the length of the run.

There are some DNC systems currently in use in which a single, general-purpose computer can control up to 256 N/C machines, with up to five axes each. In addition to the advantages obtained by faster, simpler, and more flexible programming, other major features of DNC are:

1. Elimination of the tape reader problems, resulting in the availability of more production time. Programmer productivity is also improved. Substantial savings in materials and equipment can be realized.

2. The need for special tape-punching typewriters is eliminated, as is the tape itself. The time needed for program prove-out can often be reduced as much as 65%. More new programs are possible with less programmers.

3. DNC systems equipped with CRT permit a visual display which simplifies the "conversation" with the computer.

4. The key control unit may be isolated in a data-processing room safely away from an adverse shop environment. One computer can simultaneously program many N/C machines. Machines can be installed as far as 2500 ft away from the computer and central control.

5. Machine productivity is improved (reportedly as much as 10 to 15%) because of the comparative ease in optimizing programs. Smaller lot sizes become practical, or more profitable.

6. The present DNC generation of N/C systems is considered nearly the ultimate in data-processing capabilities. Fixed-head disks provide permanent storage for N/C programs and system logic-sensing devices, which automatically compensate for desired amounts of metal removal and preferred limits of machine accuracy. In addition, features are available which greatly expand the machine-tool control functions that were not possible in previous control technologies.

COMPUTER-AIDED MANUFACTURING

CAM is a sophisticated extension of DNC in which the computer not only directs the machine movements but assists in data requisition. CAM is considered to be the ultimate in total manufacturing capability. Some CAM systems are so complete that once loaded with all operational and production parameters, they will manage and operate the entire manufacturing network continuously without any intervention. Management information is continuously monitored, updated, and concisely recorded on printouts. Status reports are instantly available to production systems managers. Production control data are furnished on scheduling, inventory levels, machine and tool maintenance, shipment, labor, in-process and final quality control, parts inspection, and other relevant manufacturing information. CAM systems are also capable of recording important around-the-clock time data such as setup time, downtime or nonproductive time, and run time. In most companies, the term "logged-on" time represents time that is not available for use (e.g., the system was not connected, the shift was not manned, etc.).

The economic pressures for continued development and increased use for computer-based manufacturing technology are strong. It has been estimated that in conventional manufacturing operations, a typical part spends about 5% of its manufacturing life on a machine tool and the other 95% sitting on pallets or waiting in "tote" pans. In contrast, an efficient N/C system can be programmed to cut metal about 75% of its average cycle time. DNC and CAM systems have gradually progressed through a series of developmental steps, with each step a proven accomplishment before progressing to the next. Any production process or machine that requires continuous motion control of any type is a potential application for DNC or CAM.

A *machining center* consists of a single, but sometimes two, machine tools equipped with automatic tool changers and capable of performing a wide range of operations on a common workpiece. Fully automatic operations, such as milling, facing, slotting, boring, drilling, and tapping, can be combined in a single location, as in Figure 21-9, in a machining center. Workpiece handling time is decreased because there is no movement of the part from one machine to another. Some machines are equipped with a shuttle system of two worktables that can be rolled into and out of the machine. While work is being done at one table, the next part can be mounted on the second table. Then, when machining is complete, the first table is moved out of the way and the second part is moved into position. Direct savings are available to manufacturers because of an increased production schedule and a reduction of in-process inventory costs.

278

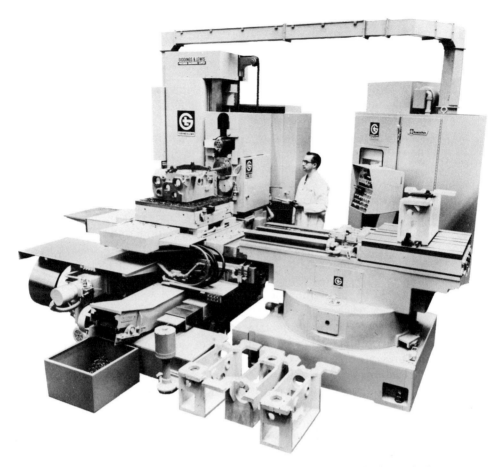

FIGURE 21-9 This precision machining center is capable of close tolerance horizontal boring, drilling, and milling in a short time interval. It features independent electric servo drives and infinitely variable speeds and feeds. The tool changer, rotary indexer and automatic work changer optimize chip production time. (Courtesy, Giddings & Lewis-Bickford Machine Company)

This bold and relatively new concept is an important departure from conventional manufacturing techniques. While there are some non-N/C versions of machining centers, most machines are normally numerically controlled. Initially, most machining centers were dominated by I.C. controls but, more recently, DNC and CAM systems have entered the field to take advantage of the potential for even greater applications which are possible because of the numerous special features which can be incorporated into the design of these astonishing machines. Machining centers are equipped with automatic tool changers, rotary work tables, automatic gaging facilities, and with a wide variety of other options, accessories, add-ons, and so on.

ADAPTIVE CONTROL

A recently developed and highly promising technique of on-line optimization of N/C manufacturing processing is called *adaptive control*. The process employs various types of sensor instrumentation to automatically pick up data in process directly from the workpiece or from the cutting tool. The purpose of adaptive control is to increase the productivity of the machine tool by maintaining the highest possible metal-removal rate consistent with the production of acceptable parts and an economical tool life. An important fact to remember is that adaptive control is obtained without operator intervention. Of prime interest in this stage of development is the measurement and monitoring of tool-tip temperature, tool vibration, and spindle torque. The information is sent back to the control unit so that the necessary corrections can be continuously made to compensate for unpredictable machining conditions.

It should be understood that the full potential of adaptive control has not yet been reached. Except for the rare instance when nearly ideal or predictable cutting conditions may exist, it has not yet been possible for programmers to predict or to simulate all of the actual cutting conditions. At the present stage of development, the technique is principally restricted to applications dealing with cutting speeds and feed rates. Cutting speeds and, to a large extent, feed rates depend, among other factors, upon the hardness of the workpiece material. In adaptive control the input, measured from heat sensors, vibration sensors, or force sensors, is used to measure changing conditions inherent in the chip-making process, particularly at the time when a cutting tool begins to wear and lose the keenness of its cutting edge.

Continuing work in creative statistical techniques and in the development of valid process models, as a means of advancing the technology of adaptive control, goes on in both the metal-cutting and welding fields. Increased productivity in the order of 20 to 30% (and even higher) has been reported, but, as in all other emerging technologies, there are still many problems to be worked out. The development of adaptive controls will continue to be a natural evolution as work progresses toward finding ways to use the machine tool more efficiently within its limitations.

ROBOTS

Robots are operatorless work-handling devices which are programmable. The largest users of industrial robots in the United States today are the major automotive builders. In Japan, there are presently over 1500 robots in operation, with over 80 manufacturers of robot devices. In most applications, robots have demonstrated they can handle heavier loads more accurately and at higher speeds than their human counterparts. The urgency to reduce costs in manufacturing industries has led to an intensive study of the robot's potential use. Figure 21-10 shows the various components and layout for an industrial robot installation.

FIGURE 21-10 Various components and layout for the installation of an industrial robot (Courtesy, AMF Incorporated)

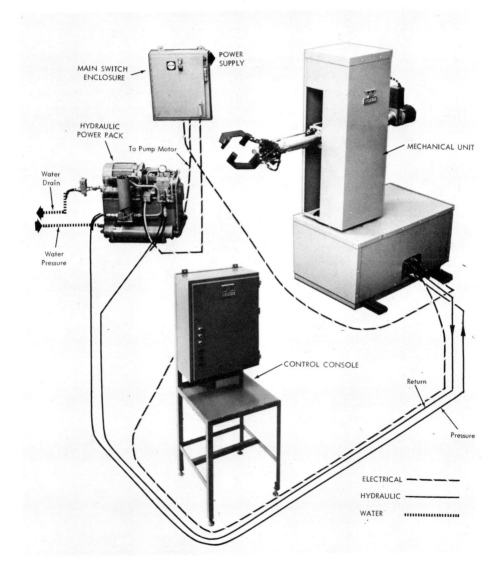

Robots are available with handling capacities that range from minisized parts of a few ounces (approximately 75g) and with a reach of a few inches (75 mm, for example) to huge robots capable of handling payloads of up to 350 lb (approximately 160 kg) having a reach as long as 20 ft (or about 6 m). Some of the larger models of robots can move at speeds of up to 3 fps.

Robots are easily programmable. Figure 12-11 illustrates a six-program-controlled articulation for an industrial robot. Programs can be stored on magnetic-tape cassettes. Some are produced with sophisticated memory systems having the capability of being able to take alternative actions under changing conditions. Other systems are in use with the same overall utility but with a limited work-handling flexibility and are largely reduced to routine jobs involving single "put-and-take" motions. It is possible to "teach" the robot to do a specified job by switching its control system to a recording mode and manually moving the "hand" of the robot through the operation to be performed. Once recorded, the movement and actions will be faithfully repeated, endlessly and precisely, time after time, in synchronization with the external equipment with which the robot works.

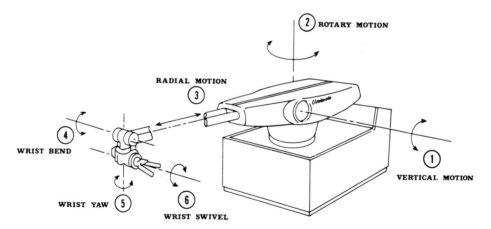

FIGURE 21-11 Articulations of a six-program controlled industrial robot (Courtesy, Unimation Inc.)

Industrial robots are classified as *general-purpose* automation devices. They are presently available as competitively priced, off-the-shelf mass-produced items. Robots have an articulated mechanical arm to which can be attached a variety of hand-like fingers or grippers, vacuum cups, or a tool such as a welding gun or impact wrench. The fingers can be made to fit the given workpiece configuration and interlocks can be quickly located and adapted to a new job. *Special-purpose* automatic equipment, on the other hand, is considerably less flexible to specific job demands. A lead time as long as 6 to 7 months or longer may be required to design, build, and make the necessary performance adjustments for satisfactory operation. In addition, certain installation costs may be necessary in altering the machine controls and replacing them with actuators or relays or in changing over other features of the machinery.

Robot capabilities have become commonplace in many industries. Some robots have operated successfully over 3 million cycles. Among the jobs they perform are:

ASSEMBLY OPERATIONS: *Loading and unloading parts on DNC machines:* lathes, chuckers, drilling machines, broaches, grinders, and multiturret machines.

MATERIAL TRANSFER: automotive assembly, automotive parts, glass (Figure 21-12), textile, ordinance, appliance manufacturing, molded products, heat treating, paper products, plating, conveyor and monorail loading and unloading, palletizing and depalletizing.

DIE CASTING: unloading (see Figure 21-13), quenching, trimming, die care, insert loading, palletizing.

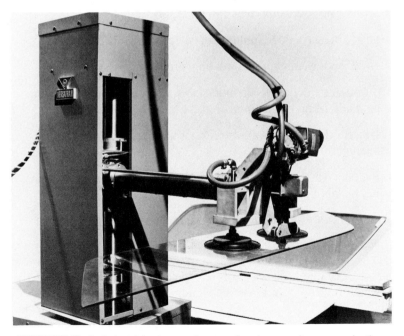

FIGURE 21-12 An industrial robot used to transfer glass panels on a production line (Courtesy, AMF Incorporated)

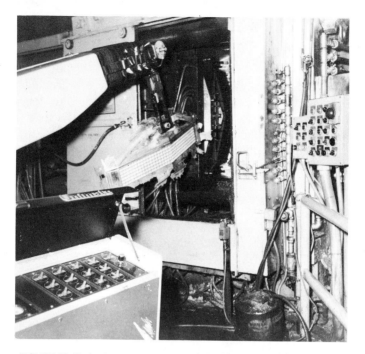

FIGURE 21-13 An industrial robot removing an automotive grille from a die casting machine (Courtesy, Unimation Inc.)

FORGING: drop forge, upsetter, roll forge, presses.

STAMPING PRESSES: load–unload, press-to-press transfer.

WELDING: spot welding, press welding, arc welding.

PLASTIC INJECTION AND COMPRESSION MOLDING: unload, trim, insert loading, palletizing, packaging.

INVESTMENT CASTING: wax "tree" processing, dipping, manipulating, transferring.

SPRAY-COAT APPLICATIONS: Mold release, undercoats, finish coats, highlighting, frit application, sealants.

FURNACE TENDING: loading and unloading parts.

Important factors to consider when selecting an application for a robot are (1) speed of movement, (2) weight of workpiece, (3) accuracy with which a workpiece can be positioned, (4) manipulative ability, and (5) effectiveness with which the robot employs its memory capacity. The selection criteria are based upon (1) whether the process orients the part to aid in pickup by the robot, (2) nonuniformity of the workpiece or other characteristics that make the process unsuitable for automation, and (3) whether the workpiece is difficult for the robot to handle, as in the case with fabric and leather.

284

Industrial robots are extremely versatile and their application to more and more jobs is broadening. Robots are helping to eliminate the risks involved to workers in hazardous operations, as on metalworking presses, or in preventing possible injury, as when lifting heavy parts. They are ideally suited to parts-handling and assembly operations in hostile, unpleasant, or unilluminated environments. Some industrial robots can automatically position parts to an accuracy of ±0.005 in. (0.13 mm), although a more universal positioning standard of accuracy is in the order of ±0.050 in. (1.27 mm).

REVIEW QUESTIONS

21.1. Give the main advantages and disadvantages of APT.

21.2. What is the principal advantage of ADAPT?

21.3. What is the principal advantage of UNIAPT?

21.4. Explain why the advent of I.C.-oriented control brought new N/C builders into the market.

21.5. Prepare a sketch illustrating how five axes of machine motion might be applied to manufacturing operations on a machine part.

21.6. Explain the essential differences between point-to-point and continuous-path types of numerically controlled machine tools.

21.7. Which system is the simplest and the least expensive? Why?

21.8. Compare the relative advantages of automatic methods of data input to manual methods.

21.9. What is the function of the planner or process engineer in parts programming?

21.10. Compare a closed-loop system to an open-loop system.

21.11. Which system results in the most accuracy? Why?

21.12. List the principal advantages of DNC.

21.13. Of what practical advantage is a CRT editor to N/C operations?

21.14. What factors govern the maximum number of machine tools that can be controlled by a single DNC computer?

21.15. Describe a system known as CAM.

21.16. Define a "machining center."

21.17. List the principal advantages of a machining center.

21.18. Define "adaptive control."

21.19. What are the essential differences between general-purpose industrial robots and special-purpose industrial robots?

21.20. What are the principal advantages of industrial robots?

PROBLEMS

21.1. Using the layout shown in Figure 21-P1 (a) convert the dimensions rounded to the nearest hundreth of an inch to the preferred form for jig boring the holes by numerical control, using the setpoint location as shown in Figure 21-P1 (b).

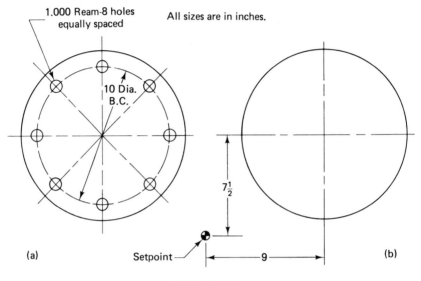

FIGURE 21-P1

21.2. Write a short paper describing current nonmetal cutting uses of numerical control.

21.3. Define the term *software program.*

21.4. List some important advantages of numerical control machining with automated tool changing capabilities.

21.5. Write a short paper describing the current "state of the art" of DNC machining systems.

BIBLIOGRAPHY

BARON, C. H., *Numerical Control for Machine Tools,* New York: McGraw-Hill Book Company, 1971.

*BERRY, S. A., "A New Look at Numerical Control," *The Bendix Corporation,* Detroit, Mich.: Industrial Controls Division, Jan. 1971.

BOLZ, R. W., *Understanding Automation,* Novelty, Ohio: Conquest Publications, 1975.

BROSHEER, B. C., "NC Speeds Contouring," *American Machinist,* Jan. 8, 1973, p. 55.

"CAD/CAM, Tool for Metalworking," Society of Manufacturing Engineers, conference and exhibit (chart), *American Machinist,* Nov. 12, 1973.

"Computer Controlled Production," *Industrial Engineering,* July 1974, p. 34.

"Control Systems—Brains of the Manufacturing Revolution," *Production's Manufacturing Planbook,* Feb. 17, 1975, p. 131.

DEGROAT, G., "NC for Five or Six Pieces?" *American Machinist,* May 27, 1974, p. 118.

*FEDEROV, A. and E. F. SINDELAR, "Numerical Control," *Machinery,* Aug. 1972.

"Future of Robots Look Bright," *American Machinist,* July 19, 1973, p. 61.

HATSCHEX, R. D., "Computer Control: Inevitable?" 1973 IEEE Machine Tools Conference, *American Machinist,* Nov. 26, 1973, p. 49.

HATSCHEX, R. L., "Controls for Cutting," Westinghouse Machine Tool Forum, *American Machinist,* July 9, 1973, p. 66.

*HEROUX, N. M. and G. MUNSON JR., "Robots Reduce Exposure to Some Industrial Hazards," *National Safety News,* Jan. 1974.

*KOHL, R., "Open Loop Numerical Control," *Machine Design,* June 15, 1972.

Machinability Answers: New Use For the Computer, ME73FE032, Dearborn, Mich.: Society of Manufacturing Engineers, Feb. 1973.

MALLEY, L. N., "Mini-DNC Goes to Work Controlling Four 4-Axis Turret Lathes, *American Machinist,* Aug. 6, 1973, p. 45.

"NC Ultrasonics," *American Machinist,* Jan. 22, 1973, p. 53.

Numerical Control in Manufacturing, Dearborn, Mich.: Society of Manufacturing Engineers, 1975.

PATTON, W. J., *NC: Practice and Application,* Reston, Va.: Reston Publishing Company, Inc., 1972.

RAYMOND, M. R. *Manufacturing Automation: Where to Begin,* ME72DE021, Dearborn, Mich.: Society of Manufacturing Engineers, 1972.

*"Robot Power Spreads All Over the World," *Iron Age,* Nov. 11, 1974.

SCHAFFER, G., "Guide to Machine Control: Programmable Controllers," *American Machinist,* Sept. 7, 1973, p. 120.

"Tactile Robots Are Coming," *American Machinist,* Aug. 6, 1973, p. 43.

*Abstracted with permission.

22

CHEMICAL MILLING

Chemical milling is also known as photochemical machining, chemical machining, photofabrication, or photoetching. It is a well-established process used to shape metals by removing unwanted metal by a controlled chemical attack. In principle, chemical milling originally evolved from the well-established process of photoengraving, but the techniques, materials, controls, and purposes presently in use differ markedly from those of its forerunners. An early application of chemical milling was in producing printed circuit boards. The aerospace industry, also an early pioneer, developed etching techniques leading to processing shallow pockets on wing and fuselage skin sections. Chemical milling is now an important, economically sound, and efficient production tool, with broad applications in many industries.

DESCRIPTION OF THE PROCESS

There are two categories of chemical milling: *blanking* and *contour machining*. The basic difference in the two types of chemical milling is that blanking is a process used to etch *entirely* through a metal part, while contour machining is a method of *selectively* etching an area to some desired depth.

FIGURE 22-1 Typical parts produced by chemical blanking (Courtesy, Chemcut Corporation)

Chemical blanking is a process of cutting or "stamping" out parts from flat thin metal or foil sheets. Various chemically blanked parts are shown in Figure 22-1. Contour machining is a method of removing metal from the surfaces of formed or irregularly shaped parts such as forgings, castings, extrusions, or formed wrought stock. Metal may be removed from the entire surface of the part or from selected areas of the part. A contour machined part is shown in Figure 22-2.

FIGURE 22-2 A contour machined part

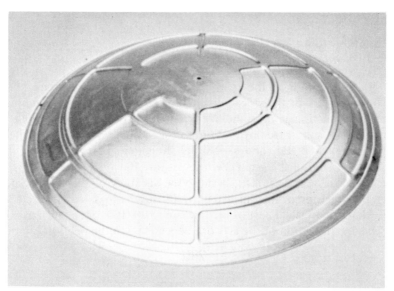

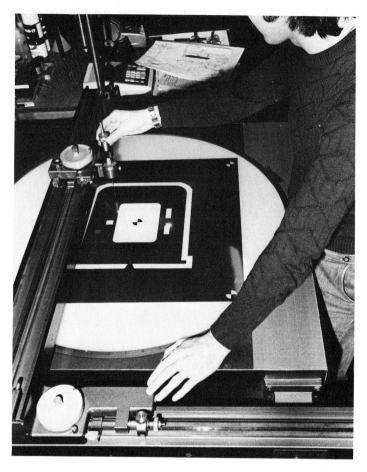

FIGURE 22-3 Artwork is cut on rubylith film with a Coordinatograph machine at several times the actual size to tolerances of ± 0.001 in. (Courtesy, Photofabrication Engineering, Inc.)

Chemical Blanking

The principal steps for chemical blanking are as follows: preparation of artwork, photographing the artwork, cleaning the raw sheet stock, coating one or both sides of the stock with photoresist material, photographically transferring the image onto the metal plate, exposure of the negative to a powerful light source, removal of photoresist to develop the image, etching away unwanted metal, and separating individual parts from the sheet.

The first step in the sequence, the preparation of the artwork, is perhaps the most important step, because the accuracy of the drawing determines the precision of the finished part. Most drawings are made to a scale larger than the finished part (sometimes up to 20, 100, or as much as 200 to 1). The magnification of the artwork depends upon the relationship of the

290

tolerances required on the part to the variables throughout the process. One such variable is artwork error. If, for example, the maximum error on artwork is ±0.010 in. (0.25 mm) and the specifications on the part call for a tolerance of ±0.001 in. (0.03 mm), the artwork should be laid out 10 times its actual size in order to maintain this single variable within acceptable limits. The actual drawing medium may be paper, polyester drafting film, or polyester or glass-base scribing film. Peel-off rubylith film, used with a coordinatograph, shown in Figure 22-3, generally provides the best artwork for chemical milling. Printed circuits are commonly made by the application of tapes, pads, and other adhesive symbols to clear transparent film.

The next step is to photograph the drawing and make a master negative reduced to the precise size of the actual part, as shown in Figure 22-4. In some companies, this negative is called a "photo tool." By using a step-and-repeat camera process, a master negative can be produced, consisting of multiple images of one or more parts. If the part being manufactured is small and the quantity is large, the fabricator will want to get more than one part out of a single sheet of metal (also shown in Figure 22-4). To do this, it is necessary to hold the finished parts to the metal until etching has been completed. This is normally done by extending one or more thin lines which link one part to its adjacent part and to the surrounding metal border. The lines, called "tie-ins" or "tabs," are normally made as few and as small as possible consistent with the quantity of parts being etched.

FIGURE 22-4 A master negative is made from the photograph of the multiple part drawing (Courtesy, Chemcut Corporation)

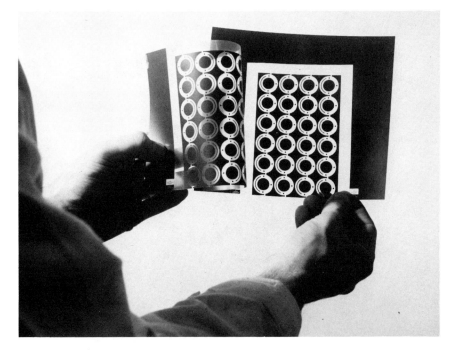

The raw sheet stock from which the parts are to be etched is thoroughly cleaned to eliminate impurities. The metal surfaces are then coated with an emulsion of photosensitive resist. The coating may be applied by spraying, brushing, roller coating, flow coating, or by dipping. Several types of resists are available. The choice of the resist depends upon the size or shape of the part, the etchant, the material to be etched, and the method of application. As with all steps in the process, extreme care must be exercised. It is imperative that the resist be applied in uniform layers. In most cases, the coating on the metal sheets is dried and hardened by baking in an oven to a maximum of 250 °F (approximately 120 °C) for several minutes. The master negative is then placed against the coated workpiece sheet in a vacuum printer and a powerful source of ultraviolet light radiation passing through the negative is used to expose the photosensitive resist. For most kinds of resists exposure to light causes polymers in the resist to cross-link in order to withstand the softening effects of the following step, the solvent bath. Each side of the sheet can be printed individually or the two sides can be exposed simultaneously by using special mirror-image master equipment.

The polymerized image, now printed on the coated workpiece sheet, is then developed by immersing the entire sheet into a tank containing an organic solvent bath solution. In some companies, a vapor and spray method is used to apply the solvent. The action of the solvent readily dissolves the exposed photoresist coating.

FIGURE 22-5 A single chamber, conveyorized, chemical etching machine (Courtesy, Chemcut Corporation)

The part is formed by manually feeding the dry sheets into a conveyorized spray etching machine, shown in Figure 22-5. A variety of metal attacking acids are used. The depth of the cut is a function of etching time, while the temperature and agitation rates usually determine the speed of the process. Most blanking operations are arranged so that the etchant can simultaneously act on both sides of the metal.

The final part is obtained when all of the metal is etched away except that portion of the metal sheet on which the protected or etch-resistant polymerized image is printed. As the sheets come out of the conveyor, each with their multiple parts still attached, they are dipped in a neutralizer, which stops the etching process. The sheets of parts are then dried and, when necessary, the resist is removed. Finally, the individual parts are separated from the sheet by cutting the tie-ins or tabs that connect the adjacent parts.

Contour Machining

Contour machining consists of two different etching techniques: selective or nonselective. The steps in *selective* contour machining begin, as in chemical blanking, with thoroughly cleaning the part to ensure good adhesion of the mask that will be used to protect the surfaces that are not to be etched. If the entire part is to be chemically reduced in size by *nonselective* contour machining, it is unnecessary to apply a maskant. Following degreasing or solvent cleaning, liquid masks are applied in a uniform thickness to the parts by brushing, spraying, dipping or flowcoating. In some applications, a special type of tape maskant may also be applied. The type of maskant material used and the size and shape of the part will dictate the method of masking. However obtained, the resulting film is dried and hardened to produce a tough etch-resistant mask that will adhere well and yet can be easily cut and peeled away in subsequent operations. Next, a template corresponding to the desired etch pattern is placed on the mask and used by the worker to guide the knife or scribing tool as it is carefully worked through the mask. The finely cut lines in the mask are usually marked with a wax pencil so that they can be easily seen after the template is removed. The mask is peeled off to expose the areas to be etched. The mask is closely inspected for pinholes or other defects. Any areas that are suspect are repaired with patching material or with tape.

The photoresist technique previously described for chemical blanking may be used, in some cases, instead of the masking and scribing technique just explained for selective contour machining. The part to be contour-machined is first coated with photoresist and then the image is photographically transferred to the selected area. After removing the photoresist from the unwanted areas with a solvent, the contours of pockets may then be etched. This procedure is used to an advantage for shallow contouring of relatively small parts that require close dimensional accuracy or fine detail. Etching for contour machining is usually done by immersing the parts in a tank, although a spray etching process is sometimes used. Most fabricators test the rate of chemical attack by the etchant on a given part by peeling off small sections of the mask in a surplus area or by using specially prepared test panels. An etching solution of caustic soda is often used for aluminum while acids are used for steel, magnesium, or titanium alloys. After etching, a vapor degreaser "breaks the bond" of the mask so that it can be easily peeled from the part. It is standard practice to wash the part in a series of hot and cold rinse solutions to clean off deposits of dissolved metal "smut" and to remove all trace of the etching chemical. As a final step, the part is brightened by dipping it into a tank containing a chromic deoxident followed by rinsing and drying in an oven.

Tapering

Taper etching may be accomplished by controlling the rate of immersion and withdrawal of a workpiece from the etchant bath in a tank. The tapered section on the workpiece is gradually formed by the chemical action of the etchant as it removes metal over differing periods of time. No maskant is required.

PRODUCT APPLICATIONS

The principal applications for chemically milled parts exist in aerospace, aviation, automotive, electronic and instrument-making industries.

Chemical Blanking

The groundwork for *chemical blanking* was developed during the 1940s by companies whose backgrounds were strong in chemistry and photographic technology. The main interest in the process was in the fabrication of microelectronic devices. Soon afterward, the tremendous potential of chemical blanking was adopted by nonelectronic industries. In many cases, it was adopted as an inexpensive way to manufacture experimental equipment and prototypes. Ultimately, the benefits of the process led to its use in the production of custom-made items. At this time, virtually any company that uses precision metal parts made essentially from flat sheets represents a valid market for this technology. Representative parts, such as screens, reticles, disks, grids, laminations, shims, dividers, cams, brushes, and dials, are to be found in photooptical mechanisms. Computers, tape recorders, telephone systems, television sets, cameras, electric shavers, radar and RF antennas, electric motors, timers, and medical instruments are examples of products that contain these parts.

Contour Machining

Contour machining has achieved a firm position as an important method of metal removal. The process may be classified into three general types of work-related applications: (1) reducing the overall weight of parts, (2) selectively machining shallow (but sometimes deep) cavities or pockets, and (3) tapering sheets, plates, formed parts, or extrusions.

Specific process applications of contour machining in the aviation industry, which is estimated to be the largest user, include the production of special surface configurations on aircraft wing and fuselage sections. Another heavy user of contour machining is the aerospace industry (for forming contoured pockets on surfaces of bulkheads, skin panels, etc.). Other production examples include machining special geometries on the contours of parabolic radar reflectors, gyro housings, heat exchangers, and in making custom hardware. Contour machining is often used to produce decorative etched surfaces on various flat or contoured items, ranging from elevator doors to ashtrays, plaques, signs, panels, and a host of other custom products involving engraving a design partway into a metal object. The artwork for instrument dials, metal tags, or nameplates can, in many cases, be made simply by applying opaque rub-on letters to a film.

PROCESS SELECTION FACTORS

Some parts are so complex, the material so thin, or the degree of temper and metallurgical properties such that chemical blanking may be the only practical process to produce them. Chemically milled parts have no burrs or deformed edges and, because no mechanical action takes place, there are no induced stresses or adverse effects upon the physical properties of the metal. Unlike some pressworking operations, for example, no costly tooling is required. Production cycles are relatively short. Most companies can deliver prototype parts within 48 hours. Short production runs (less than 10 sheets, for example) can usually be delivered in 7 to 10 days. Blanked parts can be made as flat as the original stock from which they were manufactured, and, since prehardened materials can be used, posthardening distortion, particularly of very thin and intricate shapes, is eliminated.

Contour machining can often simplify the manufacture of a part, particularly for large complicated shapes on preformed parts which would involve prohibitive costs of tooling and machining if produced conventionally. The process is particularly useful when used to selectively machine (etch) pockets or odd-shaped cavities. Surprisingly, complex patterns can be chemically machined by etching to a predetermined depth on flat or contoured surfaces. Cuts of different depths can be produced on large areas by progressive removal of different sections of masking tape at appropriate intervals. Metal may be removed from both sides of a part simultaneously.

295

The method requires an accurate register in applying the maskant pattern on the opposite sides of the workpiece. The process makes practical the design of lightweight, integrally stiffened parts because of the ability to etch away selected areas. Surfaces of the parts have raised outlines resulting from selectively etching away the areas around the outlines. The process often is used to eliminate the need for attaching by riveting or spot-welding, additional stiffeners that might normally be required for structural stability.

It is possible on a production basis to chemically mill titanium parts with a surface roughness value of 15 to 20 μin. Typical surface roughness values of aluminum alloy parts average out to be slightly in excess of 100 μin.

In some cases, the overall weight of formed parts such as castings, forgings, or extrusions may be reduced by nonselective etching. Parts are entirely immersed in the etching solution. No protective masking is needed, because the entire surface area of the parts is etched away.

Parts can be produced from virtually any metal or alloy. While chemical milling is most commonly used in processing aluminum and its alloys, parts are frequently processed from copper, iron, carbon steel, stainless steel, and nickel as well as the newer alloys such as beryllium nickel and René 41.

The complexity of the part is not an important consideration, and, in fact, the greater the complexity, the greater is the potential for the process. Design changes can be made at little cost since only the artwork need be altered.

Possible Limitations

Metal Thickness in Chemical Blanking: As a practical matter, blanked parts are usually restricted to a maximum thickness of 0.060 in. (1.52 mm).

Depth of Etch in Contour Machining: Although cuts up to 2 in. (50.8 mm) deep have been successfully made in plate stock, the following general guidelines should be used:

Extrusions	0.150 in. (3.81 mm) max. depth/surface
Forgings	0.250 in. (6.35 mm) max. depth/surface
Sheet and plate	0.500 in. (12.7 mm) max. depth/surface

The depth of etch or the amount of metal actually removed is a function of the time during which the part is immersed in the etching solution. Because of variations in grain size and grain structure on some castings or on weldments, the results obtained from etching are often highly variable.

Tolerances: Many factors influence the final part size. It is impractical to make an absolute statement concerning a tolerance that will apply in all cases. As a general rule, however, the closest practical tolerance on the width of an opening on a chemically *blanked* part is ±10% of the thickness of the metal. A reasonable production tolerance for a chemically *milled* part is ±0.002 in./0.100 in. (±0.05 mm/2.54 mm) of depth. To this must be added the actual raw stock tolerance prior to chemical milling. Finally, depending upon the method of processing a given part preceding chemical milling, the following additional tolerance should be added:

For a part formed by stretch forming, add ±0.001 in. (0.03 mm).

For a part formed by machining, add ±0.005 in. (0.13 mm).

For a part formed by forging, add ±0.015 in. (0.38 mm).

Internal Corners: Sharp internal corners cannot be produced on chemically milled parts. Figure 22-6 illustrates the general rule in which the minimum corner radius is equal to the depth of the cut.

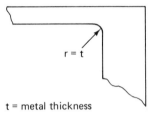

r = t

t = metal thickness

FIGURE 22-6 Internal corners

External Corners: External corners remain sharp.

Safety Precautions: Etchant vapors are very corrosive. It is necessary to isolate etching equipment from other manufacturing equipment. Special care must be used in handling the chemicals in this process and in disposing of the gas that is generated.

Production Rate: Compared to conventional machining processes, the metal-removal rate for chemical milling is considerably slower and, therefore, from this point of view, the process is not considered competitive. Chemical milling removes metal at the rate of approximately 0.001 to 0.002 in.[3] (approximately 0.016 to 0.032 cm[3]) per minute. The process is limited to relatively short production runs. Die stamping or comparable methods should be considered in lieu of chemical blanking for large production runs. Contour machining is used to best advantage for small to medium-size runs.

Inclusions and Surface Defects: Defects in the work metal are often accentuated by the etchant and, as a consequence, poor-quality parts may be obtained. There is a tendency for the etchant to attack surface imperfections at a faster rate than at the surrounding grain boundaries. Tool marks, scratches, or other indentations should be removed before processing by chemical milling. Unless removed, these defects will result in a workpiece with pinholes and pitted or ragged edges. Fine-grain material etches most smoothly.

PRODUCT DESIGN FACTORS

In the chemical milling process there is an almost unlimited design flexibility. The mechanical properties of the work metal and the configuration of the workpiece have no effect on the etching rate.

Minimum Hole Diameter and Slot Widths: As a general rule, the smallest hole size that is practical to produce by chemical blanking is 1.5 times the metal thickness, as shown in Figure 22-7.

Minimum Land Width: The minimum land width, also shown in Figure 22-8, should be twice the depth of cut but not less than 0.125 in. (3.18 mm).

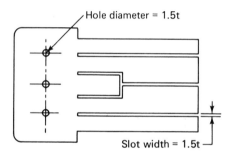

t = metal thickness

FIGURE 22-7 Minimum hole diameter and slot widths

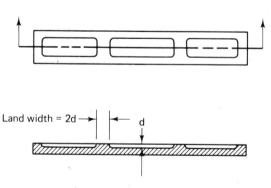

FIGURE 22-8 Minimum land width

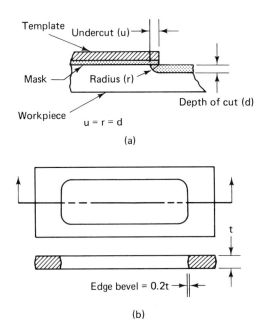

FIGURE 22-9 Effects of undercutting or edge bevel:
(a) undercutting; (b) edge bevel

Undercutting or Edge Bevel: The bevel is illustrated in Figure 22-9. Since the etchant uniformly attacks the metal (except at inclusions or defects), a radius is produced under the resist at each edge approximately equal in size to the depth of the cut. The amount of undercutting increases with etching time. Undercutting problems encountered in chemical blanking are essentially the same as those normally encountered in contour milling. Figure 22-10 illustrates how this condition may result in the formation of pockets in which gases may be entrapped. Unless the parts are mechanically agitated during etching, a rough surface may be produced due to the effects of out-of-control machining. An etch factor that relates the depth of cut to the amount of undercut may be determined on the basis of experience of trial processing

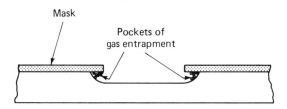

FIGURE 22-10 Formation of pockets of gas entrapment

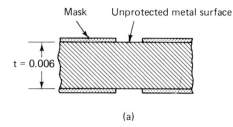

(a)

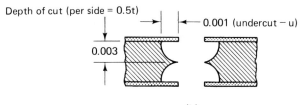

(b)

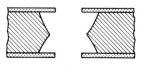

(c)

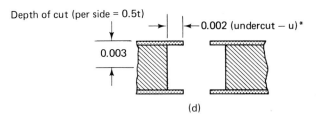

(d)

FIGURE 22-11 Effect of the etch factor in chemical machining: (a) workpiece before etching; (b) workpiece with blanking completed; (c) overetched workpiece to straighten side walls; (d) overetched workpiece with blanking completed

of prototype parts. Once the etch factor is known for a given set of conditions, the size of the edge bevel can be controlled. Rigid production control is required to consistently obtain a minimum edge-bevel size. Another way of improving straightness of side walls, by reducing the edge bevel, is to make an allowance for undercutting and etch factor on the layout of the original artwork. This technique, shown in Figure 22-11, known as "overetching," consists of varying the dimensions of the openings in the resist coating.

REVIEW QUESTIONS

22.1. Compare the important similarities and differences between chemical blanking and contour machining.

22.2. Prepare a brief outline showing the sequence of operations involved in (a) chemical blanking (b) contour machining.

22.3. Explain how the photoresist technique used in chemical blanking can be used to produce shallow contours on relatively small parts.

22.4. What are the three general types of work-related applications for contour machining?

22.5. List the principal advantages of chemical blanking.

22.6. List the principal advantages of contour machining.

22.7. Explain what is meant by "out-of-control" machining. What can be done to offset this condition?

22.8. Define the term "etch factor."

22.9. Explain how the method of "overetching" will often improve size control.

PROBLEMS

In determining the economic break-even point (Q) between conventional stamping and chemical milling (assuming the part can be made equally well by either means) one company uses the following formula:

$$Q = \frac{D + S - A}{\dfrac{R_1}{U_1} = \dfrac{R_2}{U_2}}$$

Where: D = cost for conventional punch and die; S = cost for die setup and maintenance; A = artwork and photographic costs (material and labor); R_1 = hourly rate to have parts chemically milled; R_2 = to have parts punched; U_1 = number of parts per sheet; U_2 = parts punched per hour. (A negative result always indicates that chemical milling will be more economical regardless of quantity. Because the formula is intended for first orders, the D and A terms are omitted for repeat orders.)

Calculate the economic break-even point (Q) using the data given in the table below:

	D	S	A	R	R_2	U_1	U_2
22-1.	775	100	100	4.50	4.50	40	120
22-2.	170	20	26	4.50	4.50	10	100
22-3.	1500	25	92	5.00	4.50	20	160
22-4.	4500	300	75	4.50	7.00	48	700
22-5.	1000	50	170	5.00	5.00	144	1500
22-6.	Repeat Order	100	Repeat Order	7.00	5.00	36	1250

22.7. Compare the cost of producing 1000 coated silicon steel motor laminations using the following processing costs. *Chemical milling:* Artwork costs at $8 per hour = 2 hr; photographic costs at $10 per hr = 4½ hr; process time at 0.06 min. each at $6.50 per hr; process material cost at $9.50/1000. *Conventional pressworking methods:* tooling costs at $1400; processing and material cost at $18/1000.

BIBLIOGRAPHY

"Chemical Machining," *Machining Data Handbook,* 2nd edition, Dearborn, Mich.: Society of Manufacturing Engineers, 1972, pp. 719–725.

"Chemical Machining," *Metals Handbook,* Vol. 3, *Machining,* 8th edition, Metals Park, Ohio: American Society for Metals, pp. 240–248.

*"Photochemical Machining Offers Precise Method for Forming Metal," *Product Engineering,* Feb. 1974 (copyright Morgan-Grampian, Inc., 1974).

*SPEAR, D. R., "New Status for Chemical Milling," *Manufacturing Engineering and Management,* Sept. 1972.

SPRINGBORN, R. K., Editor, *Non-traditional Machining Processes,* Detroit, Mich.: ASTME, 1967, pp. 80–102.

*Abstracted with permission.

23

ELECTRICAL DISCHARGE MACHINING

Electrical discharge machining, the process normally referred to as EDM, came into industrial use shortly after World War II. Its initial applications were in "tap-busting," the electrical erosion of broken taps in parts and die sections too valuable to discard. It was soon discovered, however, that the process of electrical erosion could be controlled to machine cavities and holes. It was observed that the shape formed by this cutting is virtually a direct conforming image of the tool. Thus began the advance of EDM as an accepted metalworking machine tool. EDM is now unquestionably recognized as an important precision machine-tool forming process for producing internal shapes on workpieces, traditionally the most difficult type of operation.

DESCRIPTION OF THE PROCESS

Electrical discharge machining removes metal by the eroding action of small electrical sparks. The energy source for these sparks can be either an intermittently discharged capacitor or an electronically switched dc power source. Normally, the power supply is connected so as to give the tool (electrode) negative polarity, and the workpiece positive polarity. In operation, both the electrode and workpiece are immersed in a dielectric fluid. As the negatively charged electrode approaches the positively charged workpiece, an

303

electrical strain builds up across the dielectric until it ionizes. Electrons then flow from the electrode, through the ionized path in the dielectric, to the workpiece. These electrons further ionize the dielectic, causing the electron flow to greatly increase, thus initiating an electronic avalanche—a spark. The spark, striking the workpiece at extremely high speed, melts and vaporizes a small amount of material, leaving a crater in the workpiece surface. The vaporized metal, floating in the dielectric as a metallic cloud, condenses into small particles that must be quickly flushed away. It has been estimated that a temperature on the order of 15,000 to 30,000 °F (approximately 8300 to 16,600 °C) is encountered at the tip of the tooling electrode.

Electron flow continues until the potential between the electrode and the workpiece falls to a point where the discharge can no longer be maintained, and the dielectric once again becomes deionized. The voltage once again builds up, and the process repeats itself. Described in this manner, the process would appear to be extremely slow; however, there are thousands of sparks per second, so the rate of material removal is substantial.

As the workpiece is eroded, the electrode workpiece gap tends to increase. For optimum machining efficiency, a servodevice controls the movement of the electrode to maintain a constant gap, which, depending upon conditions, may range from 0.0005 to 0.020 in. (0.013 to 0.51 mm). If a direct short circuit occurs between the electrode and the workpiece, the servosystem acts to quickly retract the electrode before the electrode and the workpiece are seriously damaged.

In theory, any conductive material can be cut by electrical discharge machining, but the effectiveness of the process varies widely, depending upon factors such as the workpiece material and part-finishing requirements. Workpiece hardness is of no concern in this process. Hard materials are cut as easily as soft ones. Hardened tool steels, difficult-to-machine exotic alloys, and carbides can thus be machined easily. The process results in completely burr-free parts.

PRODUCT APPLICATIONS

With the increasing use of hard, difficult-to-machine space-age metals, EDM's ability to machine burr-free, intricate configurations, narrow slots, and blind cavities or holes into these materials, at close tolerance, becomes more important. Since EDM does not set up the high cutting forces and mechanical strains often associated with conventional machining, the process is well suited for cutting tubing, honeycomb, and other thin-wall, fragile structures. Small hole drilling, 0.005 in. (0.13 mm) in diameter and as deep as 20 diameters, with virtually no bending or drifting, for example, is pos-

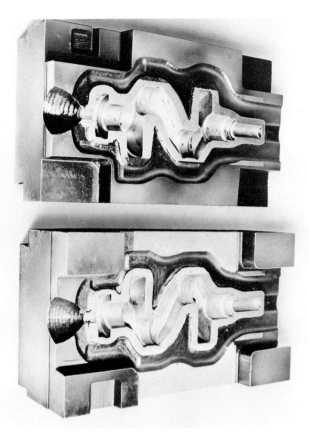

FIGURE 23-1 A large forging die finished by EDM (Courtesy, Cincinnati Milacron Company)

sible. For comparison, in conventional hole drilling, when the length of a twist drill is 10 times the hole diameter or greater, it is difficult to hold straightness. Slots as narrow as 0.001 in. (0.03 mm), recesses, reentrant features, holes in awkward locations, and even curved holes may be produced by this remarkable process.

One well-established application of EDM is in machining die cavities and molds used for die casting, plastic molding, wiredrawing, extrusion, compacting, cold-heading, and forging. In this widespread area of application, the EDM process is known as "die sinking." Figure 23-1 illustrates forging die halves made by the EDM process.

Another important application of EDM is in the metal-forming field to produce punch, trim, or stamping dies. This category of work requires that a precisely sized through hole be made, often irregular in shape, in a solid carbide die. Examples of closely fitting punches and trim die sets are shown in Figure 23-2. Electric discharge machining is ideally suited to the manufacture of dies of this type, as well as for progressive or compound dies, which are a combination of through hole and cavity dies.

305

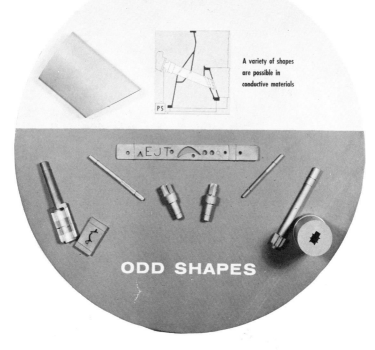

FIGURE 23-2 Odd shapes of precisely sized holes may be produced by EDM (Courtesy, Cincinnati Milacron Company)

Another use for EDM, which has proved to be economical in many companies, is for salvage work on worn forging dies. Copper bar stock is often used as the electrode material. The electrode is first rough-forged on a master die. In a final step, closer tolerances are obtained by coining (a cold-working process of forcing metal to flow within a die) the electrode.

Other special EDM applications are shown in Figure 23-3. The process is often substituted in cases where tool deflection, breakage, and/or chatter by conventional methods has proved to be a problem on parts. Although not widely used at present, EDM can be used to produce very attractive decorative finishes on metal parts. Under certain conditions, the EDM process is competitive with other finishing methods, such as etching or engraving. In some companies, EDM has become a well-established process for making prototype and short-run parts.

As a production process, EDM is particularly adaptable for specialized operations as in the machining of small holes, orifices, or slots in diesel-fuel-injection nozzles, or in aircraft turbine engines, air brake valves, and so on. In many cases, hole shapes other than circular are produced. A single-wire electrode is used to produce burr-free holes which are consistently held to 0.0002 or 0.0003 in. (0.005 to 0.008 mm) in straightness and roundness.

306

Fragile or solid parts
accurately pierced without
damaging cutting forces

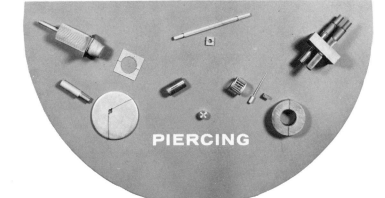

(a)

PIERCING

FIGURE 23-3(a) and (b) Special EDM applications (Courtesy, Cincinnati Milacron Company)

Machining the hidden
hole . . Valve body inter-
porting

(b)

VALVE PORTING

307

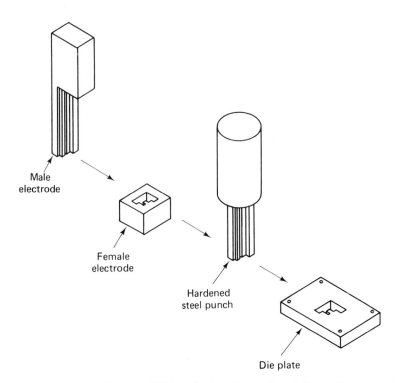

Male
electrode

Female
electrode

Hardened
steel punch

Die plate

FIGURE 23-4 Sequence of steps in EDM production of a punch and die set. The conventionally machined electrode is used to produce a female electrode and die. The punch is EDM
machined using the female electrode

The EDM electrodes, fed from a continuous spool, are less expensive than microdrill bits. EDM has also been successfully used on a production basis to produce perforated parts.

The versatility of the EDM process was recently demonstrated in the manufacture of electrical switch parts when a copper-tungsten electrode was used to machine another copper-tungsten electrode (Figure 23-4). The punch and die required for the production of the parts were made up from rectangular sections having a critical thin rectangular projection 0.030 in. (0.76 mm) thick and 0.200 in. (5.08 mm) long. This complicated job was reduced to a relatively simple EDM operation by using the male electrode to machine the shape of the female electrode.

EDM FUNDAMENTALS

Physical Setup

A schematic diagram showing the main components of the mechanical system and a typical electrical circuit for electric discharge machining is shown in Figure 23-5. The electric current usually ranges from 0.5 to 400 amperes at 40 to 400 volts dc. The capacitance may range from 0.0004 to 400 microfarads (mfd).

308

Power Supplies

Compact, solid-state power supplies are increasing in use because of improved reliability and the superior machining results made possible through precise control of the parameters. Circuits are readily available to control the current or voltage which can generate almost any shape as a single or repetitive pulse. They can also be used to sense and/or respond to various inputs.

Multiple-lead power supplies are also becoming popular, primarily because of the cost reduction possible when using one machine for several different operations or several machines with one power supply. Modular power-supply systems currently available offer good versatility and economy. Small plug-in modules can be connected to the main control and power console as needed.

Dielectric Fluid

A "dielectric" is a material that does not conduct electricity. It is pumped through the tool at a pressure of 50 psi (3.5 kg/cm^2) or less. The dielectric fluid used for this cutting process performs several functions: (1) it aids in keeping the electrode and the workpiece cool, (2) it washes the

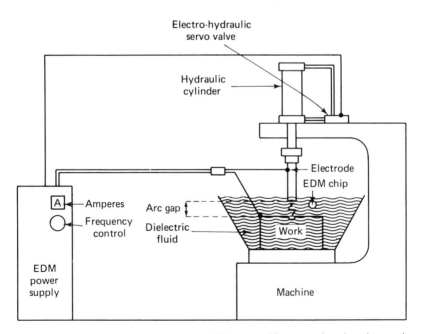

FIGURE 23-5 Schematic diagram of the EDM process (Courtesy, American Iron and Steel Institute)

debris out of the cut, and (3) when ionized, it aids in stabilizing the cut by maintaining a path of finely divided electrically conductive material in the gap. The usual petroleum-based hydrocarbon dielectric fluids tend to increase the cutting rate by reducing the vapor pressure. Since most dielectrics are flammable, a low vapor pressure is desirable from the standpoint of safety. Kerosene, mineral oil, or a mixture of these are also frequently used. Other dielectric fluids, such as silicon-base oil; polar fluids, such as ethylene glycol; and water-miscible compounds are occasionally used for special applications. Dielectric fluids must not be hazardous to operators or corrosive to equipment.

All dielectrics tend to become contaminated by the metal particles eroded from the workpiece. Since these particles are electrically conductive, they cause fluctuations in the breakdown voltage at the electrode/workpiece gap. Unless these eroded particles are flushed away, erratic, uneven cutting occurs and short circuits are likely to develop across the gap. Various methods are used to achieve proper flushing. This is accomplished by standard accessories on the machine itself, which flush by pressure or suction. In some cases an auxiliary flushing device is used. A filtration system is also ordinarily required to control the contamination of the dielectric fluid.

Electrode Materials

The prime requirements of any electrode material is that it must be electrically conductive and maintain a good tool-to-workpiece wear ratio. The choice of electrode materials for electric discharge machining is as critical as the selection of the correct cutting tool for a conventional machine. As a result of broad research and development programs, much has been learned about the optimum cutting requirements (relating to desirable properties, size, and proper contour or shape of tooling electrodes). In the final analysis, the economics of manufacturing the tool electrode to the required geometry is usually the most formidable obstacle confronting EDM users. A recent study of companies who regularly use EDM revealed electrode expenses account for 50 to 60% of the total job costs.

Electrode tool materials perform with varying degrees of success on different workpiece materials. The selection of a particular tooling electrode material depends primarily upon the specific cutting application and upon the material being machined. Graphite or copper have been found to be the best materials for general use. Alloys of zinc and tin are also commonly used. Copper-tungsten and silver-tungsten are often selected for cutting small holes or slots. Other important factors which must be considered when selecting an electrode material are the availability, the cost, and the practical limitations inherent in processing the tooling electrodes to the desired form.

In addition to the materials previously mentioned, a number of other metals, mixtures, and alloys may be used as electrode materials. New electrode materials are continually being developed with the objective of improving workpiece metal-removal rates and finishes and increasing the ease with which the electrode may be formed at reduced costs. Some of the newer materials include various combinations of metal alloys, bonded metallics and bonded nonmetallics, as well as bonded metallic and nonmetallic mixtures.

Electrode Wear

One drawback of EDM is the wear that occurs on the electrodes as the cut is being made. The wear ratios are known and predictable. Figure 23-6 illustrates the various kinds of electrode wear. When through holes are produced, electrode wear is usually of little concern since, even after the tip of the electrode has worn, the hole in the workpiece can continue to be opened to its full depth by simply ignoring the worn, rounded end and advancing the electrode through the workpiece. Electrode wear is most troublesome when complex contours must be maintained on the electrode to create intricate blind workpiece cavities.

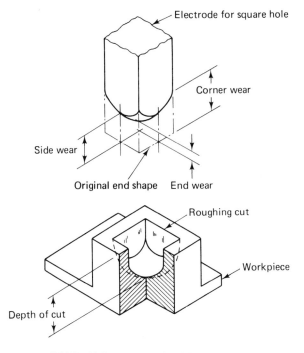

FIGURE 23-6 Examples of EDM electrode wear

A brass electrode wears at the ratio of 1:1. The volume–wear ratio equals the volume of the workpiece removed divided by the volume of the electrode consumed. As a general rule, an acceptable wear ratio for most other metallic electrodes is about 3:1 or 4:1. Graphite, with the highest melting point of any known material, 6300 °F (approximately 3500 °C) may yield wear rates ranging from 5:1 up to 50:1. Most electrode wear occurs on corners, which is about double that of end wear. The sharper the corner, the greater the wear.

No-Wear (or Low-Wear) Electrodes

An alternative EDM technique, used exclusively in place of conventional EDM for bulk metal removal, is the "no-wear" (or "low-wear") operational mode. The technique employs a reverse polarity mode in which the graphite or copper electrode receives a positive charge and the workpiece the negative charge. No-wear EDM results in a smaller electrode/workpiece gap and a higher temperature at the tool surface. Because the polarity is reversed, some of the molten metal ejected from the workpiece strikes the electrode surface and adheres to it. The ideal condition exists when the volume of the thin film of metal that clings to the electrode equals the sum of the metal volume lost through electron erosion. In this case, the tool-wear ratio is zero and it is at this phase that the no-wear mode reaches the point of greatest advantage. If an excess of vaporized or molten metal builds up on the electrode, the cutting efficiency is affected and the deposit may have to be removed. Because of the virtual absence of machining wear on the electrodes, the need for redressing or replacement becomes virtually eliminated.

Metal-removal rates in the no-wear mode are ordinarily 10 to 20% less than the rate achieved by conventional EDM roughing techniques. Furthermore, since no-wear EDM is exclusively a roughing technique, the process does not result in the production of as smooth a surface as in the conventional mode. For maximum effectiveness, most EDM users employ the no-wear operational mode to make roughing cuts close to the desired final requirements and then change to the conventional mode to obtain the final finish.

Electrode Manufacture

Tooling electrodes are made by most of the conventional machining processes as well as by casting, etching or chemical milling, electroforming, stamping, coining, or by power metallurgy. The surface finish of the electrode must be equal to or superior to that required on the finished part. Figure 23-7 illustrates examples of graphite electrodes and the corresponding dies that were produced.

(a)   (b)

FIGURE 23-7 (a) A graphite electrode and die block for a roller skate truck (Courtesy, Cincinnati Milacron Company); (b) complicated graphite electrodes and the corresponding extrusion die produced from the electrodes (Courtesy, L.C. Miller Company)

Kinds of Electrodes

In certain applications, tubes and wires may be used as electrodes. During the workpiece erosion process, any forces which are generated are normally quite negligible, so that electrode sections may be remarkably thin and complex in form. One or more small holes extending along the entire length of the electrode are generally formed so that the flow of the dielectric fluid may be directed down through the hollow tool and onto the workpiece. The small projecting cores left on the workpiece by the flow holes are later broken off and finished.

An economical technique often used to fabricate an electrode for cutting workpiece cavities is to attach by brazing or conductive epoxy cement, a tip of suitable material of desired shape to the end of an electrode mounting shank. It is important in this case that electrical conductivity be maintained. In another case, the actual punch of a punch-and-die set may be used as the tooling electrode. A blank of appropriate electrode material is applied as a tip to the end of the hardened punch. With the tip securely fastened in position, the desired punch contour is then produced by grinding. The tip is removed from the punch following its use as an electrode in the EDM operation.

Electrode Installations

One of the simplest electrode setups is where one electrode operating in a single head is used, as in the machine shown in Figure 23-8. Only one spark at a time can be fired by a single electrode. Each lead from a machine can supply only one electrode. The single electrode is custom-made and shaped to conform to the configuration of the required cavity or the through-hole. In actual practice, some electrodes when made for prototype production or die work must be replaced several times in making a single cut. As many as two to six or more electrodes of identical design, but each of slightly differing size, are required to finish some complex molds. The principal intent of this technique is to erode the maximum amount of metal at the fastest possible rate consistent with the dimensional and surface-finish tolerances of the final workpiece cavity.

Another common practice, known in some companies as "staging," is to design a single electrode having a series of steps or stages which in turn rough, semifinish, and finish the workpiece opening. The size of each stage is calculated to enable that stage to efficiently remove a predetermined amount of metal. In this manner, it is possible to systematically remove large masses of metal under controlled machine settings.

FIGURE 23-8 A typical electrical discharge machine (Courtesy, Easco-Sparcatron, Inc.)

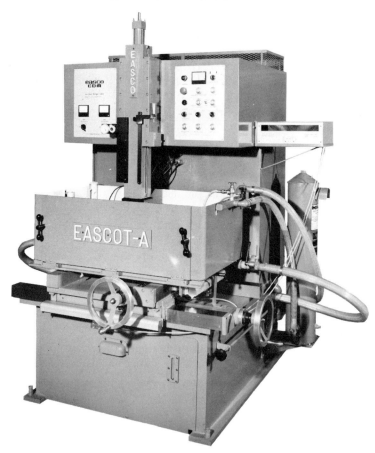

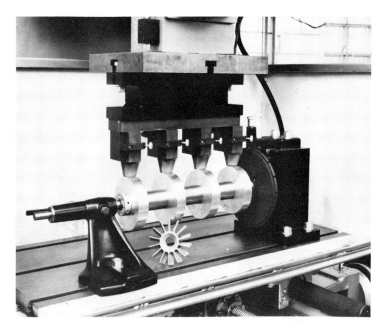

FIGURE 23-9 A multi-lead EDM system combines a four-lead power supply and special fixturing to simultaneously machine four parts in production quanities (Courtesy, Cincinnati Milacron Company)

Multiple EDM heads are also used. Machines are available that permit electrical discharge machining to occur simultaneously from opposite ends of an 8-ft (2.43-m)-long mold. Each EDM head operates independently of each other. One end of the machine is being used for finish machining. An example of a multilead EDM application is shown in Figure 23-9.

Other machines currently in use have as many as eight or more parallel heads, each equipped with an indexing turret electrode holder. Several identical electrodes can be put to work gang-machining several parts at once. The individual cutting is slow, but the total removal rate in the batch operation can be quite high. Setups such as this were originally devised to produce repetitive EDM operations on aircraft jet engine components. This arrangement permits the operator to change electrodes while the machine is cutting. While the electrode in the lower position on each turret is cutting, the upper electrode is out of the dielectric fluid and insulated from the electrical circuit.

Another variation of EDM uses a wire traveling from reel to reel instead of using a shaped tool for the electrode. The schematic diagram shown in Figure 23-10 illustrates the principle of a moving-wire electrode attachment. The wire is gradually advanced between the reels to compensate for the wear that occurs at the point of cutting. This innovative variation in the EDM process amounts to what is essentially an ''electronic bandsaw,'' which cuts precision slots as shown in Figure 23-11 by moving at a constant speed across

315

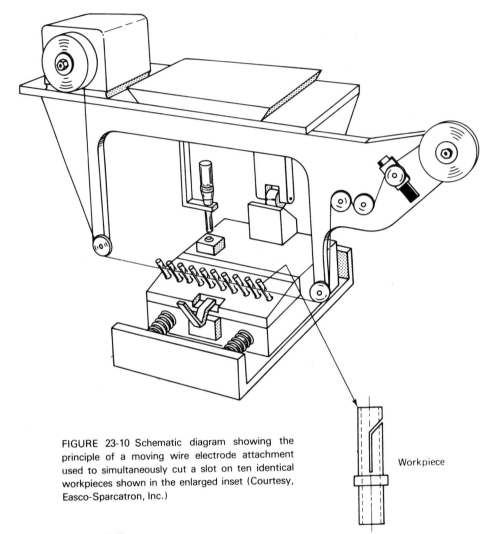

FIGURE 23-10 Schematic diagram showing the principle of a moving wire electrode attachment used to simultaneously cut a slot on ten identical workpieces shown in the enlarged inset (Courtesy, Easco-Sparcatron, Inc.)

Workpiece

FIGURE 23-11 A 0.010 in. wide slot is cut partially through a workpiece by a traveling wire electrode moved according to a programmed form to produce the required cut (Courtesy, Easco-Sparcatron, Inc.)

the workpiece. Initially, this type of equipment was used as little more than a slicing maching for tubing, thin-walled structures, and honeycombing. Now, some companies have added numerical control-positioning features so that complex two-dimensional shapes can be cut. EDM can thus be put to work on simple profiling jobs without the benefit of special electrodes. It is often possible to stack the workpieces in cases where the relatively slow cutting speed of 1 in.^{2}/hr (6.3 cm^2/hr) is objectionable on an otherwise one-piece basis.

Another ingenious design feature consists of a rotating electrode in the form of a tube, disk, or other special shape. Somewhat analogous to grinding, the rotating electrode has the advantage of a large effective surface area. As a consequence, the wear on the electrode is considerably less than that for conventional electrodes. Electrodes such as these can be mounted on the machine in the peripheral or planer mode.

PROCESS SELECTION FACTORS

Metal Removal Rate

The *metal removal rate* is the volume of metal removed from the work-piece during 1 hour per ampere of cutting current. Since the rate of removal is dependent upon current density, the removal rate is directly proportional to an increase in amperage. It is often possible to design and manufacture a part in such a way that the amount of stock that must be removed by EDM is comparatively small. Figure 23-12 illustrates a large forging die cavity which has been roughed out by use of a hand-controlled die sinker. This method is commonly employed prior to electric discharge machining operations, particularly when a large volume of metal must be removed. Another technique employed to rough out excess stock is by conventional machining methods or by using numerically controlled machines. It is also sometimes possible to produce a slightly undersized die cavity by investment casting. In this case, the die is finished to final size and form by EDM electrodes.

FIGURE 23-12 A large forging die is roughed out by conventional machining prior to finishing by EDM (Courtesy, Cincinnati Milacron Company)

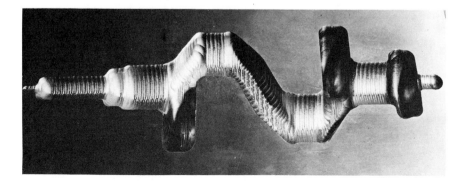

As a general rule, single-lead machines are limited to the rate of about 1 in.³ (16.39 cm³) of metal removal per hour. In spite of this relatively slow removal rate, the process continues to be particularly adaptable for the machining of one-of-a-kind dies where a delay of a few days or even a week for machining may not be objectionable.

The cutting rate may be increased by firing a larger spark. Unfortunately, unless controlled, a larger spark produces a rougher surface, with the result that dimensional tolerances also suffer. Multilead machines are available which, in effect, "spread out" the electrical energy over many spark discharges instead of just one. Thus, by this process, it is possible to put higher power levels to work without the consequence of producing rough finishes or poor tolerances. Dramatically improved metal removal rates, on the order of 25 to 50 in.³ (410 to 820 cm³) per hour have been reported on some multi-lead machines.

Overcut

Overcut is the size difference or gap between the workpiece and the electrode. It constitutes a factor that must be considered in the design of an electrode. The amount of overcut is predictable and can be controlled to within 0.002 to 0.005 in. (0.05 to 0.13 mm) per side. The amount of overcut is uniform at any position along the face of the electrode regardless of its configuration or size. Figure 23-13 illustrates the principle of overcut. In this case the sparks eject metal particles corresponding to the exact shape or contour of the electrode plus the amount of the overcut. The higher the energy level of the spark, the higher the voltage and the larger the overcut.

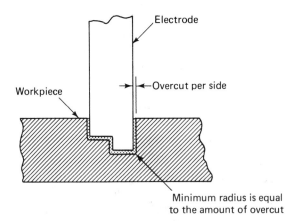

FIGURE 23-13 The overcut or the gap between the electrode and the workpiece is indicated by the shaded areas

Possible Limitations

Materials: While not often a serious limitation, it should be stated that the material for both the electrode and the workpiece must be electrically conductive. Only metals and alloys can be worked by EDM. Nonmetallics such as plastics, ceramics, or glass cannot be machined.

Electrode Wear and Overcut: In spite of the present state of the art, it continues to be a difficult problem for tool designers to determine an optimum size for electrodes, for which compensation has been made within acceptable limits, for both wear and overcut on a given job. Rapid tool wear is a costly factor, and in some companies this aspect may be the deciding factor in determining whether or not to use EDM.

Metal Removal Rates: Various categorical statements have been made from time to time in commercial literature to the effect that EDM is faster or better than conventional die sinking machining techniques. The accuracy of this statement depends upon how well the tool designer is able to control the many factors in the process. While EDM has many unique advantages, it is not ordinarily a mass-production process. Stock removal rates are comparatively slow and not competitive with milling. The process should be applied where results obtained by faster methods are not acceptable, when the total time by conventional methods is unacceptable or when closer accuracy is desired.

Heat-Affected Layer: Another possible disadvantage of electrical discharge machining is the rehardened, highly stressed zone produced on the workpiece surface by the heat generated during cutting. Any molten metal not expelled is resolidified to form a hard skin on the work surface. This condition is shown in Figure 23-14. In the shallow extreme, the depth of this rehardened or recast zone, also called the "white layer," usually varies from 0.00015 to 0.005 in. (0.004 to 0.13 mm). New surface layers are produced by high-frequency low-amperage sparks. When low-frequency, high-amperage sparks are used to achieve high metal removal rates, the heat-affected layers may be as deep as 0.015 in. (0.38 mm).

The brittle white layer can cause serious problems. Cracking may occur on the workpiece after it has been placed in service, particularly in applications where the part is subjected to shock and cyclic loads. The rehardened layer can have an adverse effect upon part performance. These effects can be minimized by finishing the part at low metal removal rates, by using a finishing technique such as lapping or grinding to remove the rehardened layer, or by retempering after machining has been completed. In some specific product applications, the rehardened layer has proved to be an asset.

319

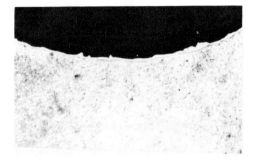

(a)

(b)

FIGURE 23-14 This photomicrograph at (a) shows the recast zone or white layer formed during EDM machining of a part made of AISI 4340 steel. The enlargement at (b) shows the area marked in greater detail (Courtesy, American Iron and Steel Institute)

Perfectly square corners cannot be formed by EDM. However, EDM can form sharper corners than those possible by conventional machining. The minimum radius is equivalent to a distance equal to the overcut, as illustrated in Figure 23-13.

PRODUCT DESIGN FACTORS

Precision: Tolerances on the order of ±0.002 to 0.005 in. (0.05 to 0.13 mm) are routinely obtained and with extra care, tolerances of ±0.0001 to 0.0005 in. (0.003 to 0.013 mm) are possible.

Taper or Draft: The opening cut in the workpiece is always larger than the cutting tool. Unless special precautions are taken, a slight taper or draft is always present on the sides of holes and cavities. In most cases, the taper can be held to less than 0.002 in. (0.05 mm) per inch of depth per side.

Surface Finish: Factors that affect the workpiece finish are (1) the dielectric constant of the spark gap, (2) the magnitude of the charge built up in the capacitor, and (3) the nature of the workpiece material being eroded. Surface roughness is the result of a multitude of individual craters or pits superimposed on each other, caused by individual machining sparks, as illustrated in Figure 23-15. Under normal cutting conditions, an EDM surface finish resembles a nondirectional matte surface and will range from 50 to 150 μin. Under closely controlled conditions, surfaces as smooth as 10 μin. have been generated on workpieces, but such finishes can only be accomplished over small areas and with very low metal removal rates. Plastic molding dies, requiring a "mirror" finish, can be produced by EDM to around 100 μin. but must then be hand-finished to the desired smoothness. Considering the lengthy time period needed, it is uneconomical to use an EDM machine to obtain an ultrafine surface finish.

FIGURE 23-15 Characteristics of an EDM surface finish

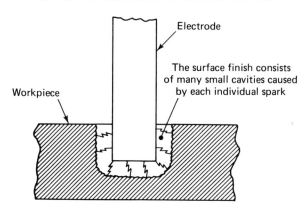

Electrode

The surface finish consists of many small cavities caused by each individual spark

Workpiece

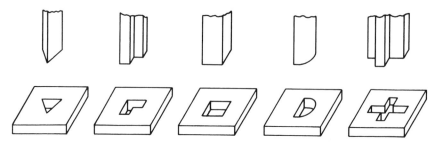

FIGURE 23-16 Noncircular holes such as these can be readily produced by EDM in one machining operation

Configurations—Through-Holes: One of the superior features of EDM is its ability to produce cuts that are either impossible or difficult to produce by conventional machining. Noncircular holes, for example, such as those shown in Figure 23-16, can be routinely produced in any electrically conductive material, entirely free from burrs, by EDM. Holes can be cut at shallow entrance angles, as shown in Figure 23-17. Fragile parts may be machined without damage or distortion from the tool itself or from the fixtures that may be needed to position and hold the part while being machined.

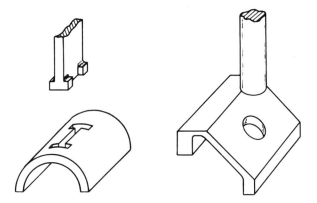

FIGURE 23-17 Holes can be cut at shallow entrance angles that would be difficult or impossible to produce with a conventional drill bit

Complex Cavities: A leading concern in the design of electrodes is to avoid conditions of opposing flow of dielectric fluid and to prevent pockets where debris may become trapped. Figure 23-18 illustrates how more than one electrode may be used for a multilevel cavity to aid in cooling the work areas and in flushing away the eroded metal particles. Steep vertical cavity walls tend to increase electrode wear.

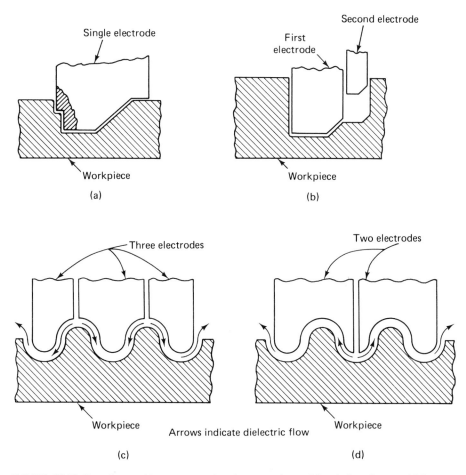

FIGURE 23-18 Complex cavities may require the use of specially designed or multiple electrodes: (a) shaded area indicates greatest wear; steep vertical walls accelerate electrode wear; (b) complex cavities with separate levels may require multiple electrodes for adequate dielectric flow; (c) poor condition; (d) improved condition; dielectric flow as in (c) and (d) must avoid stagnation or opposing flows that trap debris

REVIEW QUESTIONS

23.1. Briefly outline the process by which a controlled electric spark may be used to create a desired cavity in a metal product.

23.2. Explain what would likely occur if a constant gap were not maintained between the electrode and the workpiece.

23.3. Define the term "diesinking."

23.4. Explain the function of the dielectric fluid.

23.5. Explain why electrode wear is usually of little consequence when producing through hole cavities.

23.6. Contrast the alternative no-wear EDM technique to conventional EDM operation.

23.7. Explain how the use of various electrode tips may result in increased process economy.

23.8. Why would several electrodes of slightly different size often be used in finishing some complex molds?

23.9. Describe the process known as "staging."

23.10. Explain how the use of multiple EDM heads improves process efficiency.

23.11. Why is it of prime importance to initially rough out the general size and shape of a desired cavity prior to the EDM operation?

23.12. Describe the wire-electrode method of EDM cutting.

23.13. The cutting rate may be increased by firing a large spark. What are the attendant disadvantages of doing this?

23.14. What is the significance of overcut?

23.15. What basic consideration must a workpiece material satisfy to be considered for the EDM process?

23.16. Explain why EDM is not classified as a high-volume production process.

23.17. Explain how the heat-affected layer, often a disadvantage of the process, can, instead, be advantageous in certain cases.

23.18. Give the leading advantages of the EDM process.

PROBLEMS

23.1. One EDM company uses the following formula to determine the "gain factor" which rates a multi-lead over single electrode machining:

$$K_n = \frac{t_1}{t_n} N$$

where: K_n = multi-K factor, t_1 = machining time for one electrode, t_n = machining time with N electrodes, and N = the number of electrodes.

Calculate the multi-K factor using the following data: (a) one electrode will drill a hole in 11 seconds while 10 electrodes will drill 10 holes in 27 seconds; (b) one electrode will drill a hole in 7 minutes while 5 electrodes will drill 5 holes in 5½ minutes.

23.2. What would be the leading cause of workpiece cracking under the following conditions: cutting tungsten carbide workpiece material using copper tungsten electrodes with a 100 ampere power supply?

23.3. What would be the effect upon the surface finish of a given workpiece if the current level was reduced from 100 to 50 amperes using the workpiece material and the electrode listed in the preceding problem?

23.4. A manufacturer chose EDM to produce 31,000 holes of 0.021 in. diameter in a 0.010 in. thick stainless steel filter tube. The distance between adjacent holes was held to within 0.006 in. The drilling time required per tube was 40 hr. Give some reasons why you feel the manufacturer was justified in specifying EDM over other hole-making processes.

23.5. Careful metallurgical investigation of a group of hardened tool steel parts worked by the EDM process has indicated adverse metallurgical effects. It was noticed that a thin highly stressed untempered martensite layer developed in the heat affected zone. What action would you recommend in order that the parts be salvaged?

BIBLIOGRAPHY

Big Boost For EDM, ME72DE025, Dearborn, Mich.: Society of Manufacturing Engineers, 1972.

"Dial Type EDM Machine for Production," *American Machinist,* Sept. 17, 1973, p. 59.

*"EDM's Versatility Begins to Surface," *Iron Age,* Sept. 13, 1973.

"Electrical Discharge Machining," *Machining Data Handbook,* 2nd edition, Dearborn, Mich.: Society of Manufacturing Engineers, 1972, pp. 701–706.

"Electrical Discharge Machining," *Metals Handbook,* Vol. 3, *Machining,* 8th edition, Metals Park, Ohio: American Society for Metals, pp. 227–232.

Electrical Discharge Machining as Applied to the Practical Man, Publication M-2123, Cincinnati, Ohio: The Cincinnati Milicron Company, Meta-Dynamics Division, formerly the Cincinnati Milling Machine Co.

*"Electrode Materials for EDM," *The Tool and Manufacturing Engineer,* Jan. 1967.

*KELSO, T. D., "EDM—Production Tool?" *The Tool and Manufacturing Engineer,* Nov. 1969.

*KOHL, R., "Electrical Discharge Machining," *Machine Design,* Nov. 11, 1971.

KRABACHER, E., "Forging Ahead with EDM," *Manufacturing Engineering and Management,* Aug. 1973, p. 28.

KRASYUK, B. A., *Electrospark Machining of Metals,* Vol. 3, New York: Consultants Bureau, Inc., 1965.

"Multiple Setup Divides EDM Time," *American Machinist,* Mar. 5, 1973, p. 53.

*Abstracted with permission.

24

ELECTROCHEMICAL MACHINING

Simply stated, electrochemical machining, or ECM, is the process of removing metal from a workpiece by a reverse plating action. ECM can be used to entirely form or blank a desired external contour, to selectively shape one or more external surfaces, to produce through or blind circular or odd-shaped holes, or to machine an almost infinite array of cavity shapes into workpiece surfaces at various depths. While the ECM process is a relatively recent development, it is, nevertheless, generally accepted as an important stock-removal and surface-producing process.

DESCRIPTION OF THE PROCESS

The basic electrochemical action is an application of Faraday's laws. Figure 24-1 illustrates the principles involved in the ECM process. The positive electrode (anode) is the workpiece. It is connected to a high-frequency dc power supply. The negative electrode (cathode) consists of a hollow cutting tool. The anode and the cathode are energized by low-voltage direct current. An electrolyte (a water solution that conducts electricity) constantly flows in a space that is carefully maintained between the electrodes. As the current flows, it removes electrons from the surface atoms of the workpiece and causes the resulting ions to migrate toward the hollow cutting tool. Unlike

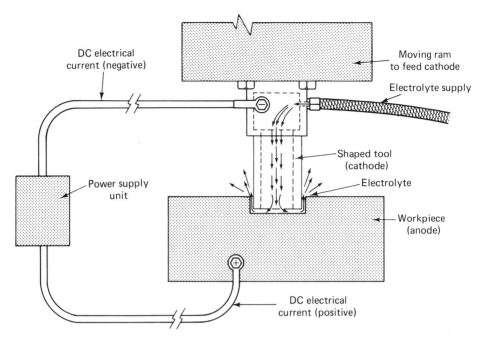

FIGURE 24-1 The ECM process

electroplating, however, ECM does not deposit or plate metal on the tool or cathode. Instead, a swiftly moving electrolyte sweeps the deplated ions out of the gap between the workpiece and the tool before plating can occur. In effect, the surface atoms of the workmetal are removed a layer at a time.

The tool movement is limited to a single axis and is advanced toward the workpiece by an automatically controlled servomechanism. The shape of the workpiece is formed by electrical and chemical energies which, in combination, etch away the unwanted metal under the tool until the final configuration conforms to the reverse shape of the tool.

PRODUCT APPLICATIONS

ECM had its beginnings as a process for certain machining operations on gas-turbine aircraft engine components. The process was selected because it could produce a burr-free finish and because there was no danger of any metallurgical damage to the workpiece. Since then, a substantial number of applications have been found for ECM outside the aircraft industry, encompassing an impressive range of product applications. While the gas-turbine industry continues to be a leading user of the process, companies specializing in airframe component fabrication, die making, and the manufacture of general machine parts are becoming more and more aware of ECM capabilities.

327

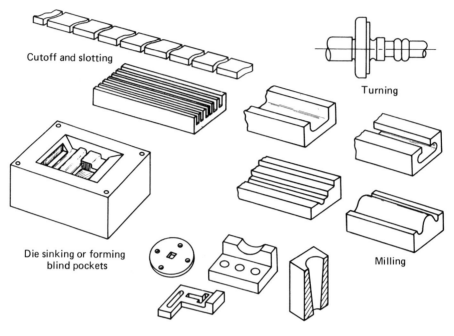

Cutoff and slotting

Turning

Die sinking or forming blind pockets

Milling

Trepanning and hole-sinking

FIGURE 24-2 Examples of ECM operations

Because of its versatility, ECM has greatly extended product design possibilities. It is now practical to combine in one operation configurations that formerly required a number of different setups using traditional methods. As an example, operations such as drilling, milling, turning, boring, shaping, planing, and sometimes grinding can be combined and performed economically and quickly. Figure 24-2 illustrates a number of examples of typical ECM operations. The major appeal of ECM to users is in producing simple to complex cavities, as in die sinking, special marked or embossed surfaces, blind holes, through-holes, square and other irregular holes, and complex external shapes. The process is often selected as a substitute for machining operations that have proved to be too difficult on conventional machines. It is also used for processing refractory or other tough, high-temperature alloys. Steel mills are using ECM equipment for cutting test blocks and for sawing ingots to various desired lengths.

Figure 24-3 illustrates several typical parts that have been processed by ECM operations. Other product applications include rough machining of massive forgings; operations on complex-shaped turbine wheels with integral blades; honeycombing aircraft panels or parts with complex grid patterns; jet engine blade airfoils and cooling holes; operations on rock boring bits,

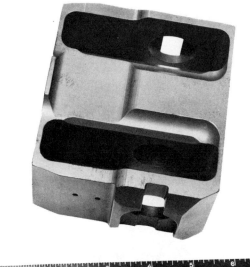

(a)

(b)

FIGURE 24-3 Typical examples of ECM production parts: (a) two blind pockets machined by EDM (Courtesy, Vought Corporation); (b) various parts and a typical ECM tool used to produce the equally spaced slots (Courtesy, Vought Corporation)

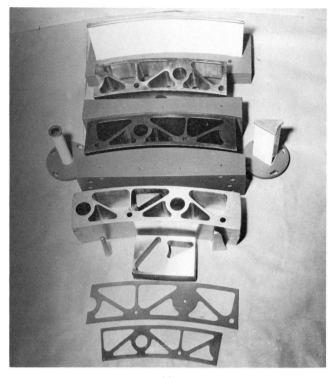

(c)

(d)

FIGURE 24-3 (c) ECM tooling and Rene 41 bulkhead sections machined by ECM (Courtesy, Vought Corporation; (d) these sprocket teeth were produced by the ECM tool shown (Courtesy, Vought Corporation)

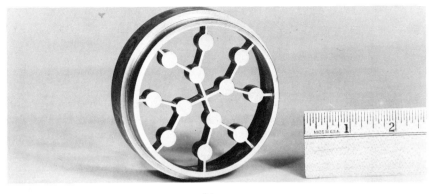

(e)

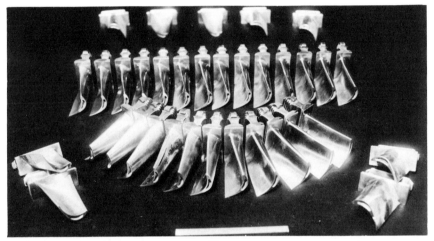

(f)

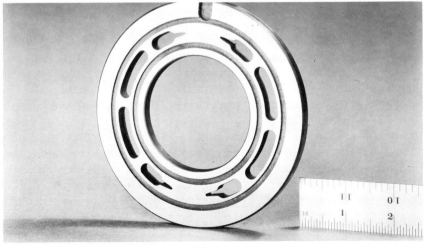

(g)

FIGURE 24-3(e) this zirconium grid is part of an isotope holder for a nuclear reactor. It was made from a solid blank (Courtesy, Anocut, Inc); (f) aircraft gas turbine blades mass produced by ECM from oversize forgings (Courtesy, Anocut, Inc.); (g) the oval holes, including the fishtail configuration, were made in this hardened steel valve plate in one ECM pass (Courtesy, Anocut, Inc.)

transformer cores, pump impellers, gears; contour-forming intricately shaped blanks for machine parts; and in die making of all kinds. Another important application for ECM is in high production, extremely rapid deburring operations, particularly in locations on components that are inaccessible by conventional machining methods. In this way, costly hand-finishing operations may be eliminated.

Finally, some companies have reported significant savings when ECM processes have been adapted to salvage operations on worn machine parts or on dies.

ECM FUNDAMENTALS

Physical Setup

Both vertical- and horizontal-type ECM machines are available in a number of sizes. They are classified according to the direction of the ram or quill feed in which the electrode tool is held. Figure 24-4a shows a vertical-type ECM machine and Figure 24-4b shows a horizontal-type ECM machine. As with conventional machine tools, depending upon the special work requirements, electrochemical machines are built in both standard and special models. ECM machines are characterized by a rugged frame built to withstand the action of the forces caused by the high-pressure flow of the electrolyte, which also supports the quill and table in smooth and precise alignment during machining. In operation, when the electrolyte is pumped at 200 psi (14 kg/cm²), for example, and directed between the hollow tool and the workpiece, a machining area of 10 × 10 in. (254 mm × 254 mm) would receive about 20,000 lb (9000 kg) of force. Rigid workpiece holding fixtures are required for ECM operations to offset rather substantial electrolyte pressures and to protect against the corrosive and electrolytic action inherent in electrochemical machining. Corrosion-resistant materials are used for all components that contact the electrolyte.

Machine sizes are governed by the metal removal rate and ampere capacity because, in general, the rate of metal removal is directly proportional to the flow of current. Machines are available with either single or multiple tool and work stations. Tool and work positioning is often simplified by electronic digital readouts at the control systems. Depending upon the nature of the special work requirements, tool feed mechanisms can be furnished to provide either straight or rotary motion. Symmetrical contours on workpieces can be produced on a special-purpose ECM machine having a lathe-like action, where either a flat or a profiled electrode tool is used to shape the rotating workpiece.

When compared to conventional machining processes and, in fact, to many of the other new machining methods, the unique electrochemical

(a)

(b)

FIGURE 24-4 Typical ECM machines: (a) a 20,000 ampere vertical machine (Courtesy, Anocut, Inc.); (b) a general view of ECM equipment. A horizontal cavity sinking machine is shown on the left with the operating console on the right (Courtesy, Anocut, Inc.)

requirements for consistently achieving high-quality control are considered by most authorities to be far more demanding. Successful results in ECM depend heavily upon the user's ability to develop a uniform system in which an equilibrium cutting condition is achieved. Proper balance over the electrode tool feed rate, cutting gap, current density, electrolyte composition, flow rate, and temperature must be continuously monitored by trained personnel. However, once the cutting requirements for a specific machining application have been determined and stable conditions across the machining gap have been achieved (usually by a series of test runs), it is not unusual for a single operator to attend to two and sometimes three ECM production machines.

Most ECM machines are equipped with an assortment of both manual and automatic controls, including switches, thermocouples, and regulating valves, to cope with the varying machining conditions. A leading difficulty in the present ECM systems is that most of the operating parameters are independently controlled, each operating within a specified tolerance band and usually within narrow limits. The newer ECM machines, presently in advanced stages of development, will be equipped with sophisticated adaptive control features. These controls will be capable of analyzing and interpreting each of the parameters simultaneously and automatically making the necessary adjustments in one or more parameters to correct for the deviations.

Power Supply

A large portion of the total cost of an ECM machine is represented by the power-supply unit that is required to convert ac electrical power to the required low dc voltage. The amount of dc power required is determined by the size of the work being performed. While some power-supply units are as small as 50 to 100 amperes, most machines may be readily obtained in the 1500-, 3000-, 5000-, or 10,000-ampere range. Other units, on giant ECM machines, are rated up to the highest at 40,000 amperes. The power-supply requirements for electrochemical machining are dictated by a necessity for a rigidly maintained condition of constant voltage. The range of machining voltage on most machines is from 5 to 30 volts dc. Voltage-regulation circuits are available that accurately maintain constant voltage across the machining gap at any pre-selected value from zero to maximum.

Power-supply units also include cutoff circuits to stop the supply of power to the machining gap, thus protecting the tool and workpiece from potential damage in the case of an arc or a short circuit. Systems are currently available that will react within 2 milliseconds (ms) to initiate power-supply shutdown. Electrochemical machine manufacturers also provide overcurrent or "rate-of-change" devices, which will respond in 10 ms or less, to anticipate arcs or short circuits. Spark control and detection systems are normally set at values higher than job requirements and operate any time the current varies from the preset values.

Current Density

For purposes of estimating current requirements, most process engineers recommend using a current density of 1000 amperes for every square inch of work metal to be removed. It is interesting to note that current density required to machine a large surface area is actually less than when machining a small surface area. Also, the maximum permissible electrode tool feed rate is inversely proportional to the size of the machining area for a given machine capacity. The current density drops rapidly with the distance from the electrode tool face to the workpiece. Fortunately, parts with large cavity surface areas—dies, for example—generally do not require close sidewall tolerances. As will be explained in the section Process Selection Factors in the paragraph entitled Surface Finish, the quality of the surface finish on the workpiece is affected by the current density.

Feedrate

Most ECM machines have a feedrate system with a capability of 0.020 to 1.0 in. (0.51 mm to 25.4 mm) per minute. In order to maintain optimum cutting conditions, the feedrate must be held constant. It is unnecessary to feed the electrode into the work when machining very shallow cavities if the gap size can be tolerated with no loss of accuracy. The need for tool feed is also eliminated for burr-removal operations.

The Electrolyte and Its System

The ECM electrolyte should have good electrical conductivity because it must serve as the vehicle for carrying the current between the electrodes. It also removes the sludge of metal oxides or hydroxides that form during the deplating process and prevents overheating of the electrodes. Neutral salts, in preference to highly corrosive and extremely dangerous acids or alkalis, are used as electrolytes. A water-base saline solution, usually sodium chloride, is most commonly used, principally because of its low cost. In addition, the current efficiency can usually be held to close to 100% when machining most metals with a sodium chloride electrolyte. Other special-purpose solutions, such as sodium chlorate or sodium nitrate, are occasionally used, sometimes in combination with special additives.

The electrolyte solution is pumped between the workpiece and the tool at a pressure of up to 350 psi (24.6 kg/cm^2). For most work applications a swift flow rate of 100 to 200 gallons/min is required. Extremely close quality control of the electrolyte is necessary for good machining results. The electrolyte system must be closely "balanced" to provide a constant supply of clear electrolyte at the required temperature, pressure, and flow for the job. Dual-compartmented stainless steel or fiberglass-lined tanks with capacities up to 500 gallons are available on most machines. The contaminated electrolyte, collected in the work area pan, drains to one side of the tank. From there it is pumped to a centrifuge for clarification. The clean electrolyte returns by gravity flow to the "clean" side of the tank, following a thorough cleaning from the continuously operating centrifuge. In-line filters are also used on some systems. The temperature of the electrolyte in the clean tank is thermostatically controlled by electric heaters together with a heat exchanger or evaporative condenser. Most ECM systems maintain temperature control over a range of from 95 to 150°F (approximately 35 to 65 °C) within ± 2 °F of the desired settings.

The Electrode Tool

In ECM, there is no tool wear. Once prepared, a single ECM electrode can produce thousands of parts. No erosion takes place because there is no physical contact between the two electrodes—the workpiece and the tool. A machining gap of between 0.001 and 0.030 in. (0.03 and 0.76 mm) is available on most machines, but for the majority of applications, a gap of 0.001 to 0.005 in. (0.03 to 0.13 mm) is used.

When compared to tools used for conventional mechanical-type machining, manufacturing requirements of the ECM electrodes, in most cases, are considerably more critical. Important characteristics of a good electrode tool are as follows: (1) it must have good thermal and electric conductivity, (2) its cross-sectional shape must be sufficiently large to conduct a machining current in the quantities required, (3) it should be of a material that can readily be machined, (4) the material should resist the chemical action of the electrolyte, (5) it must be sufficiently rigid and stiff to withstand distortion or vibration caused by the high-pressure electrolyte flow, (6) it must be made to exacting dimensional standards, since the accuracy of the workpiece configuration is a direct function of the electrode tool, and (7) the cutting surfaces of the electrode tool should be polished. The surface finish of the workpiece depends to a large extent upon the condition of the face or frontal surface of the electrode tool. Tool defects such as nicks, scratches, or burrs show up as corresponding blemishes on the workpiece.

Except for very simple, straightforward shapes, the design and construction of the electrode tool often involves numerous small compromises in accuracy and configuration in order that a successful tool might be obtained. In most critical ECM machining applications, in spite of the present high level of expertise, the development of the desired tool is largely empirical, often requiring a series of "cut-and-try" steps before achieving optimum cutting performance.

Close cooperation in tool design usually exists between the buyer and seller of ECM machines. Some machine makers furnish complete tooling services, which may include designing and building the electrode tool as well as providing technical assistance in solving problems involving work-holding fixtures. Buyers are ordinarily given the option of extensive tooling-service assistance by manufacturers, particularly in cases when a machine is purchased to make a specific part. Also available are a host of engineering design companies that have been formed to provide complete tooling and fixturing services, not only for ECM, but for conventional and other nontraditional machining processes. Because of the unique nature of the kind of work performed, most electrode tools are individually designed for their intended application. Unlike many cutting tools used in other processes,

FIGURE 24-5 Three basic methods used to provide electrolyte flow between the electrodes: (a) straight-flow tooling; (b) reverse-flow tooling; (c) cross-flow tooling

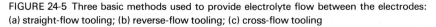

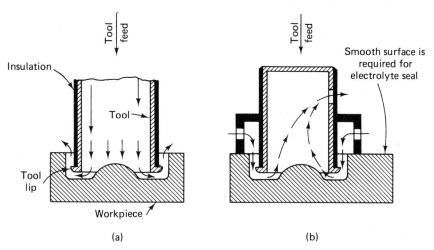

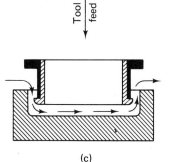

there are few "standard" cutting tools. There are, however, a limited number of electrode tool shapes which may be used to perform simple machining operations, such as hole sinking or sawing. These are generally available as off-the-shelf items from some manufacturers.

The simplest form of ECM tool is the one-piece, tube-type electrode made from titanium, copper, brass, or stainless steel. Some electrode tools have copper-tungsten tips which are attached to the electrode body by brazing or with a conductive epoxy. One of the most important considerations in designing ECM electrode tools is in the formation of electrolyte flow holes. A fundamental requirement is that the electrolyte must flow adequately and uniformly over the entire cutting area. Figure 24-5 illustrates three basic methods used to provide electrolyte flow between the electrodes: *straight flow, reserve flow,* and *cross flow.* Each of these methods is selected upon the basis of how well it accommodates the flow of the electrolyte to the shape of a specific operation being performed on the workpiece.

Except for a small tool lip, shown in Figure 24-5a, the outside surfaces of the electrode tool are insulated to minimize side cutting. Materials most commonly used for insulation include vinyl, Teflon, epoxy, high-temperature varnish, or enamels. The principal methods of applying the coatings are by spraying or dipping. A thickness of about 0.002 in. (0.05 mm) is commonly achieved.

PROCESS SELECTION FACTORS

Workpiece Materials: Practically any material that conducts electricity can be machined by the ECM process. There are very few ECM production applications for materials other than the "difficult-to-machine" alloys such as steels with high nickel and chromium contents, titanium, or beryllium. Metals such as aluminum, brass, copper, or mild steel are machined so easily by less expensive, conventional methods that they are rarely formed by ECM.

Cutting Rate: The metal removal rate in this process is directly proportional to the quantity of current used. The ECM removal rate is expressed as volume of metal removed per second. Both the theoretical removal rate and electrode tool feed rates can be calculated in advance. Machining rate is unaffected by workpiece hardness and toughness. Metal is dissolved at approximately the same rate for soft metals as for hard metals.

The ECM process is adaptable to machining almost all metals with equal ease and without regard to the traditional concerns for machinability. In this respect, the ECM and EDM processes are identical. An average metal removal rate for all metals is about 1 in.^{3}/min for every 10,000 amperes of available current.(It should be repeated that machines are currently available that will deliver up to 40,000 amperes.) It is not practical to predict more accurate metal removal rates, because of the influence of the shape of the electrode tool, the work-metal density, chemical balance, and other important factors. Compared to EDM, for example, the cutting rate is a great deal faster; in fact, ECM compares favorably to all other nontraditional stock removal processes.

Tolerances: Most ECM machine makers report that under proper tooling conditions, it is possible to produce parts with through-hole tolerances of ±0.001 in. (0.03 mm) and tolerances of ±0.005 in. (0.13 mm) for cavity-type work. A tolerance of ±0.0005 in. (0.013 mm) for each type of cutting just described is not uncommon under carefully controlled cutting conditions.

Surface Finish: The quality of the surface finish depends largely upon the current density and is generally predictable within a reasonable range. To a lesser extent, the surface finish may also be affected by the type of alloy being machined, the feed rate, gap dimension, and conditions of the electrolyte. The best obtainable finish is formed at the cutting face of the electrode tool, where the current density is highest. Finishes of 8 to 20 μin. are not uncommon on workpieces of nickel-chromium alloys, for example. Subsequent polishing operations are seldom required. Actual cases have been reported of ECM finishes down to 4 μin.

Current density drops off rapidly with the distance from the electrode tool face. Low current density results in an etched or matte-textured surface which is particularly noticeable at the sidewalls of a cut.

Thermal Effects: The resistance of both the workpiece and the electrolyte to the current flow is insignificant. Under properly controlled conditions, there can be no damaging effects to the workpiece from heat. Parts are completely free of residual stresses. ECM machining produces no significant change in the microstructure of the work metal.

Possible Limitations

Machine Tool Costs: A high capital expenditure should be expected for an ECM machine. A 10,000-ampere machine, for example, normally represents a basic investment of over $100,000. Added to this must be the cost of certain necessary accessory equipment and factory facilities, such as water, steam and electrical systems, and installation costs. The process should be seriously considered only where ECM clearly offers advantages over the metalforming methods, particularly for complicated configurations in hard and tough workpiece materials. Also, a sufficient volume of work must be available to justify the capital investment.

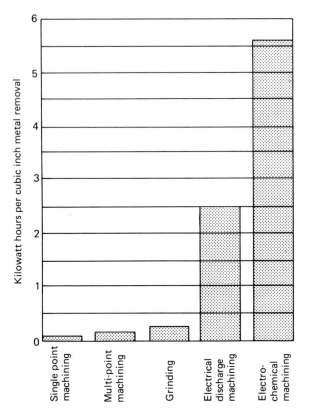

FIGURE 24-6 Power consumption of ECM compared to other machining processes

Power Requirements: In contrast to conventional, as well as to most non-traditional, machining processes, ECM consumes a disproportionate amount of electricity. Current flow in the order of 200 amperes up to 5,000 amperes /in.² (6.3 cm²) is not uncommon on some machine models. Figure 24-6 contrasts the unit power consumption requirements for ECM versus other machining processes.

Process Hazards: A cloud of corrosive spray or mist is usually associated with many ECM operations, particularly when an electrolyte solution of sodium chloride is used. Special precautions must be taken to prevent the occurrence of potentially harmful conditions, which may not only seriously affect the operators or other personnel working nearby but contaminate other machines and items in the adjacent area. For this reason, most ECM installations are safely located in isolated areas away from other machine tools, inventory supplies, or product storage facilities. Facilities such as mist collectors, exhaust systems, and sensing devices to detect leakage are standard on most ECM machines.

A potential fire hazard exists when a sodium chloride solution is allowed to soak into materials such as cloth, paper, or wood. When dry, these substances become highly combustible.

Parts Cleaning: It is recommended that the corrosive residue on workpieces be thoroughly removed as soon as practical after ECM processing. In some plants, this operation is done automatically. Most nonferrous metals and alloys may be simply rinsed in water. Steel and cast iron parts are generally cleaned with an alkaline solution and, in some cases, they may also be given a protective coating.

Malfunctions: Both the electrode tool and the workpiece are subject to damage by arcing if contact is accidentally made between the anode and the cathode. Malfunctions caused by poor spark control are most likely to occur during the preliminary setup period and while trial runs are being conducted.

Workpiece Properties: The rate of metal removal will vary slightly with the chemical properties of the workmetal. The results obtained are dependent upon the quality of the metal. Nonmetallic inclusions, for example, can lead to less than desirable results.

PRODUCT DESIGN FACTORS

Accurate results in electrochemical machining are extremely difficult to predict unless a significant number of pilot test runs have been conducted on specific product applications. Overcut, taper, and corner radius always accompany this process. These conditions are unavoidable. Size deviations, within reasonable limits, can generally be determined in advance for most standard ECM operations. To a large extent, the amount of overcut, taper, or corner radius depends upon the design of the electrode tool, penetration rate, and the gap voltage.

Overcut: Machining voltage greatly affects overcut. The overcut or sidewall tolerance may range from 0.001 to 0.030 in. (0.03 to 0.76 mm). Various precautionary measures, such as adjusting the feed rate, are often taken during production runs to control the overcut to within ±0.005 in. (0.127 mm).

Taper: As a general rule, side-wall taper may be held to 0.001 in./in. (0.001 mm/mm) of depth of the hole or cavity.

Corner Radius: A radius of 0.007 to 0.009 in. (0.18 to 0.23 mm) should be expected in *internal* corners. A minimum radius of 0.002 in. (0.05 mm) can be held on *external* corners.

Roundness: Roundness can be maintained to with 0.0005 in. (0.013 mm).

Runout: Runout is held to within 0.001 in./in. (0.001 mm/mm) of hole depth.

Workpiece Size Capacities: There are many impressive and well-documented product applications showing the versatility of the size range of work which has been accomplished by ECM. Operations such as circular and odd-shaped "drilled" holes as small as 0.030 in. (0.76 mm) up to several inches (about 75 to 100 mm) are possible. With proper tooling, a number of holes can be drilled simultaneously. The aspect ratio (length to diameter) can be as high as 180 : 1 and holes as deep as 32 in. (about 810 mm) have been produced on superalloy materials. Examples of die cavities produced with one pass of the electrode tool measuring as large as 50 in.^{2} (315 cm^2) are not uncommon.

Stock Configuration: As the process engineer prepares for the design of the raw stock, casting, or forging to be electrochemically machined, there are two main considerations affecting the initial configuration. *First,* the rough contour of the stock should conform as closely as possible to the profile of the final desired shape. Generally, the cutting face of the electrode tool is shaped to a mirror image of that shape. In operation, as the tool is advanced toward the workpiece, the machining rate is governed by the current density at the nearest point of the tool to the workpiece. With correctly applied voltage, the only significant current flow to occur initially is at this nearest or "high" point. To properly function, there must be no decrease of current density across the gap between the tool and the initial point on the workpiece. As the tool advances, the effective machining area increases, the current increases, but the current density and the voltage remain constant. When each high point on the workpiece surface has been removed, all positions along the finished surface of the workpiece are essentially the same distance from the tool. At this point, the tool contour is duplicated and the operation is complete.

A *second* important design consideration involves stock allowance, which must be *uniformly* provided so that the resulting quality of surface finish produced will be consistent across the entire area. The need for a sufficient "stock envelope" over the finished workpiece contour is shown (to a greatly exaggerated scale) in Figure 24-7.

Electrode Tool Design: Except for simple shapes, the tool will not be an exact mirror image of the shape being machined around, onto, into, or through the workpiece. Great care must be taken to design an electrode whose geometry deliberately differs by some discrete amount from that of the final shape to be machined into the workpiece. For example, the angle

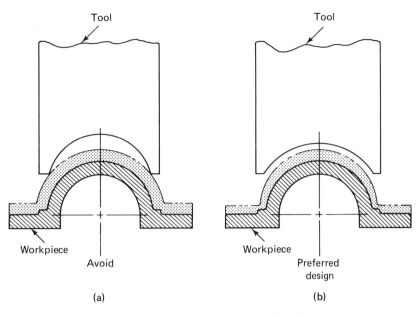

Workpiece   Workpiece

Avoid   Preferred design

(a)   (b)

Shaded area indicates "stock envelope"

FIGURE 24-7 ECM design must allow sufficient "stock envelope" over finished contour on castings and forgings

formed on the electrode tool used to produce the 30° cone-shape cavity in Figure 24-8a would actually be made to compensate at *less* than 30°. Another example where the electrode size would deviate from the desired final cavity is in the manufacture of hemispherical ball seats shown in Figure 24-8b. The shape of the electrode tool used to produce this shape would not be a true diameter.

Only the face or frontal portion of the tool actively removes metal. In practice, the tool continuously corrects the previously made cut as the depleted metal gradually forms a deeper and deeper cut onto the workpiece. Electrode tools are designed so that when the final depth is reached, the final size and finish of the cut are also reached. Tool dimensions also differ from the nominal dimensions on the workpiece to compensate for overcutting.

Sidewalls: Expensive tooling is required when the sidewalls vary more than 20° from the direction of the electrode tool feed, as shown in Figure 24-9. Parts designed with steep sidewalls are easier to machine because of the uniform electrolyte flow.

Undercuts: Because of the highly complex tooling requirements, it is rarely practical to incorporate undercuts into shapes formed by ECM. Figure 24-10 shows one method that can be used in a case where, in spite of the additional cost, it is absolutely necessary to produce a dovetail undercut.

343

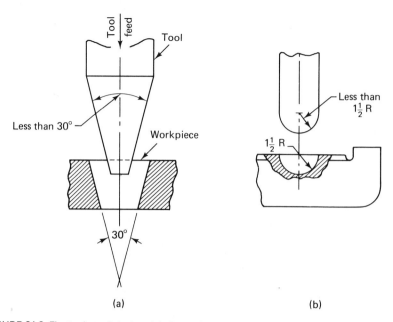

FIGURE 24-8 Electrode tool design: (a) the tool angle is made compensatingly less than the final angle of the tapered hole; (b) the shape of the tool is not a true diameter

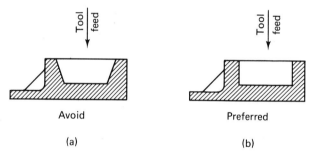

FIGURE 24-9 Sidewalls should be as steep as possible: (a) uniform sidewalls are difficult to produce if the sidewalls slope more than 20° from the direction of tool feed; (b) uniform sidewalls easily produced because of uniform electrolyte flow

FIGURE 24-10 Tooling required for undercuts: (a) required shape; (b) first operation; (c) second operation; parts held as in (b) and (c) generally require complex tooling

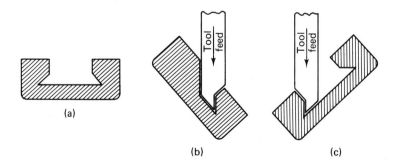

REVIEW QUESTIONS

24.1. Compare the similarities and differences of ECM and EDM.

24.2. List the principal advantages of the ECM process.

24.3. Explain why an ECM machine is characterized by a rugged frame.

24.4. Upon what important factors do successful results in ECM depend?

24.5. Describe the safeguards available with power-supply units to offset the damaging effects of a short circuit caused by the tool contacting the workpiece.

24.6. Describe the function of the electrolyte in an ECM system.

24.7. What important factors must be considered when "balancing" an electrolyte system?

24.8. Explain why there is no electrode tool wear in the ECM process.

24.9. What effect would a nick or a dent on an electrode have upon the resulting workpiece surface in the ECM process? Why?

24.10. Unlike the standard tools and cutters available for milling, most electrode tools in ECM are individually designed for specific applications. Why is this so?

24.11. Explain how the electrode tool is insulated to minimize side cutting.

24.12. How does the removal rate for hard metals compare to soft metals?

24.13. Prepare a table showing the relative removal rates, tolerances, and surface finish obtained by ECM and EDM.

24.14. Describe some of the possible process hazards associated with ECM and discuss the usual precautions taken to offset them.

24.15. Describe how "overcutting" is minimized.

24.16. Explain why it would be advantageous for the rough contour of a workpiece to be shaped by ECM to conform as closely as possible to the final finished shape.

24.17. Explain why the stock allowance must be uniformly provided on a part to be processed by ECM.

BIBLIOGRAPHY

*AARON, T. E., and R. WOLOSEWICZ, "Electrochemical Machining," *Uncommon Metalworking Methods,* Cleveland, Ohio: Penton Publishing Co., 1971, p. 44.

CHALLIS, H., "Electrochemical Machining," *Engineering,* Sept. 1974, Sup. i-xii.

*CROSS, J. A., and D. C. L. FAUST, "Electrical Energy Utilized for Chemical-Machining," paper presented at 26th Annual Machine Tool Electrification Forum, Pittsburgh, Pa.: Westinghouse Electric Corporation, June 12–13, 1962.

DEBARR, A. E., and D. A. OLIVER, *Electrochemical Machining,* New York: American Elsevier Publishing Company, Inc., 1968.

*"ECM—Electrochemical Machining," Special Report 609, *American Machinist,* Oct. 23, 1967.

"Electrochemical Deburring," *Automation,* Oct. 1974, p. 14.

"Electrochemical Grinding," *Machining Data Handbook,* 2nd edition, Dearborn, Mich.: Society of Manufacturing Engineers, 1973, pp. 711–714.

"Electrochemical Machining," *Machining Data Handbook,* 2nd edition, Dearborn, Mich.: Society of Manufacturing Engineers, 1972, p. 703.

"Electrochemical Machining," *Metals Handbook,* Vol. 3, *Machining,* 8th edition, Metals Park, Ohio: American Society for Metals, pp. 233–239.

SPRINGBORN, R. K., Editor, *Non-traditional Machining Processes,* Detroit, Mich.: ASTME, 1967, pp. 39–65.

*Abstracted with permission.

25

LASER BEAM MACHINING

Discovered in 1960, laser beam machining (LBM) has already reached the production machine status. "Laser" is an acronym derived from the initial letters of the words comprising "Light Amplification by Stimulated Emission of Radiation." Although the laser is used as an amplified light in some applications, it is mainly used as an optical oscillator for the purpose of converting energy into an intensive beam of optical radiation light. Any object absorbs light of a certain wavelength and reflects the rest. A material is machinable or weldable by the laser depending upon whether its wavelength is more or less compatible with the laser wavelength. The energy of the laser beam is concentrated onto a small area, causing the material to vaporize. Of special interest in this chapter are laser machining operations, which include drilling, scribing, cutting, and shaping. Laser beam welding is often technically effective and economically justified, particularly in situations where existing welding techniques do not work. All discussion regarding laser beam welding (as opposed to machining) will be reserved for Chapter 42.

DESCRIPTION OF THE PROCESS

Types of Lasers

Currently, there are four types of lasers: *solid-state, gas, liquid,* and *semiconductor.* All have different characteristics and are used according to the job they do best. Of special interest in materials processing are the industrial lasers: solid-state and gas.

Solid-State Lasers: These include the ruby laser, which was the first workable laser to be built. Ruby lasers (crystalline aluminum oxide or sapphire) are somewhat restricted in applications as a result of their relatively short wavelength, 0.69 micron (μm). Other recent additions to the solid-state category are the YAG (yttrium–aluminum–garnet) lasers, with a near infrared wavelength of 1.06 μm and neodymium-in-glass, also called neodymium-doped (or Nd-glass) lasers. Solid-state lasers are fabricated into rods and their ends are finished to close optical tolerances.

Gas Lasers: These may use argon or another gas but, more popularly, carbon dioxide (CO_2). While the CO_2 laser with a 10.6-μm wavelength is used to process many different metals, it is particularly adaptable to the machining of nonmetals which have very high energy absorption at the same or nearly the same wavelength. Because of the high reflectivity of metals at the CO_2 wavelength, most of the energy delivered is reflected and lost, and thus the cutting efficiency is low. It was found that by firing repetitive high-power pulses of the carbon dioxide laser beam, it was possible to penetrate the metal surface, even though only a small fraction of the radiation was absorbed. Furthermore, once the surface began to vaporize, the reflectivity dropped and the energy was absorbed effectively. The CO_2 laser is especially adaptable for working plastics that absorb readily at 10 μm.

Laser Capability

A laser is a device with the capability of producing a beam of light. Laser light can be a very intense source of power, so powerful, in fact, that light may be produced that is more than seven magnitudes brighter than the light emitted by the sun. A typical setup for laser beam machining, illustrated in Figure 25-1, can be explained in a simplified manner as follows. An initial source of 250 to 1000 watts of electrical power is used to activate tungsten-halogen or krypton-arc flash lamps located near the solid-crystal rod or the gas medium. Since ordinary light is completely incoherent (having a spectrum of wavelengths and many phases), mirrors located inside the optical oscillator or discharge tube are used to reflect and to focus the light on the laser medium. On some laser equipment, the light is focused onto the laser medium by mirror-finished, gold-plated, double-elliptical reflectors. Figure 25-2 shows some of the features of a discharge tube with its cover removed. The light energy is temporarily captured and stored. Using an "optical pumping" process, the radiation from the light source excites the electrons in the storage medium and raises them to a higher energy level. During the interval when the atoms relax back to their normal energy level, the released energy

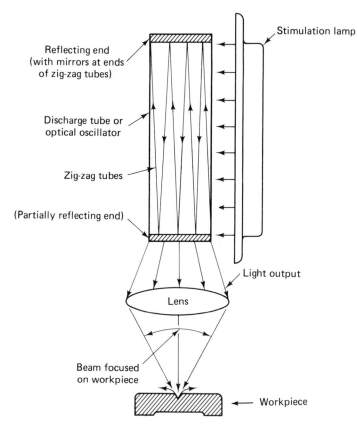

Reflecting end
(with mirrors at ends
of zig-zag tubes)

Stimulation lamp

Discharge tube or
optical oscillator

Zig-zag tubes

(Partially reflecting end)

Light output

Lens

Beam focused
on workpiece

Workpiece

FIGURE 25-1 A typical set-up for laser beam machining from a high power CO_2 laser

FIGURE 25-2 A discharge tube with its protective cover removed (Courtesy, Ferranti Electric, Inc.)

emerges as a stream of protons flowing almost completely parallel to the axis of the medium. Machining operations are accomplished by focusing the highly amplified light beam through a lens at a selected spot on the workpiece.

In summary, this phenomenal device generates, amplifies, and emits an intensely concentrated beam of coherent light energy. Once this beam of highly collimated light (*collimated* meaning: alignment of the optical axes in an optical system to the reference mechanical axes) has been released from the laser device, a controllable heat source is available which serves as the cutting tool. The collimated laser beam is focused through a prism to provide a beam spot of the precise size needed to produce a given hole diameter or kerf width, or bead size, as in welding. In LBM, both metals and non-metals are penetrated by a combination of melting and vaporizing events.

Laser Energy Modes

Two energy modes are possible, depending upon the laser medium: *pulsating* and *continuous.* The only mode possible with ruby and neodymium glass lasers is the *pulsating* mode, during which giant pulses at the rate of 1000 to 10,000 pulses per second "chip away" to generate discrete holes or lines on the workpiece. This pulsing-mode capability, called Q-switching, is available as a subsystem component on most laser equipment. The Q-switched mode develops very high power in a beam of short duration with minute amounts of energy. Briefly, the action is accomplished by temporarily reducing the Q (or quality factor) of the laser resonator which is actuated by a radio-frequency signal. With the RF signal *on,* the optical beams become scattered and, as a result, lasing is inhibited, thus allowing energy to be temporarily stored in the laser rod. Switching the RF power *off* restores the high Q-state and produces a Q-switched high-power laser pulse. The cycle is electronically synchronized during the cutting action at selected repetition rates. The Q-switched laser can be turned off on command within 250 nanoseconds (ns), thus almost instantaneously terminating the cutting action. Both YAG and CO_2 lasers emit photons in a *continuous* beam, which are used for cutting as well as for welding.

Focal Length

Lens manufacturers calculate the proper design for the lens used in LBM and fabricate them to a specific focal length. Most beams are focused, for efficient operation, on the workpiece at a distance of between 1 and 6 in. (about 25 to 150 mm). In sheet-metal cutting applications, the focal length is usually 1.5 or 3.0 in. (about 38 to 76 mm). The laser beam is easily focused to a spot as small as 0.001 in. (0.03 mm) in diameter.

Gas Jet Assist

A recent development which is now included on some LBM systems is a device for applying a continuous jet of air, oxygen, or inert gas, such as nitrogen, which is used to assist the machining process. The gases perform three functions, depending upon their types: production of an oxidizing atmosphere, maintenance of an exothermic reaction, and material removal. To more clearly explain the effects of the three functions, we shall use, as an example, oxygen as the gas and steel as the workpiece material. In the first function, the gas jet oxidizes the surface and acts to reduce the reflectivity, resulting in greatly increased light absorption. In stainless steel, for instance, the surface reflectivity may be reduced by as much as 85% by a gas jet assist. As the steel workpiece heats up from the laser beam, the jet of oxygen acts to produce a molten oxide sufficient to sustain and then increase the cutting action. It has been found that 80% of the heat required to do any cutting is provided by this important function, the exothermic reaction. Estimates have placed the temperature at the focus point of the laser beam to be about 75,000 °F (about 42,000 °C). Finally, the third function, material removal, is accomplished by the gas jet as it carries away the slag and removes the unvaporized steel from the path of the beam.

Laser Equipment

The construction of laser devices varies according to the intended use from one manufacturer to another, but each performs in essentially the same manner. Most manufacturers of laser equipment have machines that permit manual, semiautomatic, or fully automatic operation. Some LBM machines resemble a torch cutting tool (perhaps more in function than appearance). Figure 25-3 shows several types of laser machines. A typical CO_2 laser machine consists of two major components: a service module, a single unit that contains all equipment required to drive and control the laser head; and the laser head assembly. In the service module are the gas flow system, power supply, and the controls. The laser head assembly, which includes the laser tube and mirrors, is connected to the service module by an umbilical cord containing electrical and cooling lines. A separate refrigerated closed-circuit water cooling system is available as an option. Complete laser equipment generally includes a microscope for examining the workpiece, a beam-focusing device, and a movable workholding fixture.

Hand-held, portable self-contained laser systems are currently available weighing only 9 lb (about 4 kg). One such unit, shown in Figure 25-4, may be used in a wide range of applications. The laser head can be fired in a hand-held or in a stationary position. The output energy on some models is

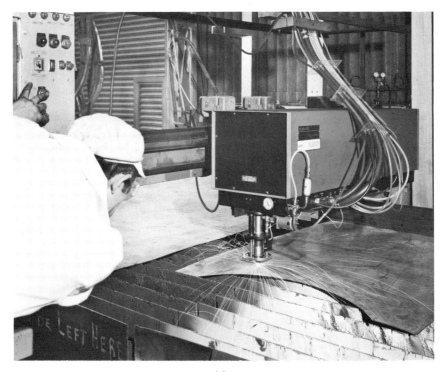

(a)

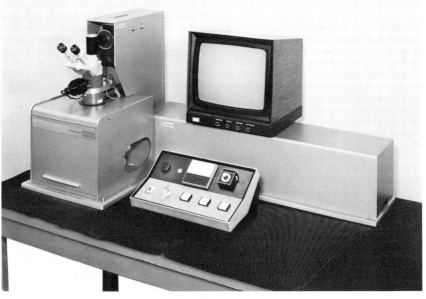

(b)

FIGURE 25-3 Types of industrial laser beam machines: (a) a CO_2 laser used to cut shapes from sheet steel (Courtesy, Ferranti Electric, Inc.); (b) a machine designed for micro-precision metal removal, scribing, welding, or drilling. Frequency doubled output can provide spot diameters down to 2 microns (Courtesy, Holobeam, Inc.); (c) this machine incorporates a high continuous power YAG laser (Courtesy, Holobeam, Inc.)

(c)

FIGURE 25-3 *(Continued)*

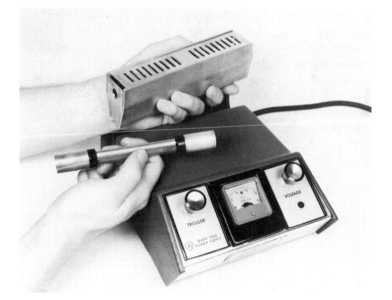

FIGURE 25-4 A portable laser system completely self-contained and permanently aligned for immediate use on a wide range of applications (Courtesy, Laser, Inc.)

353

1 joule, with an operating life of 100,000 shots or more. Desk size and even larger LBM models are also available with the capability of production hole drilling in sizes up to 0.0394 in. (1 mm) in diameter. All the necessary logic required to interface with numerically controlled, computer-controlled, and mechanical feed systems is normally included with the unit. Many laser systems, after being proved-out in process applications, may be operated and maintained by regular production personnel with minimal training.

PRODUCT APPLICATIONS

Lasers for machining purposes will certainly never completely replace some of the older, more traditional machining techniques in use today. However, because of the uniqueness of their applications, lasers have taken a commanding place in the machining market. Essentially, industrial lasers are principally confined to the category of work that might be called "micromachining," and in many industries the use of lasers for certain on-line production operations has become a well-accepted practice. As the state-of-the-art improves, the number and methods of applications also multiply. There is a growing list of applications in which the use of the laser is the only practical way some jobs can be done. LBM has found successful applications in communications, chemistry, and the military, as well as in many component-producing industries. Figure 25-5 illustrates some typical examples of industrial LBM applications.

Drilling represents an extremely large potential field of application for LBM. A complete listing of laser drilling applications would be virtually impossible to compile. The following examples will show the range of applications: holes in rubber baby bottle nipples, vacuum holes in electro-formed dies; holes in thin-wall fiberglass tubes, in ruby watch bearings, and in expanded foam parts; relief holes in pressure plugs, retainer-wire holes in brass nuts, holes in nylon buttons, in aerosol spray or particle-emission nozzles, and in surgical and hypodermic needles; flow holes in oil or gas orifices; and drilled-and-*sealed* holes in Zircaloy-4 nuclear fuel rods.

The versatility of the laser as a machining tool can be further illustrated by the following list of other applications.

1. In the garment industry: the cutting of patterns from single-ply or multilayer stacks of fabrics can be fully automated and accomplished at a rate of up to 2,400 in. (61 m) per minute.
2. Dynamic balancing of gyrorotors, watch or clock balance wheels, or precision rotating parts: high-speed balancing is accomplished by vaporizing away excess metal as the part rotates on the balancing machine. Scribing (or micromachining) lines 5 to 10 μm wide on microcircuit wafers: the cutting rate approaches 2 to 6 in. (about 50 to 150 mm) per second on materials including silicon, sapphire, and alumina.

FIGURE 25-5 Typical examples of LBM applications (a) tin plated steel cans 0.007 in. thick are "trued" to size with oxygen assisted continuous YAG laser at 18 in./sec. (Courtesy, Holobeam, Inc.); (b) laser ceramic cutting produces a smooth, fused square edge leaving no flakes to break off at a later date. Precision straight, contoured and circular cuts are possible (Courtesy, GTE-Sylvania); (c) small washers are cut with ease from 0.020 in. thick tungsten at 80 in./min. Kerf width is less than 0.030 in. (Courtesy, GTE-Sylvania); (d) metal serializing with continuous wave YAG laser (Courtesy, Holobeam, Inc.); (e) laser application for precision register trimming (Courtesy, Holobeam, Inc.); (f) a stainless steel orthodontic dental brace welded with a pulsed YAG laser (Courtesy, Holobeam, Inc.)

(a)

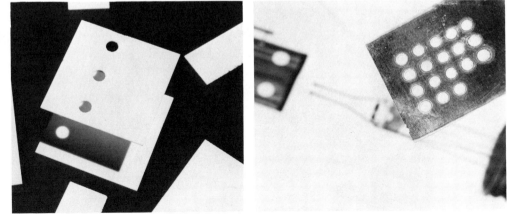

(b)   (c)

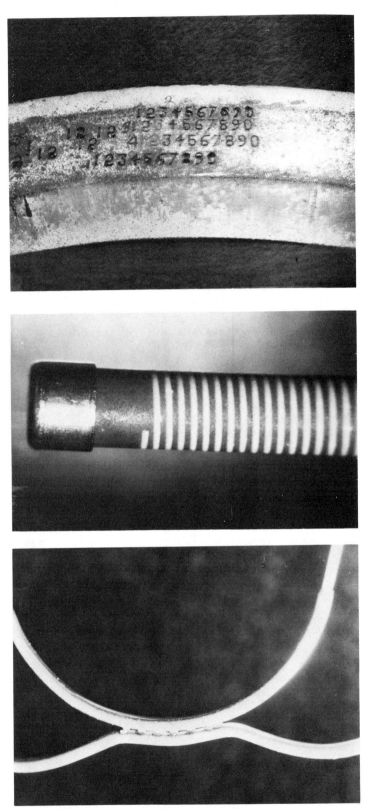

FIGURE 25-5 (*Cont'd*)

3. Insulation stripping of wire: bundles of braided wire used in aerospace and computer installations can be laser-stripped without uncoiling each wire. The metal remains uncontaminated; no nicks, cuts, or scratches are produced.

4. Surface hardening: one company uses the laser beam to case-harden surfaces of cast-iron ring grooves, camshafts, and diesel engine valve seats. Figure 25-6 shows the transformed structure of heat-treated gray cast iron.

5. Trimming flash from plastic parts: resistor "trimming" has been one of the most successful automated applications of lasers. "Trimming" in this case refers to changing the resistance of a carbon resistor by striking it with a laser beam.

6. Photo-etching: fast, accurate, and inexpensive.

7. Measuring and testing: laser beams have been used to check machine-tool indexing toolheads to accuracies of ± 0.00001 in. (0.000254 mm). It is often necessary to precisely align a series of mechanical parts in such a manner that some reference point on each part lies on the same basic straight line. Examples include aircraft structure component assembly jigs, assembly of turbine generator parts, or installation of the various components of production welding fixtures. Laser beam alignment devices are available which will operate along distances ranging from a few inches (about 75 mm) up to 100 ft (approximately 30 m) to accuracies within 10 μm per foot or 0.001 in. in 100 ft (0.03 mm in about 30 m). Straightness and flatness of machined surfaces such as beds and ways on large machine tools may be checked by laser beam techniques. Accuracy requirements in the installation of long lead screws may be readily maintained by laser equipment. A recently perfected

FIGURE 25-6 Heat treated gray cast iron. The picture shows a complete transformation of pearlite to martensite and bainite. Slight melting is also evident. Magnification is 1600X (Courtesy, GTE-Sylvania)

laser measurement technique is the development of a device that precisely measures small deep circular holes for diameter, roundness, locations, angularity, and taper.

8. Flaw detection: it is possible to detect fatigue cracks, inclusions, or other defects on a variety of materials by a process known as "laser holographic interferometry."

9. Marking or engraving: numerically controlled or computer-controlled data in the form of coded information or special numbers may be applied to various workpiece materials.

10. Cutting operations: battery plates of lead-alloy composition can be cut as fast as 20 in. (or about 500 mm) per minute. Precious stones may be cut and shaped by laser methods.

11. Impurity removal: in the jewelry industry, objectional black carbon deposits in diamonds may be removed by drilling holes as small as 0.001 in. (0.03 mm) in diameter up to depths of 0.050 in. (1.27 mm).

Nonindustrial Applications

There are three major areas of experimental medical applications of lasers:

1. Experimental medicine and surgery: laser injury of the atrioventricular node results in cardiac block similar to that which may follow coronary infarction. Bone fractures may be "spot-welded" with the high-temperature CO_2 laser.

2. Microbiology: a laser beam fine enough to drill a hole through a wire no larger than a human hair is used to remove part of the mitochondrion from a living myocardial cell, and to alter a single chromosome, both without killing the cells.

3. Analytic and diagnostic procedures: lasers are already being used to monitor blood oxygen during bypass surgical procedures. Thanks to recently developed "tunable" lasers, very rapid *in vivo* qualitative and quantitative analyses of circulating alcohol, acetone, urea—even bacteria—may soon be practical.

Lasers have also been successfully used in clinical applications in dermatology (to remove birthmarks, tatoos, melanomas), opthalmology (photocoagulation of affected vascular branches in diabetic retinopathy), "spotwelding" of retinal perforations and tears to prevent detachment of the retina, and surgery (ablation of brain tumors with the so-called "laser knife").

Another nonindustrial application of laser beams of interest is in perforating as many as 220 minute holes in contact lenses so that fluids can circulate freely in the eye, permitting the eye itself to "breathe" beneath the plastic lens.

PROCESS SELECTION FACTORS

Cutting Quality: The most important parameters that affect cutting quality are laser power, feed rate, nozzle design, and the type and conditions of the gas jet assist. Edge conditions are discussed later in this chapter under the heading Possible Limitations.

Tolerances: Dimensional accuracy in the order of ±0.001 in./in. (0.001 mm/mm) can be obtained under closely controlled conditions.

Corner Radii: A minimum corner radii of 0.010 in./in. (0.010 mm/mm) should be expected.

Materials: It is significant that *all* metals and nonmetals may be vaporized by this process. High-strength, exotic, and refractory alloys such as Inconel, René 41, Hastelloy, tantalum, zirconium, and titanium may be machined. Other representative metals include stainless steel, aluminum and brass. Prominent nonmetals include diamond, ceramic, vinyl, rubber, plastics, wood, cloth, and paper.

Aspect Ratio: The ideal depth-to-diameter ratio, the aspect ratio, is about 4 : 1, but holes as deep as 12 times the diameter have been successfully laser-drilled. As shown in Figure 25-6, holes may be conveniently drilled at an angle to a surface or on curved surfaces without special preparation. Laser-beam-drilled holes may often be produced in inaccessible locations.

Range of Operations: LBM is especially adaptable and principally used for relatively small materials-processing applications, such as cutting, trimming, scribing, piercing, drilling, or other delicate material removal operations similar to milling or shaping.

Minimum Hole Size: Hole diameters of "micro" size can readily be produced.

Burrs: Laser-drilled holes are burr-free.

Workpiece Deformation: Laser machine tools never touch the material being worked, and, as a consequence, there is no possibility of deforming the workpiece from tool pressure. Any distortion due to heat buildup is minimal and, except in special cases, may be neglected.

Production Rate: Currently, there are LBM units in operation that process millions of parts per week in practically every area of manufacturing.

Cutting Rate: Table 25-1 shows typical cutting rates for a 1000-watt CO_2 laser. The kerf width ranges between 0.002 in. (0.05 mm) and 0.005 in. (0.13 mm), depending upon the workpiece material and the thickness involved.

359

TABLE 25-1 Typical CO_2 laser cutting rates

Metal thickness	Stainless steel	Cutting rates (in./min.)		
		Aluminum	Galvanized steel	Titanium
0.020	750	800	250	—
0.032	650	—	—	—
0.040	550	350	100	250
0.062	450	200	50	150
0.080	325	100	—	100
0.125	200	—	—	—

Compared to friction sawing, for example, laser cutting speed is estimated to be as much as 30 times faster.

Drilling Rate: Less than 1 ms is required to drill each of these holes:

Material	Thickness	Hole diameter
Tungsten	0.020 in. (0.51 mm)	0.020 in. (0.51 mm)
Ceramic	0.101 in. (2.57 mm)	0.050 in. (1.27 mm)
Brass	0.010 in. (0.25 mm)	0.250 in. (6.35 mm)

Possible Limitations

Edge Condition: LBM, when used in cutting applications, usually results in a series of striations or ridges along the workpiece edge. This condition is the result of metal "freezing" as the heat source passes. The ridges are at a right angle to the surface. There is, also, a small amount of slag present on the bottom edge, but most is blown away by the gas. The depth of the ridges is dependent upon the feed rate. As a general rule, the condition of the cut edge is not detrimental to the appearance of the workpiece product and, in fact, the cut edges are generally superior to those obtained by torch, plasma arc, or nibbling.

Removal Rate: When comparing the volume of stock removal to any other machining process, the LBM removal rate of 0.0004 in.^{3} (0.00656 cm^3) per hour is by far the slowest. Also, power requirements tend to be higher. As a general rule, LBM consumes about three times the amount of power to remove the same volume of tungsten as it does aluminum. Important factors affecting the removal rate are the specific heat and heats of fusion and vaporization for each work material.

Out-of-Roundness: Hole roundness is usually a problem, but it can be controlled to fall within specified tolerances for small holes less than 0.030 in. (0.76 mm) in diameter, for example.

Taper: As a result of the blasting effect of the laser, sidewalls of holes produced by LBM may taper from entry to exit. The taper measured along one side may be as much as 0.050 in./in. (0.050 mm/mm) on holes deeper than 0.010 in. (0.25 mm).

Production Rate: In some specific cases, because of the requirements for precise positioning of the cut on the workpiece or other factors, the production rate may be relatively slow. However, recent advancements in work-holding and positioning fixtures, and improvements in other processing techniques as well, have made LBM more competitive. It is currently possible, for example, to preprogram and automatically drill small holes in thin stock at the astonishing rate of 50 per second.

Maximum Hole Size: As a practical matter, *circular* hole-drilling operations are restricted to holes not over 0.125 in. (3.18 mm) in diameter in stock thickness not to exceed 0.500 in. (12.7 mm).

Heat Effect: LBM operations result in a small heat-affected zone on workpieces, which, except in special cases, may usually be disregarded. One leading manufacturer of LBM equipment reports that the heat-affected zone "seems to be limited to about 0.001 in. (0.03 mm) or less on either side of the cut."

PRODUCT DESIGN FACTORS

Materials Selection: In transferring laser beam energy into the workpiece material, many factors related directly to the material are of concern. In fact, the effectiveness of the laser as a machine tool is a function of the workpiece surface itself. Because of the influence of the material parameters upon the ultimate job success, designers often confer with technical staffs in laser system companies. Many laser firms operate laser laboratories and invite potential customers to send in materials for testing. Such services give the potential user all the information that is needed to determine whether or not the laser is the answer to the user's production requirements without the necessity for investing thousands of dollars in laser equipment. Many manufacturers of industrial lasers also offer job-shop services.

Flammable Materials: Some materials burn, char, and bubble when exposed to the intense thermal energy released by the laser, and as such are not suitable for laser processing. Certain epoxy-laminated materials, phenolics, and some vinyls are difficult to process by laser beam.

Safety Hazards: Of principal concern is the damage potential to the eyesight of the high-intensity light of the laser. The visible ruby laser beam is extremely dangerous to the eye, since it can penetrate the cornea and do retinal damage. The infrared gas laser beam is, of course, not visible and, because of this, is considered even more dangerous. OSHA (Occupational Safety and Health Act) has adopted the ANSI 2-136.1 1973 standard (Safe Use of Lasers) and has incorporated it into the broader OSHA regulations. In addition to the necessity for wearing protective eyewear equipment, safeguards such as special laminated plastic viewing windows and plate glass filters are also necessary. When considering all of the possibilities, there is little danger in LBM if proper precautions are taken.

Cost: In spite of a downward trend in costs for capital investment and operation, LBM equipment continues to be more expensive than comparable conventional equipment. CO_2 laser beam machines range in cost from about $25,000 for a 70kW to 1.5 megawatt, 33 watt system to well over $100,000 for the more sophisticated high-power models. Solid-state lasers cost somewhat less than these amounts.

REVIEW QUESTIONS

25.1. In what form are solid-state lasers fabricated?

25.2. Explain why the CO_2 laser is particularly effective for machining nonmetals.

25.3. Prepare a brief outline explaining how laser light energy is used as a cutting tool.

25.4. What is the essential difference between the two laser energy modes?

25.5. Explain the three functions of a gas jet assist.

25.6. List the important advantages of LBM.

25.7. Compare the dimensional accuracy achieved in LBM to EDM and ECM.

25.8. Compare the removal rate achieved in LBM to EDM, ECM, and chemical milling.

25.9. Describe an important limitation of LBM which occurs on the sidewalls of holes.

25.10. Give some important precautions that must be observed when operating laser beam equipment.

BIBLIOGRAPHY

BELFORTE, D. A., "Laser Metal Cutting," *Journal .of the Fabricator,* Vol. 4, No. 3, May/June and July/August, 1974.

CHARSCHAN, S.S., editor, "Lasers in Industry," New York: Van Nostrand Reinhold Company, 1972.

COOK, N. H., *Manufacturing Analysis,* Reading, Mass.: Addison-Wesley Publishing Company, Inc., 1966, pp. 127–133.

HARRY, J. E., *Industrial Lasers and Their Applications,* New York: McGraw-Hill Book Company, 1974.

IRVING, R. R., "Metalworking Lasers: Their Time Has Come," *Iron Age,* Sept. 9, 1974, p. 51.

"Laser Beam Machining," *Metals Handbook,* Vol. 3, *Machining,* 8th edition, Metals Park, Ohio: American Society for Metals, p. 225.

"Laser Makes and Fills Holes," *American Machinist,* Mar. 19, 1973, p. 49.

Lasers for Industry, Catalog, Sturbridge, Mass.: Laser, Inc.

LAVOIE, F. J., "The Blue-Collar Laser," *Uncommon Metalworking Methods,* Cleveland, Ohio: Penton Publishing Co., 1971, p. 35.

Model 971 High Energy Laser Processor, Catalog, Mountain View, Calif.: GTE Sylvania-Electronic Systems Group.

MORTIMER, J., "NC Laser Cuts Steel, Wood and Plastics," *Engineer,* Nov. 14, 1974, p. 13.

Non-traditional Machining Processes, Dearborn, Mich.: Society of Manufacturing Engineers, 1975.

SPRINGBORN, R. K., editor, *Non-traditional Machining Processes,* Detroit, Mich.: ASTME, 1967, pp. 147–158.

*"There's More to Lasers Than Light," *Iron Age,* Oct. 14, 1971.

WEINER, M., "Lasers Begin to Shine as Processing Tools," *Automation,* Dec. 1974, p. 30.

*Abstracted with permission.

363

26

ULTRASONIC MACHINING

Ultrasonic machining (USM) is essentially an abrasion or erosion process. Its principal applications are in milling, drilling, surface grinding, threading, or lathe-type operations. On hard, brittle materials such as ceramic, germanium, and glass, unlike the ECM process, electrical conductivity of the workpiece material is unimportant. The process is regarded as competitive *only* when an operation cannot be practically and economically performed on conventional machining equipment. The ultimate value of USM lies in the ability to do work that cannot be practically accomplished in any other way.

DESCRIPTION OF THE PROCESS

Techniques

There are two basic techniques currently in use. The *first* employs a nonrotating, but vibrating, metal tool. The tip of the tool, which never touches the workpiece, is bathed in an abrasive slurry–water mixture which serves as the cutting medium. The abrasive grains, activated by the subtle but persistent vibratory action of the tool, continuously work to erode away small particles of the workpiece. This technique is sometimes referred to as "impact machining." A *second* technique involves a rotating—vibrating diamond-tipped cutting tool which cuts in a way similar to a twist drill or

364

milling cutter. No abrasive slurry is used. The cutting action is entirely the result of the sharp cutting edges on the tool itself as it intermittently contacts the workpiece.

The principle of operation of the first technique in which the nonrotating tool is used is explained as follows. The cutting tool vibrates with small amplitudes of about 0.0025 to 0.003 in. (0.06 to 0.08 mm) and at high frequencies of electrical energy, customarily up to 20,000 to 30,000 hertz. Frequencies in this range produce a sound too high to be heard by the human ear. A stream of slurry, or abrasive powder (normally boron carbide, silicon carbide, or aluminum oxide), suspended in a liquid medium, is directed at the tool tip. The particles in the liquid are impacted off the tool face and are driven against the workpiece with an accelerated force, estimated to be approximately 150,000 times the force of gravity. Actual cutting conditions are brought about by the bombardment of these abrasive particles on the workpiece.

In the second technique, the principle of operation involves a rotating diamond tool having a form that resembles a grinding wheel, toothed cutter, or twist drill to which are imparted ultrasonic oscillations along its axis. The cutting tool has a shape conjugate to the required hole or cavity on the workpiece. The action is basically a modified grinding process. Unlike the technique that employs the nonrotating tool, no abrasive slurry is used. The diamond-tipped tool used in this process barely contacts the workpiece during the intervals between vibrations. Each cutting grain on the tool has lateral motion as well as its normal circumferential motion. This lateral motion does two things: (1) it reduces friction so that the tool cuts more freely, and (2) it cleans the residue away so that the wheel cuts aggressively.

Resonance

All objects may be made to vibrate, but the greatest amplitude of vibration for each object occurs at a specific vibratory frequency. When an object is vibrating at this greatest amplitude, it is said to be in *resonance* or that it has reached its *resonance point.* In operation, the ultrasonic tool must be driven at a frequency that causes it to resonate, or the amplitude of the vibrations will be too small to impart proper cutting force to the abrasive particles. While tooling engineers can design an ultrasonic tool to resonate well within the frequency range of the frequency generator with which it is used, the actual resonance point must be determined after the tool is made and installed in the machine. At this time, the frequency generator on the machine is dial-adjusted until the tool resonates or vibrates through its maximum amplitude.

Several techniques are used to determine if a tool is resonating. One technique employs a microscope with a cross-hair reticle so that the amplitude of vibration can be measured. Another employs a dial guage that measures cutting speed. Still another technique is to use a small electric meter to measure the current flow in the transducer coil, which varies sharply when the tool reaches its resonant point. In some instances, experienced ultrasonic machine operators use a simpler method—merely listening to the audible subharmonics being generated in objects on or around the machine, which are activated by the ultrasonic motions of the tool. When these subharmonics are loudest, the tool is resonating.

USM Machines

All ultrasonic machines, in spite of other wide differences, have facilities that permit the tool to be fed down onto the workpiece and a worktable capable of positioning and controlling movement in three directions. One type of USM machine which employs a nonrotating tool is shown in Figure 26-1. It consists of a machine frame with a slurry pot, a power-supply unit, an acoustic tool head to which is attached the cutting tool, and a circulating

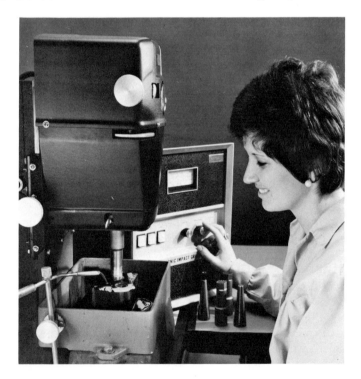

FIGURE 26-1 A pedestal unit ultrasonic machine with a slurry abrasive work tray. The operator is adjusting the drive unit to the right (Courtesy, Raytheon Company)

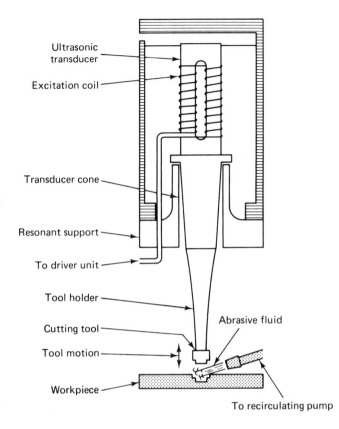

FIGURE 26-2 Schematic diagram shows an ultrasonic transducer or "transmission line" which converts electrical power into mechanical power (Courtesy, Raytheon Company)

pump. The machine frame, with its adjustable table, resembles a conventional milling machine. A major working part of any USM machine is the acoustic head, in which the main function is to convert electrical energy into mechanical energy, resulting in tool vibrations at ultrasonic frequencies. A typical "transmission line" for sending sound-wave energy to the cutting tool tip is shown in Figure 26-2. The action may be briefly explained as follows. A high-frequency current developed by the audio-frequency generator, in the power unit, is fed into a magnetostrictive ultrasonic transducer. The transducer consists of an excitation coil wound around a laminated nickel core. Many metals, particularly nickel, contract and expand in response to the influence of a rapidly alternating current. These vibrations are linearly transmitted and further amplified at a Monel connecting body, which in turn relays the oscillations to the cone-shaped tool holder. Essentially, all these parts collectively react along one axis as a single, continuous elastic body carrying and magnifying the vibrations (by resonance) to the tip of the cutting tool. USM machines are equipped with a feed mechanism that controls the linear motion of the acoustic head. To insure the accuracy needed in working to specified tolerances, the feed mechanism has slides that are both precise and sensitive.

Figure 26-3 illustrates a typical rotary ultrasonic machine. The main components are a drive-mechanism (pneumatic or electric motor) power unit, an acoustical head with a rotary transducer, and a diamond-tipped tool. No circulating pump or slurry pot is needed. The cutting-tool spindle operates at infinitely variable speeds over the range 0 to 5000 rpm. The speed of rotation is a dialable function, with most jobs being dialed at ap-

FIGURE 26-3 A universal ultrasonic machine tool for machining of glass, ceramics, ferrites, and other hard, brittle materials (Courtesy, Branson Sonic Power Company)

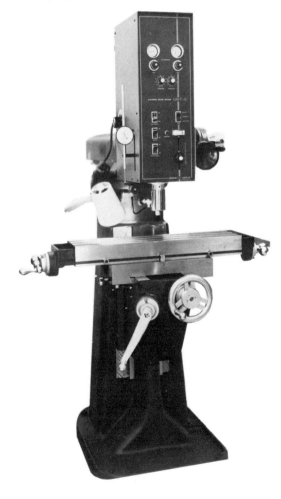

proximately 3000 rpm. The link between the spindle and the tool tip is a cone-shaped tool holder or "horn" (shown also in Figure 26-2, page 367. The end of the horn is fitted with a threaded adaptor to which the diamond tool is fastened. A tight fit between the two is essential to provide a good acoustical coupling and thus prevent the dissipation of energy in the form of heat. The weight of the tool is a critical factor in the horn design. Currently, the practical limit is about 1 ½ oz (approximately 42 g).

Also available from one manufacturer is a multipurpose ultrasonic machine with a special head which can be installed with a rotating diamond tool and can also be locked and used with nonrotating metal-cutting tools. In the later case, slurry may be fed to the tool by hand or with a spatula or squeeze bottle.

In some instances, USM machine manufacturers mount a machining head on special machine bases supplied to them by the customer, or they may deliver only the head, which the customer will mount on one of his own machines. USM heads have been mounted and used successfully on numerically controlled milling machines, for example. In another case, a user mounted an ultrasonic head on a flame cutter in place of the torches normally used. Ultrasonic heads have also been adapted to lathe-tool-post setups for specialized lathe-type manufacturing operations.

USM Cutting Tools

Figure 26-4 illustrates an ultrasonic machine and a close-up of the non-rotating ultrasonic horn used to produce the design shown on the glass ashtray. Figure 26-5 shows mother-of-pearl cufflinks machined with the tool head shown.

Nonrotating tools are of two-piece construction and usually consist of a tool holder to which a specially shaped toolface is attached by soldering or brazing. A high-quality bonded joint is required to prevent deterioration of the amplitude of vibration across the joint. The tool holder may be machined from titanium alloy, Monel, aluminum, stainless steel, or various other metals. Important factors to be considered when selecting the material for tool holders include conductivity, how well a material can be brazed, and fatigue properties. Diamond tool faces may be used as shown in Figure 26-6, or they may be produced from a variety of metals, including tool steels, annealed stainless steel, cold-rolled steel, brass, and copper. USM tools must be ductile and tough rather than hard. Important factors to consider when selecting a tool-face material are hardness of the workpiece material, quantity of parts to be produced, and the tolerances required. Compensation for tool-face wear and erosion resulting from the harsh action of the abrasive particles may be made by machining or undercutting the area surrounding the actual cutting surface off the tool face to a suitable depth.

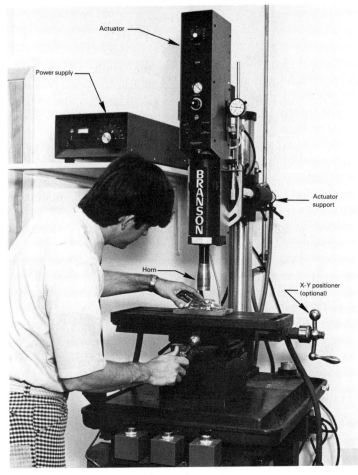

(a)

(b)

FIGURE 26-4(a) Picture shows operator producing an engraved glass ash tray on an ultrasonic machine (Courtesy, Branson Sonic Power Company); (b) close-up of the ultrasonic horn and the design on the ashtray (Courtesy, Branson Sonic Power Company)

FIGURE 26-5 Toolhead and cufflinks produced on an ultrasonic machine tool. Gemstones such as jade, agate, spinel, onyx, and others can be easily tooled by USM (Courtesy, Raytheon Company)

FIGURE 26-6 Diamond tools are available in a variety of shapes and sizes for ultrasonic drilling, milling and machining operations. Core drills are available in standard diameters up to 1 ½ in.; miniature drills from 0.020 to 0.062 in. diameter; and machining wheels up to 1 ½ in. diameter (Courtesy, Branson Sonic Power Company)

For example, in Figure 26-7 the original cutting-surface tolerance on USM tools can often be restored by regrinding as the wear occurs.

Rotating diamond tools are made in a variety of shapes and sizes. Miniature drills in sizes ranging from 0.020 to 0.062 in. (0.51 to 1.57 mm) in diameter may be conveniently obtained from suppliers as well as standard diameters of core (hollow) drills up to 1½ in. (38.1 mm). All tools are tuned to the proper frequency and are guaranteed by the manufacturer to withstand vibrations during normal use. Tool wear is negligible since no abrasive slurry is used.

In ultrasonic machining, a fast, smooth cutting action is possible. There is a minimum of friction, resulting in only a slight temperature increase between the tool and the workpiece. Low friction is principally due to light tool pressure of the longitudinal motion of the tool, normal circumferential motion, and the fact that adjacent surfaces are either barely touching or not touching at all during the intervals between vibrations. A coolant is used to produce a self-cleaning effect, which eliminates binding and loading the diamond matrix, resulting in efficient cutting action. Some USM systems employ sensing devices in which the ultrasonic power load may be used to sense and adapt to conditions at the cutting-tool tip.

FIGURE 26-7 Because the wear rate is greater along sharp cutting tool edges and on small radii, periodic re-sharpening is necessary. A tool's usefulness can be greatly extended by leaving ample stock for re-sharpening requirements

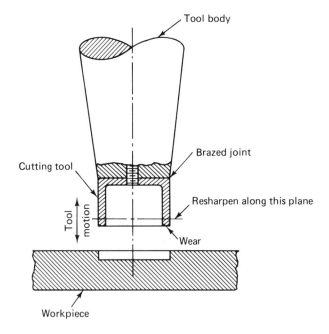

Workholding Methods

Workpieces must be held as rigidly as possible. Vibration must be prevented to achieve maximum efficiency in penetration. For nonrotating ultrasonic machining, the workpiece is securely clamped to the worktable inside the slurry pot and then positioned under the tool. Small or thin workpieces are often cemented to a glass or metal base, which, in turn, is clamped to the worktable. After processing, the cement is dissolved, thus disattaching the parts from the base. When USM is used to produce holes, the workpiece is usually rotated to improve uniformity. Some workholders have provisions for moving the part back and forth as in a slicing application, for example. In addition, indexing fixtures are also available which permit a series of machining operations on parts at staggered or at regular intervals.

PRODUCT APPLICATIONS

As in the case of most new processing techniques, users are reluctant to be identified or to provide much information or precisely how they are using ultrasonic machining. Most users feel that the process has given them a competitive edge that they would prefer not to discuss. USM is a process that has proved to be useful for production work as well as for prototype work.

Materials used in high-temperature applications in laboratories, in aeronautics, and in the missile field are ideal candidates for specialized ultrasonic machining processes. The process is commonly used in the electronics industry for the production of very small diced germanium crystals used in the manufacture of semiconductors and transistors. Another application for USM is in machining ceramic fuel-injector nozzles, which give longer life than those made from steel and are not subject to corrosive attack at high temperatures. Another important application of USM is in shaping solid or cored carbide seats in dies used for extruding metals or in wire drawing and sizing. USM has also been effectively used to produce special ornamental effects on glass products.

One particularly innovative application of the process is in the production of external threads on nonmetallic parts by holding the part to be threaded in a motor-driven rotary chuck. Incorporating a special lead screw, the chuck is raised or lowered a distance equivalent to the pitch for every revolution. Figure 26-8 illustrates a variety of USM work-related applications.

FIGURE 26-8 (a) Small hole drilling and threading alumina (Courtesy, Branson Sonic Power Company); (b) holes drilled in ferrite by USM (Courtesy, Branson Sonic Power Company); (c) carbide die blocks produced by USM (Courtesy, Raytheon Company); (d) various examples of parts machined by USM in hard, brittle materials such as alumina, ceramics, ferrites, porcelain, glass, and boron-tungsten laminates (Courtesy, Branson Sonic Power Company)

(a)

(b)

(c)

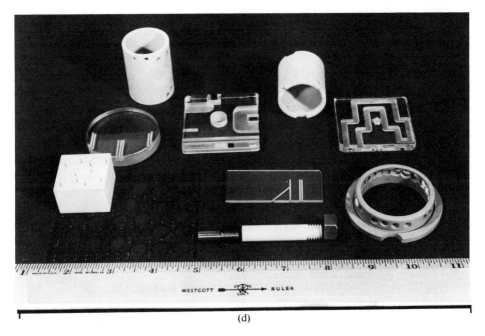

(d)

FIGURE 26-8 (Cont'd)

PROCESS SELECTION FACTORS

Workpiece Effects: Because USM is essentially an erosion process, with little or no heat involved, the metallurgical, chemical, and physical properties of the workpiece material are not altered in any way. Because of the low friction and tool pressure—the force rarely exceeds 10 lb (4.54 kg)—components may be drilled or otherwise processed without chipping. Delicate, brittle components may be machined without a danger of cracking or of creating thermal-stress problems.

Surface Finish: The process can produce finishes of 10 μin. depending upon the workpiece material, size of the abrasive particles, tool amplitude, and finish of the tool face. An optimum finish on a given surface may be produced by using two or more cutting tools while gradually reducing the abrasive grit size. It is possible to obtain a 15 μin. surface finish, for example, by using an 800 abrasive grit, and a 250 grain size produces a 25 μin. surface finish. The use of a fine grit results in a slower rate of stock removal. As in conventional machining, USM tools are designed for roughing, semi-finishing, and finishing operations.

Tolerances: Precision depends to a large extent upon the size and finish of the tool that is used. Tolerances within ± 0.0005 in./in. (0.0005 mm/mm) can be obtained using special precautions, but a tolerance of ± 0.001 in./in. (0.001 mm/mm) is considered more practical. In a recent report describing a related research project, it was disclosed that four 0.095-in. (2.413-mm)-diameter holes were required through the sides of a 7½-in. (190.5-mm)-

375

square block of glass. The accuracy of intersection of the holes drilled by USM techniques at a depth of 6½ in. (165.1 mm) was attained within 0.005 in./in. (0.005 mm/mm).

Aspect Ratio (Diameter-to-Depth Ratio): As a practical matter, an aspect ratio of 5:1 can be routinely achieved. With special precautions, however, it is possible to do even better. One exceptional case has been reported where an aspect ratio of 1:200 was achieved whereby 0.062-in. (1.57-mm)-diameter holes were drilled into solid glass to a depth of 12 in. (304.8 mm). The procedure involved using three different drills, each penetrating the material to a successively deeper dimension. A starting hole was first drilled about ½ in. (12.7 mm) deep. The hole was then drilled with a second drill to a depth of 6 in. (152.4 mm), followed by a third drill to the final depth. Hollow diamond core drills were used, through which flowed a steady stream of water. The drills used in this application rotated at 4500 rpm.

Sharp Corners: A finishing tool must be used if sharp corners are required. Corners have a tendency to chip at the exit end of holes unless the workpiece is cemented to a backing material (using the method described previously).

Intricate Detail: The clarity and accuracy of fine detail required in coining or in fine engraving operations may be improved by adjusting the vibratory stroke of the cutting tool to a minimum setting. Unnecessary damage to the cutting-tool face may also be prevented by shortening the stroke.

Removal Rate: Important factors that affect the volume of material removed per unit of time are the size and nature of the abrasive grit used, and the hardness density and brittleness of the workpiece material. As an example, the removal rate is the highest when coarse abrasive particles are used on brittle materials. Slow material removal rates are generally associated with high wear rates.

Hole Sizes: Drilling operations can be accomplished on workpieces in sizes ranging from 0.040 to 1½ in. (1.02 to 38.1 mm) in diameter, but holes as small as 0.003 to 0.006 in. (0.08 to 0.15 mm) in diameter have been successfully produced on experimental work. Larger holes (over 1½ in. in diameter, for example) can be cut by trepanning or by rotating the workpiece.

Drilling Rate: Table 26-1 shows typical cutting rates for USM drilling.

Maximum Machining Size: At the current stage of development, the USM process is limited to cuts on workpieces of not more than 3 in. (76.2 mm) across. For larger cuts, which demand a larger tool, it is better to use two or more tools in successive setups, with the tool faces correlated so that one cut

TABLE 26-1 Typical USM cutting rates-drilling
(Courtesy, Branson Sonic Power Company)

Material	Diameter of drill (in.)	Depth of cut	Approximate time required (sec.)
99.9 Alumina	0.042	0.250	110
	0.125	0.250	14
	0.250	0.250	11
	0.375	0.250	16
	0.500	0.250	18
Glass	0.042	0.500	75
	0.250	5.000	130
Ferrite	0.080	0.250	23
	0.125	0.250	19
	0.250	0.250	13
Boron	0.500	0.700	48
composite	0.250	0.500	19
	0.125	0.500	26

takes up where the other one leaves off. In this case, it would be necessary to reposition the workpieces between successive cuts.

Work Materials: While USM can be applied to ductile materials such as soft steel, copper, and brass, it is best suited to machining operations on hard, brittle materials that are not practical (or possible) to process by other methods. In general, USM is not recommended on steel that is softer than Rockwell C45. Ultrasonic machining can be used for metals and nonmetals, conductors or nonconductors.

Possible Limitations

Removal Rates: USM is not competitive with conventional machining operations on the basis of stock removal rates.

Sidewall Taper: It is not possible to ultrasonically machine parallel walls in deep holes or cavities. As an example, the taper may exceed 0.005 in./in. (0.005 mm/mm) of depth on some operations. Sidewall taper exists because of two conditions: first, there is tool wear, which is greatest at the lower end and along the sides of the tool face. Second, a condition known as "secondary impact" occurs in the gap between the tool and the workpiece. These conditions are represented in the shaded areas illustrated in Figure 26-9. The amount of taper associated with a given operation may generally be determined by trial runs. In most cases, taper may be almost entirely eliminated by using a finishing tool.

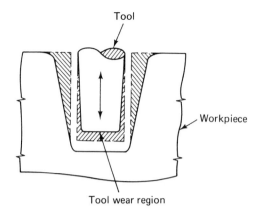

Tool

Workpiece

Tool wear region

FIGURE 26-9 Sidewall taper

Hole Depths: Nonrotating tools are limited to hole-penetration depths of about 2 in. (50.8 mm) unless special provisions can be made for maintaining the circulation of the vital abrasive slurry. The ratio of penetration can be improved in cases where holes may be preformed by coring or by other methods prior to USM processing.

Overcut: The minimum diameter of a circular hole produced by non-rotating USM tools is equal to the tool diameter plus twice the size of the abrasive particles. For example, a hole produced by a ⅛-in (3.18 mm) tool using a 240 abrasive grit size [0.0025 in. (0.064 mm)] results in a hole 0.125 in. (3.18 mm) plus 2 × 0.0025 in. (2 × 0.064 mm) = 0.130 in. (3.30 mm) in diameter.

REVIEW QUESTIONS

26.1. Contrast the cutting action of a nonrotating vibrating metal tool to that of a rotating–vibrating cutting tool.

26.2. List some abrasive powders commonly used in the process known as "impact machining."

26.3. Explain the two major functions obtained by imparting ultrasonic oscillations along the axis of a rotating diamond tool.

26.4. Explain how resonance is related to efficient ultrasonic cutting.

26.5. Prepare a simple sketch showing how on a typical ultrasonic machine the direction of the tool is fed with respect to the work and the three worktable directions of motion. Label fully.

26.6. Give three examples of metals used for cutting tips or toolfaces on non-rotating tools.

26.7. List some methods of applying tool faces to tool holders.

26.8. Give some important considerations when selecting a tool-face material.

26.9. Explain why only low frictional forces are normally encountered in ultrasonic machining.

26.10. Explain how the use of a coolant increases cutting efficiency.

26.11. Explain the major advantage of cementing small or thin workpieces to a second material and then clamping the assembled components to the machine worktable.

26.12. Give the important factors that control the quality of surface finish obtained by USM.

26.13. Compare the limits of accuracy obtainable in USM to that of AJM, EDM, ECM, and LBM.

26.14. Give the leading advantages of USM.

26.15. Explain why sidewall taper should be expected for USM.

BIBLIOGRAPHY

ARMAREGO, E. J. A. and R. H. BROWN, *The Machining of Metals,* Englewood Cliffs, N.J.: Prentice-Hall, Inc., 1969, pp. 383–397.

BABIKOV, O. I., *Ultrasonic and Its Industrial Applications,* New York: Consultants Bureau, Inc., 1960, 224 pp.

BLITZ, J., *Fundamentals Of Ultrasonics,* Woburn, Mass.: Butterworth Publishers, Inc., 1963, 214 pp.

*DALLAS, D. B., "The New Look of Ultrasonic Machining," *Manufacturing Engineering and Management,* Feb. 1970.

DRAPER, A. B., and B. W. NIEBEL, *Product Design and Process Engineering,* New York: McGraw-Hill Book Company, 1974, pp. 460–565.

GOLDMAN, R. C., *Ultrasonic Technology,* New York: Van Nostrand Reinhold Company, 1962, 304 pp.

*KAZANTSEV, V. F., L. O. MARKAROV, L. P. ROZENBERG, and L. P. YAKHIMOVICH, *Ultrasonic Cutting,* New York: Consultants Bureau, Inc., 1964.

*KOHL, R., "Machining with a Soft Touch," *Machine Design,* Aug. 24, 1972.

Non-traditional Machining Processes, Dearborn, Mich.: Society of Manufacturing Engineers, 1975.

SPRINGBORN, R.K., editor, *Non-traditional Machining Processes,* Detroit, Mich.: ASTME, 1967, pp. 24–36.

"Ultrasonic Machining," *Machining Data Handbook,* 2nd edition, Dearborn, Mich.: Society of Manufacturing Engineers, 1972, pp. 717–718.

"Ultrasonic Machining," *Metals Handbook,* Vol. 3, *Machining,* 8th edition, Metals Park, Ohio: American Society for Metals, pp. 249–251.

*Abstracted with permission.

27

ELECTRON BEAM
MACHINING

Applications of electron beam technology other than welding are mostly in a state of infancy. It appears that much progress is still needed in research and technology in machining applications before it can be considered a well-established, mature manufacturing process. Although not widespread, high-volume operations such as drilling and perforating and the somewhat lower-volume operations of slotting, scribing, and engraving using electron beam equipment are currently being performed on commercial products. Of particular interest is the adaptability of the process to superalloys and refractory materials, as well as to nonconducting materials such as ceramics or quartz. As knowledge grows and new effects are discovered and perfected, the flexibility and high speed of electron beam machining technology will doubtless result in considerable improvement in productivity and manufacturing economy.

DESCRIPTION OF THE PROCESS

An electron beam may be defined as a stream of electrons all moving with about the same velocity and in the same direction so as to form a beam. Electron beam machining (EBM) is a thermoelectric process in which electrons are accelerated to ultrahigh velocities of nearly three-fourths the speed of light.

380

Material is instantly removed by a melting and vaporizing action caused by concentrating a high-velocity narrow stream of electrons on a precisely limited area upon the workpiece. Upon impact with the workpiece, the kinetic energy of the electrons in the beam is converted to powerful heat energy. The beam power is continuously variable. It can be pulsed on and off at rates up to at least 10 kHz, with typical pulse widths from 10 to 80 μs. EBM work is performed in a high-vacuum chamber to eliminate the scattering of the beam of electrons as they contact the gas molecules on the workpiece.

A major factor in the efficiency of the process is in conducting the heat away from the workpiece at the beam focus point. The cutting action is accomplished by a pulsing technique. This consists of repeatedly impinging the electron beam upon the workpiece for a few milliseconds and then turning it off for a certain period of time. Careful control must be exercised over the heating and cooling cycle locally at the cut to prevent welding the surrounding metal. Short pulses are necessary to prevent ceramic material from cracking during machining.

FIGURE 27-1 Inside the electron gun (Courtesy, Sciaky Bros. Inc.)

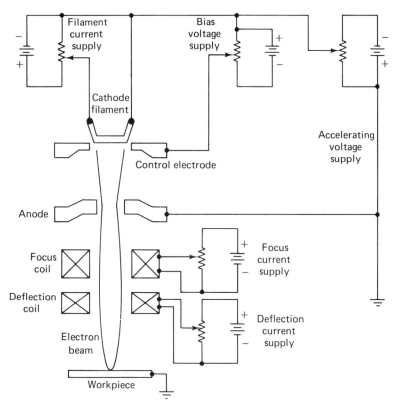

The essential difference between EBM and EBW (electron beam welding) is that in machining, the beam energy is higher, the focusing sharper, and the pulse duration shorter. Also, instead of just melting the metal as in EBW, electron beam machining *vaporizes* the metal.

A typical EBM system, schematically shown in Figure 27-1, employs a heated tungsten-filament cathode. The beam power of the electron gun is the product of the control electrode and the anode voltage. The control electrode regulates the beam *current,* and the anode voltage regulates the velocity of the electrons in the beam. Magnetic lenses or focusing coils concentrate the beam to a precise spot on the workpiece. The spot size can measure as small a diameter as 0.001 in. (0.03 mm), but more typically between 0.005 in. (0.13 mm) and 0.015 in. (0.38 mm). To illustrate the tremendous concentration of beam power density, a 0.001-in.-diameter spot can attain an energy density of up to 7 billion watts/cm^2. The beam spot size can be regulated by adjusting the gun-to-workpiece distance (standoff), and, by using a special optical tracing device, the beam may be directed anywhere in an area within a ¼-in. (6.35-mm) square on the workpiece or moved within

FIGURE 27-2 The optical viewing system (Courtesy, Welding Products Dept., Linde Div., Union Carbide Corp.)

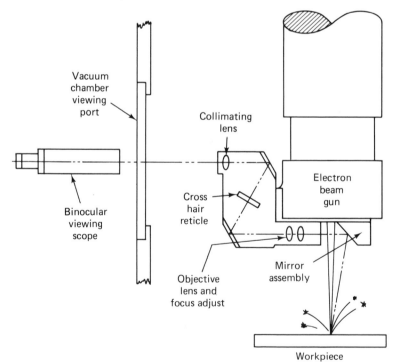

that area through an endless variety of patterns. A minimum vacuum of 10^{-4} Torr (1 standard atmosphere = 760 Torr) is essential for generating and transmitting the electron beams. Such a vacuum can easily be obtained with standard commercial equipment. Finally, the beam may be precisely positioned on the workpiece by numerical-control or computer operations.

Because the beam impact point is focused on such a small spot, the precise position of the sharp point of the beam on the workpiece is of critical importance. Figure 27-2 illustrates a method by which an operator may accurately locate the beam and observe the operation taking place by means of an optical viewing system. Essentially, the system works by superimposing a cross-hair on the image of the work surface and then collimating and relaying the combined image to a viewing telescope outside the vacuum chamber. The beam spot may be viewed up to 40 times magnification. There is no mechanical contact between the tool and the workpiece. The distance from the electron gun to the workpiece, or standoff, is usually about 4 in. (101.6 mm).

Different power settings for various cutting applications are necessary. Adjustments must be made which conform to a wide range of changing conditions, depending upon the work material and the size and shape of the cut. As an example, an increased beam current in longer pulses and at higher frequencies is needed when machining alloys with high melting points or thick stock.

PRODUCT APPLICATIONS

Electron beam operations are principally restricted to micromachining applications on thin materials. A recently perfected micromachining technique called *electron beam lithography* is now used in the manufacture of electronic devices such as field-emission cathodes, computer memories, and integrated circuits. Researchers believe that this technology can be used to make smaller devices, with better resolution, than any other current process. It has already been used to produce an ''object'' 0.5 μm wide, about the width of a single wavelength of light.

Perhaps the most typical application for EBM at the present time is in precise high-speed drilling of circular or profile-shaped holes. Drilling time for holes as small as 0.005 in. (0.13 mm) in diameter in thicknesses up to 0.050 in. (1.27 mm) in any material is almost instantaneous. The time required to drill holes larger than 0.005 in. increases but is still relatively short. A recent development called *transpiration cooling* requires drilling up to 30,000 holes in each turbine blade, an operation that is economically feasible only with a high-speed drilling tool such as the electron beam. Other production examples where EBM drilling is currently being used are in making holes in mixer plates, combustion chamber rings, and other gas turbine parts, as well as in metering or flow orifices or rough wire dies.

Perforating is another example of the capability of EBM. The process was recently selected as a method to perforate holes in a glass fiber spinning head made from a heat-resistant superalloy. Another recent application was in perforating "breathing holes" in artificial leather at a rate of 5000 holes per second. Still another outstanding example of EBM capability in perforating is in the production of 400,000 holes 80 μ in. in diameter in each square inch (or 6.45 cm^2) of a 0.001-in. (0.0254-mm)-thick metal foil.

Slotting and related milling operations are economically practical with EBM technology. One company uses EBM to cut 0.005-in. (0.13-mm)-wide slots into 0.030-in. (0.762-mm)-thick aluminum oxide wafers at a rate of 24 in. (610 mm) per minute. Scribing and engraving of thin films are other practical applications of EBM. The process has been successfully used to dice silicon chips for the semiconductor industry. Salvage-type work, as in the removal of small-diameter broken taps, is another use of EBM technology.

The electron beam has proved to be an ideal thermal tool with which to heat-treat metallic workpieces. Electron bombardment for steel hardening results in a controlled surface hardness in selected areas on the workpiece. In this process a precisely defined localized area is rapidly heated to an appropriate temperature. Other areas on the workpiece absorb the heat and rapidly "self-quench" as the heating cycle terminates. The advantage is that small geometrically defined areas such as edges can be heat-treated without affecting other areas of the workpiece.

PROCESS SELECTION FACTORS

Hole and Slot Tolerances: Very precise and fine cuts are possible. Typical EBM tolerances are about 10% of the hole diameter and slot widths.

Hole and Slot Sizes: Micromachining operations of thin materials and small-hole-diameter hole drilling are the principal applications of EBM. Holes as small as 0.0005 in. (0.013 mm) in diameter may be EB-drilled. Multiple slots as narrow as 0.001 in. (0.03 mm) have been produced and spaced as close as 0.005 in. (0.13 mm).

Aspect Ratio: Holes 10 to 20 diameters deep may be readily drilled. The maximum aspect ratio (depth-to-diameter ratio) is about 100:1 using multiple pulsing.

Machining Locations: Positions of machining cuts on workpieces such as holes, profile shapes, and slots can be held to ±0.0005 in. (0.013 mm) or better with work-handling devices installed within the vacuum chamber.

Materials Applications: All metals and alloys are readily machinable by EBM. In addition, EBM is especially adaptable to machining non-metallics, such as ceramic, plastic, and glass.

Effect of Materials Properties: EBM is essentially independent of material properties such as hardness, brittleness, or electrical conductivity.

Surface Finish: EBM produces holes and cuts with relatively clean surfaces. Surface finishes ranging from 20 to 100 μin. have been obtained.

Possible Limitations

Taper: Holes and slots produced by EBM in materials less than 0.005 in. (0.13 mm) thick exhibit little or no wall taper. Holes in greater depths have a taper of 2-to-4° included angle. The minimum hole diameter occurs on the opposite or the beam exit side of the workpiece. A sidewall taper of 1 to 2° should be expected for slots exceeding 0.005 in. (0.13 mm) deep.

Material Spatter: A small amount of recasting and material spatter usually accumulates on the side of the cut where the beam enters. This layer may usually be removed without difficulty by abrasive cleaning.

Hazardous X-rays: Not unlike commercial X-ray equipment, electron beams in an EBM system generate X-rays when they impact a metal surface. Protective measures must be incorporated into the design of EBM equipment which assure that radiation transmitted to the surrounding area environment remains below the safe level of 0.25 mR/hr. Periodic inspections on equipment are necessary to verify the integrity of the various protective measures, which include suitable wall thickness, lead content of the glass in the windows, and the use of labyrinth seals.

Removal Rates: The thermal properties of the workpiece material and the power level of the EBM equipment control the stock removal rates. Electron beam machining may have no equivalent for some micromachining applications. For example, one company uses EBM to machine 0.005-in. (0.13-mm)-wide slots in 0.010-in. (0.25-mm)-thick material at a rate of 10 to 24 in. (254 to 609.6 mm) per minute. At this stage of technological development, the removal rate for EBM is only 0.006 in.^{3} (0.098 cm^3) per hour. In contrast, conventional milling can reach 250 in.^{3} (about 4100 cm^3) per hour. The process should be considered only for special work applications where no better process can produce the required operation.

Productivity: The process can be somewhat cumbersome and expensive because of the requirement that it be operated entirely in a high-vacuum environment that may require pumpdown times up to 30 minutes or more. Batch loading of small parts increases production rates.

Heat-Affected Zone: Very high temperatures are reached at the focal point of the beam, which results in a heat-affected zone often penetrating to a depth of 0.010 in. (0.25 mm). To minimize potential thermal damage to the workpiece, pulses of shorter duration are used which reduce the heat flow and minimize local recasting and other potential surface-damage effects.

Cost: An important limitation is in the high cost of commercial equipment. The high temperatures necessary for achieving practical removal rates require the use of high beam current. Costly machines with accelerating voltages up to 150,000 are necessary to produce a power output of 100 watts.

PRODUCT DESIGN FACTORS

Parallelism: The edge of slot walls can be held parallel to a tolerance of 0.002 in. (0.05 mm) (see Figure 27-3).

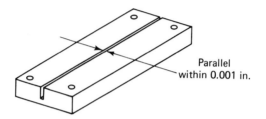

Parallel
within 0.001 in.

FIGURE 27-3 Parallelism

N/C or Computer-Aided Control: Because the electron beam has very low inertia, it can be steered rapidly from point to point across the workpiece. Electromagnetic beam deflection systems can be economically adapted to N/C or computer control. All the parameters of electron beam processing —beam power, beam focus, precise beam positioning, beam pulsing, and workpiece speed—are quite amenable to N/C or computer control.

Hole Entrant Angles: The electron beam can drill a hole at any angle between 20° and 90° to the workpiece surface, as illustrated in Figure 27-4. This capability is presently being utilized to produce holes in numerous components of gas turbine engines. Examples of hole-drilling applications are shown in Figure 27-5.

FIGURE 27-4 Hole entrant angles

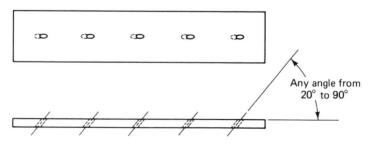

Any angle from
20° to 90°

(a)   (b)

(c)

FIGURE 27-5 The electron beam is ideal for drilling these holes in a gas turbine engine mixing plate. The angle between the hole axis and the surface is approximately 30° (Courtesy, Farrel Company-Div. of USM Corporation); (b) holes in a gas turbine engine part are drilled with an electron beam. Holes are at right angles to the surface (Courtesy, Farrel Company— Div. of USM Corporation); (c) imitation leather used for shoe uppers is perforated to allow the material to "breathe". Holes are approximately 0.005 in. diameter (Courtesy, Farrel Company-Div. of USM Corporation)

Machining Configurations: The diameter of an EB-drilled hole depends upon the beam diameter and the energy density. If holes larger than the beam diameter are required, the following methods may be used. One method to produce large circular holes is by trepanning, shown in Figure 27-6. Cutting in this manner is accomplished by using magnetic deflection coils and by rotating the workpiece on a turntable. A computer-controlled electron beam may be used to traverse the workpiece and create a hole of virtually any size and shape as long as it has vertical walls.

A method for producing odd-shaped cuts or for drilling precision grids and multiple uniform or nonuniform spaced holes in parts employs a "flying spot scanner." This system optically copies or traces patterns over a ¼-in.² (1.58-cm²) area from negatives made by photographing the original artwork. Light emitted from a cathode ray tube passes through transparent lines on the negatives and is picked up by a phototube. The signal is relayed to a

387

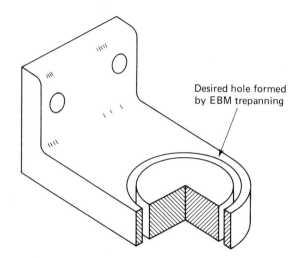

FIGURE 27-6 Large diameter holes may be produced by trepanning

logic system that sets off the electron beam. The pattern may be automatically reproduced at a 10:1 reduction. This method may also be used to etch copper gravure plates.

Numerically controlled machining operations show considerable promise in adaptability to this emerging technology. Standard EBM equipment is available with mechanical facilities for manually moving the worktable so that large areas on a workpiece may be machined.

Depth of Cuts: Most holes or slots are cut entirely through the workpiece, but bottoming cuts may be produced. This is accomplished by shutting off the power when the correct depth has been reached. The maximum practical depth for holes and slots is limited to less than ½ in. (12.7 mm).

REVIEW QUESTIONS

27.1. Explain why electron beam machining is considered to be in a state of early development at the present time.

27.2. Define an electron beam.

27.3. Describe how the process of EBM may be used to remove required amounts of material from a workpiece.

27.4. Explain why EBM work is performed in a vacuum chamber.

27.5. Describe the essential differences between EBM and electron beam welding.

27.6. By what means is the electron beam directed to a precise location on the workpiece?

27.7. Give some techniques that may be used to regulate the beam spot size.

27.8. Explain the function of an optical viewing system.

27.9. At the present stage of development, what classification of work is best suited for EBM?

27.10. Explain the major advantage of using an electron beam for heat-treating applications.

27.11. Give a statement that establishes the versatility of EBM regarding the range of a material that may be appropriately processed.

27.12. Prepare a sketch illustrating a condition resulting from material spatter and suggest one or more ways of removing material spatter.

27.13. Outline precautions ordinarily taken to offset potential dangers from hazardous X-rays associated with this process.

27.14. Explain why the productivity of EBM is regarded as comparatively slow and suggest ways that may be utilized to improve production rates.

27.15. Explain why there might be a significant advantage in using EBM to produce multiple holes 0.010 in. (0.25 mm) in diameter at an angle of 25°, for example, to a workpiece surface.

27.16. Describe a method for producing odd-shaped cuts or precision grids on workpieces by EBM.

27.17. Give the maximum practical depth for producing holes by this method. Suggest a reason why a depth limitation exists for EBM.

BIBLIOGRAPHY

BELLOWS, G., *Nontraditional Machining Guide,* Cincinnati, Ohio: Metcut Research Associates. Inc., 1976.

COOK, N. H., *Manufacturing Analysis,* Reading, Mass.: Addison-Wesley Publishing Company, Inc., 1966, pp. 138–140.

"Electron Beam Machining," *Machining Data Handbook,* 2nd edition, Dearborn, Mich.: Society of Manufacturing Engineers, 1972, pp. 727–728.

"Electron Beam Machining," *Metals Handbook,* Vol. 3, *Machining,* 8th edition, Metals Park, Ohio: American Society for Metals, pp. 253–254.

*HEGLAND, D. E., "Electron Beams Come Down to Earth," *Automation,* Mar. 1976.

*KOHL, R., "Machining with a Soft Touch," *Machine Design,* Aug. 24, 1972.

Non-traditional Machining Processes, Dearborn, Mich.: Society of Manufacturing Engineers, 1975.

SPRINGBORN, R. K., editor, *Non-traditional Machining Processes,* Detroit, Mich.: ASTME, 1967, pp. 135–145.

*Abstracted with permission.

28

FORGING

Forging is defined as the process of plastically deforming metals or alloys to a specific shape by a compressive force exerted by a hammer, a press, by rolls, or by an upsetting machine of some kind.

The process of forging is considered to be the oldest of all metalworking production processes. Its beginning extends as far back in history as primitive man, who hammered his metal tools, utensils, and weapons into crude shapes from the heat of an open fire. The equipment in use today serves as a logical extension of the smithy's hammer and anvil. Particularly significant changes in forging techniques have taken place, especially within the past decade.

DESCRIPTION OF THE PROCESS

Forging Temperatures

While some metals can be forged at room temperature, the majority of metals are made more suitable for forging by heating. Forging at elevated temperatures improves the plasticity of the metal while reducing the required forging forces. Ferrous materials are forged from 1700 to 2500 °F (927 to 1372 °C); from 1100 to 1700 °F (594 to 927 °C) for coppers, brasses, and bronzes; and from 650 to 900 °F (344 to 483 °C) for aluminum and magnesium alloys. Each specific metal composition has its own plastic range. Some alloys have a fairly wide forging-temperature range, but for others the range is very limited. Alloys that do not possess a plastic range are considered to be unforgeable.

The *conventional* forging processes described in this chapter are intended to apply only to hot-forging methods. However some of the *special* forging processes described later in the chapter include cold-forming techniques.

Structure

Forged parts are characterized by a fibrous structure, a highly advantageous condition, which is generally attributed to the effects of hot working. The fibrous structure is produced only in wrought metals. When metal is forged, both strength and ductility increase significantly along the lines of flow, resulting in a characteristic pattern of flow that is unbroken and follows the contour of the part. The main objective of good forging design is to control the lines of metal grain flow so as to put the greatest strength and resistance to fracture where it is needed on a part, thus ensuring a part with a high strength-to-weight ratio. Figure 28-1 illustrates the cross section of an etched forging which shows the typical lines of grain flow.

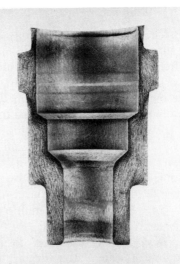

FIGURE 28-1 Properly developed grain flow in hot forging closely follows the outline of the component (Courtesy, National Machinery Co.)

Hammering and pressing, the two main methods of forging, differ only in their relative speed of pressure application. Under special conditions, depending upon the type of equipment, the two methods are considered to be interchangeable. *Hammering* causes the metal to change shape by repeated flows of a freely falling ram. *Pressing* changes the shape of a metal bar or billet by the squeezing action of a slowly applied force.

Open or Closed Dies

Forged parts may be produced in either *open* or *closed* dies. Open-die forging is a method that progressively applies local compressive forces on different parts of the metal stock. Simple forged shapes may be obtained by either hammering or pressing a material that is held between two flat surfaces or between dies with V-shaped or half-round cavities. Parts produced by a closed-die forging process are formed by applying force to the entire surface, causing the metal to flow into a die cavity that has been cut to a specific shape. More complex and accurate parts are produced with closed dies and with increased production rates.

Forging Processes

The deformation of a metal into a specific shape can be accomplished by a surprising number of forging processes. Modern forging techniques are often used to combine two or more methods in the production of a given workpiece. For example, a part may be preformed (a metal blank whose geometry is a rough approximation of the final shape) by a conventional forging hammer, with the final forged shape completed on a forging press. It is not unusual in some forging shops to produce the initial shape of a part by preforming a part on a steam hammer or between forging rolls, further developing its configuration in upset forging dies, followed by one or more bending operations, and finally completing the part in a forging press. Some parts begin with a billet that is drop-forged and then completed by press forging. Such innovative multiple forging sequences usually lead to greatly increased forging potential.

Basic forging processes normally consist of the following: smith or hammer forging, drop forging, press forging, machine or upset forging, and roll forging. There are several other variations of the forging processes, some recently developed and others traditional, which, for purposes of simplicity, will be classified as *special* forging processes. These techniques include ring rolling, orbital forging or rotaforming, no-draft forging, high-energy-rate forming, cored forgings, wedge rolling, and incremental forging. A considerable amount of confusion exists in the current technical literature, arising from the terms that are used to denote some of the more recent developments in forging. It is expected that standardization will eventually occur as each process reaches a broader level of acceptance.

CONVENTIONAL FORGING PROCESSES

Smith Forging

Smith forging (or *flat-die forging*) consists of forming the shape of a heated metal part by applying repeated blows of a hand-held hammer. A flat die or an anvil is used. The desired shape of each metal piece is maintained by the smith during the forging process as the desired length and cross section are adjusted manually by positioning and turning the part on the flat surface of the anvil. The quality of the forging is totally dependent upon the skill of the smith. Smith forging is not classified as a production process. Only relatively simple shapes in small quantities are forged by this technique. It seems reasonable to assume that this old and acceptable method of metalworking will continue to be used in its present form for many years to come.

Hammer Forging

Hammer forging differs from smith forging in that the heated billet is shaped by the impact of a steam- or air-operated hammer (see Figure 28-2). Larger, heavier work is possible since higher pressures may be exerted on the workpiece. Hammer-forged parts have a more uniform structure.

FIGURE 28-2 A high strength fitting is being forged in impression dies in a forging hammer (Courtesy, Forging Industry Association)

Mechanical manipulators are used to hold and handle work that is too heavy to be positioned by hand. The horizontal face of the hammer (upper die) and the anvil (lower die) are both flat. Special tools, attached to long safety handles, are used by the hammer man. These consist of various punches and specially shaped dies for producing holes, chisels for notching or cutoff operations, and devices that help grip the parts while forging special configurations. Similar to smith forging, parts produced by hammer forging are restricted to reasonably simple shapes in small quantities. Except for special preforms, the process is not generally classified as a production-forging method.

Drop Forging

Drop forging is a process of forming metal parts by hammering a heated bar or billet into aligned die cavities. Dies are made in sets or halves; one half of the die is attached to the hammer and the other to the stationary anvil.

FIGURE 28-3 A die set. Note the guide pins which aid in producing forgings with a minimum of mismatch and make it possible to install the die set as a unit (Courtesy, National Machinery Co.)

FIGURE 28-4 A power drop hammer utilizes either air or steam power through a differential porting arrangement (Courtesy, Chambersburg Engineering Company)

Parts made in this way cannot be formed by a single blow of the hammer. The final shape of a drop-forged part is progressively developed over a series of steps from one die impression to the next. Figure 28-3 illustrates a typical closed impression die set which contains a sequence of three cavities of successively different shapes. Each impression gradually distributes the flow of metal and changes the shape of the workpiece as it is transferred from one impression to the next between hammer blows. Some die sets consist of groups of cavities that increase the production rate to two, three, or more pieces at a time.

Steam hammers used in drop forging are similar to those used in hammer forging. A typical steam hammer works rapidly, with over 300 blows per minute, and they range in capacities from 500 to 50,000 lb (225 to 22,500 kg). Steam pressure is used to raise the ram as well as to control the striking force on the workpiece. Figure 28-4 illustrates a power drop hammer. A variation of the steam hammer is the *gravity drop hammer*. In this machine, steam that is used to raise the ram is suddenly released to allow the ram to free-fall by gravity.

Another form of hammer, the *board hammer,* employs a unique design consisting of hardwood boards and steel rollers whose combined action raise the ram to permit it to fall by gravity. Unlike the previous open-die processes, it is practical to produce drop-forged parts in quantities that may vary in special cases from only a few to millions of duplicate pieces. The varieties of shapes and the size range of drop-forged parts are almost limitless.

Depending upon the size and complexity, parts may be drop-forged at the rate of 10 to 500 pieces per hour. For parts weighing 1 to 5 lb (0.45 to 2.3 kg), tolerances of ± 0.006 to 0.030 in. (0.15 to 0.76 mm) across the parting line may be obtained; for parts weighing in the vicinity of 100 lb (45 kg), a tolerance of ± 0.030 to 0.090 in. (0.76 to 2.29 mm) is often possible.

Press Forging

Press forging is an important production process by which parts are made by plastically deforming a metal blank into one or more die cavities by a slow squeezing action. Unlike the foregoing processes of hammer and drop forging, no impact blows are struck. An operator has little influence over the quality of press-forged parts, since the operator's principal duties consist of loading and unloading the parts. The parts are formed by a single stroke in closed-impression die sets. Maximum pressure is built up, with increasing intensity, from the start to the end of the stroke, resulting in maximum penetration and in improved grain flow throughout the entire forging. Completed forgings are manually ejected from the die cavity.

Most press-forged parts are initially *preformed* on other machines, such as forging rolls, upsetters, benders, or on special machines. The press is utilized only for the blocking and final forging operations. A recent trend in preforming is to use pieces that have been cut from machined, extruded, or rolled shapes, which often results in considerable cost savings.

Press forgings are formed in large vertical presses that may be either mechanically or hydraulically operated. Figure 28-5 illustrates the largest mechanical press in the world. Mechanical presses are capable of delivering from 25 to over 100 strokes per minute and range from about 100 to 12,000 tons (approximately 90 to 11,000 metric tons) capacity. Such presses are generally limited to nonferrous alloys weighing under 30 lbs (13.5 kg). Hydraulic presses are generally larger and have varying capacities, extending up to 50,000 tons (45,000 metric tons).

Pressforging is a considerably quieter operation than drop forging. Because of the need for less draft, somewhat more accurate forgings may be produced. The structural quality of press forgings is generally considered to be at least equal, and, in many cases, superior, to drop forgings. Forging presses may also be used for secondary operations—to restrike certain parts, previously made by other processes, for the purpose of correcting dimensional

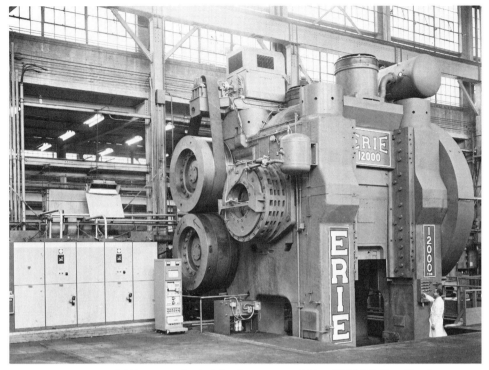

FIGURE 28-5 A 12,000 ton mechanical forging press weighing 2,750,000 lbs (Courtesy, Erie Foundry Company)

size and improving the surface quality. Little metal flow is involved in either operation, and flash is not formed (see Chapter 35). Special trimming presses are available which are used to remove or trim excess metal from forgings.

Impact Forging

Impact forging, also known as *automatic closed-die forging,* consists of mechanically positioning and holding a precut heated billet, usually a pre-form (a rough approximation of the final shape), between opposing die sets which are attached to air-operated cylinders or rams mounted in a horizontal plane. The metal stock is formed to shape by the counterblow action of the die halves as they simultaneously strike the opposite sides of the billet, which is held in place by a programmed manipulator. The operation is remarkably free of excessive shock or vibration. The energy is absorbed as the billet deforms equally from both sides. The process may be completely mechanized as in Figure 28-6, or it can be limited to a simple, automatic single-blow forging cycle.

FIGURE 28-6 A fully automated impact forging machine that combines heat, feed, forge, and control into one continuous operation (Courtesy, Chambersburg Engineering Company)

The types of forgings produced in impactors are generally comparable to those produced on gravity hammers. Impact forging, a relatively recent innovation, may be regarded as a refinement of press forging, because each die impression is used only for one stroke, to form the part. On the other hand, impact forging may be considered an outgrowth of drop forging, since the final forging results from plastic deformation caused by a severe impact. As in most of the other closed-die forging processes, a thin projection of excess metal, called *flash,* which may later be trimmed, extends around the forging at the parting line.

Compared to press- or drop-forged parts, impact forgings exhibit improved structural characteristics and physical properties. Closer part tolerances are generally possible, which result in cost economies from reduced machining time. The process is adapted to the production of parts in large quantities. Parts may be completely automated, starting with the raw, precut stock and ending with the ejected forging.

Upset Forging

Upset or *machine forgings* are produced in double-acting mechanical presses similar to that shown in Figure 28-7, which operates in a horizontal plane. Upsetting is performed by the impact blow of a punch against the end of a piece of stock inserted into a shaped die. Earlier models of upsetters, sometimes called "headers," were originally developed for cold-heading nails, rivets, and small bolts and continue to be an important method of producing these products today. Chapter 34 discusses the technology of cold heading. Modern upsetters squeeze the desired shape of a heated metal workpiece into die cavities. It is often the method selected when certain parts are required with an increased volume of metal, which must be gathered up at the center or only at one end. Such parts include gear blanks, valve stems, axles, couplings, and so on.

FIGURE 28-7 An upset forging machine with an automatic handling mechanism for upsetting the flange on an automotive axle shaft (Courtesy, Ajax Manufacturing Company)

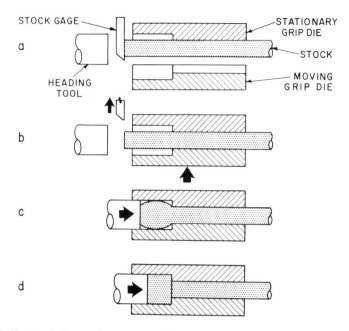

FIGURE 28-8 Typical operating sequence in upsetting on a forging machine (Courtesy, Forging Industry Association)

The operation cycle shown in Figure 28-8 consists of placing heated metal bar stock in a stationary die. A moving-grip die closes against the stationary die, gripping the stock tightly. The heading die with its attached punch moves forward and presses against the workpiece, displacing or upsetting the stock until it conforms to the shape of the die impression. The heading die then releases the workpiece and returns to its original position. Depending upon the workpiece configuration, the die cavity may be entirely contoured in the grip die, the punch, or partially shared by both. The final step may include a punching or shearing operation. The dies and punches shown in Figure 28-9 were used to form the axles at the left. Upset forgings

FIGURE 28-9 Sequence of upset forging operations on an automotive axle shaft (Courtesy, Ajax Manufacturing Company)

generally do not require a trimming operation. Upset-forging machines can handle bar stock ranging in diameter from ¼ in. (6.35 mm) to stock as large as 10 in. (254 mm).

Roll Forging

Roll forging is a process that reduces the cross section of short lengths of bar stock for products such as rifle barrels, levers, axles, and leaf springs, and for making simple preform shapes for further processing. Roll-forming deforms the material by changing the cross-sectional shape or by reducing the diameter or thickness while increasing the stock length. Figure 28-10 illustrates a roll-forging operation. Pairs of semicylindrical rolls are used with the active die surface occupying only a portion (usually half) of the roll circumference. The operator holds the heated bar stock with tongs, and the stock is manually inserted between the roll dies. The shape formation begins as the revolving rolls grip the stock, squeeze it, and roll it back toward the operator. As the first pass is completed the rolls open, the bar is removed and inserted between another set of grooves for the next pass. The process is progressively repeated using smaller die impressions until the workpiece reaches the desired size and shape. Both tapered and straight work can be roll-forged. Surfaces may be obtained which are both smooth and free of scale pockets. An improved fiber structure is developed on parts made by roll forming by the progressive squeezing and hot-working action of the roll dies on work blanks.

FIGURE 28-10 Forging roll with a three impression die is an efficient means of shaping a workpiece (Courtesy, Forging Industry Association)

SPECIAL FORGING PROCESSES

Ring Rolling

A special process, which is gaining increasing application in the forging industry, is the rolling of rings. The ring-rolling machine is essentially a vertical "two-high" mill with one of the rolls driven. Figure 28-11 illustrates the following typical rolling sequence. The starting material is a thick-walled ring that has been formed by forging, or by some another suitable preparatory operation. As the wall thickness decreases, the ring diameter increases while the stock is held between the driven and idling roll. In operation, the idling roll is gradually moved nearer the work roll so that continuous deformation and wall thinning results. Suitably arranged rollers secure the proper position of the ring during rolling, and provision is made for supporting the weight of the ring on a base plate or guide.

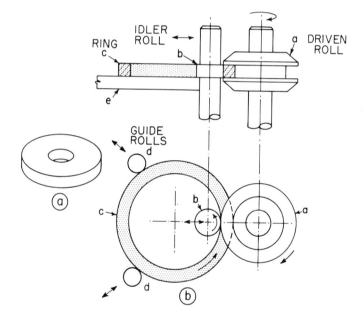

FIGURE 28-11 Typical sequence of a ring rolling operation
(Courtesy, Forging Industry Association)

In principle, there is no limit to the size of the rings rolled. In practice, they range from small rings used as roller-bearing sleeves to rings of 25 ft (approximately 7.5 m) diameter with face widths of 80 in. (approximately 2 m). Various profiles, as illustrated in Figure 28-12, may be rolled by suitably shaping the driven or idling rolls.

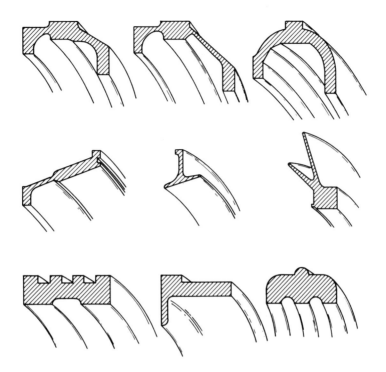

FIGURE 28-12 Representative contours machined from rolled rings

Orbital Forging

Orbital forging (or *rotaforming*) is a recent forging development in which metal workpieces are quietly and accurately cold-formed. Figure 28-13 illustrates a model of an orbital-forging press. Figure 28-14 shows the basic die motion of a typical orbital-forging press (certain features have been exaggerated for clarity). The upper die, which is inclined at a slight angle from the vertical, imparts a high-frequency circular rocking motion across the top surface of the workpiece. At the same time, the lower die, powered by a hydraulic ram, moves the workpiece upward, pressing against the orbiting upper die. The forging cycle ends when the hydraulic ram contacts the preset stop. The ram is then lowered, and a hydraulically powered rod moves up to eject the finished part. Metal may be upset by gradually deforming it either outward or downward to form flanged parts with indented or crown shapes, disks, or even asymmetrical parts, as shown in Figure 28-15. Types of deformation by orbital forging are illustrated in Figure 28-16. The forging pressure is reported to be only 5 to 10% of that needed by conventional cold-forming equipment, because only a relatively

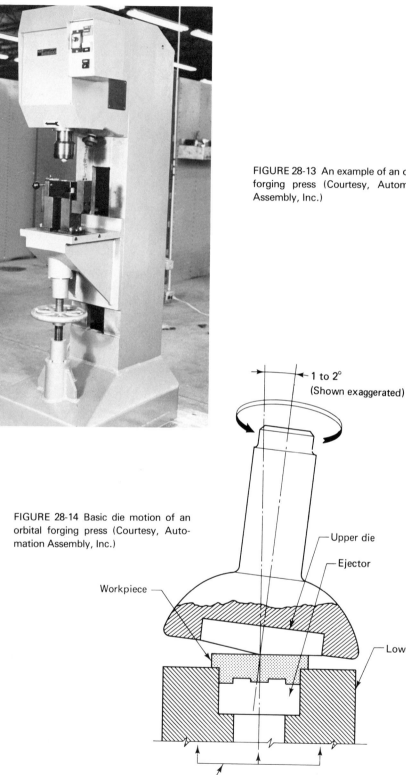

FIGURE 28-13 An example of an orbital forging press (Courtesy, Automation Assembly, Inc.)

FIGURE 28-14 Basic die motion of an orbital forging press (Courtesy, Automation Assembly, Inc.)

1 to 2°
(Shown exaggerated)

Upper die

Ejector

Workpiece

Lower die

Hydraulic ram

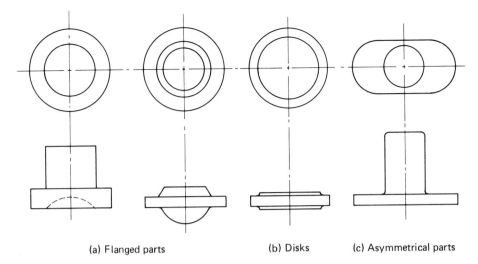

(a) Flanged parts   (b) Disks   (c) Asymmetrical parts

FIGURE 28-15 Representative shapes of orbitally forged parts

FIGURE 28-16 Examples of types of deformation by orbital forging:
(a) upsetting; (b) vertical deformation; (c) extrusion

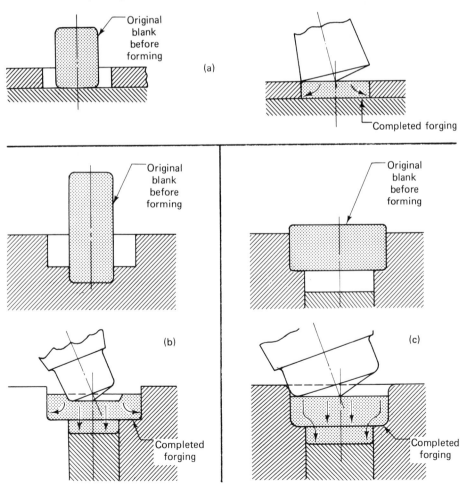

small portion of the die actually contacts the workpiece. In comparison to parts made by an impact hammer, for example, the material deformation is slower and more gradual. The process results in smooth surfaces which are typical of those produced by cold working. Also, orbital forging requires smaller and lower-cost equipment.

The type of tool motion used in orbital forging depends upon the nature of the deformation needed. Any of the four motions shown in Figure 28-17 can be "dialed in." In all cases, the blank is deformed by the rocking tool motion. The upper die does not rotate; a given point on the die always contacts the same point on the workpiece.

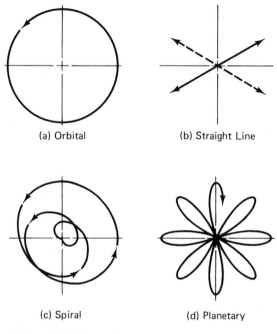

(a) Orbital   (b) Straight Line

(c) Spiral   (d) Planetary

FIGURE 28-17 Four tool patterns used in orbital forging
(Courtesy, Automation Assembly, Inc.)

The unique tool motions are particularly effective in forming parts that are relatively thin with respect to their diameter. Such parts, because they require significant lateral movement of metal, are difficult to cold-form by conventional hydraulic presses. Bevel and face gears, stepped-diameter circular parts, and similar pieces as large as 3¼ in. (82.55 mm) in diameter can be orbital-forged. Materials that can be formed by orbital forging include all metals that can be cold worked.

No-Draft Forging

No-draft forging, also called *precision* or *close-tolerance forging,* is an important and recent variation of traditional press forging and, in fact, is considered the ultimate in present forging techniques. Early applications of no-draft forging centered principally in the aerospace field, but recently other industries have gradually become aware of its advantages. Because the process yields a forging that conforms more closely to required finish sizes and shapes than is possible with conventional methods, there is a substantial saving in machining costs and material waste. Many parts can be installed into service "as-forged" or with only a limited amount of subsequent machining. A particularly unique advantage is that only a minimum amount of end grain is exposed on a no-draft forging and, even in cases where parts require a small amount of machining, the grain flow is usually undisturbed. End-grain exposure often brought on by heavy machining cuts, may lead to problems of stress corrosion and cracking. The aerospace industry is, of course, particularly concerned with preventing part failures of this kind. No-draft forgings can be made with physical characteristics not obtainable in the same material by any other processing method.

The process consists of preparing an accurately sized blank by an extensive sequence of preforming steps, using various production methods. The final die cavity is made to exacting standards and, as the name of the process implies, the die walls have little or no draft. Several other refinements are also necessary. Increased control over preheating both the dies and the blank, descaling, lubricants, die materials, and the use of special die inserts, punches, pins, and so on, are also important requirements. Complex-shaped turbine blades or spiral-teeth bevel gears are examples of parts made by this process. Because of the high development and tooling costs, no-draft forgings are justified only when large quantities of close-tolerance, complex-shaped parts are required. Because of the need for specialized forging requirements, the production output ranges between one-third and two-thirds that of conventional forging. No-draft-forging techniques are currently used in the production of aluminum, magnesium, and copper products. Development work is now being carried out in titanium and some steels.

High-Energy-Rate Forging

High-energy-rate forging is a variation of press forging in which parts are produced within a span of a few milliseconds by applying extremely high pressures. The process was developed in an attempt to reduce the size of the larger-capacity forging presses and, at the same time, increase forging production. Simple preforms are usually prepared by some other

FIGURE 28-18 A forging produced on a nitrogen operated high energy rate forging machine: (a) stock; (b) completed forging (Courtesy, Forging Industry Association)

forging process, or, in some cases, raw stock consisting of a slug or billet is used, which is the case in Figure 28-18. The final forging is produced by a single blow of a high-velocity hammer whose short stroke is actuated by releasing sudden energy from explosives, hydraulically compressed gases (such as nitrogen), or electrohydraulic sources. In some cases, deformation velocities may reach upward of 700 feet per second (fps). At the current stage of development, the process seems best adapted to the production of reasonably symmetrical parts, principally because of an apparent inability of the machines to withstand eccentric loading. Aluminum alloys and high-strength, high-temperature steel parts may be forged. More applications are being found for parts that formerly were difficult or impossible to forge due to problems encountered as the metal cooled and stiffened. In comparison to other forging processes, parts can be made with an improved surface finish, to closer tolerances, and generally with improved structural characteristics.

Cored Forgings

Except in special cases, the forging processes discussed thus far in this chapter have applied to *solid* forgings. Typical examples of *cored* forgings are shown in Figure 28-19. A special technique has been developed to produce parts with cored openings utilizing a sequence of punch actions in combination with the motions of the dies of a forging press. This action requires only a single stroke of the press to produce cavities not possible with standard forging procedures. The entire forging cycle takes only a few seconds. Physical properties of cored forgings are equal to or better than conventional solid forgings. The primary advantages of cored forgings are reduced metal cost and reduced machining and finishing time necessary to complete the part. The process leads to the economical production of high-volume parts.

Cored forgings are produced on mechanically operated special vise-type crank presses, with vertically split dies that are closed during the operation and which open to release the finished forging. Figure 28-20 illustrates a typical cored-forging setup. Movable punches or core pins form the holes. Cored holes can be formed at right angles only and in a single plane. The main difference between the cored and the solid forging procedures is the location at which the pressure is applied. In solid forging, pressure is exerted on the billet by the closing die halves in a hammer-blow motion. In cored forging, the main pressure is exerted by the top core pin. Special engineering expertise is required to successfully produce parts by this process. Die design is very important. As in the production of no-draft forgings, the small amount of flash produced necessitates careful calculation in the weight of the billet.

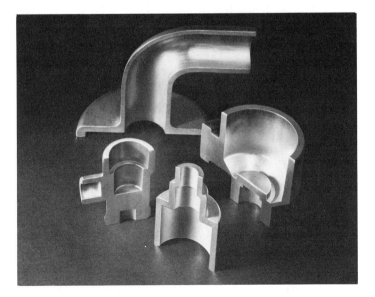

FIGURE 28-19 Representative examples of cored forgings
(Courtesy, Bridgeport Brass Company)

(a)

(b)

FIGURE 28-20 The picture shows the original blank (a) and the final cored forging (b) of the finished part. Note the core pin which was used to form the hole (Courtesy, Bridgeport Brass Company)

The major design advantages are as follows:

1. Parts can be produced with parallel walls without draft on both inside and outside wall dimensions.
2. Long, deep cores are possible.
3. As many as four cored cavities can be generated in the same plane.
4. General tolerance limits of ± 0.005 in. (0.13 mm) can be held.

Cored forgings can be produced in a wide range of nonferrous metals, including copper, brass, silicon bronze, and aluminum. Some parts have been forged of several different metals using the same dies.

410

OTHER FORGING TECHNIQUES

Many other new forging innovations are currently being developed which are of considerable interest to design and process engineers who continually search for ways to improve production capabilities. Considerable research in forging variations is constantly taking place in an effort to extend forging versatility.

Wedgerolling

Wedgerolling (also called *crossrolling* or *transverse roll forging*) is a method of accurately forming surfaces on shafts with tapers, steps, and shoulders. The process seems to have a potential which, when fully developed, may favorably compete with parts currently made by roll forging and screw machining, particularly in high-volume quantities. While some aspects of the process resemble classical rollforging, wedgerolling should not be considered one of the same. Figure 28-21 shows a typical arrangement for wedgerolling a simple shaft. The process consists of automatically feeding preheated metal bars or billets between two rolls. Special wedge-shaped dies are mounted in t-slots milled into the surface of the rolls. The desired shape of the part is progressively produced during one revolution or less of the rolls. The rolling or squeezing action displaces the metal and causes the part to elongate along its center line. Each wedge tool can reduce the diameter of a bar by a ratio of 2:1. Large reductions can be accomplished over two or more stages of wedge tools. The grain flow closely corresponds to the profile of the finished part. There is little or no scrap produced.

FIGURE 28-21 A typical arrangement for wedgerolling a shaft

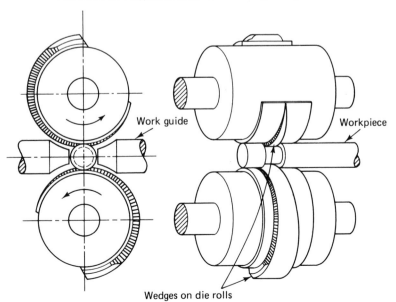

Work guide

Workpiece

Wedges on die rolls

The process is remarkably fast, with a normal cycle time of 4 to 6 seconds, which results in an output of 600 to 900 pieces per hour. Because of the high equipment costs, the economic advantage is generally limited to high production runs of no less than 100,000 pieces.

Depending upon the intricacy of shape, tolerances of ±0.003 to 0.004 in. (0.08 to 0.10 mm) on diameters and ±0.004 to 0.006 in. (0.10 to 0.15 mm) on length are possible. Bars up to 11 in. (279.4 mm) long and billets up to 13 in. (330.2 mm) long can be worked in diameters that range from ½ to 1¾ in. (12.7 to 44.5 mm).

The principal uses of wedge-rolled products are for automotive parts applications, including steering arms, drive pinions, tapered axle shafts, connecting rods, and various kinds of transmission shafts. A second important use of wedgerolling is for preparing preforms for other forging processes. Material savings of 20% have been reported using this process.

Incremental Forging

Challenges brought on principally by aerospace requirements have spurred research programs directed toward finding ways to get more performance out of existing forging equipment. Aircraft designers have long been interested in large one-piece parts, because they avoid the structural inefficiency normally associated with welded, riveted, or bolted joints. Large titanium forgings ranging from from 10 to 14 ft (approximately 3 to 4 m) long and an aluminum forging that measures over 16 ft (4.9 m) long have recently been produced. Under the constraints of conventional technology, parts of this size would require presses far more massive than any presently available.

In *incremental* forging, different areas of the forging are worked into shape, one at a time. Since only a limited area is worked, the forging equipment can be much smaller than conventional closed-die equipment. This technique makes it possible to form gigantic parts on presses of quite modest size. The process seems simple enough in concept, yet making it a workable production process has required significant research and refinement in forging methods. A major problem is to prevent the part from cooling below the forging temperature as it moves from incremental step to step. Many alloys cannot be reheated fully, because bringing them back to the original forging temperature can destroy the thermomechanical work that is one of the prime benefits of the forging process. Two techniques have been developed to offset the detrimental effect of reheating:

1. When the workpiece cools below its forging temperature, it is progressively reheated to a lower temperature.
2. Heat loss is minimized by using insulating blankets and other coatings to cover the billet after it is brought up to heat.

Long, slender parts lend themselves to incremental forging. Heat-conservation techniques, perfected for incremental forging, are being applied to forge normal-size parts in new hard-to-work alloys. Some of the new high-temperature jet-engine alloys, for example, must be worked at temperatures considerably below normal forging temperatures. It is now possible to forge parts with complex shapes and high mechanical properties in many of the new alloys.

PRODUCT APPLICATIONS

Forgings are generally specified when maximum strength and reliability are required. It is reported that over 50% of the forgings produced in the United States are used in the automotive or aerospace industries. Other heavy users of forgings, however, include marine equipment manufacturers, the turbine industry, railway and ordinance equipment makers, tool and machinery builders, manufacturers of hardware, and builders of construction, mining, and agricultural equipment. Figure 28-22 illustrates the range of sizes that can be forged. Figure 28-23 illustrates other representative forged parts.

FIGURE 28-22 Pictures show the range of sizes for forged parts: (a) tiny forgings for a desk calculator weigh less than an ounce (Courtesy, Forging Industry Association)

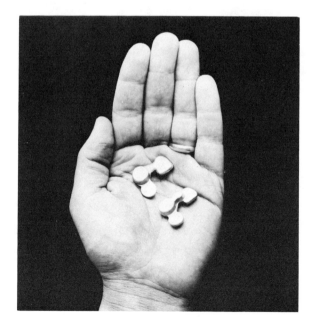

(a)

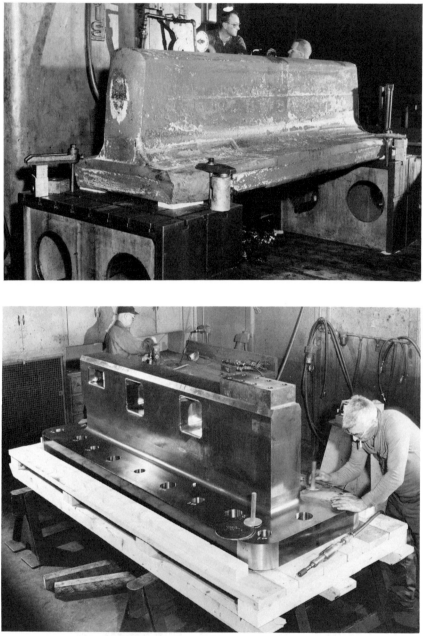

(b)

FIGURE 28-22 *(continued)* (b) Picture shows a giant forging in both rough and finished form for an anchor rail used in a space vehicle launching complex (Courtesy, Wyman-Gordon Company)

(c)

FIGURE 28-22 *(continued)* (c) Picture shows the machined anchor rail in place at a space vehicle launching complex (Courtesy, Wyman-Gordon Company)

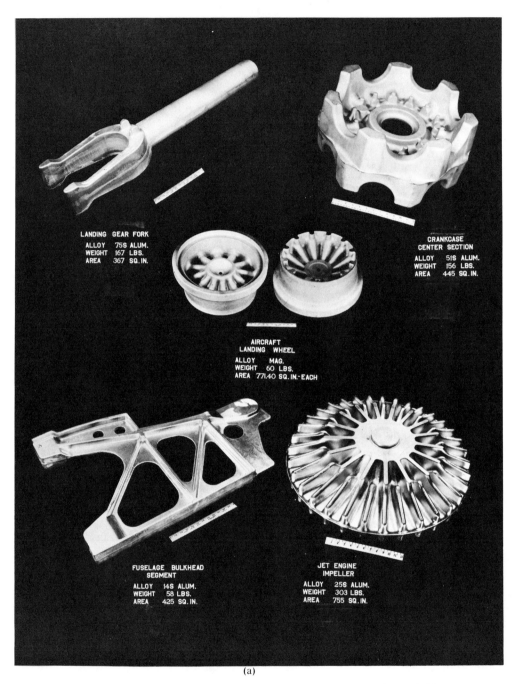

FIGURE 28-23 Representative forged parts: (a) aluminum alloy and magnesium parts (Courtesy, Wyman-Gordon Company)

(b)

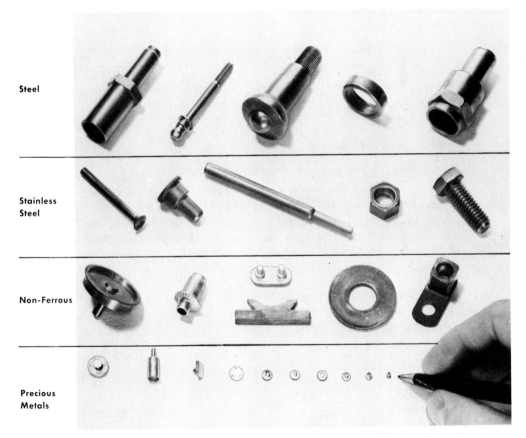

FIGURE 28-23 (continued) (b) an alloy steel press forging and the completed machine part (Courtesy, Forging Industry Association); (c) various shapes of forgings in a range of materials (Courtesy, National Machinery Co.)

PROCESS SELECTION FACTORS

Forgeability is defined as "the relative ability of a material to deform without rupture."

Materials: While all ductile materials can be forged, forgeability of a metal at the forging temperature depends upon the crystallographic structure, the melting point, yield strength, strain rate, and die friction. The following materials are ranked in order of increasing forging difficulty:

1. Aluminum alloys	10. Titanium alloys
2. Magnesium alloys	11. Iron-base superalloys
3. Copper alloys	12. Cobalt-base superalloys
4. Carbon and low-alloy steels	13. Columbium alloys
5. Martensitic stainless steels	14. Tantalum alloys
6. Maraging steels	15. Molybdenum alloys
7. Austenitic stainless steels	16. Nickel-base superalloys
8. Nickel alloys	17. Tungsten alloys
9. Semiaustenitic PH stainless steels	18. Beryllium

Advantages of Forging

Strength: Forging reduces the risk of part failures. Forging develops the quality of a specific type and grade of metal to its greatest potential. It is the only basic method of fabrication that improves the metal. When compared to sand castings, forgings are from 1 to 1½ times stronger. The process yields parts with high strength-to-weight ratios and is particularly appropriate for parts that are subject to fluctuating stresses caused by sudden shock loading.

Metal Conservation: Forging to close tolerances avoids waste of metal. Strong, thin-walled parts may be produced without sacrificing important physical requirements.

Finish: A smooth, fine-textured "hardware surface" is possible—often requiring no further finishing operations. Figure 28-24 illustrates a connecting rod after trimming and cleaning.

Machining Time: Forgings can be made to close tolerances, which reduces machining time. No-draft forgings with parallel walls permit the use of positive chucking methods which facilitate fixturing and subsequent machining.

FIGURE 28-24 An example showing the quality of finish obtained on a forged steel part (Courtesy, Forging Industry Association)

Tool Costs: Since less machining is required, fewer tools are required. The absence of sand or hard spots on forgings permits free machining and extends cutting-tool life.

Grain Structure: Parts can be made that are gas-, air-, or watertight. Forgings have consistent characteristics.

Fewer Rejects: Forgings are free from porosity, surface inclusions, or concealed flaws and other defects. There is less chance of work spoilage.

Design Potential: Irregularly shaped or complicated parts are possible. Many subassemblies may be combined and forged in one piece, which helps to avoid the relative unreliability of joints. There is a wide range of forgeable metals.

Production Rate: High rates of production are possible.

PRODUCT DESIGN FACTORS

With the advent of the fixed-stroke drop hammer, newer developments in forging presses and recent advancements in some of the newer forging technologies, the responsibility for the success of the final forging has been transferred from the operator to the die designer. Die design is a difficult and complex task. Because of variations in materials, part geometry, final component requirements, and various economic factors, it is virtually impossible to list rigid rules governing the design of all forging dies. There are, however, certain "ground rules" which should be carefully considered in achieving practical, economical part designs.

Final die design is usually accomplished as a result of close cooperation among the designer, the metallurgist, and the forging engineer. Forgings with unusually complicated shapes, consisting of holes, pockets, recesses, and bends, require a highly skillful blend of forging design and craftmanship.

Parting Line: This is the plane of separation between the die halves. The location of the parting line is generally the first step in the design of a forging die. If kept as nearly as possible in one centrally located plane of the part, die sinking, forging, and subsequent trimming operations are simpler and less costly. Symmetrical impressions in both the top and bottom dies tend to distribute the loading on the dies more evenly. Figure 28-25 relates typical cross-sectional shapes to the selection of parting-line locations.

Draft Angles: Except for the no-draft process, *draft angles* must be provided on the die walls to facilitate removal of the forging from the dies after forging. Most forge shops can work to a 3° draft angle for aluminum and magnesium, 5° for steel, and 7° for some of the tougher alloys. As a general rule, the deeper the cavity, the greater the draft angle. In any event, the

FIGURE 28-25 Parting line locations (Courtesy, Forging Industry Association)

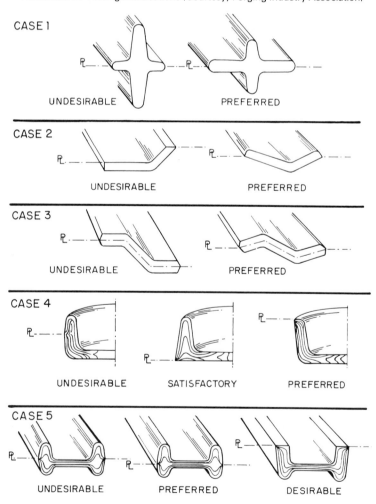

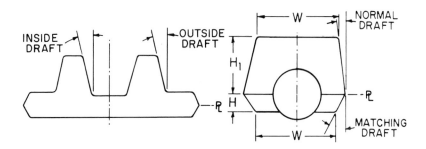

FIGURE 28-26 Locations of typical draft angles
(Courtesy, Forging Industry Association)

draft angle should be kept uniform so that finish machining on the die can be done with a minimum number of cutter changes. Figure 28-26 shows preferred locations for draft angles.

Corner Radii: These (on forgings) are formed by the *fillets* in the die. Die fillets should be uniform so that the same-size round-nose cutter can be used. A minimum die-fillet size of ⅛ in. (3.18 mm) is recommended to offset the tendency of stress concentration effects, which may lead to die failure. Die cavities with large fillet radii are easier to fill with metal and require less forging pressure than those with small fillets.

Fillet Radii: These (on forgings) are formed by the *external* corners of the die. Fillet radii on forgings should be at least ½ in. (12.7 mm) and, if possible, at least 2 to 3 times larger than the radii on external corners.

Webs: Webs measuring less than ³/₁₆ in. (4.763 mm) thick are rarely practical. A web, ordinarily the thinnest portion of a forging, cools first and when it goes below the forging temperature, the forging pressure required increases rapidly.

Ribs: These should be proportionately low and wide and limited in height to eight times the width. The minimum recommended thickness for ribs is essentially equal to that for webs. The relative sizes of both elements are influenced by the type of metal being forged and the geometry of the part. Figure 28-27 shows examples of four basic types of rib designs.

Shrinkage Allowance: This is defined as the amount of contraction of a given metal during cooling after forging. Die cavities are made correspondingly larger according to shrinkage values for various materials.

Finish Allowance: Some forged parts require surface conditions or accuracy limits that may be obtained only by subsequent machining operations.

FIGURE 28-28 Multiple forged parts showing flash before trimming (Courtesy, National Machinery Co.)

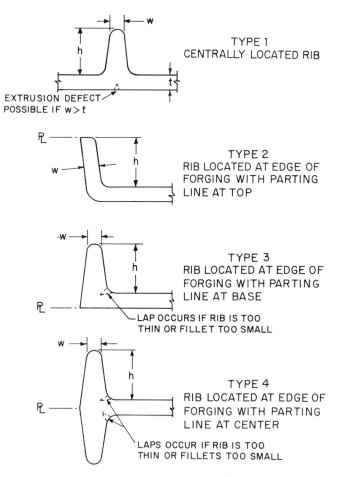

FIGURE 28-27 Examples of four basic types of rib designs (Courtesy, Forging Industry Association)

Depending upon the material and the relative size of the part, from $\frac{1}{32}$ in. (0.794 mm) in special cases to $\frac{1}{8}$ in. (3.18 mm) of extra stock is provided on the forging for this purpose.

Forging Tolerance: This is the sum of all the tolerances caused by process variables. These are principally due to die wear and mismatch (misalignment of the dies, etc.). Effects of these variations have been explained in considerable detail in the *Forging Industry Handbook.*

Surface Condition: Standards of surface measurement of forgings have never been clearly established. It is impractical to use specifications associated with surface measurement in units of microinches (μin). Discontinuities on forged surfaces that result from the effects of heat, handling, die surface, and metal flow are completely random, as opposed to the regular patterns generated by machine tools.

Finishing: Most forgings are constructed so that some excess metal is forced out into a relieved area between the dies when the dies are closed and the impression is filled. The excess material, called *flash,* is represented by the metal surrounding the parts in Figure 28-28. The relieved space around the impression is called a *flash gutter.* Prior to shipment, the surplus flash is removed by trimming, sawing, or machining. Surfaces on forgings are often covered with a scale that is formed during the forging operations or by heat treating.

Forgings may be cleaned by shotblasting, pickling, or tumbling, depending upon the metal or finish requirements.

R E V I E W   Q U E S T I O N S

28.1. Give two advantages of forging at an elevated temperature.

28.2. Define "plastic range."

28.3. Explain the importance of strength and ductility upon a forged part.

28.4. State a main objective of a good forging design.

28.5. Explain the important differences between the two main methods of forging: hammering and pressing.

28.6. Contrast open-die forging methods with closed-die forging.

28.7. What is a preform? What is its leading advantage?

28.8. Explain why smith forging is not classified as a high production process.

28.9. Explain why hammer forging is restricted to producing reasonably simple shapes and in small lots.

28.10. Explain the procedure and the reasons for "progressively developing" a part in the drop forging process.

28.11. Contrast the methods used in press forging to those used in drop forging.

28.12. Explain why it is sometimes necessary to "restrike" certain forgings.

28.13. List the major advantages of impact forging.

28.14. Explain the basic differences between upset forging and impact forging.

28.15. Give the principal advantages of roll forging.

28.16. What basic stock shape is required prior to ring rolling?

28.17. Give the principal advantages of orbital forging.

28.18. Define the term "no-draft" forging and give what you consider to be the main advantages of the process.

28.19. In what important ways does high-energy-rate forging differ from press forging?

28.20. Give the principal advantages of the process of cored forging.

28.21. List the essential differences between wedgerolling and roll forging.

28.22. List the advantages of incremental forging.

28.23. Under what conditions would forgings be specified for a particular component application?

28.24. Define "forgeability." Discuss reasons why this condition is of critical importance.

28.25. Prepare a sketch of a machinist's ball-peen hammer. Indicate the ideal location of the parting line.

BIBLIOGRAPHY

Aluminum Forging Design Manual, FDM-15, New York: The Aluminum Association, 1976.

*Bridgeport Forgings—Pressure Formed Non-ferrous Metals, Catalog, Bridgeport, Conn.: The Bridgeport Brass Company, 1965.

*CHEPKO, F. E., "Forging Today," Special Report 596, *American Machinist,* Nov. 21, 1966.

*COYNE, J. E., and J. D. McKEOGH, "What's New in Forging?" *Machine Design,* Dec. 23, 1971, pp. 39–44.

DOEHRING, R. C., *New Developments in Cold and Warm Forging,* MF72-526, Dearborn, Mich.: Society of Manufacturing Engineers, 1972.

*DREGER, D. R., "Noiseless Cold Forging," *Machine Design,* Jan. 10, 1974.

*DREGER, D. R., "Non-impact Forging," *Machine Design,* Oct. 4, 1973.

"Forging," *Metals Handbook,* Vol. 5, *Forging and Casting,* Metals Park, Ohio: American Society for Metals, pp. 1–105.

*Forging Industry Handbook, Cleveland, Ohio: Forging Industry Association, 1970.

*GOUGH, P. J. C., "Forging," *Engineer's Digest,* Vol. 34, No. 12, Dec. 1973.

"High Energy Rate Forging," *Metals Handbook,* Vol. 5, *Forging and Casting,* 8th edition, Metals Park, Ohio: American Society for Metals, pp. 99–104.

HINES, C. R., *Machine Tools and Processes for Engineers,* New York: McGraw-Hill Book Company, 1971, pp. 73–93.

*"Hot Shaped Parts—Forging," *Machine Design Metals Reference Issue,* Feb. 14, 1974.

"Ring Rolling," *Metals Handbook,* Vol. 5, *Forging and Casting,* 8th edition, Metals Park, Ohio: American Society for Metals, pp. 105–112.

"Roll Forging," *Metals Handbook,* Vol. 5, *Forging and Casting,* 8th edition, Metals Park, Ohio: American Society for Metals, pp. 95–98.

"Rotary Forger Shapes Refractory Alloys," *American Machinist,* Mar. 5, 1973, p. 64.

*Abstracted with permission.

29

POWDER FORGING

This innovative and relatively new forging process is fast gaining universal acceptance over conventional forging techniques. Essentially, powder forging combines the advantages of the well-established powder metallurgy process (see Chapter 30) with those of forging to produce high-strength, accurate parts at a lower cost. This achievement requires strict process control from start to finish.

DESCRIPTION OF THE PROCESS

Powder forging (also called *powder metal forging* or *hot forming*) starts with a preform, instead of the usual solid metal forging billet. The preform, consisting of impurity-free powders, is made on a standard powder metallurgy compacting press (see Figure 30-2) and is sintered in a conventional powder metallurgy furnace. Sizing and re-pressing operations are often performed to improve the dimensional accuracy. Coining is frequently necessary to increase the density of the part by reducing the voids. Unlike conventional forged parts, which normally require a succession of blows in a series of dies, the heated powder preforms are forged in one operation by a single blow of a set of preheated closed dies.

A typical hot-forming press is shown in Figure 29-1. Combinations of mechanical and hydraulic actions are used to create the very high forging pressures that are required for this process. Prior to forging, the preforms

FIGURE 29-1 A 400-ton double action hot forming press
(Courtesy, Cincinnati Incorporated)

are usually coated with graphite, which prevents the effects of oxidation and provides a lubricant. Figure 29-2 shows a coated preform and the resulting spur gear produced by powder forging. Lubricants are also periodically sprayed on the die cavities. Because the preforms are made to a size that may closely correspond to the end-product size, there is little, if any, material waste. Figure 29-3 shows a spur gear and the corresponding preform that are typical of the size used to produce the gear. Parts may be produced to close dimensional tolerances and with a surface finish of such quality as to entirely eliminate the need for subsequent machining. A high-strength gear may be formed by conventional forging methods followed by machining to final size. The disadvantage in the example is that extensive and costly machining is required to obtain the finished gear. Also, the degree of accuracy and the ability to make complicated forged shapes by conventional forging methods must usually be compromised. By contrast, the same gear could be made at a lower cost and to a closer tolerance by the power metallurgy process, but the physical properties, impact strength and fatigue limit, for example, would not be as high.

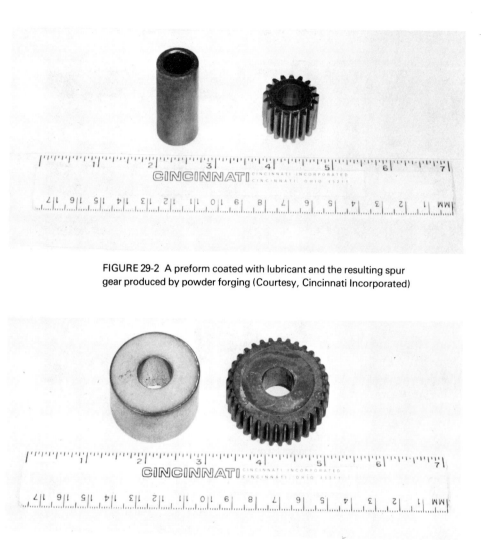

FIGURE 29-2 A preform coated with lubricant and the resulting spur gear produced by powder forging (Courtesy, Cincinnati Incorporated)

FIGURE 29-3 A preform which closely approximates the size of the final part (Courtesy, Cincinnati Incorporated)

PRODUCT APPLICATIONS

The automotive industry is the principal user of this process at the current stage of development, although considerable interest has been expressed by agricultural and recreational equipment manufacturers. Examples of powder-forged production parts include gears and pinions, bearings, pistons, special flanged parts, connecting rods, clutch knobs, and, in general, other similar load-bearing parts. Figure 29-4 illustrates some typical powder-forged parts.

428

FIGURE 29-4 Typical parts produced by powder forging
(Courtesy, Hoeganaes Corporation)

PROCESS SELECTION FACTORS

Cost: Considerable savings over other manufacturing methods have been reported because there is less need for stock removal. As a consequence, costly tool design and machining time can be reduced or entirely eliminated with a minimum of stock waste.

Properties: Powder forging unites the best features of conventional powder metallurgy with those of forging. In general, powder-forged parts have properties that are comparable with parts made from conventional wrought materials. While the billets used in traditional forging may contain undesirable segregates, the grain structure of powder metal preforms is generally more uniform. Because of the manner in which metal powders are blended, as compared to billet production by melting, each small powder particle contains all the desired balanced chemistry. Up to 99.5% of the theoretical density of a given metal can be obtained by powder forging.

Heat Treating: High-quality structural parts made from high-purity steel powder can be heat-treated, case-hardened, or plated if they have been compacted to a theoretical density of at least 85%, or the equivalent of about 6.7 grams/cm^3. Iron parts may be carburized.

Production Rates: Powder forging can be an economical method for short-run production jobs. It has been reported that in some cases substantial savings are possible for productive runs of less than 10,000 parts when compared to alternative, existing manufacturing methods. A practical production rate for reasonably complex parts is about 1000 parts per hour, with

429

1800 parts per hour possible for simpler configurations. The process is well adapted to high production quantities of parts (on the order of 20,000 or more).

Materials: While both ferrous and nonferrous metals can be used, the greatest application is in the production of steel parts. Powder preforms of new and complex superalloys can be obtained in considerably less time than is possible in the production of forging billets, which use the conventional melt-and-cast technology. The development of this process has led to the production of metal alloys, which, heretofore, were considered to be impossible to make in billet form.

Configurations: Part shapes are governed by the usual design restrictions commonly associated with producing metal powder parts. In general, more complex parts with improved detail are possible by this process when compared to conventional forging techniques.

Part Size and Weight: Forging machines are currently available for producing parts up to 8 in. (203.2 mm) in diameter by 4 in. (101.6 mm) in height. Most parts currently being produced weigh less than 4 to 5 lb (approximately 1.8 to 2.3 kg). Forge shops can maintain a net weight variation of one part to another to within ¼%.

Surface Finish: The powder forging process is capable of yielding parts with surfaces that are usually equal to or better than a machined surface. Many parts may be installed in-service "as forged."

Tolerance: For runs of up to 20,000 parts in the same die, a tolerance of ±0.001 in. (0.03 mm) has proved to be practical. Die wear increases with production quantity and, accordingly, more generous tolerances should be expected.

PRODUCT DESIGN FACTORS

There are no universal or clear-cut rules that can be used to guide the design of parts for the powder forging process. As a practical matter, a completely different set of design considerations must usually be established for each new part. When evaluating the feasibility of powder forging a part that has previously been made by another manufacturing process, it may be necessary to completely redesign the part.

Most designers agree that powder forging design should begin with the preform dies. The shape of a preform may either closely resemble the end product or it may be only a rough approximation of the final shape. Designers always strive for simple, straightforward preform shapes. Such shapes usually allow free material flow in all directions to completely form the part. Of critical importance also is the weight control of the preform, since the preform must completely fill the die. As a general rule, the preforms are designed so that the metal will be channeled toward critical, high-strength or load-bearing locations for the highest possible part density. Maximum metal flow also helps to ensure maximum strength and uniformity of the part.

It is suggested that *product design factors* relating to forging (Chapter 28) and powder metallurgy (Chapter 30) be studied in conjunction with this chapter.

REVIEW QUESTIONS

29.1. In what ways do the preforms used in powder forging differ from those used in conventional methods of forging?

29.2. Explain how sizing and coining each contribute to an improved preform.

29.3. Explain the essential difference between powder forging and press forging.

29.4. List the principal advantages of the powder-forging process.

29.5. Compare the complexity of part shapes possible by powder forging to shapes produced by conventional forging techniques.

29.6. Explain why it is advisable to use only simple, uncomplicated preform shapes.

29.7. Regarding the as-cast tolerance, which of the casting processes summarized in Chapter 11 compares most favorably with the tolerance given for powder forging?

BIBLIOGRAPHY

ANTES, H. W., "Processing and Properties of Powder Forgings," in J. J. Burke and V. Weiss, (eds.) *Powder Metallurgy for High Performance,* Proceedings of the 18th Sagamore Army Materials Research Conference, Syracuse, New York: Syracuse University Press, 1972, pp. 171–190.

BELDEN, B. B., and R. F. HALTER, "PM Parts with the Strength of Forgings," *Machine Design,* July 12, 1973, pp. 116–121.

BROWN, G. T., "Powder Forging: Properties Related to Sintered and Wrought Steels," *Metallurgia and Metal Forming,* Vol. 41, No. 6, June 1974, pp. 172–173 and 175–178.

COYNE, J. E. and J. D. McKEOUGH, "What's New in Forging," *Machine Design,* Dec. 23, 1973, pp. 39–44.

"Design for Powder Forging—The Route to Strong Precision Parts," *Design Engineering,* Feb. 1975, pp. 41–42.

DIAKOKU, T., F. MASAYAMA, and T. MATSUMOTO, "Studies on Powder Forging," Proceedings of the Annual Meeting, Japan Society of Powder and Powder Metallurgy, Okayama, Japan: November 26–28, 1974.

DIVAKAR, S., "Powder Forging," *Metallurgy Engineering IIT,* Bombay, India, Vol. 8, 1977, pp. 31–35.

FISCHMEISTER, H. F., L. OLSSON, and K. E. EASTERLING, "Powder Metallurgy Review: Powder Forging," *Powder Metallurgy International,* Vol. 6, No. 1, 1974, pp. 30–39.

GALE, K. "Powder Forging Makes the Best of Both Worlds," *Engineer,* Vol. 235, Sept. 7, 1972, pp. 54–55.

GRIFFITHS, T. J., R. DAVIES, and M. B. BASSETT, "Compatibility Equations for the Powder-Forging Process," *Powder Metallurgy,* Vol. 19, No. 4, 1976, pp. 214–220.

JARRETT, M. P. and P. K. JONES, "The Powder-Forging Process and its Potentialities," *Engineer's Digest,* Vol. 34, No. 3, March 1973, pp. 41–43.

JOHNSON, H. V., *Manufacturing Processes: Metals and Plastics,* Peoria, Ill.: Chas. A. Bennett Co., Inc., 1973, pp. 453–463.

JOHNSON, R. K., "Putting Metal Powder Parts to Work," reprinted from *Appliance Engineer,* Vol. 7, No. 1 (c), Elmhurst, Ill.: Dana Chase Publications, Inc., pp. 371–375.

JONES, P.K., "Forging of Powder, Metallurgy Preforms," Vol. 6, *New Perspectives in Powder Metallurgy, Fundamentals, Methods, and Applications,* Metal Powder Industrial Federation, New York 1973, pp. 19–34.

KHOL, R., "Forged Powder Metal," *Machine Design, Uncommon Metalworking Methods,* Cleveland, Ohio: Penton Publishing Co., 1971, pp. 52–56.

LINDBERG, R. A., *Processes and Materials of Manufacture,* 2nd edition, Boston: Allyn and Bacon, Inc., 1970, pp. 356–359.

PILLIAR, R. M., W. J. BRATINA, and J. T. McGRATH, "Fracture Toughness Evaluation of Powder-forged Parts," *Modern Developments in Powder Metallurgy,* Vol. 7, 4th International Powder Metallurgy Conference, Toronto, Ontario, July 15–20, 1973, pp. 51–72. Available from Metal Powder Industrial Federation, Princeton, N.J., 1974.

"P/M Forging," Geneva Ill.: Burgess-Norton Manufacturing Co., 1973.

WICK, C. H., "Forging Powder Metal Preforms," *Machinery,* August, 1970, Reprint 7642, Cincinnati Incorporated, Article 5.

YAMAMURA, T., K. TATSUMO, and Y. SASAKI, "Powder Molding Techniques, Powder Forging Techniques, and the Facilities," *Research and Development,* Vol. 24, No. 2, April 1974, pp. 58–72. In Japanese.

YUASA, K., "Experimental Manufacture of Gears by Powder Forging," *Komatsu Technical Report,* Vol. 22, No. 73, 1976, pp. 61–71.

30

POWDER METALLURGY

Powder metallurgy (P/M) is a well-established production process by which parts are made by compressing metal powders in a mold. Strength and other properties are added to the parts by subsequent sintering (heating) operations. The process may be used to produce porous parts as well as parts with high-density structures.

DESCRIPTION OF THE PROCESS

Basically, the process consists of three steps; the *preparation, blending,* and *mixing* of the metal powder; its *compaction* in a die to a prescribed density; and finally, *sintering* in a specially controlled atmosphere furnace.

Metal Powder Preparation

Pure metal powders may be used and, because molten material is not involved in the process, a variety of alloy combinations can also be used. The most commonly used raw materials are iron-base and copper-base powders. Other powders, such as stainless steel, aluminum, tin, nickel, titanium, chromium, graphite, metal oxides, and carbides, are used less extensively. Refractory metals such as tungsten, tantalum, and molybdenum are impractical to melt and cast. They are, however, economically fabricated by the P/M process. P/M may also be used as a method to produce non-metallic parts made from graphite or silicon. A further classification of

433

powders, consisting of special combinations of metals and ceramics, called *cermets,* are also adaptable to this process.

Metal powder manufacturers work to certain specifications, which, in turn, determine the ultimate characteristics and physical properties of the compacted parts. Such specifications commonly include requirements governing particle size and shape, and other qualifying characteristics affecting powder flow, compressibility, purity, and apparent density. The nature of these characteristics guide the powder manufacturer in choosing which of the many methods available should be employed to produce the powder.

In general, the most common powder production methods are reduction, atomization, electrolysis, shotting, mill grinding, and milling or crushing. *Reduction* of refined ore or sponge iron is a common method used to produce iron, nickel, cobalt, molybdenum, and tungsten powders. *Atomization* is a method that is used principally for brass and bronze, but it may also be used to produce zinc, tin, lead, and aluminum powders. The process consists of interrupting a fine stream of molten metal with a jet of air, inert gas or steam. Varying sizes of particles are formed as the molten metal solidifies. Copper and iron powders may be produced by *electrolysis.* The process is similar to electroplating, in which a metal deposit is allowed to build up on metal plates suspended in a tank filled with electrolyte. After a controlled period of exposure to a dc current, the plates with their metal coating are removed, dried, and stripped of the deposit. The material is generally milled or otherwise broken down to the desired particle fineness. *Shotting* results in the production of relatively large spherical particles and is adaptable for producing most metal powders. In this process, molten metal is poured through a small opening. The droplets are transformed into small particles as they are agitated and cooled in water. *Mill grinding* may be used to produce metal powders from a large block of material or as a method to further reduce the particle size of iron or steel produced by some other powder production method. *Milling* or *crushing* is accomplished by subjecting brittle material to crushing or impact loads applied by crushers, stampers, and rotary or ball mills. This method is capable of producing extremely fine particle sizes. Other, less commonly used powder production methods include granulation, machining, precipitation, condensation of metallic vapors, thermal decomposition of carbonyls, and high-velocity impacting.

Mixing or Blending

It is generally unnecessary to mix the powder unless the distribution of the particle size is unsatisfactory or in cases where a lubricant is to be added. Lubricants promote the flowability of the powder, which is considered to be its most important characteristic. Lubricants also reduce die wall friction during the subsequent pressing operation, and aid in the ejection of the compacted part. Stearic acid, lithium stearate, or powdered graphite are the principal lubri-

FIGURE 30-1 Powder metal is mixed or blended in tumbling barrels (Courtesy, Burgess-Norton Mfg. Co.)

cants used. The mixing is mechanically accomplished in a tumbling barrel as in Figure 30-1 and may be done either wet or dry. Wet mixing is accomplished by adding water or a solvent to the dry powder to reduce the possible hazards of dust and explosion. Alloyed powders are produced by combining a homogeneous mixture of carefully weighed and blended powders. Mixing and blending are necessary steps when combining metallic and nonmetallic powders.

Compacting

The compacting process, called *briquetting,* starts with the automatic filling of the die cavity with the required amount of blended powder (Figure 30-2). In the next step, punches simultaneously enter from the top and bottom of the die to compress the metal powder and lock the particles together. In most cases, P/M parts are pressed cold. Metal powder does not behave or readily flow as a fluid or semiplastic material. The resistance to flow builds up as the compacting pressure increases; double-action punches are usually used to offset this tendency. The final step consists of the ejection of the finished part by the rising action of the lower punch. At this stage, the somewhat fragile briquette is known as a *green compact.* Figure 30-3 shows a compacted part being ejected to make ready for another operating cycle.

435

FIGURE 30-2 Inset—die cavity is filled with powder prior to the downward squeezing action of the upper punch which forms the shape of the spur gear similar to the one shown in the inset (Courtesy, Kux Machine Division-The Wicks Corporation)

FIGURE 30-3 A compacted part being ejected to make ready for another operating cycle. Note punch in raised position (Courtesy, Aluminum Company of America)

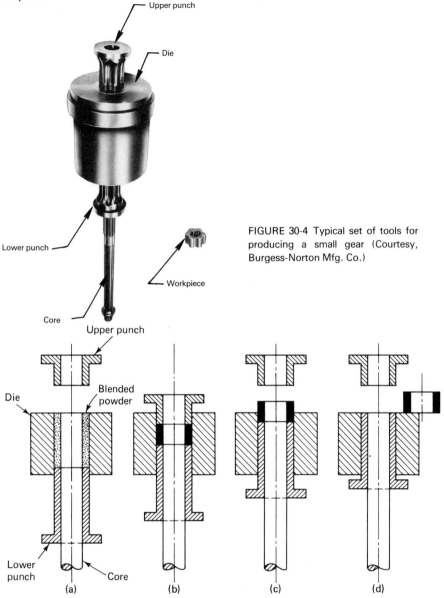

FIGURE 30-4 Typical set of tools for producing a small gear (Courtesy, Burgess-Norton Mfg. Co.)

FIGURE 30-5 The operating cycle for compacting a simple part: (a) the empty die cavity is filled with blended powder (b) both the upper and lower punches simultaneously press the metal powder into the die (c) the top punch is withdrawn and the green compact is ejected from the die by the lower punch (d) the green compact is pushed out of the pressing area to make ready for another operating cycle (Courtesy, Metal Powder Industries Federation)

Compaction by a single punch, except in the case of a small part whose diameter-to-length ratio is nearly equal, often results in a part with non-uniform density. The density on the end of the part nearest the punch face is at maximum while the density at the center and at the opposite end is considerably less. The resulting compact may have relatively large voids between the adjoining particles. Figure 30-4 shows a typical set of "tools" for producing a straight cylindrical part. Core rods for producing holes as well as specially designed punches for making various stepped sections can also be incorporated into the die set.

Figure 30-5 illustrates a typical operating cycle for a simple punch and die arrangement for compacting a metal powder part. It should be noted that the top and bottom punches determine the vertical configuration of the part while the die-wall configuration forms the lateral dimensions of the part.

While most presses operate in a range 5 to 50 tons (4.5 to 45 metric tons), the maximum press capacity is 300 tons (270 metric tons). Some are mechanically operated with a very fast automatic action and are used for high production runs. Figure 30-6 illustrates a 200-ton (180 metric ton) mechanical top-drive compacting press. Special rotary table presses are used also. These are equipped with multiple die cavities, each having a corresponding set of telescoping top and bottom punches. The final briquette is progressively formed as the table indexes from station to station. Large parts, or those which require higher pressures, may be compacted on hydraulic presses, shown in Figure 30-7.

FIGURE 30-6 These 200-ton mechanical top drive compacting presses have a 6 in. depth of powder fill (Courtesy, Burgess-Norton Mfg. Co.)

(a)

FIGURE 30-7 Two models of hydraulic powder metal presses (Courtesy, Kux Machine Division-The Wicks Corporation)

(b)

Sintering

The final step in the production of most P/M parts consists of heating the green compacts to within 60 to 80% of the melting point of the principal constituent. It is important to keep the temperature below the melting point so as to prevent actual *casting* of the metal. The sintering time varies with different metals and may range from 20 minutes to 1 hour or more. Batch- or continuous-type furnaces are used for sintering. A typical high-volume continuous-type furnace is shown in Figure 30-8. In this case, the wire mesh conveyor shown in Figure 30-9 carries the compacts through the controlled-atmosphere furnace. The lubricants that were originally blended with the powders are permitted to burn off in a special chamber before the parts reach the high-heat zone of the furnace.

It is also essential to reduce the tendency of surface oxide films to form. This is done by introducing a protective atmosphere—gases, such as hydrogen or dissociated ammonia—that react with the scale on the metals and cleans away the contamination as well as controls the carburization or decarburization of iron and iron-rich compacts.

FIGURE 30-8 Powder metal compacts are sintered at roughly 2000°F in endothermic gas (Courtesy, Burgess-Norton Mfg. Co.)

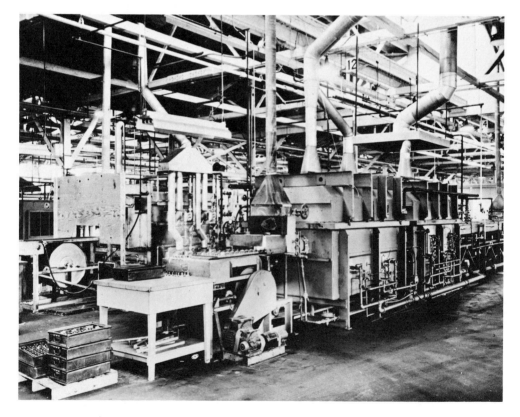

FIGURE 30-9 Iron textile ring blanks entering a sintering furnace
(Courtesy, Metal Powder Industries Federation)

Sintering is an essential step in the development of the ultimate strength of the final part. The most important purpose of sintering is to establish, by intimate contact, a diffusion bonding of the atoms of the particles. Most materials remain solid during sintering. A liquid phase may not necessarily occur. In the case of cemented carbides (carbides plus a cobalt binder), for example, sintering increases the density of the part due to the action of the cobalt, which dissolves small fragments of the carbide and promotes fusion and bonding. Particles of iron, copper-base, or iron-base parts do not fuse, however, and are strengthened solely by the action of the atoms as they diffuse and bind to one another. The sintering temperatures help to promote these reactions as well as to dissipate trapped gases locked in the voids.

Depending upon such variables as the particular powder composition, particle configurations, compacting pressure, and so on, a sintered part may measure either undersize or oversize. This tendency of shrinkage or of growth may be offset or compensated by machining the die cavity to a predetermined size according to the density of the green compact.

FIGURE 30-10 An appliance gear which has been infiltrated with copper (Courtesy, Delco Moraine)

Infiltration

Infiltration is a secondary P/M step, principally used for iron-base compacts, which increases the density, strength, and hardness of a part. It consists of placing a metal blank or slug directly on or under a previously sintered part and then reheating it to a temperature slightly below the original sintering temperature. Only blanks or slugs of low-melting-point metals such as copper, copper alloys, or brass are used. The metal blanks melt and flow or infiltrate from above, or are absorbed by capillary action from below into the pores of the presintered part. Strength may be increased from 70 to 100% by this step. A recently developed extension of this step now makes it possible to accomplish sintering and infiltration simultaneously. However, maintaining dimensional control of the parts is difficult. An example of a part that has been copper-infiltrated is shown in Figure 30-10.

Impregnation

Impregnation, also a secondary P/M step, is a process by which lubricants are added to porous bronze or iron composition bearings, gears (as in Figure 30-11), and pump rotors. Wax, grease, or oil may be used as lubricants. Various impregnation methods are used to fill the pores. The simplest method is to immerse the parts in a tank of heated oil. Vacuum pressure is another method commonly used.

FIGURE 30-11 A crankcase oil pump gear which has been oil impregnated (Courtesy, Delco Moraine)

Sizing

Certain critical features of P/M parts are often processed as a secondary operation to a more precise size and with a better surface finish than was obtained in the sintered condition. *Sizing* is a cold-working operation. One method of sizing parts is shown in Figure 30-12. The procedure consists of holding the part in a simple fixture so that an accurately machined mandrel or tool can be forced through a hole, slot, or some hollow feature of the part. Another method is to "extrude" or force the part through die openings in a special die, resembling that used originally for compacting the part.

FIGURE 30-12 Critical dimensions on aircraft pistons are sized to insure the required tolerances (Courtesy, Burgess-Norton Mfg. Co.)

443

Coining

Coining, like sizing, is also a cold-working process. It consists of repressing P/M parts by use of high pressures in dies made especially for this purpose. Increased densities on the order of 95% may be obtained by mechanically reducing the voids that remain between adjoining particles. Densification also improves the surface quality and precision of the part. Raised or indented letters or numbers may be formed on outer surfaces of parts by using high-pressure embossing–coining dies. Coining is not practical on carbide materials.

PRODUCT APPLICATIONS

Product applications for powder metal parts are virtually unlimited in scope. While commercial applications for the process extend back to World War II, a number of new and interesting adaptations are being developed almost on a daily basis. Figure 30-13 illustrates some typical product applications. The variety of applications may be conveniently divided into the various categories that follow.

Hydraulic pump gear

Rotor in agricultural spray pump (stainless steel)

Operating cam mechanism—vending machines (5-piece assembly)

Gearotor assembly

Sector gear

Spur-bevel gear— garden tractor transmission

Thrust plate

Golf cart differential gear

FIGURE 30-13 Typical powder metallurgy production parts (Courtesy, Burgess-Norton Mfg. Co.)

444

Porous Parts: P/M is unique in that it is the only practical method of producing parts with controlled porosity. Both metallic and nonmetallic structures can be obtained with a uniformly spaced network of pores. Of considerable importance is the production of special screens and filters used in controlling the flow of gases and liquids. Such parts are made with porosities as high as 80% of the total volume. Less-porous structures are created for self-lubricating bearings, slide strips, expansion plates, gears, cams, and so on. Parts of this kind are sintered with a porosity of up to 40% of the total volume followed by impregnation with a suitable lubricant.

Structural and Machine Parts: Some parts, especially those used in the aerospace industry, often have exacting heat-resisting, corrosion-resisting, or hardness requirements. Special combinations of alloyed metal powders are readily obtainable to meet these in-service demands. In fact, powders may be combined to produce parts from materials that are not formable by any other method. In some cases, parts can be made of materials too hard or brittle to process by conventional methods. Since the process is adaptable to the production of complicated shapes to a precise part size and superior finish, machining operations are frequently unnecessary. P/M gears and a wide variety of other parts, such as windshield wipers and switches, are used in the automotive industry. Hardware items or assorted parts found on many household appliances, industrial equipment, power tools, business machines, farm and garden equipment, and on various machine tools are frequently produced by powder metallurgy.

Cemented Carbides: Cutting tools and dies made from tungsten carbide continue to be an important P/M product. Thin-wall wear-resistant carbide bushings are also included in this category.

Friction Parts: Various combinations of metals and ceramics are used to produce brake and clutch linings, friction disks, and for other special friction parts, such as rollers, spacers, and sliders.

Electrical Parts and Magnets: Other important users of P/M products are the electrical or electronic fields for parts such as motor or generator brushes and various other electrical contacts. A combination of graphite and copper powders is generally employed for such parts. Of considerable importance also are magnets and cores composed of iron in combination with aluminum, nickel, and cobalt powders. Various radio and electronic production components are made by this process.

Other Miscellaneous Uses: Thermit welding powder, arc welding electrodes, and nuclear fuel elements are manufactured by the P/M process. Self-lubricating bearings are produced by blending combinations of certain hard and soft dissimilar metal powders. Another unique application is in the production of laminated special-purpose parts with various layers of materials securely and permanently interlocked.

PROCESS SELECTION FACTORS

Part Quantities: Powder metallurgy producers universally agree that a minimum production run of 10,000 parts is needed to offset the tooling costs. As in almost any other manufacturing process, a lower-quantity run may be justified in special cases where savings may be obtained by eliminating expensive machining operations or if the unique material specifications for a given part cannot be met by any other practical method. An important consideration is that the process is adaptable to automated production methods.

Production Rates: Some parts may be manufactured in quantities of 500 to 1000 pieces per hour.

Relative Costs: Powder metallurgy dies are subjected to extremely high pressures, which may range from 5 to as much as 300 tons per square inch (or about 700 to 42,000 kg/cm²). Furthermore, powder compaction, especially for carbide materials, imposes severe wear on the die walls. Dies that are designed for long production runs, in the order of 150,000 to 200,000 pieces, for example, tend to be rather massive, resulting in high material costs. Die costs increase as the complexity of P/M parts increases. High powder material costs may usually be offset by an almost total absence of scrap. In many cases, part costs are reduced by eliminating machining. The complete assembly of press tooling usually consists of dies with special inserts, together with corresponding punches and core rods. Dies are made of high-alloy steel or, where abrasion is especially severe, tungsten carbide. Hardened-steel- or carbide-tipped punches are required. Core rods may be coated with carbide, applied by flame or plasma spraying, to resist abrasion.

Labor costs are comparatively low. Relatively unskilled workers can operate the production compacting and sintering equipment.

Materials: Powder metal parts of very high purity can be produced. Various combinations of materials commonly used include iron, iron–copper, iron–carbon, iron–copper–carbon, brass, bronze, aluminum, magnesium, stainless steel, nickel, silver–nickel, and nickel alloys. In addition, various mixtures of metals with nonmetallic materials (such as refractory oxides) are used for many applications.

Properties: Mechanical properties of P/M parts are determined largely by how the part is compacted. Increasing the density of a compact made from iron, steel, copper–steel, or nickel–steel powder invariably increases the tensile strength, hardness, and usually the elongation. Iron-base powder blends seldom are compacted in the cold-pressing stage to much more than 85 to 90% of the density of the parent metal. Factors limiting cold densification of P/M compacts include press capacity and compressibilities of powder blends. Repressing by coining may improve the density to about 95% of the theoretical density. It is possible to produce parts with densities as high as 99.5% by extending conventional P/M techniques by a process known as *hot densification.* To do this, a *preform* that closely resembles the finished part is heated to an elevated temperature and compacted in a finishing die with a single stroke of the press. Another method of increasing the density of a P/M part is by powder forging, which is explained in Chapter 29.

Part Size and Weight: These are regulated by the capacity of available presses and the compressibility characteristics of the various powder combinations. When comparing the size of P/M parts to some of the larger parts which are produced by sand casting or by forging, the limitations of the process become apparent. P/M is principally restricted to the manufacture of comparatively "small" parts. An example of a "mini-size" part is the ball in a ballpoint pen, which weighs only a fraction of a gram. Figure 30-14 shows other typically small production powder metal parts. Parts as thin as 0.030 in. (0.76 mm) can easily be produced. On a somewhat larger scale, however, parts as large as 4 in. (101.6 mm) in diameter by 6 in. (152.4 mm) in length can be produced in presses with up to a 100-ton (90-metric-ton) capacity. Still larger parts, up to 10 in. (254 mm) in diameter by 8 in. (203.2 mm) in length, have been successfully produced on presses with 100- to 300-ton (90- to 270-metric-ton) capacities. Parts weighing as much as 200 lb (90 kg) and more have been successfully mass-produced by P/M. An important advantage of the process is that consistent production control can be maintained to equalize the weight of one part to another.

Configuration: In general, parts must have a uniform cross section which extends throughout the entire length. Cross-sectional shapes such as cylinders, squares, and rectangles are the preferred configurations. The ideal part geometry is achieved when the length is relatively short in comparison with the diameter or width. Parts whose length-to-diameter ratio is not more than 3:1 are the easiest to manufacture.

Surface indentations or projections can easily be formed on the tops or bottoms of parts, and flanges or lateral projections can be formed at either end but not both. Splines, gear teeth, axial holes, counterbores, straight knurls, slots, and keyways can be formed easily.

Tolerances: For *sintered* parts, the tolerance in the direction across the part (horizontally) is ±0.001 to 0.002 in./in. (0.001 to 0.002 mm/mm). The

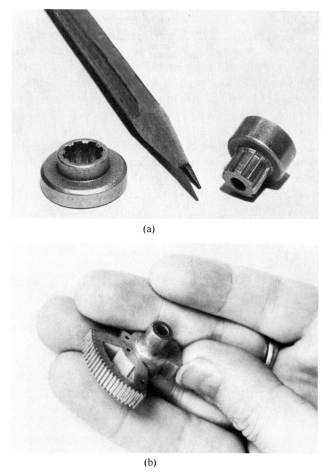

(a)

(b)

FIGURE 30-14 Examples of small powder metal production parts:
(a) (Courtesy, Chrysler Corporation-Amplex Division); (b) (Courtesy,
New Jersey Zinc Co.)

length (or die fill) tolerance is 0.004 in./in. (0.004 mm/mm). Concentricity for holes up to 1 in. (25.4 mm) in diameter is 0.003 in. (0.08 mm) T.I.R. (total indicator reading) with 0.001 in. (0.03 mm) for each additional inch in length. For *sized* or *coined* parts, the tolerance in the direction across the part as well as for the length is ±0.0005 in./in. (0.0005 mm/mm). The tolerances for concentricity cannot usually be improved by sizing or coining.

Heat Treating: Except in special cases where heat treatment may adversely affect the accuracy of the part, most ferrous parts may be hardened by quenching and tempering, case hardening, carbonitriding, or tuftriding.

Some Limitations

Strength: As a general rule, wrought, cast, or forged parts may exhibit superior physical properties over similar P/M parts. Residual oxides, which cannot be removed by normal sintering operations in some metal powders such as aluminum, titanium, zinc, and tin, may yield parts with inferior properties. As a practical matter, P/M parts are selected for particular in-service applications under conditions of predictable light to medium loads. If the P/M process proves to be the only practical method for producing a given part, however, improvements in mechanical properties can usually be obtained by altering the metal powder combinations.

Powder Storage: Some powders tend to deteriorate when stored over prolonged periods of time. Furthermore, aluminum, zirconium, and magnesium powders, for example, may present a fire or explosion hazard.

Density: For any given powder combination, there is a practical density limit to which the powders may be compacted by this process. Annealing operations are usually employed to increase the compressibility of some powders, particularly the harder metal powders. Annealing is often necessary to offset the effects of workhardening, which usually accompanies powder pressing.

PRODUCT DESIGN FACTORS

Unlike the forging or casting processes, in which parts are formed in three dimensions, the P/M process is somewhat restricted by the two-dimensional formation of part shapes. Because metal powders do not flow (even under high pressure) as do fluid or semiplastic materials, part complexity can only be incorporated along the axis of compression and not in a third dimension across the die. Since it is impractical to apply pressure from the sides of the die, the lateral flow of powders in a die is nonexistent. As a result, parts cannot be produced with reentrant forms such as threads, peripheral grooves, undercuts, diamond knurls, or overhangs. In this particular respect, the limitations attendant upon this process resemble those in other pressworking processes. Other similarities in the design of P/M tooling to forging, as well as to some of the casting processes, include precautions that must be observed for part ejection, the necessity of strong tooling, and requirements for fillets and rounds on the respective parts.

Section Thickness: Sidewall or flange thickness, shown in Figure 30-15, should not be less than 0.030 in. (0.76 mm). It is unwise to concentrate a large load on a punch over a small area. A practical minimum wall thickness for a bushing should be about 0.060 in. (1.52 mm), as shown in Figure

449

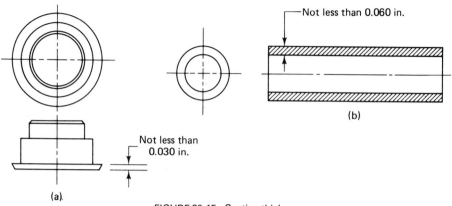

FIGURE 30-15 Section thickness

30-15. Extremely large flange diameters (as compared to a given body diameter, for instance) are not practical. A suggested ratio of flange diameter to body diameter is given in Figure 30-16.

Cross-Sectional Shapes: As has been mentioned previously, the process is most ideally suited to the production of cylindrical, rectangular, or irregular cross-sectional shapes, particularly those which do not have variations in form.

Undercuts and Annular Grooves: Undercuts and annular grooves cannot be formed on the sides of a part. These must be accomplished as secondary machining operations (see Figure 30-17).

Avoid narrow deep splines as in Figure 30-18a.

Slots: Slots deeper than one-fourth the axial length of the part require multiple punching action, which inevitably results in higher production costs. A 4° taper on each side of the slot will strengthen the tool and permit easier ejection of the part.

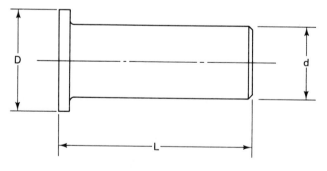

When L is greater than 0.750 in., D should
not exceed 1.5d

FIGURE 30-16 Relative diameters

450

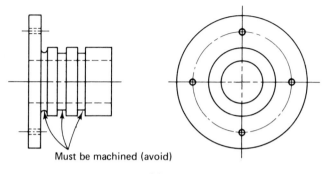

Must be machined (avoid)

(a)

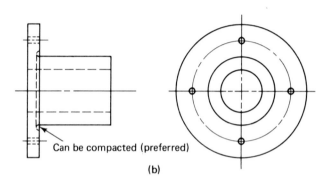

Can be compacted (preferred)

(b)

FIGURE 30-17 Undercuts and annual grooves

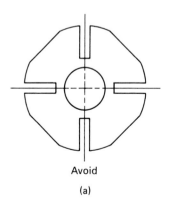

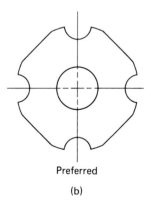

Avoid   Preferred

(a)   (b)

FIGURE 30-18 Splines

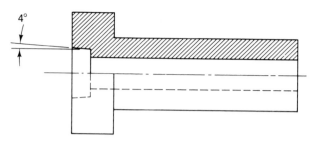

Counterbored holes

FIGURE 30-19 Counterbores

Holes: Holes along the axis of compression should be no less than 0.040 in. (1.016 mm) in diameter and, generally, no deeper than 3½ in. (88.9 mm). Holes may be circular, square, fluted, or odd-shaped, although circular holes are the easiest to produce. As in the case of slots, counterbored holes (Figure 30-19) should also have a 4° wall taper. Blind holes may be produced using special tooling unless the part geometry prevents convenient ejection of the part from the die. Cross holes cannot be produced.

Fillets and Rounds: Avoid sharp internal and external corners. A good practical minimum radius for fillets and rounds, shown in Figure 30-20, is 0.010 in. to 0.015 in. (0.25 to 0.38 mm). Sharp corners can weaken parts and dies. Corner radii permit uniform powder flow in the die.

FIGURE 30-20 Fillets and rounds: (a) avoid; (b) preferred

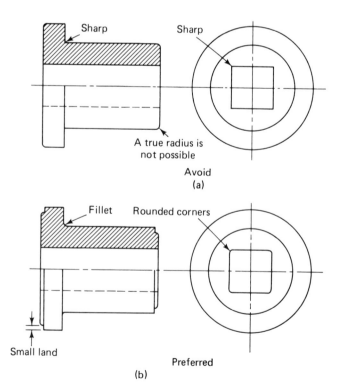

Feather Edges: Feather edges should be avoided.

Threads: It is not practical to compact internal or external threads on powder metal parts.

Stepped Cylinders: Each step should be from $\frac{1}{16}$ to $\frac{1}{8}$ in. (1.588 to 3.18 mm) greater in diameter than the preceding step, as shown in Figure 30-21.

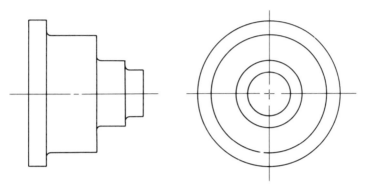

FIGURE 30-21 Stepped cylinders

Gears: The root diameter should be at least 0.100 in. (2.54 mm) greater than the hub diameter.

Cylindrical and Spherical Parts: Figure 30-22 illustrates how the design of a special cylindrical part must be changed by adding a small, $\frac{1}{32}$-in. (0.794-mm), flat surface. In a similar way, spherical parts (Figure 30-8b) must be altered to offset the inability of metal powders to uniformly flow toward the sides of the dies.

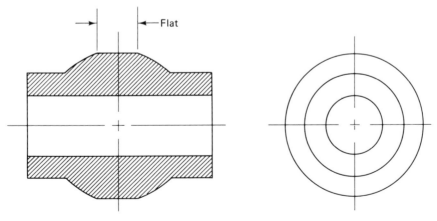

FIGURE 30-22 Cylindrical and spherical parts

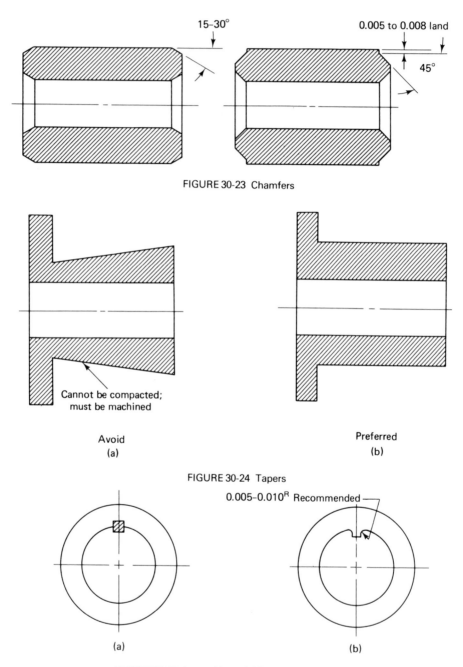

FIGURE 30-23 Chamfers

Avoid
(a)

Cannot be compacted;
must be machined

Preferred
(b)

FIGURE 30-24 Tapers

(a)

0.005–0.010^R Recommended

(b)

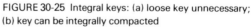

FIGURE 30-25 Integral keys: (a) loose key unnecessary;
(b) key can be integrally compacted

Chamfers: Chamfers of 45° are preferred. This design technique will give added strength to the tooling. The addition of the 0.005- to 0.008-in. (0.13- to 0.20-mm) land, shown in Figure 30-23, eliminates the need for sharp corners on tools, resulting in reduced downtime in tool maintenance.

Tapers: It is not possible to produce reverse tapers, as shown on the part in Figure 30-24.

Integral Keys: Keys can be produced as in Figure 30-25. A 0.005- to 0.010-in. (0.13- to 0.25-mm) radius is recommended.

REVIEW QUESTIONS

30.1. Prepare a brief outline describing the essential elements of the powder metallurgy process.

30.2. Define a "briquette."

30.3. Explain why sintering is performed at a temperature below the melting point of the powder metal part being formed.

30.4. Discuss why sintering is considered such as essential step in developing the ultimate strength of the final part.

30.5. Describe the process of infiltration.

30.6. How does the process of impregnation differ from that of infiltration?

30.7. What is the minimum number of parts needed to offset powder metallurgy production costs?

30.8. What single factor in the powder metallurgy process generally tends to offset high powder materials and die costs?

30.9. What are the principal limiting factors which govern the size and weight of parts that may be produced?

30.10. Contrast the tolerances obtained for sintered and for sized or coined parts with those possible for powder-forged parts.

30.11. Discuss some of the important *limitations* of this process.

30.12. List what you consider to be the main advantages of this process.

30.13. Prepare a sketch of three different parts, each with a variety of representative features that are impractical to produce by powder metallurgy.

PROBLEMS
(See appendix for useful formulas.)

30.1. Calculate the press tonnage capacity and the die fill depth for the parts shown in Figure 30-P1(a-c).

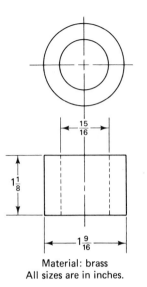

15/16

1 1/8

1 9/16

Material: brass
All sizes are in inches.

(a)

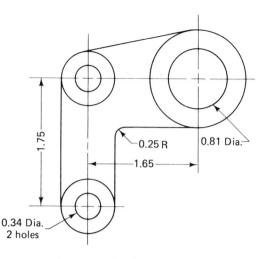

1.75

0.25 R

0.81 Dia.

1.65

0.34 Dia.
2 holes

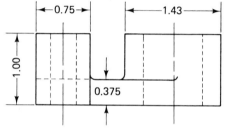

0.75

1.43

1.00

0.375

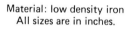

Material: low density iron
All sizes are in inches.

(b)

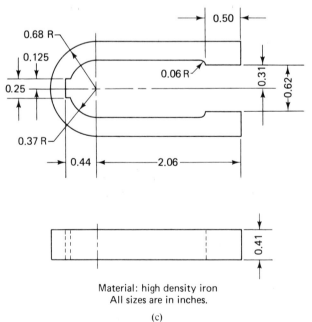

0.50

0.68 R

0.125

0.06 R

0.25

0.31

0.62

0.37 R

0.44

2.06

0.41

Material: high density iron
All sizes are in inches.

(c)

FIGURE 30-P1(a-c)

30.2. Indicate the reasons why the preferred designs shown in Figure 30-P2 (a-e)
will result in the most suitable and the lowest cost powder metallurgy part.

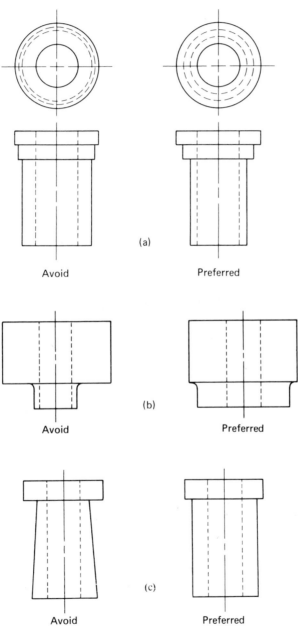

FIGURE 30-P2 (a-c)

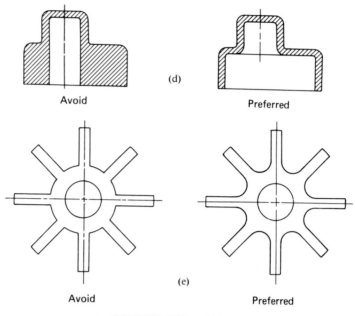

(d)

Avoid   Preferred

(e)

Avoid   Preferred

FIGURE 30-P2(d) and (e)

BIBLIOGRAPHY

*BELDEN, B. B., "PM Parts with the Strength of Forgings," *Machine Design,* July 12, 1973.

HAUSNER, H. H., *handbook of Powder Metallurgy,* New York: Chemical Publishing Co., Inc., 1973.

HAUSNER, H. H., and W. E. SMITH, *Modern Developments in Powder* Metallurgy, Vols. 6, 8, and 9, Princeton, N.J.: Metal Powder Industries Federation.

HIRSCHHORN, J. S., *Introduction to Powder Metallurgy,* Princeton, N.J.: American Powder Metallurgy Institute.

New Perspectives in Powder Metallurgy, Vol. 6, Princeton, N.J.: Metal Powder Industries Federation.

Powder Metallurgy Equipment Association, *Powder Metallurgy Equipment Manual,* revised edition, Princeton, N.J.: Metal Powder Industries Federation, 1974.

Powder Metallurgy Parts Association, *Powder Metallurgy Design Guidebook,* Princeton, N.J.: Metal Powder Industries Federation.

"Production of Powder Metallurgy Parts," *Metals Handbook,* Vol. 4, *Forming,* 8th edition, Metals Park, Ohio: American Society for Metals, pp. 449–465.

Standards and Specifications for Metal Powders and Powder Metallurgy Products, Princeton, N.J.: Metal Powder Industries Federation.

*Abstracted with permission.

31

PRESSWORKING

Pressworking includes a wide variety of chipless processes by which work-pieces are shaped from rolled metal sheets. A general term, *stamping,* is used almost interchangeably with pressworking. Stamping is normally understood to include the general headings of *cutting, bending* or *forming, squeezing,* and *drawing.* Each of these headings has been further subdivided and will be described in considerable detail in successive chapters.

DESCRIPTION OF THE PROCESS

Stampings are produced by the downward stroke of a ram in a machine called a *press.* The ram is equipped with special punches and moves toward and into a die block, which is attached to a rigid bed. The punches and the die block assembly are generally referred to as a "die set" or, more simply, as the "die." Pressworking operations are usually cold-worked at room temperature or in some cases hot-worked at an elevated temperature that does not exceed the annealing temperature of the metal.

Presses

The main function of a stamping press is to provide sufficient power to operate the die and the movements necessary to close and open the die. The press also maintains the alignment of the punches and dies. Some simple

459

(a)

(b)

(c)

FIGURE 31-1 Various metalforming presses: (a) a 60-ton single crank OBI press-air clutch flywheel model (Courtesy, Rousselle Corporation); (b) a 100-ton straight-side press (Courtesy, Rousselle Corporation); (c) a battery of nine stamping presses produces 30,000 catalytic converters a day. The line includes four 1800-ton presses with 13 die stations and four 1200-ton presses with nine stations (Courtesy, Verson Allsteel Press Company)

single-action presses can be operated by hand feeding the metal stock and by manually tripping the dies on a one-to-one basis. Most production stamping presses, however, are equipped with multiple station dies and are designed for high-speed, completely automatic stock feeding and stamping operations. Some presses are restricted to single-purpose types of operations, such as bending, coining, or punching. Figure 31-1 illustrates a variety of presses. Multipurpose presses are available and, depending upon the design and construction of the dies, are capable of performing varied stamping operations, such as cutting, forming, and punching. Figure 31-2 illustrates a numerically controlled punching and nibbling machine.

Press Classification: Presses are classified in various ways. Some manufacturers list presses according to the type of work for which the press has been designed. Other manufacturers group press types according to the method of power transmission, which may include simple manual operation or mechanical, steam, or hydraulic power operation. Frame and bed designs constitute another way that may be used to classify stamping presses. Finally, press types may be listed according to the action or to the number of rams (single, double, triple); methods of ram operations (crank, eccentric, toggle, knuckle-level); and the position of the ram guides (vertical, horizontal, inclined).

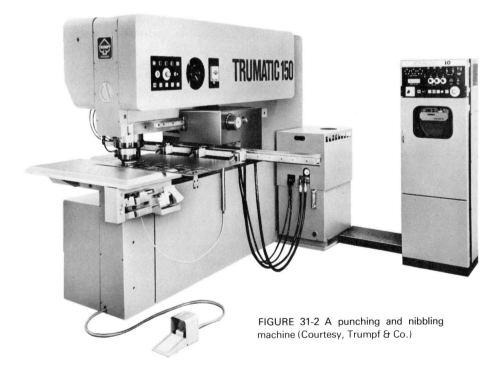

FIGURE 31-2 A punching and nibbling machine (Courtesy, Trumpf & Co.)

JIC Press Identification System: The Joint Industry Conference (JIC) system of identifying press characteristics is in general use. In a typical sample, S4-750-96-72, the press is identified by the S as a single-action model (D is used for double-action, T for triple-action, and OBI for open-back inclinable); by the 4 as having four-point suspension; by the 750 as being rated at 750-ton (670-metric-ton) capacity; and by the 96 and 72 as having a bed measuring 96 in. (2.45 m) left to right and 72 in. (1.83 m) front to back. Any other press can be so identified by substitution of appropriate numerals for the number of suspension points, tonnage rating, and bed dimensions.

The JIC also recommends that a metal tag be attached permanently to the press, stating the stroke length, shut (or maximum die opening) height, kind and length of adjustment, strokes per minute, and bed size and machine weight.

Press Selection Factors

The overall workpiece size, the stock thickness and material, and the nature of the individual operations as they relate to final as-stamped configuration are important factors that must be considered in selecting the type of press to use for a given job. Once these initial considerations have been carefully evaluated, the fabricator can usually determine the type of tooling, the power requirements of the press, and the approximate production rate at which a given workpiece may be produced. In the chapters to follow (cutting, bending, etc.), which explain important elements of the various stamping operations, recommended classifications of presses will be given for most operations. It will be stated, for example, that a *crank-type* press is almost exclusively used for *punching* or *notching* operations.

High-Production Presses: Some presses are made especially for high-production applications. While a machine is rated on its ability to operate at the highest number of strokes per minute, this is not the only determining factor. A high-production press must also be able to operate continuously over a period of many hours, almost unattended and with a minimum amount of wear and vibration.

High-production presses include dieing machines, multiple-slide machines, transfer presses, and "high-productivity" presses. *Dieing machines* are used principally for blanking operations and, in some cases, for forming and drawing. These single-action underdrive presses are set up with progressive dies for long-run operation. Unlike conventional press action, the dies are pulled together rather than pushed together. Large guide rods and bushings are used which keep the die halves in proper alignment, resulting in a prolonged die life. *Multiple-slide machines* are fully automatic and are used principally for producing high quantities of small parts from metal strips or wire.

Two principal types of high-production *transfer presses* are used: One type is a wide-bed, straight-side, transfer press which has a horizontal slide that permits the movement, or transfer, of the workpiece blank from one die to the next. The press is adaptable to the use of either single-operation or compound dies. A press of this type is used to produce parts that cannot be manufactured on a die setup, such as deep drawing, restriking, or cam piercing and trimming. A second type of transfer press is the eyelet machine, which uses straight punches, in a series of single stations, which may be individually repaired or replaced without affecting the rest of the setup. Eyelet-type transfer presses are used for a wide variety of operations, such as bottom stamping and marking, threading, beading, knurling, piercing, and slotting. The coiled stock is fed into the machine and the operations are progressively accomplished as the blank is moved from station to station. Such presses may have from 7 to 15 stations and, depending upon the press design, may have a capacity of from 15 to 75 tons (13.5 to 67.5 metric tons). They operate at up to 250 strokes per minute and produce parts as large as $3\frac{7}{8}$ in. (98.43 mm) in diameter. Draws as deep as $3\frac{1}{8}$ in. (79.38 mm) are practical.

FIGURE 31-3 High productivity presses: (a) a 5-ton press, flywheel type with mechanical clutch (Courtesy, Rousselle Corporation); (b) a 150-ton straight-side double crank press (Courtesy, Perkins Machine Company)

(a)   (b)

High-productivity presses are used for relatively light stamping operations. Such presses are equipped with precision automatic stock-feeding devices and can operate continuously at speeds equal to or greater than conventional presses. High-productivity presses range in design from the 5-ton (4.5-metric-ton) type, shown in Figure 31-3a, to the more rigid straight-sided presses shown in Figure 31-3b, of 150-ton (135-metric-ton) capacity, for heavier work.

Lubrication

In most blanking operations, no lubrication is required. There is usually a light residue remaining on the stock, which was deposited at the mill. Some fabricators routinely apply a light coating of mineral oil to the stock before blanking. It is important that moving parts of the punches and dies be adequately lubricated.

Lubricants are of two main types: water-base and oil-base. The final selection of a specific type of lubricant for a given forming operation is largely dependent upon the severity of the press operation. Lubrication may be applied by sprayer, roller, drip-feed, dip, flood, swab, or brush.

PRODUCT APPLICATIONS

Parts produced by pressworking operations can be as small as a shoe eyelet or as large as the end of a freight car. Compared to other metalworking processes, pressworking techniques offer an almost unlimited choice of metals and design versatility and can be produced in extremely large quantities. Metal stampings are lightweight, strong, and have a superior strength-to-weight ratio. It is estimated that the average household contains products in which there are over 100,000 pressworked items.

Skillful designers are often able to redesign parts previously made by forging or by casting, with significant savings in time, labor, and materials. Because it is practical to produce parts to close limits of accuracy, interchangeability is assured.

Pressworked parts are used for internal components on business machines, machine tools, household equipment, aircraft and small engines, and for locks, various other hardware items, and countless other functional applications. It is estimated that approximately half the weight of an automobile consists of pressworked parts. Formed parts are widely used as containers of various kinds, ranging from household pots and pans, to pails, buckets, and bins. Sheet-metal forms are also widely used for heating, exhaust, and ventilating equipment, medical and food processing equipment, for buildings and structures, household appliances (stoves, refrigerators, freezers, washers, and dryers), bathroom and plumbing and electrical articles, highway

FIGURE 31-4 Typical pressworked parts (Courtesy, Carlstrom Pressed Metal Co., Inc.)

vehicles, farm equipment, office furniture, and for many other applications too numerous to mention. Figure 31-4 illustrates some typical sheet-metal pressworked parts.

PROCESS SELECTION FACTORS

Workpiece Size Capabilities: Factors that govern the maximum size at which a part may be stamped are regulated by the tonnage rating of the press and the bed area. The tonnage rating of a press is defined as the maximum force that a press can apply. Some hydraulic presses are capable of applying up to 50,000 tons (45,000 metric tons). Most general-purpose stamping operations are performed in production presses which may be rated as low as 15 to 1000 tons (13.5 to 900 metric tons), however. Higher pressures and heavier equipment are needed for cold-working than for hot-working operations.

465

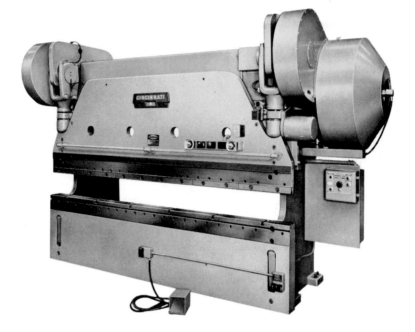

FIGURE 31-5 A 225-ton automatic cycle press brake which can accommodate workpieces 10 ft. long (Courtesy, Cincinnati Incorporated)

As a general rule, parts with a thickness range of 0.020 to 0.750 in. (0.51 to 19.05 mm) may be successfully cold-pressed. Parts as thick as 3½ in. (88.9 mm) have been successfully worked at elevated temperatures. Fabricators classify *small* stampings as those up to 3 in. (76.2 mm) in the longest dimension; *medium* from 3 to 11 in. (76.2 to 279.4 mm), and *large* as roughly equivalent to the size of an automobile roof.

Press brakes are available for forming various cross-sectional bends in steel workpieces up to ⅝ in. (15.88 mm) thick by 20 ft (approximately 6 m) wide. One such press brake is shown in Figure 31-5. Figure 31-6 illustrates a variety of stock die sets available to form the most common bends shown in the inset. Press shears, shown in Figure 31-5b, can cut, in a single stroke, aluminum sheets as thick as ¾ in. (19.05 mm) by 10 ft (approximately 3 m) wide. High-production turret punch presses, shown in Figure 31-6, can punch holes up to 3½ in. (88.9 mm) in diameter in steel sheets as thick as ⅜ in. (9.53 mm) at the rate of 30 holes per minute. High-production transfer presses can operate as fast as 250 strokes per minute and produce draws (shape deformations) in thin metal workpieces as deep as 3⅛ in. (79.38 mm). Large hydraulic presses are capable of forming large automotive body or appliance sections in a single stroke of the machine.

FIGURE 31-6 Stock die sets which are used to form various bend configurations (Courtesy, Cincinnati Incorporated)

Used for air bends from very shallow angles to 30° angles. The angle formed depends on the depth to which the male die enters the female die. Acute angle dies are commonly used to perform hems.

Acute angle dies

Included angle for both male and female dies allows for overbending of metal to compensate for springback. Angles from very shallow to 90° formed by adjusting the press brake ram.

Air bend dies

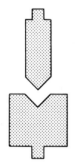

90° included angle for both male and female dies. Used for making very accurate bends with relatively sharp inside radii in comparatively light gage material, such as 12 gage and thinner.

Bottoming dies

Used for "hemming" or flattening acute angle bends. Flattening dies and acute dies mounted side by side in a press brake can produce a hem with each press stroke.

Flattening dies

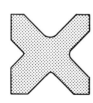

Useful for jobbing work where changes in die opening are frequently desired.

Fourway dies

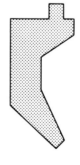

Used for making channels or special shapes when a straight sided die would interfere. Deeper "throat" beyond the die centerline increases width of return flange but reduces die capacity.

Gooseneck dies

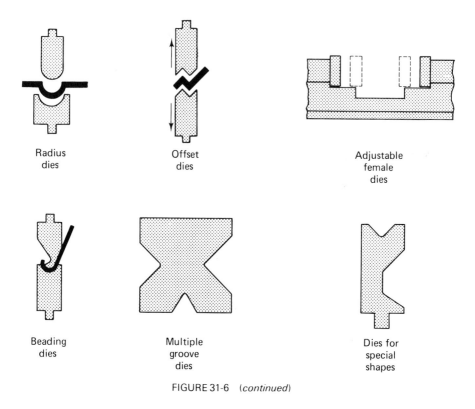

Radius
dies

Offset
dies

Adjustable
female
dies

Beading
dies

Multiple
groove
dies

Dies for
special
shapes

FIGURE 31-6 *(continued)*

Materials: Most metals, except brittle alloys, can be press-worked. Materials with a hardness value up to Rockwell C50 are workable except those in very narrow sections.

The most widely used metal for stamping operations is cold-rolled strip and sheet steel that has a carbon content between 0.005 and 0.20%. Stainless steel types 302, 304, and 305 are excellent for stamping. They should be worked in the annealed condition for maximum forming qualities. Other commonly used materials include copper and brass alloys and soft-tempered aluminum alloys. Magnesium alloys can also be press-worked but at an elevated temperature of from 450 to 600 °F (approximately 232 to 316 °C).

Precoated materials are practical to use in pressworking. These include Terne plate, lithographed stock, tin plate, and galvanized, copper-plated, vinyl-plastic-coated, cadmium-, chrome-, and nickel-plated and stainless-clad steels. It is possible to produce parts without damage to the coating in most cases without unusual tooling adaptations.

A recently developed technique of considerable interest in the pressworking field is the use of conventional metal-forming presses to form plastic sheet material. The material, known as *Azdel* (a product of the G.R.T.L. Co.,

Southfield, Michigan), consists of long glass fibers in a matrix of thermo-plastic material, usually polypropylene. The process consists of four basic steps in which the blank is heated, transferred to the press, formed, and finally ejected. Parts are reported to have high strength, light weight, and corrosion-resistant properties.

Production Rates: There appears to be no practical limit to press-working production. Small stampings may be made at upward of 18,000 strokes per hour, with as many as 20 parts per stroke. Large or complicated parts requir-ing several operations with intermediate anneals are much slower, however, and are often restricted to production runs of less than 100 parts per hour.

Some Limitations

Tolerances: In some plants, it is customary to maintain a running log that records the accuracy and condition of each press over various production runs. Such data are helpful to the designer and the process engineer who are charged with the responsibility of specifying and maintaining practical tolerance limits for production workpieces. A stamping press is rated as being in "good condition" if its accuracy is found to conform to the follow-ing limits:

> *Maximum Tolerance for Parallelism* (between slide and bed): 0.001 in./ft (0.099 mm/m) at the bottom of the stroke for all slides; 0.003 in./ft (0.263 mm/m) at midstroke for punch slides; 0.005 in./ft (0.428 mm/m) at midstroke for blankholder slides.
>
> *Stock Feed Tolerance:* ±0.003 in. (0.08 mm) at 900 in. (22.8 m) per minute.

It is sometimes possible to economically produce parts to a general toler-ance of ±0.005 in. (0.13 mm), and in special cases certain operations may be held to even closer tolerances. As a general rule, production problems are simplified when workpieces are produced to a general tolerance of ±0.010 in. (0.25 mm) or more.

Springback: When metal is bent or formed by cold working, it springs back to a slightly smaller angle after the pressure is released. Allowance is usually made in the press tools to offset this tendency. Overbending or restriking are the more common methods employed by fabricators to overcome spring-back. The condition is particularly evident when forming metals through a relatively small bend radius or through large bend angles. It is also most noticeable on thick materials or on harder materials, and especially on metals that rapidly work-harden. It is interesting to note that the extent of springback can vary in a production run of a given part because of variation in stock thickness, hardness, or temper, and in tool wear and adjustment.

Figure 31-7 illustrates how bending or forming operations on sheet metal or wire stock often result in a condition in which the metal on the outside of the neutral axis stretches while that on the inside is compressed.

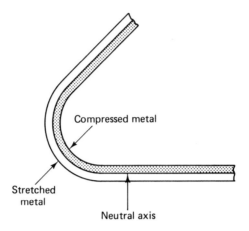

FIGURE 31-7 The neutral axis is neither lengthened nor shortened by bending

Die Marks: Die marks, caused by surface contact with the press tooling, may be noticeable on workpieces. One method commonly employed· by metal fabricators is to use urethane or rubber die pads to eliminate tool scoring on surfaces of the workpieces that contact the pads. Some fabricators insert oil-impregnated paper between the tools and the work metal, which helps to reduce the tendency of surface marking. Polishing is the usual method of removing such marks, particularly on parts formed by drawing operations.

Stretcher Strains: These impose no harmful effects on the strength of the workpiece, but for some in-service product applications, these characteristic surface markings may be objectionable. Press-forming operations on low-carbon-steel workpieces, for example, may result in surfaces that are marked with lines or "worms." Stretcher-strain lines may be visible even on painted surfaces. One method that is used to control the formation of surface marks of this nature is to anneal and then cold-work the metal sheets by temper-passing them through reduction rollers.

Scrap Loss: Most metal stamping jobs result in up to 25% stock waste. Substantial savings in material cost can often be achieved, however, by careful preliminary planning. Most metal fabricators begin a job by making a few trial layouts to find the optimum stock width which results in the least amount of scrap loss. Regardless of the layout arrangement of the blank configuration on the strip or sheet, a small amount of material must usually be provided between the adjacent blanks. Some material allowance is required to support or hold down the strip between blanking or other stamping operations. In an effort to reduce costs, many fabricators try to utilize punched scrap slugs of various shapes for subsequent product applications.

PRODUCT DESIGN FACTORS

Dies and Tooling: All pressworking operations require the use of a matching set of specially designed tools, called a punch and a die block, which may be aligned by means of guide or leader pins. When assembled, these tools are called a *die set.* Close alignment of the punch and die block is essential. In actual practice, the proper clearance is maintained by installing the complete die set as a unit of the press. After the production run has been completed, the entire unit is then removed from the press and returned to storage.

Die Classification

Dies may be conveniently classified according to their function: those used to *cut* metal and those used to *form* metal. Cutting operations include blanking, trimming, shaving, cutoff, shearing, piercing, slitting, perforating, lancing, extruding, notching, and nibbling. Forming operations include a variety of processes, which are usually grouped under the general headings of bending or forming or squeezing or drawing.

Conventional Dies: These consist of one or more mating pairs of rigid punches and die blocks. Additional auxiliary equipment may be added to increase the pressworking versatility. Conventional dies include single-operation or simple dies, compound or combination dies, progressive dies, transfer dies, and multiple dies.

Some die sets are designed to perform a single pressworking operation, which may include any of the operations listed under cutting or forming. Such dies are called *single-operation dies* or *simple dies.* One operation is accomplished by a single stroke of the press. Figure 31-8 illustrates a simple die for trimming a horizontal flange on a drawn shell. In this example, the workpiece is positioned on a locating plug. After the ring-shaped scrap from a sufficient number of trimmed shells has accumulated, a scrap ring at the bottom is severed with each stroke of the press by the scrap cutter and falls clear.

Compound or Combination Dies: These are used when two or more operations may be performed at one station. Figure 31-9 illustrates a

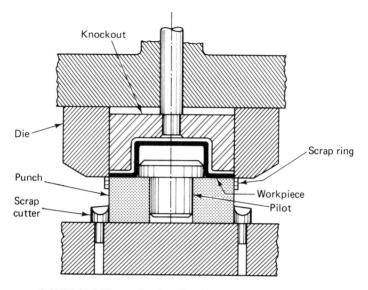

FIGURE 31-8 The tooling for trimming a horizontal flange on a drawn shell

FIGURE 31-9 An example of a part formed in a multiple compound die by one stroke of the press

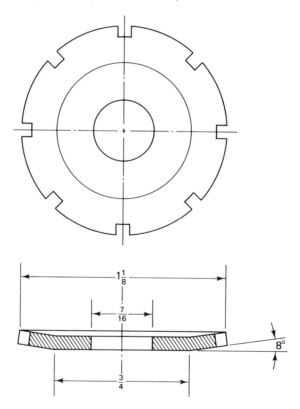

0.093 thick

part that was formed by one stroke of the press in a compound die. In this example, the dished washer was simultaneously blanked, pierced, and formed. Compound dies are more economical in mass-production operations than for a series of single operation dies, and they are usually more accurate.

Progressive Dies: These are used for high-production applications. In this case coil stock or flat strip is fed from station to station. The die performs work at some or all of the stations during each press stroke. When all the work has been completed, the work is cut off and unloaded. Figure 31-10 shows an adapter ring that was produced in a five-station progressive die. Progressive dies are generally expensive to construct. In addition, the cost of the required auxiliary equipment is high. Progressive dies are usually set up on automatic presses with a scrap cutter, feeder, straightener, and uncoiler.

Transfer Dies: These are used to produce parts which, because of their general configuration, are difficult to feed from station to station in progressive dies. Individual precut blanks are first prepared by feeding coil stock into a press. Additional cutting and forming operations on the separate workpiece blanks are then performed by mechanically feeding each blank from station to station. Like progressive dies, transfer dies and their related equipment (presses, special attachments, and feeding devices, etc.) are expensive. Their use is recommended only in cases of high-quantity production.

Multiple Dies: These are also used in mass production. Such dies produce two or more workpieces at each stroke of the press. It is possible to produce pairs of right-hand and left-hand parts, duplicate parts, or unrelated parts. Multiple dies may consist of two or more single-operation dies or multiples of compound dies. Advantages of multiple dies may include savings in material resulting from more efficient blank layout, and reduction of labor costs. The leading disadvantages are increased costs in die construction and in setup and maintenance.

Short-Run Dies: These are frequently used by metal fabricators, particularly for blanking operations. Such dies are generally employed when the production run is limited from a few hundred to about 10,000 pieces. Because short-run dies can usually be made more quickly and installed in the press with less setup time than is required for conventional dies, they are used for trial runs and as a means of expediting delivery of parts. In most cases, changes can be readily incorporated into the design of conventional high-production die sets. Inexpensive steel-rule and template dies are the principal types used for short-run applications.

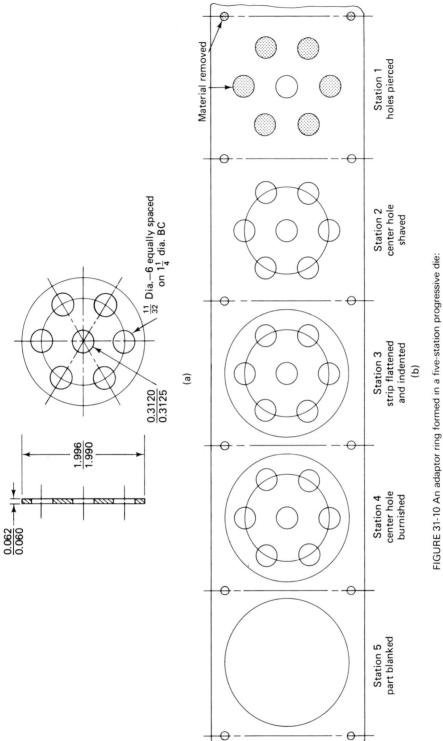

(a)

$\dfrac{0.3120}{0.3125}$

$\dfrac{1.996}{1.990}$

$\dfrac{0.062}{0.060}$

$\frac{11}{32}$ Dia.—6 equally spaced on $1\frac{1}{4}$ dia. BC

Material removed

Station 1
holes pierced

Station 2
center hole
shaved

Station 3
strip flattened
and indented

Station 4
center hole
burnished

Station 5
part blanked

(b)

FIGURE 31-10 An adaptor ring formed in a five-station progressive die:
(a) adaptor ring blank; (b) 2 ¼ wide strip

474

Die Material: Punch and die life vary with tool materials and hardness. While die materials are usually selected on the basis of the total quantity of parts to be produced, other factors, such as the type of workpiece metal, punch-to-die clearance, and the dimensional and surface finish requirements of the workpiece, must also be considered. Tool-steel dies, hardened and tempered to their highest usable hardness, are often used for low to medium production runs. However, for longer runs, carbide dies are used. Such dies have 10 to 20 times as much life per grind as do tool-steel dies.

Die Tolerances: Some pressworking operations may require more precise toolmaking tolerances than others. As a result, it is not practical to state a single tolerance that will satisfy all conditions. Cutting and forming operations on some workpieces may be as generous as $\pm \frac{1}{32}$ in. (0.794 mm) or more.

REVIEW QUESTIONS

31.1. Explain the function of the die used in a stamping press.

32.2. List some important factors that must be considered when selecting an appropriate press for a given job.

31.3. What is the principal type of forming operation performed on a dieing machine?

31.4. What are some of the leading reasons for selecting a pressworking technique as a means of producing a required part?

31.5. What is the recommended thickness range for cold-pressed parts?

31.6. Define "springback" and explain how allowance may be made to compensate for its undesirable effects.

31.7. Give a technique commonly used by metal fabricators to eliminate the possibility of formation of blemishes on workpieces caused by tool scoring or die marks.

31.8. Discuss some techniques commonly employed to reduce scrap loss in pressworking.

31.9. Name the two main work classifications into which dies are divided.

31.10. Explain the essential differences between single-operation and compound dies.

31.11. Under what conditions would transfer dies be used in preference to progressive dies?

PROBLEMS

31.1. Press manufacturers have developed a nomograph for determining the approximate press tonnage at any point on the press stroke of standard double geared presses. Find this nomograph in the technical literature and determine the percentage of rated tonnage that a press can exert using the following data: (a) a 2-in. working stroke with an 18-in. stroke length, (b) a 3-in. working stroke with a 12-in. stroke length.

31.2. When estimating manufacturing costs some pressworking companies apportion their capital equipment and original tooling costs in relation to a given number of parts produced—usually 1000. This apportionment assumes a full write-off of these costs over a 4-year period and production operation of the equipment for 4000 hours per year. Determine the machine cost per 1000 pieces when the original machine and tooling costs are $125,000 and the production rate is 50 parts per minute.

31.3. Assume the cost of plant services, utilities, and floor space at a rate of $15 per hour of machine production time. Determine the overhead cost per 1000 pieces with a production rate of 75 pieces per minute.

31.4. Explain the function of a flywheel in a press.

31.5. Suppose your plant has a crank drive press rated at 100 tons of force. A product you would like to draw requires a maximum force of 80 tons. What information do you need to know before deciding whether the 100-ton press is capable for the job?

BIBLIOGRAPHY

DALLAS, D. B., *Pressworking: The Punching Machines Have Arrived,* ME73FE018, Dearborn, Mich.: Society of Manufacturing Engineers, Feb. 1973.

EARY, D. F., *Techniques of Pressworking Sheet Metal,* Englewood Cliffs, N.J.: Prentice-Hall, Inc., 1974.

JOHNSON, H. V., *Manufacturing Processes,* Peoria, Ill.: Chas. A. Bennett Co., Inc., 1973, pp. 187–200.

"Presses," *Metals Handbook,* Vol. 4, *Forming,* 8th edition, Metals Park, Ohio: American Society for Metals, pp. 1–17.

32

CUTTING—PLATE, SHEET, AND STRIP

Cutting is a press operation that may be accomplished in a variety of related, but distinctly different ways. Regardless of the term used to describe the particular cutting operation, each method is associated with high-volume stamping of various outlines of flat workpieces.

Most cutting operations are performed in a press equipped with standard punches or with specially designed punches and dies. The punch, generally attached to the ram, moves downward through the work material and into the die to form the desired configuration. The punch, at each downward stroke, momentarily lodges in the die cavity and then rises above the stock to permit the stock to index to a new position. It is possible to combine many different cutting operations by using progressive dies.

DESCRIPTION OF THE PROCESS

Cutting operations generally consist of *blanking* (cutoff, parting, shearing, lancing, notching, nibbling); *piercing* or *punching* (slitting, perforating, extruding); and *edge-improvement methods* (trimming, shaving, fine-edge blanking).

Blanking

Almost all press operations begin with the blanking of a piece of material from flat stock. The punch penetrates the material a distance equivalent to 25 to 50% of the stock thickness, depending upon the brittleness, and the part simply fractures and breaks off. The punch continues its stroke, pushing the part into the die and below the face of the die. The sheared edges of a blanked part are characterized by a small ridge of torn metal called a "die break." Figure 32-1 shows some typical characteristics of the sheared edge of a hole in a workpiece and on the corresponding blank. In one stroke of a single-action mechanical press, blanking produces a complete outline of an individual flat workpiece. Figure 32-2 shows a typical workpiece blank which was produced during die-set tryout.

Cutoff: This is a pressworking operation by which individual flat metal blanks are produced by cutting along a line that extends entirely across the width of the stock. Special cutoff punch and die sets are designed for this

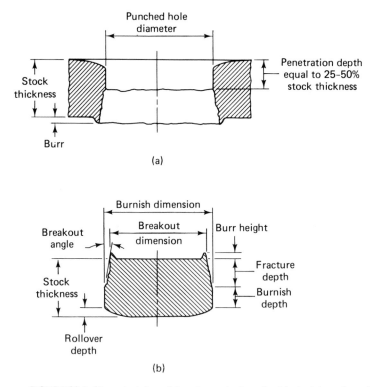

FIGURE 32-1 Characteristics of the sheared edge of a blank: (a) section of the punched workpiece; (b) section of the corresponding scrap

478

FIGURE 32-2 An example of a workpiece produced by blanking during die tryout (Courtesy, Larson Tool and Stamping Co.)

purpose. In general, the dies are relatively simple and inexpensive. As shown in Figure 32-3, the cutoff line may extend across the stock in a straight line or along a broken or curved path. The nested blanks are carefully arranged in a layout which permits the cutoff operation to produce one or more identical blanks for each stroke of the press. It is sometimes possible to eliminate scrap by alternating the positions of the blanks as shown in Figure 32-3 at c. Single-action presses, which operate at 50 strokes per minute, may be used for cutoff operations.

Parting: Unlike cutoff operations, *parting* results in some scrap production. The process consists of die cutting the required flat blank shapes by sacrificing stock which lies between the adjacent blanks. Each press stroke produces one complete part. Parting is a method employed to produce blanks with outlines too complex to permit compact nesting. Figure 32-4 shows a typical part shape that cannot be efficiently cut off without scrap production. A cutoff or parting design, shown in Figure 32-5, is justified when it is necessary to space blanks at intervals along a strip of material to avoid a tendency of stock distortion, to obtain greater accuracy of the outer contour or to allow sufficient room for bulky tool movement. As a general rule, the simplest example of a stamping tool is a parting tool, in which each press stroke produces one complete part.

479

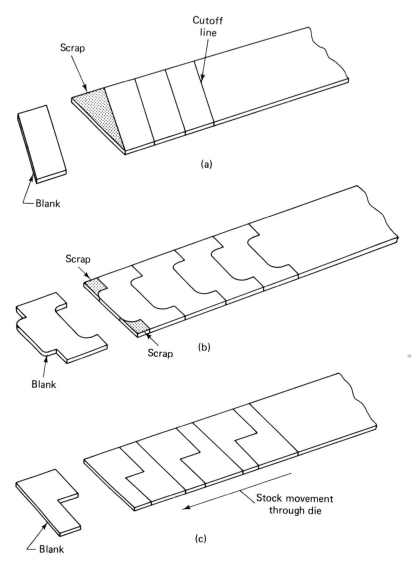

FIGURE 32-3 Nesting blanks for cutoff on strip stock

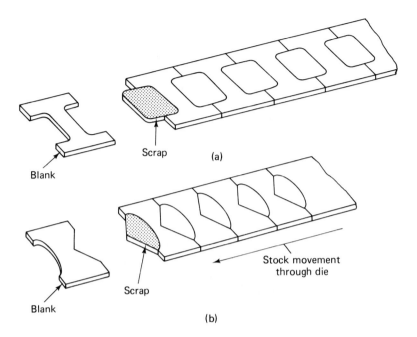

Scrap   (a)

Blank

Scrap

Stock movement
through die

Blank

(b)

FIGURE 32-4 An example of unavoidable scrap production

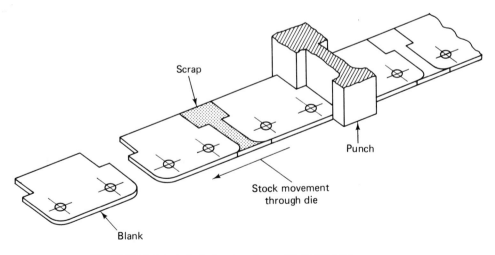

Scrap

Punch

Stock movement
through die

Blank

FIGURE 32-5 A technical advantage is sometimes obtained
by a cutoff or a parting design such as this

Shearing: This involves the cutting of flat metal forms from sheet and plate. The process may be broadly classified by the type of blade or cutter used: *straight* or *rotary shearing. Straight-blade shearing* is used for squaring and cutting flat stock to the shape and size of the required blank. The operation consists of rigidly holding stock which is severed by the action of an upper blade as it moves down past the stationary lower blade. The amount of penetration of the upper blade into the stock is governed by the ductility and thickness of the work metal. As in conventional blanking, the sheared edge is relatively smooth where the blade penetrates, with a considerably rougher texture along the torn portion. It is not possible to cut multiple layers of work sheets with a single cut. Each layer prevents the necessary breakthrough of the preceding workpiece.

Fabricators often use this process to prepare blanks prior to production punch-press operations. It is practical to shear sheet stock in lengths of 12 ft (304.8 mm) into widths as accurate as 0.0005 in./ft (0.043 mm/m) in thicknesses no greater than 0.135 in. (3.43 mm). A tolerance of ±0.001 in./ft (0.099 mm/m) is required for thicker sheets. A tolerance as large as ±0.010 to 0.020 in./ft (0.822 to 1.678 mm/m) is required when shearing plate. Squaring shears are available with mechanical, hydraulic, or pneumatic mechanisms. A mechanical power shear is shown in Figure 32-6.

FIGURE 32-6 Mechanical power shears. This machine can cut stock 6 ft. wide up to 1/2 in. thick (Courtesy, Cincinnati Incorporated)

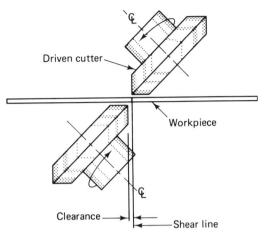

FIGURE 32-7 Conventional arrangement of cutters in a rotary shearing machine

Rotary shearing is a process by which sheets and plates are cut by means of two revolving tapered circular cutters. It is possible to make straight-line cuts as well as to produce circular blanks and irregular shapes by this method. Figure 32-7 illustrates the conventional arrangement of the cutters in a rotary shearing machine for the production of a perpendicular edge. It is also possible to produce beveled edges on workpieces by overlapping the cutters. Only the upper cutter is rotated by the power-drive system. The upper cutter pinches the workpiece stock and causes it to rotate between the two cutters.

Rotary shearing machines are equipped with special holding fixtures to rotate the work material to generate the desired circle. A straight-edge fixture is used for straight-line cutting. Rotary shears are limited to stock thicknesses of 1 in. (25.4 mm) or less. It is possible to cut circular blanks up to 10 ft (approximately 3 m) in diameter by using special attachments. Commercially available equipment is limited to the production of a minimum blank diameter of about 6 in. (152.4 mm).

Lancing: This is a press operation in which a single line cut is made partway across the work metal. Because no metal is actually removed, there is no scrap. Unlike blanking, the lanced cut does not form a closed or continuous contour. Figure 32-8 illustrates how the designer uses a lancing operation on strip stock to free metal stock subsequent to forming operations. This method is often employed as a means of forming small tabs or protrusions which may later be used for spring retainers or for assembly purposes and for louvers on certain parts. A common lancing application is the formation of small knockouts on electrical circuit-breaker panels, or on junction and outlet boxes.

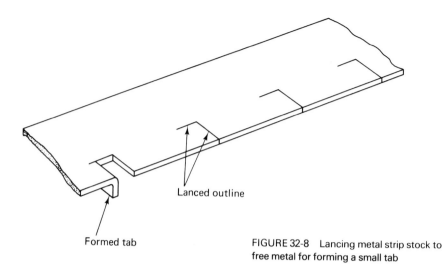

Lanced outline

Formed tab

FIGURE 32-8   Lancing metal strip stock to free metal for forming a small tab

Lancing is also used in special cases to partially cut the outline of a flat part that will be separated at a final operation in a progressive die, from the work metal, by blanking.

Triple-action presses (a press with three independent slides) are often used for lancing. The operation cannot be done satisfactorily with a die cushion and is usually accomplished while the punch slide is in motion. Designers specify a lancing operation on some parts to offset the possibility of fracture in cases where metal must be drawn or formed into a restricted area.

Notching: This is a pressworking operation by which metal pieces are removed from the edge of a blank, strip, or sheet. Depending upon the configuration of the punch used, the notched pieces may have almost any desired shape. The metal removed by notching is scrap.

Metal fabricators often notch the edge configuration of some sheet metal blanks that might be otherwise difficult to cut. A typical example of this technique is shown in Figure 32-9a. Notching is also often done prior to forming a concave radius on certain parts to help eliminate the formation of wrinkles as shown in Figure 32-9b.

Production notching (or piercing) presses are characterized by very high speeds but short punch strokes. The press used is usually the single-column, closed-back type.

Nibbling: This is an economical and versatile method of cutting out small lots of flat parts which range from simple to complex contours. The process is generally substituted for blanking when the cost of special tooling would not be justified. Figure 32-10 shows how intricate internal and external cuts can be made with a solid punch cutting at one side. One model of a nibbling machine uses standard round or triangular punches rigidly clamped to a tool adapter. (Lateral deflection of the punch would result in

484

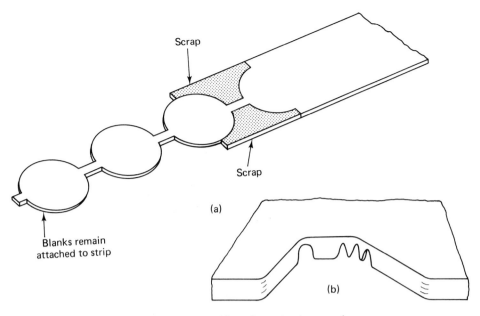

FIGURE 32-9 Notching: (a) notching prior to drawing operations;
(b) notching of flange helps eliminate wrinkles but weakens part

tool breakage as a result of collision with the die.) The small punch oscil-
lates in and out of its mating die at a rate that may range from 300
to as many as 900 strokes per minute. The metal sheets may be manually
positioned or clamped and guided by means of an electrohydraulic pulse
motor. A device called a "stripper" remains in an adjusted down position
and eliminates sheet "flutter." The shape of the desired cutout is progressively cut by the action of continuous and overlapping punch strokes which
slice the moving metal along the desired contour. There is no scrap.

FIGURE 32-10 Freehand nibbling of shaped parts with a rotary nibbling tool and solid punch.
Contours can be milled to a template clamped on the workpiece sheet (Courtesy, Trumpf
& Co.)

(a)

FIGURE 32-11 Intricate shapes may be cut in steel sheet up to 15/32 in. thick by copy nibbling: (a) working from a guide pin which follows the contour of the template on the right, naves are cut from the workpiece sheet on the left (Courtesy, Trumpf & Co.); (b) copy nibbling of a circular cutout to a template (not shown) (Courtesy, Trumpf & Co.)

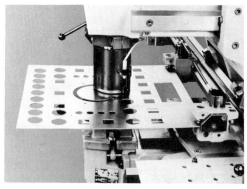

(b)

Nibbling is practical on stock up to ¼ in. (6.35 mm) thick. Numerically controlled coordinate punching machines are available which use simple and inexpensive tooling to punch or nibble an almost unlimited variety of contoured workpieces. These machines are used whenever heavy-tonnage presses and fixed tooling cannot be justified, because of material size and thickness or when only a small quantity of parts are required. The maximum punched hole diameter is approximately 4 in. (101.6 mm), and the number of tools per job is infinite. Figure 32-11 illustrates two applications of a process known as *copy nibbling.* No scribing or layout is required. A great number of parts of any shape can be quickly and economically produced with a hollow punch and guide pin.

Piercing or Punching

Piercing, also known as *punching,* is a cutting operation by which various-shaped holes are sheared in blanks. Piercing is similar to blanking except that in piercing, the work metal that surrounds the piercing punch is the workpiece, and the slug that is cut out is the scrap.

Piercing is ordinarily the fastest method of making holes in blanks, sheets, or strips. Single-operation dies are used when piercing is the only required operation or in a case where compound dies cannot be used because of design limitations imposed by holes that are too close to the edge of the stock. Compound dies are particularly appropriate for piercing operations where close accuracy of hole positions is important.

Standard punches and dies for round, square, and rectangular holes are commercially available in increments of 1/64 in. (0.397 mm). Most fabricators prefer tooling of a type that permits the piercing of all holes in the same stroke whenever possible.

Slitting: This consists of cutting strip or coiled stock lengthwise by passing the stock through spaced, circular blades. Continuous strips of stock in various widths may be obtained by adjusting the spacing of the various slitters (blades) to the desired settings. Briefly, the slitting line, or manufacturing sequence for dividing a coil into narrower coils, consists of an uncoiler for holding the original coiled stock, one or more slitters, and a recoiler that has spacers which correspond to the various required widths of the slit stock.

The stock may be driven or pulled through the slitting line. In most cases, metals less than 0.010 in. (0.25 mm) thick are driven to offset the tendency of tearing the metal. Synchronized motors are used to drive the uncoiler, the slitters, and the recoiler to maintain a constant speed. Thicker metals are pulled through the slitting line by means of a motor in the recoiler.

It is practical to slit stock in coils as narrow as 1 in. (25.4 mm) wide or less. When strip is slit to a width of less than ½ in. (12.7 mm) it is called "flat wire." Equipment is available that can produce coiled stock in widths of 90 in. (approximately 2.28 m). Steel, in most cases, is slit into coils 36 to 60 in. (0.91 to 1.52 m) wide and 0.020 to 0.125 in. (0.51 to 3.18 mm) thick.

Commercial tolerances for slit widths of hot-rolled low-carbon steel strip with a No. 3 edge (approximately square) range from ±0.005 in. (0.13 mm) for stock 0.019 in. (0.48 mm) thick or less for strip widths of 5 in. (127 mm) or less, to ±0.016 in. (0.41 mm) for stock greater than 0.019 in. (0.48 mm) thick for widths up to 12 in. (304.8 mm). Tolerances for cold-rolled low-carbon steel 0.069 in. (1.75 mm) thick or less in widths from ½ in. (12.7 mm) to approximately 24 in. (609.6 mm) range from ±0.005 in. (0.13 mm) to ±0.020 in. (0.508 mm). Coiled stock ¼ in. (6.35 mm) thick up to 20 in. (508 mm) wide, for example, is available in width tolerances of ±0.031 in. (0.79 mm). More detailed specifications for coiled steel as well as for other metals is available by consulting manufacturers' handbooks.

Burrs are usually formed when slitting coiled stock. A process known as "edge rolling" is generally employed to produce burr-free edges.

Perforating: This is a term sometimes used by fabricators to describe a process that consists of piercing multiple holes in flat-work material. The holes may take on almost any shape. The term "perforating" usually implies the production of closely and regularly spaced small holes. The tooling for perforating is usually the same as that used for piercing.

Extruding: This is a hole-making process that is performed on flat stock. It is regarded as a special piercing operation and, as such, requires special tooling. The term "extruding," in this sense, refers to the formation of a flange on a flat part by drawing stock out of a previously made hole. In general, extruding involves rather severe stretching of the metal. Extruded holes are often made on sheet-metal parts to provide added thread length for a tapped hole (for assembly purposes), to increase a bearing surface, or to produce a recess into which the head of a flat-head machine screw or rivet may fit.

One method often used to extrude a flanged hole is shown in Figure 32-12a. A punch of the desired hole diameter is forced through a small pre-pierced hole, thereby drawing its periphery. Another method, shown in Figure 32-12b, is to use a shouldered or pointed punch that both pierces the holes and flanges it.

As a general rule, a flange with a depth equal to one-half of the hole diameter may be produced without splitting the metal. Greater depths have been achieved by metal fabricators by initially forming an indent or a small cup, punching a small hole in the bottom, and extruding the flange. For best extruding results, the material should be in the dead-soft condition.

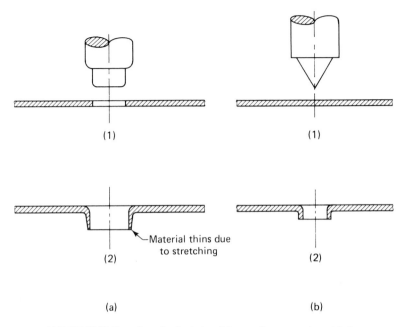

FIGURE 32-12 Extruding circular holes: (a) extruding a pre-pierced hole;
(b) combination pierced and extrusion operation

Edge-Improvement Methods

Trimming: This is similar to blanking. It consists of die-cutting unwanted excess material from the periphery of a previously formed workpiece. Trimming is frequently combined with one or more other pressworking operations, such as punching or notching, in a compound die. It is often possible to design special-purpose dies that accurately control the relationship of some other feature on a workpiece, such as a hole or a slot, to the trimmed outline of the workpiece. Deformed or uneven metal on the edges of blanked workpieces and "flash" on castings and forgings are usually removed by trimming.

Shaving: This improves the quality and accuracy of blanked parts by removing a thin strip of metal along the edges. Edges of blanked parts are generally unsquare, rough, and uneven. In addition to the rough die break texture, the edges are characterized by a "tensile burr" along the top of the cut. Only a few thousands of an inch (or approximately 0.10 mm) of metal is removed by shaving. In fact, the scrap is so thin that it resembles the chips produced in finish machining.

On very heavy stock, a double break sometimes occurs, as shown in Figure 32-13. The edge of the stock may have a burnished area on both the punch and die sides, with an irregular recess of torn material in between. Two or even three shaves may be required to improve the edge straightness and finish. It is necessary, of course, to provide a small amount of extra stock on the blanked workpiece for subsequent shaving or trimming operations. It is helpful to the fabricator to designate those surfaces which are to be shaved on the part print.

Shaving may be done as a separate operation or it may be incorporated into one station of a progressive die. In any case, the blank must be precisely located over the die or the punch because of the small amount of metal that is actually removed.

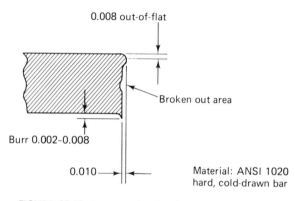

FIGURE 32-13 An example of a heavy stock blank with a double shear, producing burnished area near each side of the edge

Fine-Edge Blanking: This is a variation of conventional blanking. This process makes it practical to produce a smooth edge on a workpiece during a single press stroke. A relatively new process, it involves the use of a special V-shaped impingement ring (a holding device not used in conventional blanking), which is forced against the stock prior to blanking to lock it tightly against the die. The metal ring is positioned very closely to the outline of the part to be punched. The firmly held stock is forced to flow toward the punch, which, in effect, "extrudes" the part out of the strip without fracture. As a result, no die break shows on the sheared edge of the workpiece. Compared to conventional blanking, the die clearance is very small and the punch speed is slower.

Both ferrous and nonferrous parts may be worked with edges comparable to those found on blankings that were shaved or produced by machining. A practical surface finish range for fine-edge blanked steel parts is between 16 and 32 μin. with some improvement for aluminum and brass alloys. Clean and accurate holes of various shapes may be pierced by this method which eliminates the need for broaching and milling operations. The small burrs formed by fine-edge blanking may be removed by abrasive finishing or by vibratory finishing.

Companies that use this process have achieved tolerances as close as ± 0.0005 in. (0.013 mm) per inch on stock up to ½ in. (12.7 mm) thick. Significant cost savings, resulting from the elimination of special secondary finishing operations, are possible on steel parts such as pawls, levers, cams, and gears.

There are some practical limitations to workpiece sizes which, in addition to restrictions imposed by press capacity, are largely governed by work metal conditions such as thickness, tensile strength, and hardness. It is possible, for example, to produce 8 in. (203.2 mm)-diameter blanks from ⅛ in. (3.18 mm) low-carbon steel. Smaller parts up to ½ in. (12.7 mm) thick have been successfully blanked.

A triple-action hydraulic or a combination hydraulic and mechanical press is used for fine-edge blanking.

PRODUCT DESIGN FACTORS

It should be made clear that, in most companies, machine designers do not design the press tools that will be required for the production of the stampings they design. Most competent designers, however, visualize or anticipate the important features of the die construction so that they can estimate the approximate cost of producing the required machine parts. The selection of a given method of producing a part is, of course, heavily influenced by the design features of the part, as well as by other characteristics such as tolerance, finish, quantities, and material.

Standard Gages: Production engineers use standard-gage stock for stamped parts wherever possible. Price advantages may usually be obtained if the fabricator can combine one order with other orders, particularly in sizes that correspond to normal warehouse inventories.

Stock is usually considered "light gage" up to 0.031 in. (0.79 mm), "medium" from there to 0.109 in. (2.77 mm), and "heavy" from 0.125 in. (3.18 mm) up. Material thickness should always be specified on engineering documents in decimal parts of an inch or in millimeters rather than in gage numbers.

The thickness of sheet metal stampings may range as thin as 0.003 in. (0.08 mm) for foil to 0.500 in. (12.7 mm) or greater. The majority of stamped parts fall between 0.020 in. (0.51 mm) and 0.080 in. (2.03 mm) in thickness.

Specification of Materials: Generally, all metals for stampings should meet the following requirements: (1) comparatively low in cost, (2) high strength, (3) good surface finish, (4) dimensionally uniform, (5) uniform crystalline structure, and (6) easy workability.

Essentially, any material that can be produced as sheet or strip stock, and that will not shatter on impact, can be worked with press tools. Only brittle materials are excluded.

Burrs usually form on the die side of the stamping. With proper punch and die clearance, the burr height rarely exceeds 0.003 to 0.005 in. (0.08 to 0.13 mm). Burrs are potentially dangerous in handling and they frequently cause trouble in assembly operations or when workpieces are to be plated. Difficulties may also arise when burrs spall off from stamped parts into oil systems. Burr removal around circular holes on some parts can be accomplished by lightly countersinking. Some parts which are heavy enough not to distort, and where quality of finish is not vital, can be grouped together and agitated, thus knocking the burrs flat.

For most parts a notation on the production drawing which specifies "Remove all burrs and sharp edges" calls for tumbling barrel or vibratory finishing (explained in Chapter 39), which is quite effective and costs about the same as an additional press operation. The same drawing notation referring to a thin or a large part, however, calls for hand removal of burrs by grinding or filing. Such operations are often expensive and time consuming. Limiting the burr removal only to critical edges, specifically designated on the print, will save money, particularly when a part has to be hand-deburred.

Material Economy: The relative "as stamped" position on the work material should be carefully laid out, for the greatest material economy, to avoid unnecessary scrap. Metal fabricators avoid using very narrow stock whenever possible because it costs more per pound than correspondingly wider material. A good rule to follow is to lay out blanks in such a way as to utilize at least 70 to 80% of the strip or sheet area.

Rectangular blanks can generally be laid out more easily and less expensively than other shapes. Other shapes may be positioned side to side across the full width of the stock, as shown in Figure 32-14a, or, under certain conditions, may be blanked end to end as shown in Figure 32-14b. Circular blanks can be separated as shown in Figure 32-15a. If pairs of circular blanks are required in sufficient quantities, substantial tooling and labor savings can often result by nesting the parts as shown in Figure 32-15b. In this case, a double-blanking die is used.

Circular blanks should never be laid out as in Figure 32-16 when a subsequent cutoff operation is planned. Severe difficulty in the cutoff operation

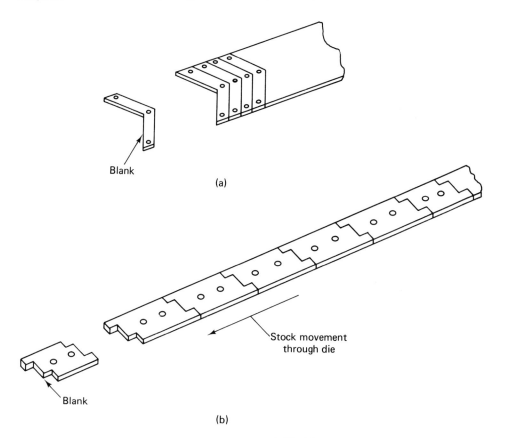

Blank

(a)

Stock movement
through die

Blank

(b)

FIGURE 32-14   Material economy in blank layout

arises due to the formation of a feather edge on the scrap skeleton between the adjacent blanks. Also, proper die register is practically impossible, and die wear is abnormal.

Rings with large holes are very wasteful of material. Stampers are constantly on the alert to utilize the scrap from one part as material for another. For example, the scrap center that otherwise would result in waste from a ring-shaped part is used whenever possible.

Odd-shaped parts are particularly difficult to lay out for the greatest economy. Parts with two opposite parallel sides can be cut off with a minimum of waste, as shown in Figure 32-17a, even if the ends do not mesh. Whenever the shape permits, blanks should be interlocked or nested (see Figure 32-3). Figure 32-18 illustrates how the outer contour of a commercial stamping was altered to permit nesting and reduce scrap loss.

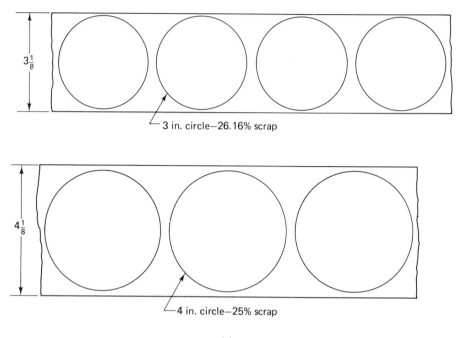

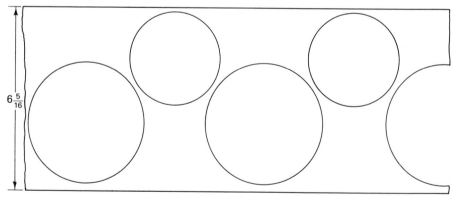

23.3% Scrap if all circles are same size
or if diameter difference is extreme,
savings may be greater

(b)

FIGURE 32-15 Material economy in blank layout:
(a) separate; (b) nested

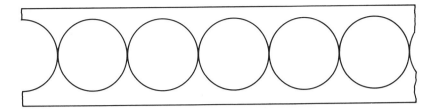

FIGURE 32-16 The layout in this example for circular blanks should be avoided

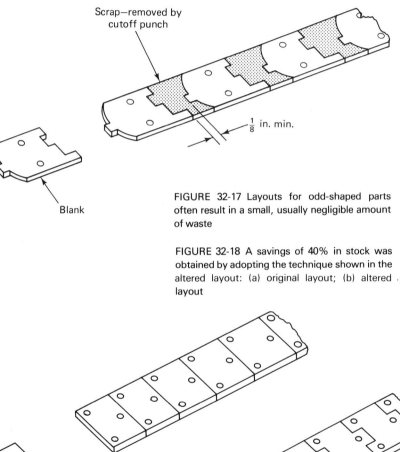

Scrap—removed by cutoff punch

$\frac{1}{8}$ in. min.

Blank

FIGURE 32-17 Layouts for odd-shaped parts often result in a small, usually negligible amount of waste

FIGURE 32-18 A savings of 40% in stock was obtained by adopting the technique shown in the altered layout: (a) original layout; (b) altered layout

(a)

(b)

Blank

Blank

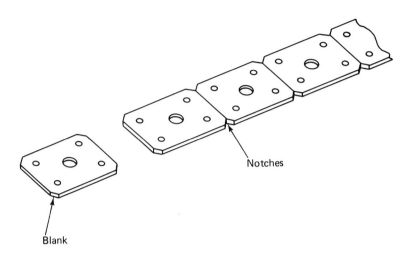

FIGURE 32-19 Blanks with chamfered ends require notching prior to cutoff

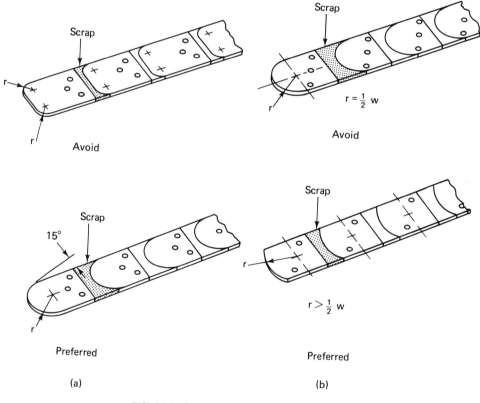

FIGURE 32-20 Layouts of blanks with rounded ends

End Shapes: Designers give careful consideration to the end shape of a stamped part. The order of preference starts with straight ends, chamfered ends, and, finally, rounded ends. Chamfered ends require notching operations as shown on the typical part in Figure 32-19. More elaborate press tools are required to obtain parts with rounded ends, however. Figure 32-20 illustrates important design considerations which affect the design of end radii.

Pierced Holes: Diameters should not be less than half the stock thickness and never less than $\frac{1}{32}$ in. (0.794 mm). Groups of holes located from formed edges, center lines, or ribs may be produced as accurately as ±0.004 in. (0.10 mm).

Careful attention must be given to hole locations to minimize bulging and distortion that usually occurs around closely spaced adjacent holes or between a hole and the edge of a part. Also, to offset abnormal punch breakage there must be sufficient stock between the die walls and the punch. For light-gage stock, round holes should be no closer than 0.062 in. (1.59 mm) to an edge or to each other. The preferred hole spacing for heavier stock is at least 0.125 in. (3.18 mm). Regardless of stock thickness, a good "rule of thumb" to follow is to maintain a minimum distance of twice the stock thickness between adjacent holes or between a hole and an edge.

Extruded Holes: Figure 32-21a illustrates some general guidelines which most designers follow when designing parts with extruded holes, shown in Figure 32-21b. Holes should be spaced a distance equal to at least four times the thickness to an edge and six times the thickness from hole to hole.

FIGURE 32-21 (a) Design parameters for extruded holes (Courtesy, S. B. Whistler & Sons)

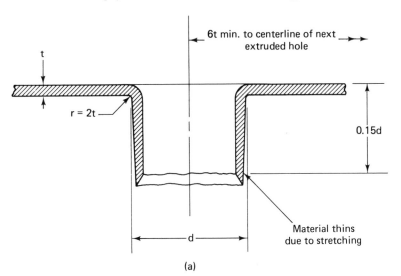

6t min. to centerline of next extruded hole

t

r = 2t

0.15d

Material thins due to stretching

d

(a)

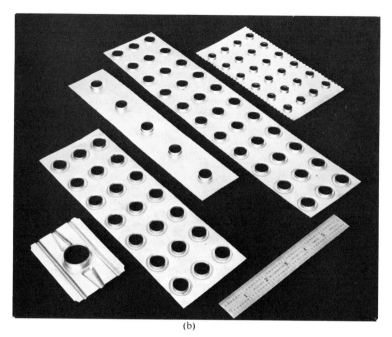

(b)

FIGURE 32-21 (b) Examples of parts with extruded holes (Courtesy, S. B. Whistler & Sons)

Slots: Standard practice is to control the location of slots on a given stamping by dimensioning to their centers. Slots with rounded ends are usually more economical to produce than those with square ends, unless standard square or rectangular punches are available.

Holes and Slots on Formed Stampings: In most cases, it is usually less expensive to pierce openings in flat stock before forming. Holes and slots which are relatively close to a bend often distort during bending. Figure 32-22 shows some recommended distances for spacing holes and slots on formed stampings. Complex shapes having one or more holes, which are usually located close to a bending radius, are produced most satisfactorily by drilling or punching the holes *after* forming.

Perforating: In general, fabricators avoid perforating holes smaller in diameter than the equivalent thickness of the material. As a result of the increased costs of maintaining the punches and the dies, holes less than 0.030 in. (0.76 mm) in diameter are rarely practical or economical to produce. A good rule to follow is to maintain the distance between adjacent holes or between holes and the edge of the stock at least equal to and preferably greater than the stock thickness. Otherwise, there is a tendency of the stock to wrinkle or distort.

Stacked Stampings: These are an interesting and innovative extension of conventional metal-stamping techniques. The process consists of conventionally producing individual flat blanks of various outlines and

498

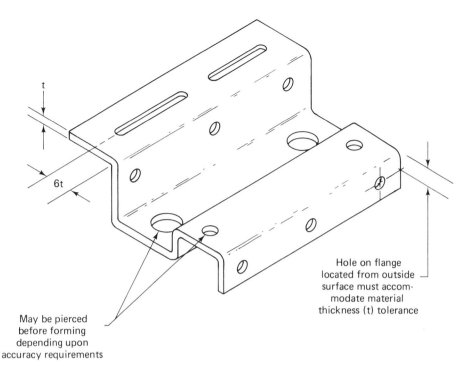

t

6t

May be pierced
before forming
depending upon
accuracy requirements

Hole on flange
located from outside
surface must accom-
modate material
thickness (t) tolerance

FIGURE 32-22 Recommendations for holes and slots

thicknesses and then stacking them, one on another. The blanks are assembled into a rigid structure by brazing, welding, riveting, or by using threaded fasteners.

Common examples of stacked stampings include washers and spacers, cam blanks, transformers, padlocks, shims, laminations, perforated sheets, link-arms, and cover plates and frames. Figure 32-23 illustrates other typical examples of stacked stampings.

FIGURE 32-23 Examples of stacked stampings

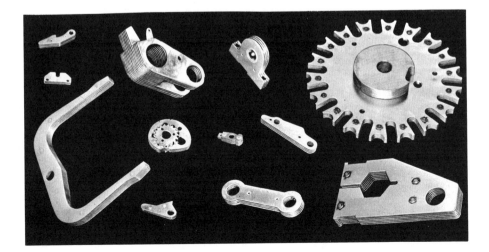

REVIEW QUESTIONS

32.1. Explain why most pressworking operations begin with blanking a piece of material from flat stock.

32.2. Prepare a sketch with adequate labels illustrating the action of a punch as it blanks out a flat workpiece from a piece of stock.

32.3. What are the essential differences between the pressworking operations of cutoff, parting, and shearing?

32.4. Describe the press operation known as lancing and give some examples showing applications.

32.5. Compare the principal reasons for notching to those of nibbling.

32.6. Under what specific conditions would a piercing operation be justified on a component of a given product?

32.7. Define the term "flat wire."

32.8. Give some valid reasons for press extruding.

32.9. Explain why it is often possible to combine a trimming operation with one or more other press operations.

32.10. Approximately how much stock is removed by shaving?

32.11. Give the advantages of fine-edge blanking.

32.12. Explain why it is generally advantageous to use standard-gage stock for press-worked parts.

32.13. List six important requirements to be considered when specifying materials for pressworking operations.

32.14. Explain why burr removal, although always costly, is, nevertheless, considered to be a necessary operation.

32.15. Prepare sketches showing the most efficient utilization of a strip of metal when laying out the following stamping shapes: (a) a letter "N" (b) a letter "T" (c) a letter "W" (d) a letter "A."

32.16. Give examples of one or more undesirable conditions likely to occur when piercing closely spaced adjacent holes.

32.17. What is the major advantage of stacked stampings?

PROBLEMS
(See appendix for useful formulas.)

32.1. Calculate the cutting force necessary to blank without shear a 6 in. dia. circle from a 0.044 in. thick sheet of low carbon steel.

32.2. Calculate the cutting force necessary to notch out the shape shown in Figure 32-P2. The material is stainless steel, 0.036 in. thick by 3 in. wide.

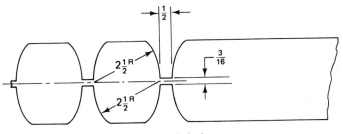

All sizes are in inches.

FIGURE 32-P2

32.3. A manufacturer is faced with the decision of using a punch press or a press brake to make a certain part. What would be the break-even point expressed in numbers of pieces using the following data? Punch Press: set-up time for 100 pieces = 1 hr with a production rate of 0.0621 pieces/min.; Press Brake: set-up time for 100 pieces = 42 min. with a production rate of 0.1072 pieces/min.

BIBLIOGRAPHY

"Blanking," *Metals Handbook,* Vol. 4, *Forming,* 8th edition, Metals Park, Ohio: American Society for Metals, pp. 31–43.

"Extrusion," *Machinery's Handbook,* twentieth edition, New York: Industrial Press, Inc., 1975, pp. 2177–2178.

"Fine-Edge Blanking," *Uncommon Metalworking Methods,* Cleveland, Ohio: Penton Publishing Co., 1971, p. 41.

*HIGGINS, C. C., "A Practical Manual for Stamping Design," *Machine Design,* Part I, July 10, 1958, and Part II, Aug. 7, 1958.

HOFFMAN, W. E., "What Can Fine Blanking Do? . . . Its Limitations," *Journal of the Fabricator,* Part I, Jan./Feb. 1974, and Part II, Mar./Apr. 1974.

Metal Stamping, Research Report, Vol. IV, Detroit, Mich.: ASTME, Aug. 1, 1960.

"Piercing," *Metals Handbook,* Vol. 4, *Forming,* 8th edition, Metals Park, Ohio: American Society for Metals, pp. 44–55.

"Shearing of Plate and Flat Sheet," *Metals Handbook,* Vol. 4, *Forming,* 8th edition, Metals Park, Ohio: American Society for Metals, pp. 265–271.

*"Stampings Offer Wide Choice of Metals, Design Versatility," *Materials Engineering,* Mar. 1972.

*STRASSER, F., "How to Stamp the Hole Thing," *Machine Design,* Nov. 1, 1973.

*Abstracted with permission.

33

CONVENTIONAL BENDING AND FORMING PROCESSES

While the terms "bending" and "forming" are often used interchangeably, these common metalworking processes are distinctly different. *Bending* consists of uniformly straining flat sheets or strips of metal around a linear axis. Metal on the outside of the bend is stressed in tension beyond the elastic limit. Metal on the inside of the bend is compressed. The metal tends to become slightly thinner at the bend. This condition is due to the tension of the metal fibers as they are plastically stretched around the outside of the bend. *Forming,* as compared to bending, involves little or no metal flow, and, as a consequence, the metal thickness and area of the original blank is virtually unchanged.

DESCRIPTION OF THE PROCESSES

Bending

Precise right-angle bends are difficult to form on high-volume production of parts particularly when harder materials and heavier-gage metals are used. Because of the possibility of cracking, parts with right-angle bends should be laid out, whenever possible, so that the bends are formed *across* the strip. Small radii should be avoided because of the possibility of weakening the part and also because additional press operations are often required. As a general rule, fabricators prefer that inside radii of bends should be equal to or greater than the stock thickness. Typical examples of sheet metal bends are shown in Figure 33-1.

FIGURE 33-1 Typical examples of sheet metal bends

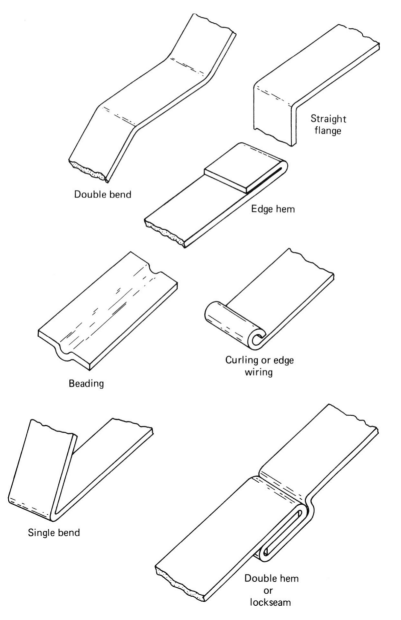

Double bend

Straight flange

Edge hem

Beading

Curling or edge wiring

Single bend

Double hem or lockseam

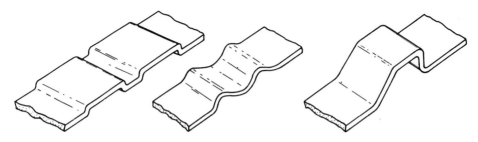

FIGURE 33-2 Various formed shapes of metal strips

Forming

The shape that is formed on the metal sheet or strip, as shown in Figure 33-2, depends upon the shape of the punch and die used to produce the workpiece. Depending upon the nature of the desired configuration, the formed shape may be gradually developed over a progressive sequence of operations, or it may be simultaneously formed. Figure 33-3 shows a variety of press-formed products.

FIGURE 33-3 Examples of various press formed products (Courtesy, Larson Tool and Stamping Co.)

PRODUCTION EQUIPMENT

Bending and forming operations are usually performed on crank, eccentric, or cam-operated presses, press brakes (see Figure 31-5), or by roll forming. Most operations on sheet-metal workpieces are cold-worked.

Press Working Equipment

Press working equipment for bending or forming is the same as that used for cutting operations. Figure 33-4 shows workpieces being formed in a battery of presses. Punch presses are used to produce high-volume parts to close tolerances and are generally restricted to relatively small workpieces, in lengths not greater than 2 ft (about 305 mm).

FIGURE 33-4 A battery of presses used to form sheet metal products (Courtesy, Larson Tool and Stamping Co.)

A wide variety of punches and dies are used in punch presses. Some of the more basic types, used to produce the forms shown in Figure 33-5, are V-dies, wiping dies, U-bending dies, rotary bending dies, cam-actuated dies, and wing dies. Largely depending upon the desired production rate and volume, but at times also upon other factors, either single-operating, compound, or progressive dies, as illustrated in Figure 33-6 may be used.

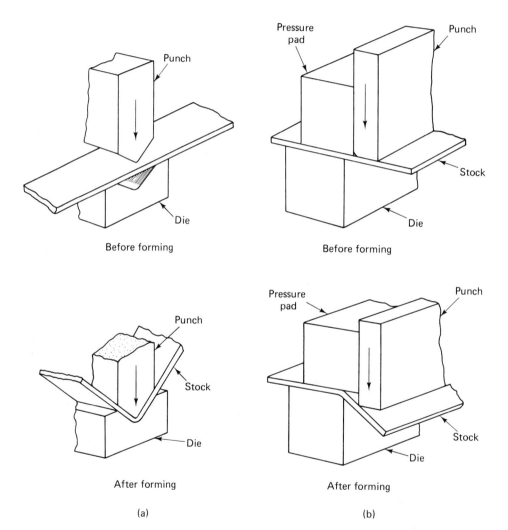

FIGURE 33-5 Typical punch and die setups used for:
(a) V-die bending; (b) wiping die bending

FIGURE 33-6 Worker checks a part produced by a set of progressive dies temporarily removed from the press. The scrap skeleton is in the foreground (Courtesy, Larson Tool and Stamping Co.)

Press-Brake Equipment

Press-brake equipment (shown in Chapter 31) is commonly used by metal fabricators to form straight and relatively narrow workpieces whose lengths exceed approximately 2 ft (about 305 mm). Also, press-brake forming operations (shown in Figure 31-6) are used for limited production quantities and when close tolerances are not required. This versatile and economical process may be used to form thin metal sheets and plates as well as for forming stock up to 1 in. (25.4 mm) thick. Press brakes are currently available that can accommodate stock as long as 24 ft (approximately 7.3 m) and in capacities up to 2500 tons (2250 metric tons). Both mechanical and hydraulic ram actions are available, which have an operating range of about 5 to 50 strokes per minute. The action of a press-brake approximates that of a slow-speed punch press. In both cases, punches and dies are used to form the desired shape in the workpiece. Perhaps the most commonly performed operation on press brakes are 90° or off set bends. Fabricators take full advantage of the remarkable press-brake versatility in forming channels, U-bends, lock seams, and for bending and curling as well as in performing blanking, piercing, shearing, lancing, straightening, embossing, corrugating, and flanging operations.

ROLL FORMING

There are two important production methods which are generally classified under roll forming: *contour roll forming* (also called cold-roll forming) and *three-roll forming.* In general, any product formed by a press brake can be roll-formed.

Contour Roll Forming

Contour roll forming is a high-production process that progressively forms various uniform cross-sectional shapes by continuously passing flat sheet or strip through a line of contoured rolls. The work material may be in the form of cut lengths or it may be fed from a coil, which is most economical. In this case, an automatic cutoff device is used to obtain desired lengths of roll-formed workpieces. As the material passes through the machine, each pair of forming rolls (called a "station") produces a partial change in the cross section to gradually transform the flat stock into the desired final shape. Figure 33-7 shows completed stock emerging from the final set of

FIGURE 33-7 An inboard roll former can accommodate various gauges of metal at speeds related to the shapes being formed (Courtesy, Maplewood — a unit of Ex-Cell-O Corporation)

FIGURE 33-8 Inboard rolling machines of this type are designed to accommodate various gauges of metal at speeds related to the complexity of the shapes being formed (Courtesy, Maplewood—a unit of Ex-Cell-O Corporation)

forming rolls on a roll-forming machine. The metal-gage thickness remains essentially constant. The number of stations or roll stands that are required on the line depend upon the complexity of the desired shape, the material used, the metal thickness and hardness, and tolerances. Some simple shapes may be formed with as few as two or three roll stations. Machines are currently available which can be equipped with as many as 40 stations. An example of an inboard-type roll-forming machine is shown in Figure 33-8.

Figure 33-9 shows a typical roll-forming sequence for producing lengths of decorative architectural trim. Figure 33-10 illustrates a sampling of the wide range of almost unlimited shapes made by this process.

Conventional contour roll-forming machines can produce long sections of workpieces on which all bends are parallel to each other and in the same longitudinal direction. The versatility of the process may be extended by adding a special fixture called a "beading shoe" to the roll-forming machine. This device permits the production of ring- or hooped-shaped parts. Stock for wheel rims for automobiles and bicycles are contour-roll-formed in straight lengths and are then bent and cut off in one continuous operation. Other applications of this process variation include various ring-shaped parts which are used for heating and ventilating equipment, machine guards, automobile headlamp rims, glass retaining rings, and so on.

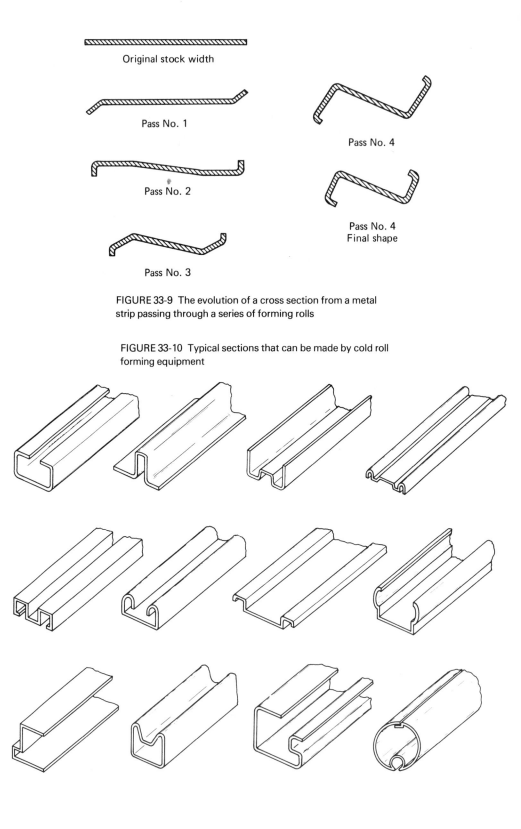

Original stock width

Pass No. 1

Pass No. 2

Pass No. 3

Pass No. 4

Pass No. 4
Final shape

FIGURE 33-9 The evolution of a cross section from a metal strip passing through a series of forming rolls

FIGURE 33-10 Typical sections that can be made by cold roll forming equipment

All the ductile metals and alloys that can be worked by any of the other forming methods can be roll-formed. Successful production runs have been made using stock as thin as 0.003 in. (0.08 mm). Parts may be made from stock up to 0.750 in. (19.05 mm) thick, but most production is in the thickness range 0.020 in. (0.51 mm) to 0.125 in. (3.18 mm). While the practical maximum width limit is about 80 in. (2.04 m), the width in the bulk of product applications is usually much less. Modern roll-forming machines are capable of continuous operating speeds of from 50 to 800 ft (approximately 15 to 243 m) per minute, with speeds between 80 and 100 ft (approximately 24 to 30 m) per minute most widely used.

Contour roll-forming is applicable to the manufacture of an impressive range of products. Shapes that are considered ideal are those that are symmetrical about a vertical center line, because they require the least amount of straightening. Products commonly formed by this process include, in addition to those already mentioned, architectural door and window trim, plaster board corners, gutters, downspouts, and siding, fenceposts, structural shapes such as Z- and C-channels, railway, truck and automotive engine, frame, and body parts, various forms of tubing and piping, parts for household and commercial appliances, rail sections for conveyors, turrets, and overhead doors.

Stainless steel and aluminum are often roll-formed for decorative and for architectural applications—often starting as prepainted, plated, anodized, polished, or preembossed stock. The gentle action of the progressive bending rolls does not usually mar the prefinished materials. A notable exception occurs on sections that have been severely bent where the finish may deteriorate and spall off.

The initial finish on stock is seldom improved by contour-rolling operations. The surface finish on hot-rolled and cold-finished steel and on brass and aluminum stock is maintained during contour roll-forming by using a flood of lubricant. Lubricant is not used for embossed metals or when forming materials with vinyl or other organic coatings.

Press-brake forming methods are competitive only if the desired sectional shape can be produced in a single press stroke. For sections requiring two or more forming strokes, contour roll forming is generally the most economical method. In spite of higher tooling costs, roll-forming production is usually justified from the standpoint of increased production capability.

Workpieces with sections up to several inches high (or wide) may be produced as accurately as ± 0.005 in./in. (0.005 mm/mm) or less on one or more dimensions. Larger and more complex shapes require increased tolerances, often as much as $\pm \frac{1}{64}$ to $\frac{1}{32}$ in./in. (0.015 to 0.03 mm/mm) for economical production. The tolerance on the cut length varies from ± 0.015 to 0.125 in./in. (0.015 to 0.125 mm/mm) depending upon the cross-sectional shape, length of the part, and speed of the line.

Three-Roll Forming

Three-roll forming, as the term implies, consists of forming various shapes by passing work metal blanks between three rolls. Depending upon such variables as the work metal composition, machine capability, or part size, a shape may be formed in a single pass or over a series of multiple passes. Figure 33-11 illustrates two basic setups for three-roll forming machines which, in most cases, are positioned in the shop for horizontal stock feed. On some pinch-type machines, all the rolls are driven while on others, only the two front rolls are powered. Pinch-type forming machines hold the work metal very firmly. They are capable of producing parts with greater dimensional accuracy than are the pyramid types. The two lower rolls on the pyramid-type machines are driven while the adjustable top roll serves as an idler and rotates on friction with the work metal blank.

In most cases, short curved sections of circular work are performed on the ends of the work metal blanks in a press brake or on a hydraulic press. Otherwise, the work blanks would have ends which, instead of being curved, are straight. Prebending is also possible by a combination of operations on the forming machine itself, which involves inserting the stock and then reversing the rotation of the rolls.

Special irregular shapes may be formed on pyramid-type machines by attaching dies to the top roll. Unlike the pinch-type machine, pyramid-type machines may be used to perform straightening operations on lengths of plates, beams, angles, and other structural forms. Parts in the shape of truncated cones are made on pyramid-type machines. One of the main disadvantages of the pyramid-type machine, unlike the pinch-type machine, is that the shape of the flat areas, which remain on the ends of the rolled metal blanks, cannot, in all cases, be satisfactorily curved. There are various techniques that may be used to minimize problems affected by the flat leading and trailing ends, but always at additional cost.

Figure 33-12 shows some typical shapes that may be produced from flat stock by three-roll forming. Some suggested applications are also given. Corrugated piping is formed from stock that is initially corrugated at the steel mill. The forming rolls are shaped accordingly to conform to the corrugations on the work metal.

Most metals are work-hardened by cold roll forming. In severe cases, as encountered in forming small-diameter cylinders, intermediate annealing is usually necessary. Cold forming is always preferred over hot forming, simply because it is less costly. A process known as *warm forming* may be used in preference to cold forming for severe forming requirements. Warm forming is usually done at a temperature that is equivalent to the tempering temperature of the metal and is often employed in cases where hot forming would alter the mechanical properties of the metal. Also, less force is required

512

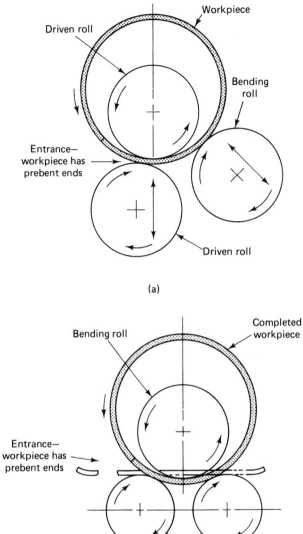

(a)

(b)

FIGURE 33-11 Two set-ups for rolling a cylindrical work-piece on roll forming machines: (a) pinch-type machine; (b) pyramid-type machine

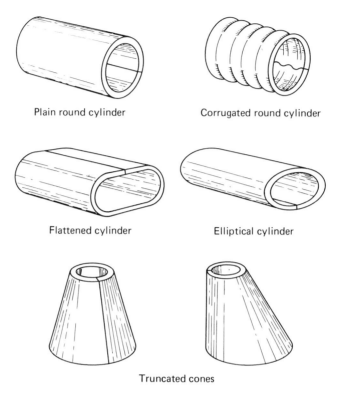

Plain round cylinder

Corrugated round cylinder

Flattened cylinder

Elliptical cylinder

Truncated cones

FIGURE 33-12 Representative shapes produced by three-roll forming

to form parts by hot forming. In addition, fewer passes are needed to roll-form a given shape.

Medium- to large-diameter parts up to $\frac{5}{16}$ in. (7.938 mm) thick may be formed at speeds up to 60 fpm in low-carbon steel plate. Heavier thicknesses are usually cold- or hot-formed from 10 to 20 fpm.

The desired size of the part must be compatible with available three-roll forming equipment. Metal sheets as thin as 16 gage (0.0598 in. or 1.52 mm) can be rolled. A practical maximum thickness is 10 in. (254 mm). Different capacities of roll-forming machines must be used as the work metal thickness increases. It is necessary, for example, to roll-form stock from 0.0598 to $\frac{3}{8}$ in. (1.52 to 9.53 mm) thick on one machine; stock from about $\frac{3}{8}$ to 7 in. (9.53 to 177.8 mm) thick on another machine; while stock 6 to 10 in. (152.4 to 254 mm) thick would require still another machine with a heavier capability. It is not practical to roll-form in a pinch-type machine a cylinder with an inside diameter of less than 2 in. (50.8 mm) more than the diameter of the top roll. Cylinders as large as 7 ft (approximately 2.1 m) in diameter in lengths of 35 to 40 ft (approximately 10.6 to 12 m) have been successfully roll-formed. In general, the maximum workpiece size is limited only by the available equipment and handling capabilities.

The operator's skill and the condition of the equipment play a vital role in maintaining any variation of out-of-roundness of production parts. Stiffening features such as beads or flanges are incorporated when possible into the workpiece design to maintain roundness. Oil drums and other such containers are common examples of this technique.

PROCESS TERMS AND DESIGN FACTORS

There are a number of important terms commonly associated with bending and forming operations that will be briefly explained and illustrated.

Seaming: Seaming (or *lockseaming*) is a bending operation that is frequently used on some products as a locking method for joining the ends of the workpiece blanks. Lockseams are commonly used to fabricate a wide variety of light-gage metal containers, such as gutters, downspouts, cans, pails, drums, open-ended boxes, tubes, paintbrush ferrules, and radio shields. Small hand-operated machines are used for low-volume workpiece production. Other types of automatic machines are available when large-quantity production is required, as in the manufacture of cans for food and beverages. Figure 33-13 illustrates various types of seams used on sheet-metal parts.

FIGURE 33-13 A variety of seams used on sheet metal parts

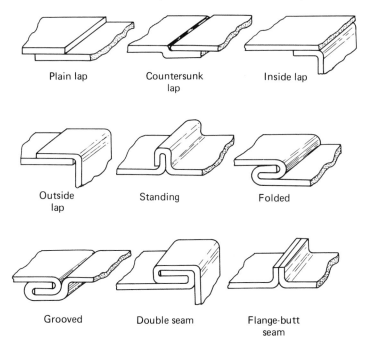

Plain lap   Countersunk lap   Inside lap

Outside lap   Standing   Folded

Grooved   Double seam   Flange-butt seam

Hemming: This is frequently done on sheet-metal workpieces to achieve a strong, smooth edge. The process consists of bending a desired width of metal back on itself. Hemming is commonly used to stiffen outer edges of thin-metal products and to safeguard the user from possible injury from an otherwise sharp and potentially dangerous edge.

Curling: As shown in Figure 33-14, *curling* consists of forming a rounded, folded-back or beaded edge on thin metal parts for the purpose of stiffening and for providing a smooth, rounded edge. Curling is not generally considered to be practical for material thickness greater than 0.031 in. (0.794 mm). The technique is principally restricted to soft and ductile metals such as steel, brass, copper, or tin. The curl diameter is limited to approximately ¼ in. (6.35 mm) for most products.

Depending upon the part shape and the required quantity, curling may be done on press brakes, single-action punch presses, roll-forming machines, or on special beading machines.

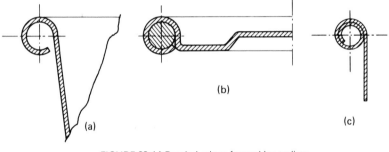

FIGURE 33-14 Beaded edges formed by curling:
(a) curling; (b) wired; (c) double curl

Curled edges on some products such as pails and pans are often further strengthened by inserting wire compactly into the beaded rim space. A common example of a product with a *functional* curled edge is a hinge.

Flanging: This is a process by which an angular lip is formed on the outer edge of a metal blank, resulting in a more rigid and stronger workpiece. The process differs from conventional angle bending in that the bending takes place on *curved* or *irregular* edges of blanks. Figure 33-15 illustrates the basic types of formed flanges, which are produced by cam-actuated flanging dies.

A flange on a concave radius severely stretches the material. The severity continues to increase as the flange width and the sharpness of the curve increase. A flange on a convex radius tends to wrinkle and distort the work-

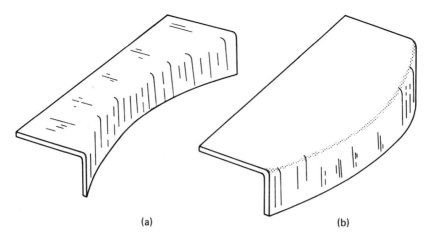

FIGURE 33-15 Two types of formed flanges:
(a) stretch flange; (b) shrink flange

piece. Notching the flange (shown in Figure 33-8) relieves the problem but may tend to weaken the support.

The radius of a flange on an irregular curve should be as large as possible coupled with the least practical height.

Draw Beads: These help to control metal flow into the die cavity and promote desirable stretching of the metal blank as it is being formed. Since the final location of draw beads is usually determined by die tryout, the manner of installing the draw beads is traditionally left to the discretion of the diemaker. Whenever possible, draw beads are placed outside the trim of the workpiece. In some cases, however, it may be necessary to incorporate draw beads as an integral feature on the final product as shown in Figure 33-16.

FIGURE 33-16 An example of a part stiffened by using a draw bead

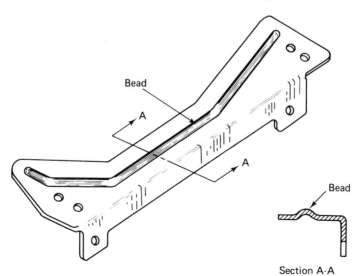

Bead

A

A

Bead

Section A-A

REVIEW QUESTIONS

33.1. What is the essential difference between bending and forming?

33.2. What is the essential difference between press-brake forming and contour roll forming?

33.3. List the factors that determine how many "stations" are required in the contour roll forming process.

33.4. Explain why it would normally be advantageous to roll-form components directly from prefinished stock.

33.5. Under what conditions is press-brake forming competitive with contour roll forming?

33.6. Explain why cold forming is preferred over hot forming.

33.7. What are the main differences between hot forming and "warm forming"?

33.8. Explain how the amount of variation caused by out-of-roundness on a circular part is controlled.

33.9. What is the purpose of hemming and curling?

33.10. Explain the major differences between hemming and curling.

33.11. Explain the major differences between flanging and conventional angle bending.

PROBLEMS
(See appendix for useful formulas.)

33.1. Calculate the force required to bend a 10 in. piece of low carbon steel bar stock ¼ in. thick by 4 in. wide on a press brake using a V-die.

33.2. Calculate the bend allowance of a 90° bend with a ¼ in. inside radius in 0.040 in. thick steel stock.

BIBLIOGRAPHY

"Press Bending" and "Press-Brake Forming," *Metals Handbook,* Vol. 4, *Forming,* 8th edition, Metals Park, Ohio: American Society for Metals, pp. 89-111, pp. 101-111.

Source Book on Cold Forming, ASM Engineering Bookshelf Series, Metals Park, Ohio: American Society for Metals, 1975.

518

34

SPECIAL COLD-FORMING PROCESSES

Ordinary manufacturing methods can produce just about any shape that a designer is likely to need. Yet, there are times when available processes cannot form the part exactly as specified on the print, or at least cannot form the part at a reasonable cost.

This chapter describes a number of cold-working processes, which, in most cases, were developed specifically to overcome major drawbacks inherent in some of the other forming processes. Many of the processes described in this chapter handle jobs that are completely beyond the capability of conventional methods, while some others offer nothing special in the way of shapes or forms, but turn out parts faster and cheaper than their conventional counterparts. The following special forming processes, listed below, do not cover all the emerging technologies in manufacturing but focus primarily on production-tested techniques presently used in industry:

Flexible die forming	Electroforming
Guerin process forming	Swaging
Peen forming	Cold heading
Magnetic pulse forming	Cold-impact extruding
Explosive forming	Thread rolling

DESCRIPTION OF THE PROCESSES

Flexible Die Forming

Flexible die forming is a method that eliminates the expense of manufacturing special precision mating male and female dies. Figure 34-1 shows the basic operation of a solid-urethane (a man-made elastomer) die pad in press-brake forming. Hydraulic presses are also used in flexible die forming. A typical setup is illustrated in Figure 34-2, in which the ram contains a flat, thick

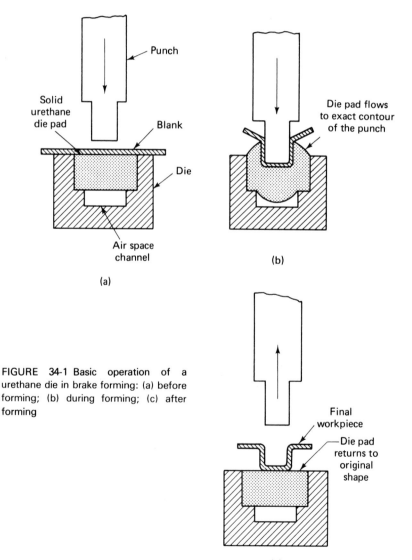

FIGURE 34-1 Basic operation of a urethane die in brake forming: (a) before forming; (b) during forming; (c) after forming

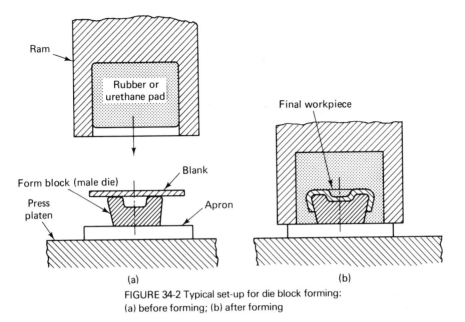

FIGURE 34-2 Typical set-up for die block forming:
(a) before forming; (b) after forming

pad of rubber or urethane. A single-form die block serves as a cutting template and is mounted on the platen. These low-cost form blocks, which are specially made for each product, can be made of epoxy, resin, wood, steel, aluminum, or cast iron. In operation, a metal blank is placed on the form-die pad and is formed to the desired shape under continuous pressure during a complete stroke of the press.

To visualize the action of this flexible material, it is helpful to understand the basic behavior of the material. Rubber or urethane die pads act somewhat like a "solid fluid," changing shape under a load while the volume remains constant. In this respect, the action is very much like water. As the totally confined material deforms or deflects, it transmits the applied force equally in all directions, exerting high, uniform, and continuous counterpressure. A property of the material often called "memory" provides for a quick return to its original shape when the load is removed. In the actual metal-forming application, the pad resists deformation at the beginning of the press stroke, thus generating high blank-holding pressure before any forming occurs. Then, as press tonnage overcomes this resistance, it deflects in every direction in which it is free to bulge. The high forming pressure that is exerted forces the material to flow around the die block, forming the desired part. Air space must be provided in the die-block cavity to permit the urethane or rubber pad to deform within allowable deflection limits.

Urethane is often selected as the pad material, in preference to neoprene and rubber, because of its superior oil and solvent resistance and better thermal stability. It also has greater abrasion and tear resistance, in combination with increased load-bearing capabilities.

The chief *advantages* in metal forming with flexible dies are that the process saves money, often results in a generally improved product, and cuts lead time. Here are some reasons why this is possible:

1. Flexible dies eliminate the expense of:
   a. Mating male and female dies.
   b. Tool steels and hardening.
   c. Grinding and fitting.
2. In some cases, the same flexible die pad can be used for several different workpiece shapes.
3. Flexible dies make it possible to work with prepolished or preplated materials and eliminate the need for protective coverings.
4. Specialists, such as tool and die makers, are not usually necessary.
5. Tools have fewer components. Forming operations can be simplified. Overforming can be accomplished without expensive cam-action tools.
6. The process is versatile. In addition to bending and forming operations, cutting (blanking, piercing) and drawing (embossing, bulging) can be accomplished.
7. Thinning of the work metal is practically nonexistent.
8. The process automatically compensates for normal material thickness variations, saving on die wear.
9. Surfaces on workpieces are not marred by the shock or draw marks commonly associated with conventional steel dies.

Important *disadvantages* of metal forming in flexible dies include the following:

1. Certain operations may require special accessory equipment (in cases where the definition may be less sharp than that obtained with conventional tooling).
2. Problems sometimes develop as a result of the tendency for wrinkles to form, especially on shrink flanges.
3. The die life of rubber or urethane pads is less than that for conventional tooling. The average life of a rubber pad is about 20,000 "hits," while urethane pads average nearly 50,000 "hits."

Guerin Process Forming

The *Guerin process* is the oldest and most basic of the several production flexible die-forming processes. It is especially preferred where quantities of differently shaped machine parts are needed, particularly at regular intervals. In this case, a number of different form blocks (or punches) may be used, each shaped for the different desired workpiece requirements. The form blocks are mounted at intervals along the press bed. Large quantities of different parts may be simultaneously produced by a single downward stroke of the upper ram of the press as the flexible die pad presses against and envelops the work metal over the various form blocks. Some parts may be blanked and formed in one operation. No lubrication is necessary.

Guerin forming is used to produce parts with reasonably simple bends and those with comparatively shallow draws. Blanking and piercing operations on aluminum and aluminum alloys are limited to stock thickness not exceeding 0.032 in. (0.81 mm). Flanges, beads, and ridges may be formed on parts up to $\frac{3}{16}$ in. (4.763 mm) thick. The process is also adaptable for forming workpieces from light-gage mild sheet steel, stainless steel sheets, and copper. Magnesium must be hot-formed.

Parts designed for flexible die forming require generous corner, bottom, and flange radii.

Peen Forming

Peen forming, also called *shot forming,* is essentially a free-forming technique where a torrent of small steel balls are hurled against a metal surface. The process is restricted to the fulfillment of a fairly specialized function in metal forming. It is principally used to form various irregular contours on aluminum sheet and plate where the required part is large but where only gentle contours are required.

Peen forming was developed during the early 1950s as a method of forming integrally stiffened wing panels for large aircraft. Because of the extreme length of the ribbed structure and additional problems imposed by the variable skin thickness, it was virtually impossible to form the panels to the required contours using conventional press-forming techniques.

Peen forming is a variation of a well-established surface-cleaning process called *shot peening,* in which a stream of metal shot is blasted against specific portions of metal parts such as crankshafts, connecting rods, and gears. The numerous small impacts in shot peening produce a uniform layer of surface compressive stress to increase the fatigue life of the metal parts. Under certain conditions, the compressive stresses also cause the workpiece to curl or warp out of shape. Peen forming harnesses this effect for a useful purpose.

Peen forming offers a simple alternative to conventional press-forming methods in that no dies or forming presses are required. Some parts may be made by simply placing the metal sheet on a table or suspending it from a support and then blasting it with shot. Other parts, depending upon the desired contour, are clamped over simple form blocks during peening. A simple change in spray pattern and intensity can change the shape of the part. The usual method of checking the accuracy of a peen-formed part is by use of a template.

Honeycomb panels such as aircraft wings and large tubular shapes may be formed by this method. Peen forming easily produces compound curvatures which are often quite difficult or economically unfeasible to produce by conventional press-forming methods. Peen forming is also adaptable to salvage operations in correcting bent or distorted parts.

Considerable experimental work in peen-forming technology is currently taking place. Uniform peening of flat sheet or plate ordinarily urges the metal toward compound curvature. It has been found, for example, that uniform peening over a square or round plate tends to produce a hemispherical curvature. Likewise, uniform peening over a rectangular shape tends to produce a somewhat ellipsoid form. It is possible, by carefully masking the part, to produce almost any shape that may be required by applying the peening in narrow bands or in selective regions. Masks are usually made of rubber.

Another important factor that affects successful achievement of the ultimate desired shape is the depth of the layer of induced compressive stress as a ratio of the total part thickness. When this ratio exceeds a certain value, which varies with specific applications, the sheet or plate warps in an unpredictable or uncontrolled manner. The selection of optimum shot size and shot velocity is also important in producing a desired curvature. In some cases, the shot is fed to a nozzle through a pneumatic hose and propelled toward the workpiece by a blast of high-pressure air. In most commercial applications, however, large quantities of shot, in highly controllable patterns, are thrown against the workpiece by a specially developed centrifugal wheel that can accept shot up to ¼ in. (6.35 mm) in diameter.

A flat steel disk 6 ft (1.82 m) in diameter and ¼ in. (6.35 mm) thick may be peen-formed to about a 6 in. (152.4 mm) hemispherical or parabolic rise with up to a 10 ft (3m) radius. In this application 0.066 in. (1.68 mm) shot is used. Using steel shot of the same diameter, the largest steel plate thickness that can be contoured is about ³⁄₁₆ in. (4.763 mm). Steel plate as thick as ½ in. (12.7 mm) can be formed with ¼ in. (6.35 mm) diameter steel shot.

Magnetic Pulse Forming

Magnetic pulse forming, also known as *electromagnetic forming,* is a production process for forming individual metal workpieces or assemblies by the force of a very intense pulsating magnetic field lasting only a few microseconds.

Electrical energy, stored in capacitors, is discharged rapidly through a forming coil. An interaction of the current, which is induced from the coil into the conducting workpiece, and the magnetic field causes a repulsion that serves as the forming force. The forming force accelerates the workpiece against a mandrel or a die. Permanent deformation on the workpiece occurs when the work metal is stressed beyond its yield strength.

High-intensity magnetic fields behave in much the same way as compressed gases. Pressures up to 50,000 psi (3500 kg/cm^2) can be uniformly created on metal workpieces. Once the correct amount of force to satisfy a specific forming requirement has been determined, the quality control on production operations may be assured.

The process is used chiefly to assemble tubular parts to each other or to other components. The automotive industry, for example, is using magnetic pulse forming to assemble steering gears, drive shafts, ball joints, and shock absorbers. Other industries have adopted the process as a method of assembling vial caps, potentiometers, instrument bellows, coaxial cable, electric motors, and many other items that would be difficult and costly to process by conventional methods. Commercial applications of this relatively new process are also being found in the manufacturing of workpieces of various shapes from flat sheets as well as in tube sizing and for piercing or shearing operations. Figure 34-3 illustrates examples of different types of work performed by magnetic forming.

Figure 34-4 illustrates three basic methods of magnetic pulse forming: *compression, expansion,* and *contour.* In each case, a different type of forming coil or "tool" is required. While there is no actual physical contact between the coil and the workpiece, the coil has sufficient strength and mass to withstand the impacts caused by the repetitive pulsating discharges of large amounts of electrical energy. While a variety of coils of standardized design are available, it is estimated that approximately 60% of the coils used are customized to particular applications. Coils can be easily changed for different applications.

In *compression* forming a contoured mandrel is inserted into the workpiece. The tubing is compressed by the sudden inward force of the forming coil that surrounds the workpiece. In *expansion* forming, the coil is inserted inside the tube or ring, over which a split collar or mating component is preassembled. Expansion of the workpiece takes place by the outward force released by the forming coil. Because of energy limitations, it is not practical to use coils with an outside diameter of less than 1 in. (25.4 mm). *Contour* forming is a method in which flat-forming coils are used to produce a uniform pressure against flat stock, to force it into a die cavity.

Good electrical conductors, such as copper, brass, aluminum, carbon steel, and molybdenum, can be formed easily. Nonconductive materials can

FIGURE 34-3 Examples of magnetic pulse forming applications: (a) swaging a rectangular tube to notched fitting; (b) attaching a rubber hose with an expanded aluminum band to an auto gas tank (Courtesy, Maxwell Laboratories, Inc.)

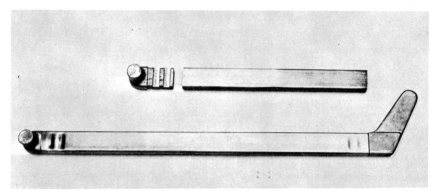

(a)

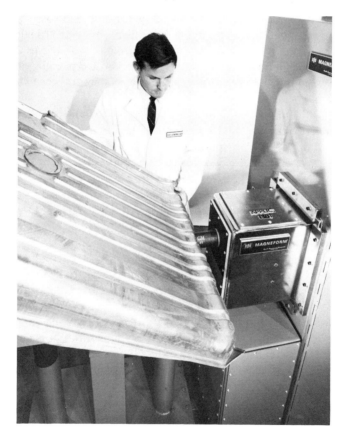

(b)

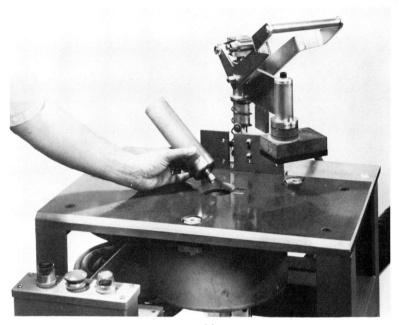

(c)

(d)

FIGURE 34-3 *(continued)* (c) assembly of end caps onto power fuses; (d) removing a 2 in. O.D. aluminum tube which has been expanded into a die to form a bead and a lip (Courtesy, Maxwell Laboratories, Inc.)

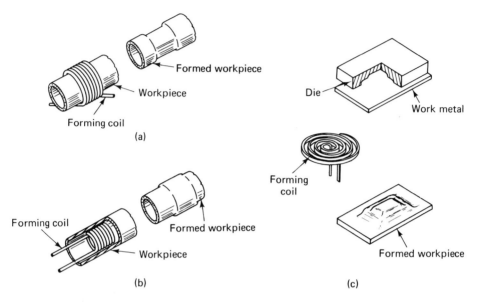

FIGURE 34-4 Three basic methods of electromagnetic forming: (a) compression forming; (b) expansion forming; (c) contour forming

be formed magnetically by placing a conductive material, called a "driver," between the coil and the workpiece. An aluminum driver, for example, can be used to form stainless-steel parts. Among the advantages of the process is that there are no tool marks on the surfaces of parts or assemblies. Joints can be produced that are free from such heat effects as contamination and deformation. Close tolerances on individual component parts are not required. Pressure and vacuum joints can be assembled without welding or brazing or use of special seals. Since the magnetic field will pass through electrical nonconductors, metal rings or tubes can be assembled to ceramic, Bakelite, and phenolic parts. Metallic bands can be compressed over rubber bands and other soft materials. The process is very rapid.

There are some important limitations of magnetic pulse forming. The magnetic field cannot be easily adapted to fit all workpiece contours. It is not possible to apply a high pressure in one area and a low pressure in an adjacent area. Nonsymmetrical parts do not lend themselves well to the process. Magnetic force cannot be applied to workpieces having slots or holes that interrupt the path of the inducted currents.

Explosive Forming

In forming parts by conventional methods, male and female dies are normally required to obtain the desired workpiece shape. *Explosive forming,* by contrast, requires only a female die form, since the energy transfer media, usually water on air, acts as a mating punch to deform the work blank. For convenience, the various explosive-forming techniques may be classified into two general categories of work: (1) *tube bulging* and (2) *die forming* of sheet and plate. In each category, the shape of the workpiece is formed by an instantaneous high-pressure shock wave that results from the detonation of an explosive. Explosive-forming techniques are used in commercial applications when a conventional process cannot do a job adequately.

Tube Bulging: Specific contours such as bulges or beads may be formed on thin-wall tubular shapes by using the energy of an explosive charge. The process, as illustrated in Figure 34-5, requires the use of massive dies to offset the severe impact loads that are imposed by the expanding gases released by the high explosives. The dies, which are split to facilitate part removal, completely enclose and confine the workpiece. The operation begins with the installation of the preform into the required position and then assembling the die halves. The explosive charge is accurately positioned inside and along the centerline of the tube and is detonated. Misalignment of the charge usually results in an unacceptable part, as a result of uneven forming pressure against the side walls.

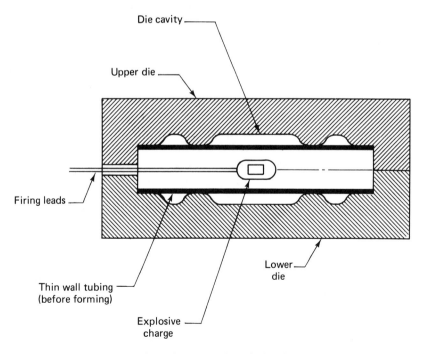

FIGURE 34-5 Explosive tube bulging

The energy source for thin-wall tube forming may also be obtained by using a so-called "low explosive." In this case, a smokeless or black powder is used. Instead of exploding, the powder rapidly burns to form a gas. The pressure produced by the expanding gas, being about 40,000 psi (2800 kg/cm²), is adequate to force the tubing into the die configuration.

Tolerances of ±0.010 in./in. (0.010 mm/mm) are readily obtainable. Tube forming by explosives is restricted to relatively small workpieces. The economic advantages of the process tend to decrease as the size of the parts increases.

Die Forming Sheet and Plate: Parts that are too large to produce by conventional forming techniques can often be successfully made by explosive forming. Explosive forming is usually not an economical method for the fabrication of small parts from sheet and plate. Maximum cost effectiveness for explosive forming is in the range 3 to 6 ft (0.91 to 1.82 m). For parts larger than 6 ft at their widest point, special problems may arise in relation to the die, such as excessive weight, difficulties in construction, and cost. The production of most concentric sheet-metal shapes by this process presents less difficulty in tooling and explosive charge placement than for nonconcentric shapes. Explosive forming of workpieces from heavy metal plate is practical because presses large enough to form heavy plate are generally not available. The process is usually considered for small quantities of large parts, particularly those requiring relatively close tolerances. When the dies are designed and constructed to assure adequate die life, the process has proved to be economical for production runs of several thousand pieces. Almost any of the common metals, except magnesium, can be explosively formed. Product applications include pressure vessels, heat exchangers, hemispheres, fan hubs, and inlet vanes, as well as a variety of components for chemical and petroleum processing equipment.

Large explosive-forming operations are normally conducted in relatively isolated areas. A mobile or fixed crane for transporting sheet and plate stock, the heavy dies, and finished work is a practical necessity. Figure 34-6 shows a typical arrangement for a "standoff" operation. In this type of operation, sometimes called hard-die forming, the explosive charge is placed some distance away from the work blank. The sequence of operations begins by loading and securely clamping the workpiece, consisting of a circular flat sheet, to a specially contoured female die. A rubber hose is attached to the vacuum outlet on the die and a vacuum pump is used to evacuate the air from under the sealed work blank. The assembly is then lifted and immersed into a large steel-lined water tank. Rugged in-ground or above-ground water tanks as large as 30 ft (approximately 9 m) in diameter and 25 ft (approximately 7.6 m) deep are often used. Finally, the explosive charge is submerged and suspended over the blank at a predetermined distance and is detonated.

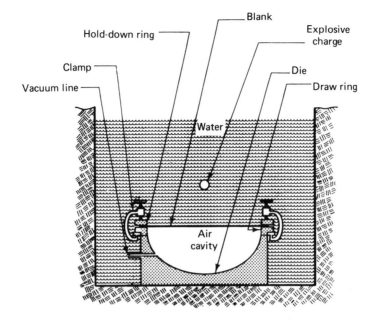

Hold-down ring

Clamp

Vacuum line

Blank

Explosive charge

Die

Draw ring

Water

Air cavity

FIGURE 34-6 A typical arrangement for a stand-off explosive forming operation. The die is submerged in a water-filled tank buried underground

The detonation velocity can be as high as 25 times that of a rifle bullet. Behind this wave, the explosive is converted, in a matter of microseconds, into a high-temperature, high-pressure system of gases that can exert extremely large forces on the medium in which it is immersed and on other objects in the vicinity.

In some cases, two or more explosives may be required to shape a workpiece to the required dimensions. Heavy dome-shape workpieces, for example, are sometimes formed in a series of two steps. In the first step, the approximate shape of the dome is explosively formed, in what might be called a roughing die, in the manner previously described. The partially formed dome is then withdrawn from the die and the flange is removed by an abrasive cutter. The second step consists of loading the deflanged preform into an accurately sized truing die, evacuating the trapped air, and returning the assembly to the tank. One or more shots are then fired to form the final outline.

A less widely used technique, known as a "contact" operation, may also be used in which the explosive charge is placed in direct contact with the workblank. Using this technique, the workpiece is forced into a female die or onto a male die by the sudden release of high-intensity, transient stress waves. The principal advantage of a contact operation is that the need for an in-ground water-tank installation is eliminated because the work can usually be done above ground.

531

Principal forming die materials, used for large production runs of hundreds of parts, are alloy steel, meehanite, and ductile cast iron. A reclaimable material known as Kirksite is often used for small production runs of 10 to 50 parts. Kirksite dies are particularly suitable for forming thin or soft, ductile copper or aluminum workpieces which do not require explosive charges. Reinforced concrete, wood, epoxy, and ice have been used for prototype work or for short production runs. Some dies are laminated for the purpose of reducing the weight and ease in handling.

Tolerances on explosively die-formed parts may be obtained as precise as ±0.010 in./in. (0.010 mm/mm) on diameters and ±0.020 in./in. (0.020 mm/mm) on contours for dome shapes ranging from 30 to 60 in. (762 to 1524 mm) in diameter.

Because of the inherent nature of the process, springback is generally less for explosively formed materials than for press-formed workpieces. Overforming, restriking, and using thicker materials are all useful ways of reducing springback. Each of the methods employed influences the design and construction features of the die.

A technique known as *free forming,* involving only a draw ring and a hold-down ring, is illustrated in Figure 34-7. No dies are required. The method is suitable for producing large parts when the workpiece contours are reasonably simple and tolerances are not severe. Only parts with axially symmetrical shapes, such as domes, may be free-formed. Domes as large as 140 in. (approximately 3.6 m) in diameter have been successfully formed in this manner. After detonation, the workpiece expands freely into the air. Since no die cavity is used to control the final workpiece contour, the ultimate accuracy of this technique depends upon the reproducibility of the workpiece materials used, their deformation-hardening characteristics, and the effectiveness of the explosive charge.

FIGURE 34-7 Free-forming a workpiece without the use of a die

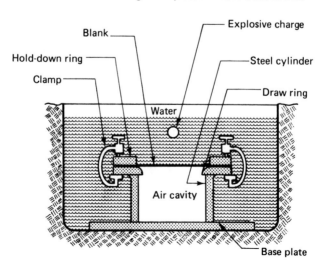

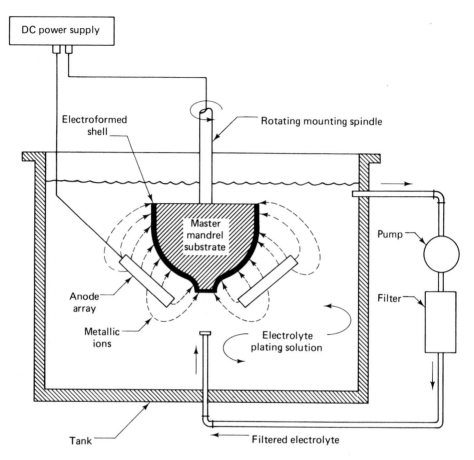

FIGURE 34-8 A typical electroforming set-up

Electroforming

Electroforming is a well-established, low-cost, high-volume process. It is used to produce specialized, complex, thin-walled components. The process uniquely combines the advantages of lightweight sheet-metal fabrications, the complex shapes obtained by casting, and the accuracy of precision machining. Figure 34-8 illustrates a typical electroforming setup. A part is electroformed, as in plating, by electrochemically transferring metal ions through an electrolyte from an anode to the surfaces on a mandrel, where the ions are deposited as atoms of plated metal. Selected surfaces on the mandrel which receive the plated metal are coated so that the deposit will not adhere to it. When the part has achieved the desired form and thickness, the plated metal shell, called the *electroform,* is removed from the mandrel and retains its as-deposited shape as a discrete component. Excess metal deposits may be trimmed and, when necessary, parts may be machined.

533

The shape of the mandrel determines the desired configuration of the required part. Most electroformed parts are produced over a positive or male mandrel. When it is necessary to produce an accurate part with a smooth surface finish on the outside surface rather than on the inside, a negative or female mandrel is used. Mandrels must be removable and electrically conductive. Depending upon the final workpiece configuration, two kinds of mandrels may be used: *permanent* and *disposable.* Permanent mandrels are used when the shape of the electroform is free from undercuts and when there is sufficient draft to permit it to be freely withdrawn from the mandrel. Disposable mandrels are required in cases where the electroform cannot be freely lifted from the mandrel.

Mandrels may be prepared by machining entirely from stock or sometimes are cast to an approximate size and shape, followed by finishing to the final dimensions. Considerable attention must be given to the dimensional accuracy and surface finish of the mandrels, since the accuracy and finish of the electroform will be only as good as that of the mandrel. Also, mandrels must be handled very carefully, since the smallest imperfection, even a small scratch, will be reproduced in the electroform. Permanent mandrels are usually made about ½ in. (12.7 mm) longer than the electroform at each end and have some provision for attaching a hanger or stem for mounting in the tank.

Stainless steel and aluminum are the most commonly used materials for mandrels. Aluminum is often selected as the material for disposable mandrels, as it readily melts, thus facilitating removal from the electroformed product. Material known by the trade name Cerrocast and a variety of low-melting bismuth alloys are also used. These products, when used for mandrels, can be readily melted by submerging them in hot water to remove them from the electroform. Other materials used for mandrels are cast alloys of nickel or brass as well as plaster, quartz, glass, wood, wax, and various plastics and elastomers. The nonconductive materials are coated with a metallic film before processing.

The process differs from electroplating in two important ways. First, in electroplating, a thin protective or decorative surface layer of metal is deposited upon a suitably prepared part. It is considered a final finishing process. Parts commonly electroplated include automotive trim, hardware, and plumbing fixtures. The adherent metal coating is usually limited to 0.0001 in. (0.003 mm) in maximum thickness. Electroforming has no such limits. While parts may be produced as thin as 0.005 in. (0.13 mm) or thinner, the process can also be used to produce parts as thick as 3 in. (76.2 mm). The second important difference is that in electroforming, unlike electroplating, which begins with a previously developed shape, the workpiece is entirely formed by building up a shell of plated metal. The process makes possible the production of parts with extremely complex and precise internal shapes, with a quality of surface finish not obtainable by other production methods. Electroforming is used in applications when it is possible to eliminate time-consuming contour machining and hand-finishing operations. In some instances, there is no other practical way to make parts, particularly those with internal forms too difficult to core or to machine.

It should be emphasized that the actual design, selection of the materials, and the preparation of the mandrel, for a given product application, is the responsibility of the electroformer. The descriptions given here and in Figure 34-9 are intended solely for the guidance of the product designer in order that he/she may adapt the design of a component to permit the realization of the full advantages of this process.

The materials most widely used to produce components by electroforming are nickel, copper, silver, and iron. The type of metal or alloy selected depends upon the nature of the various properties a given workpiece must have, such as hardness, ductility, electrical or thermal conductivity, and corrosion resistance.

Electroforming is a high-precision process and, unlike most other processes, parts are subject to less variation in size on a production basis. Accuracies of ±0.0001 in. (0.003 mm) are obtainable. There may be some variation in wall thickness, especially for irregularly shaped parts. When necessary, special-shaped anodes are used which are made to conform to the changing contour of the part. The use of "conforming anodes" ensures that the deposition of metal is more uniform on parts with intricate shapes. Thin walls up to 0.030 in. (0.76 mm) can be held to a tolerance of ±0.002 in. (0.05 mm). A tolerance of ±0.005 in. (0.13 mm) can be expected on thicker walls. To obtain uniform wall thickness, the electroformed part must be machined. Figure 34-9 illustrates important design factors that influence electroformability. Many irregularities of the deposit distribution can be minimized by the electroformer through the application of special processing techniques.

A major advantage of electroforming is that components may be produced with ultrasmooth or mirror finishes. Surface finishes of better than 2 μin. are possible. Conversely, the process is used to produce surfaces to a desired texture or roughness, examples of which are shown on the precision surface finish comparators, as in Figure 34-10. Decorative and ornate or functional finishes are electroformed on many types of household products, such as wall switch plates, sinks, and teapots. Layers or laminations of different materials can be built up on opposite surfaces of parts. The maximum size of parts that can be produced is limited only by the capacity of the available plating tanks. Some companies have plating tanks that can accept parts up to 20 ft (approximately 6 m) long. Parts over 7 ft (approximately 2 m) long have been successfully electroformed. Most electroforms, however, are in the range of 0.010 to 0.250 in. (0.25 to 6.35 mm) thick.

FIGURE 34-9 Design factors influencing electroformability (Courtesy, ASTM B-450-67 American Society for Testing and Materials)

Design factors influencing electroformability[a]

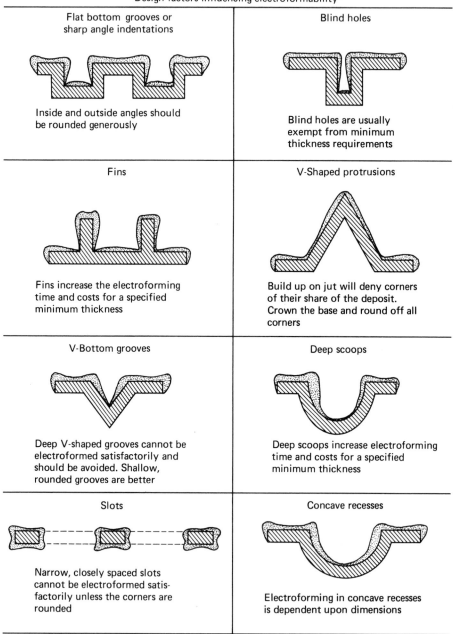

Flat bottom grooves or sharp angle indentations	Blind holes
Inside and outside angles should be rounded generously	Blind holes are usually exempt from minimum thickness requirements
Fins	**V-Shaped protrusions**
Fins increase the electroforming time and costs for a specified minimum thickness	Build up on jut will deny corners of their share of the deposit. Crown the base and round off all corners
V-Bottom grooves	**Deep scoops**
Deep V-shaped grooves cannot be electroformed satisfactorily and should be avoided. Shallow, rounded grooves are better	Deep scoops increase electroforming time and costs for a specified minimum thickness
Slots	**Concave recesses**
Narrow, closely spaced slots cannot be electroformed satisfactorily unless the corners are rounded	Electroforming in concave recesses is dependent upon dimensions

FIGURE 34-9 (*continued*)

Rings	Ribs
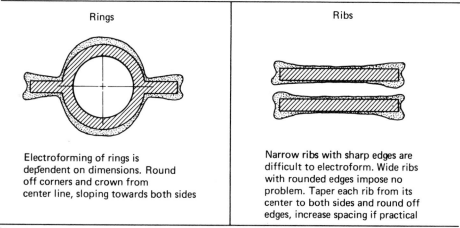	
Electroforming of rings is dependent on dimensions. Round off corners and crown from center line, sloping towards both sides	Narrow ribs with sharp edges are difficult to electroform. Wide ribs with rounded edges impose no problem. Taper each rib from its center to both sides and round off edges, increase spacing if practical

[a] Metal deposit is shown in an exaggerated fashion.

The part need not be made entirely from deposited metal. Other materials, even nonconductors, can be incorporated into the component by placing onto, over, or around separate pieces attached to the mandrel. Threaded inserts, bearing surfaces, shafts, and other such inserts—called "grow-ons"— are often incorporated in electroforms by this technique, a patented proprietary process. The nondeposited metal, in fact, often constitutes a larger portion of the final part than the electroformed metal. Some waveguides are composed primarily of machined wrought pieces, joined by a relatively small amount of electrodeposited metal.

FIGURE 34-10 A microfinish comparator on which has been duplicated by electroforming 22 specimens of machine surfaces from 2 to 500 μin. The surfaces range from lapped, ground, turned, milled and profiled surfaces (Courtesy, GAR Electroforming, Div. MITE Corporation)

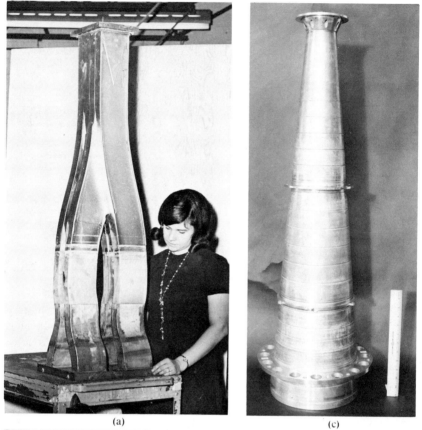

(a)   (c)

(b)

FIGURE 34-11 Representative electroformed parts: (a) a heavy nickel walled waveguide called a "combiner". It is approximately 0.200 in. thick by 6 ft. long (Courtesy, GAR Electroforming, Div. MITE Corporation); (b) nickel reflectors consisting of electroformed reflector surface epoxy-bonded to electroformed back support (Courtesy, GAR Electroforming, Div. MITE Corporation); (c) electronic feed horn—42 in. long (Courtesy, GAR Electroforming, Div. MITE Corporation)

Perhaps the greatest benefit of electroforming is the diversity of products effectively produced. One of the oldest users of the process is in the phonograph record industry, to make metal stampers by electroplating the surface of mechanically cut plastic masters. Another well-established user of the process is the printing industry, for making printing plates. Electroforming is also used to produce continuous rolls of copper and nickel foil to extremely thin gages and wide widths which cannot be produced by rolling. Other product applications include radar waveguides, aircraft wing tips and leading edges of intake scoops and propeller blades, missile bodies, lenses, mirrors, and reflectors, foundry patterns, computer cams, toys, and various kinds of dies and molds used in manufacturing rubber tires, glass, or plastics products, and for the die-casting process. Special nickel screens used in Vidicon television cameras and others for rotary screen printing of fabrics are also made by this process. Mesh sizes range from a fine mesh of 1000 (about 15 μin.) to a coarse mesh of 20 to 120 mesh. Figure 34-11 illustrates some representative electroformed components.

Swaging

Swaging is a general term that may be appropriately applied to a number of metal-forming operations. Production swaging operations are commonly performed on rotary swaging machines, and the discussion in this chapter will be confined to this process.

FIGURE 34-12   Examples of swaged shapes (Courtesy, The Torrington Company)

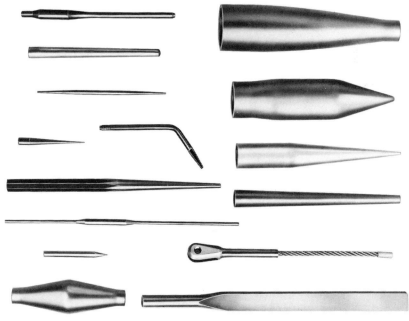

(a)

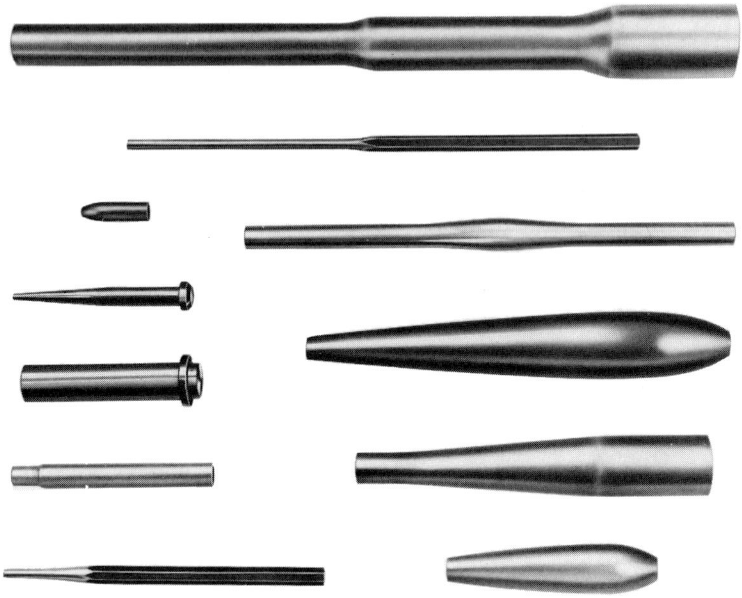

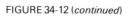

(b)

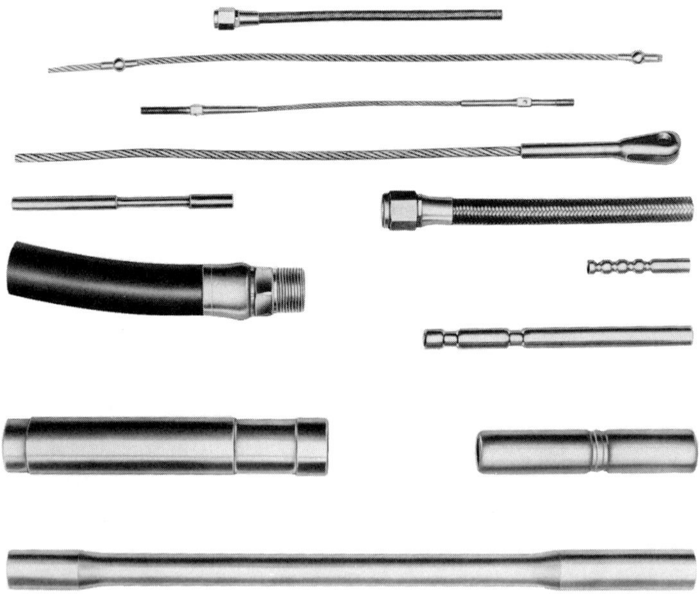

(c)

FIGURE 34-12 (*continued*)

Rotary swaging is a process of progressively reducing the cross-sectional shape of bars, rods, tubes, or wires by a large number of impacting blows with one or more pairs of opposed dies. The blows displace metal and form the blank to the shape of the dies. The final shape is round because the dies rotate around the workpiece while the operation is being performed. Swaging has proved to be an economical production method for forming shapes usually confined to a portion of the total length of a given part, by pointing, tapering, reducing, or sizing. Figure 34-12 shows some typical swaged shapes. The process is also used for various joining and fastening operations.

The process is normally limited to the forming of blanks having symmetrical cross-sectional shapes, such as rounds, squares, or hexagons, although other forms such as rectangles can be swaged using special equipment. Various types and sizes of rotary swaging machines are available. All these machines are equipped with dies that open and close rapidly to provide the hammering action to shape the workpiece.

Figure 34-13 illustrates the principle of operation for a standard two-die rotary swager used for straight reduction of stock diameter or for producing tapers on round stock. Most rotary swagers have either two or four dies. The sequence of operations begins by feeding the workblank into the forming die opening. The revolving spindle throws the rotating dies outward by centrifugal force against a series of rollers held in a "cage" surrounding the spindle. As the backers strike the rollers they rebound to force the dies inward, causing them to impact against the workpiece. After the impact, the dies open and release the workpiece, when the backers lose contact with the rollers. This condition of the backers, as they intermittently contact the rollers, may be described as a "wiping" action. The time interval between the opening and closing of the forming die halves depends upon the speed of the spindle rotation, the distance between the adjacent rolls, and the radius of the backer. A swaged part may be subjected to as many as 1000 to 5000 repeated blows or impacts per minute.

FIGURE 34-13 Principles of operation for a standard two-die rotary swager

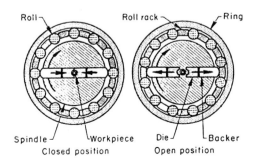

Roll   Roll rack   Ring

Spindle   Workpiece   Die   Backer
Closed position   Open position

Standard two- and four-die rotary swaging machines can be used only for reducing the ends of tubing or solid bar stock. The tooling must be changed in order to produce step reductions on workpieces. Excessive surface contaminants, such as oxides, or scale must be removed from the work-blank before feeding into the swager. This precaution is necessary to prevent the tendency of the foreign materials to load the die and interfere with other moving parts of the equipment. Another precaution taken is that swaging operations are always performed on a part *before* plating or painting.

Because of the greater ease in handling, and because a superior surface finish may be obtained, work is usually cold-swaged. Cold-swaging work hardens most metals. If a considerable amount of reduction is required, the parts may have to be annealed after one or more passes. Like other cold-working deformation processes, cold swaging normally results in a product with improved physical characteristics. Swaging at elevated temperatures is necessary with refractory metals in order to reduce the yield strength.

Under normal conditions, rotary swaging improves the initial finish of the work blank. This is principally due to the overlapping action of the dies as they periodically strike the workpiece. The resulting "burnished" appearance on swaged work is often superior to that obtained by machining.

The control of tolerances varies with the size of the parts. For example, it is possible to swage small-diameter wire to a tolerance of ± 0.0005 in. (0.013 mm) and closer while parts from 0.125 to 0.500 in. (3.18 to 12.7 mm) may require a tolerance of ± 0.002 in. (0.05 mm). Tolerances of the diameters of parts of 0.500 to 1.000 in. (12.7 to 25.4 mm) are normally obtainable within ± 0.002 to 0.005 in. (0.05 to 0.13 mm) and the tolerance on concentricity can be held within ± 0.003 in. (0.08 mm). To control both the diameter and the wall thickness of tubular shapes, special mandrels are used inside the tube. Metallic eraser caps on pencils and welding torch tips are common examples of parts that are formed over a mandrel.

While swaging is normally ideal for producing extremely small parts, the process has a practical minimum-size-diameter limitation of about 0.020 in. (0.51 mm). Machines are currently available for swaging tubes as large as 6½ in. (165.1 mm) in diameter and bars up to 4 in. (101.6 mm) in diameter. Factors that determine the maximum size of swaged work are workpiece material, the amount of reduction, and whether the operation is done hot or cold.

Except for the pointing of wire for wire-drawing operations, the process is not always economical for short-run production. The formation of steep tapers (in excess of an 8° included angle) or right-angle shoulders on parts is normally very difficult. Objectionable die marks are often noticeable on parts having various step configurations along the length of the part that was formed by two or more dies. Rotary swaging machines are extremely noisy. Most machine operators wear special earmuffs to reduce the possibility of ear damage.

The process should be specified only if swaging is the most practical method of producing the required workpiece shape. Other competitive processes, such as press forming, spinning, or machining, may be better employed to produce the given shape. Process engineers may combine swaging with some other process, such as machining, for example, to achieve increased production, decrease tooling costs, or obtain improved tolerances or surface finish.

Commercial applications of swaged parts include various styles of ratchets and sockets, a wide variety of small and large pins, such as fluorescent tube pins, guide pins, watchband pins, hinge pins, stop pins, and shoulder pins. Other parts include contacts for printed circuits, pen caps, shafts, spacers, bushings, umbrella components, mechanical pencils, a wide assortment of studs, fasteners, and eyelet type or headed parts, wheel axles for toys, shafts, screwdriver points, hose fittings, automotive exhaust pipes, and metal chair and table legs. Swaging is an important process for stepping golf-club shafts and fishing rods. As previously indicated, this versatile process may be used to assemble two or more components by joining a bushing or sleeve to a shaft, swaging of rings or terminals onto wire for use as electrical connectors, assembling fittings to cable, and for attaching various fittings to tubes.

Cold Heading

Cold heading, also known as *cold upsetting* or *axial flow forming,* is an important high-volume production method. It is one of the fastest and most economical methods of producing assembly components. The largest single use of cold heading is in upsetting heads on bolts and rivets or for nail making, but an astonishingly wide variety of other shapes, shown in Figure 34-14, can also be successfully formed by this process. Essentially, parts are formed by squeezing an unheated metal slug of a specified volume into a precise shape within a die cavity.

In cold heading, machines called "headers" are classified according to the types of dies used: *solid* or *open.* The length of the required part usually determines the type of die used. Solid-type dies are used for wire of relatively small size. Machines are also classified according to the number of forming strokes or "blows" made by the punch onto the workpiece during each forming cycle. Cold headers, using either solid or open dies, may impart only one or as many as two or three blows per cycle to produce a part. Most commercial parts are made on two-blow machines. The type of header required for a particular job depends upon the amount of material needed to form the upset contour and the location and shape of portion to be upset. The forming operation is fully automatic.

FIGURE 34-14 Typical parts made by upsetting and related operations such as extruding, piercing and trimming (Courtesy, National Machinery Co.)

In a *solid die machine,* the die is simply a hardened-steel cylinder with an axial hole through its center. The hole is a few thousandths of an inch (or about 0.08 mm) larger than the outside diameter of the stock. Parts are automatically formed by feeding wire into a stationary cutoff quill in which slugs are cut to a predetermined length. The slug is moved laterally and pushed into the die by a punch. A knockout or ejector pin, inside the die, acts as a stop to ensure that the correct amount of the blank projects out from the end of the die. One or more blows are then struck by the punch against the end of the protruding portion. The original solid cylinder is thus

544

reshaped or headed by a force that is sufficient to cause the metal to plastically flow into the die impression. Header dies may or may not include cavities for shaping the part. Four basic methods of forming a head are shown in Figure 34-15. Most simple head forming is done on the face of the die with a punch cavity shaping the head. For countersunk parts, however, the punch is flat and the cavity is entirely in the die. Some parts are upset in both the punch and the die and others between the punch and the die.

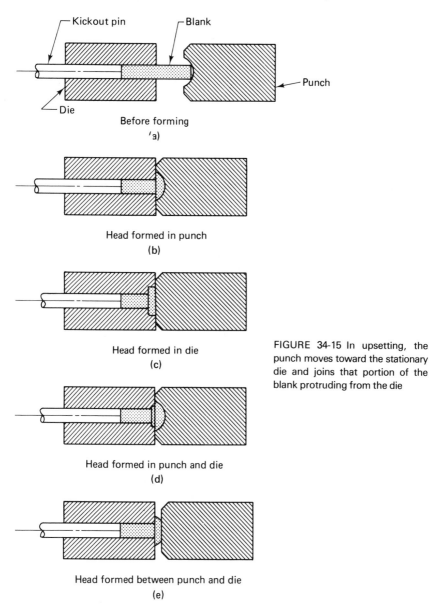

FIGURE 34-15 In upsetting, the punch moves toward the stationary die and joins that portion of the blank protruding from the die

When the heading operation is complete, the knockout pin advances to eject the finished part.

Open-die machines use dies composed of two halves with matching grooves machined to the desired configuration. The principle of operation of this type of machine resembles that of the solid die header. Coiled wire stock is fed into the open-die halves against a stop. The die halves then close to securely grip the wire and shear it to the correct length. The slug is then formed to the final part shape by one or more blows of a punch. After the heading operation is complete, the dies open and the incoming wire ejects the finished part. Fully automatic machines are available that can perform at a rate of up to 450 parts per minute.

Other types of machines, called *progressive* or *transfer* headers, may have from two to six sets of punches and dies and work by transferring stock from die to die to form complex contours. Parts are formed by multiple blows. Extruding, piercing, and trimming operations may also be combined with the cold-heading operations. Such machines are ideal for producing parts up to 1 in. (25.4 mm) in diameter in lengths up to 9 in. (228.6 mm). Figure 34-16 shows the sequence of operations in producing a typical part on a multiple-die-forming machine.

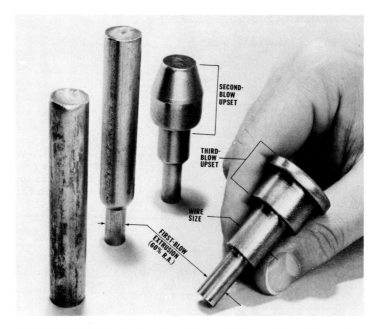

FIGURE 34-16 A typical part produced by combining extruding and upsetting on a multiple die forming machine (Courtesy, National Machinery Co.)

Another important category of cold-heading machines includes the "boltmakers," which combine heading, trimming, pointing, and threading operations in one machine. Completely finished hexagonal head and socket-head cap screws are made on boltmakers. Machines can accept material up to 1¼ in. (31.75 mm) in diameter. A production rate as high as 300 parts per minute may be obtained.

Cold heading is a chipless process that generates only from 1 to 3% scrap. Traditional forging and ultimate machining operations may result in as much as 70 to 75% material waste.

The process produces cold-worked components with improved grain structure. Metal grain flow lines in the original wire are upset to follow the configuration of the part being manufactured. Strength in critical load areas, such as head contours, collars, shoulders, and body diameters, is improved. Parts should be made with generous fillets on inside corners. Sharp corners, unless produced by a secondary machining operation, should be avoided. A good rule to follow is to use a fillet radius equivalent to 10 to 13% of the smallest adjacent body diameter. Smooth, rounded external corners are also recommended on cold-headed parts. Tapers should be avoided, but when necessary, the small diameter must be located at the bottom or inner end of the die cavity.

While cold heading is most commonly used to manufacture symmetrical parts, many different shapes can be produced by this process. By changing the original cross section from round to square, hexagonal, or some other shape, a large number of commercial applications have been made.

Burr-free parts as small as 0.020 in. (0.51 mm) in diameter by ¼ in. (6.35 mm) in length are formed by cold heading. As a general rule, conventional cold-forming operations are restricted to materials not more than 1 in. (25.4 mm) in diameter to a maximum length of about 9 in. (228.6 mm). A special-purpose machine called a *rod header* can accept parts of almost unlimited lengths. Materials 1 to 2 in. (25.4 to 50.8 mm) in diameter are generally formed at elevated temperatures.

Cold-headed parts may be produced with surfaces suitably finished so that the part may be used in service "as is." It is not unusual for the surface roughness to vary from part to part or, in fact, on different surfaces of the *same* workpiece. A high-quality surface finish is dependent upon the use of high-quality wire, the use of a suitable lubricant, and highly finished dies.

Production rates vary with part size and complexity of the machines. Small rivets, for example, can be produced at 36,000 per hour. Bolts ⅜ in. (9.53 mm) in diameter can be headed, pointed, and threaded at the rate of 15,000 per hour. The production rate may drop to only 2000 parts per hour for certain relatively large and complex parts.

Most metals, ferrous and nonferrous, can be cold-headed. The low- and medium-carbon steels are perhaps the best and most commonly used ferrous materials for this process, although certain grades of alloy steels and stainless steel are used for some products. Principal nonferrous materials for cold-heading applications include brass, copper, and nickel alloys, bronze, and aluminum alloys.

Many factors can affect part tolerances. Variables such as part geometry and size, the material, die-making practices, and the quality of the machines each determine the precision to which a given part may be produced. Parts such as nails, or other cold-headed products of this kind, do not require high precision. The main requirement is that they must be economically produced, and cold-heading fulfills this requirement.

As a general rule, lengths on upset portions, which are formed completely within a die, can be held within ±0.005 in. (0.13 mm), and a tolerance of ±0.010 in. (0.25 mm) on upset diameters is practical. The recommended tolerance on the overall length of the parts should not be less than ± 1/32 in. (0.794 mm). Closer tolerances are often possible with careful control of the variables, but at added cost. Solid dies yield parts with improved tolerances over those produced in open dies, particularly for parts in smaller sizes.

Cold-impact Extrusion

Cold-impact extrusion is a high-volume production process in which a metal slug is shaped in a die by the blow of a punch. Most extruded parts are characterized by fairly long, tubular or constant-section regions. Cold extrusion offers many advantages in the manufacture of parts, among them a savings of metal, an improvement in the physical properties of the metal, and accurate parts with a good surface finish.

The basic principle of cold-impact extruding is that the shape of the metal slug is rearranged in the manner of a viscous fluid when hit by a punch. The shape of the final part results from plastically deforming the metal slug under tremendous compressive forces, which may be as high as 300,000 psi (approximately 21,000 kg/cm²).

There are three basic methods by which parts are produced: *backward* extrusion, *forward* extrusion, and a *combination* of backward and forward. The precise method of extruding is generally selected on the basis of the shape of the required part, although other factors, such as the material, accuracy requirements, and cost and quantity, must also be considered.

The principle of *backward extrusion* is illustrated in Figure 34-17. Parts are extruded from a solid slug in a closed bottom die so that a portion of the slug flows backward over the descending impacting punch in a direction opposite to the punch travel. As the punch withdraws, the completed extrusion is removed by a device on the press called a "stripper." Backward

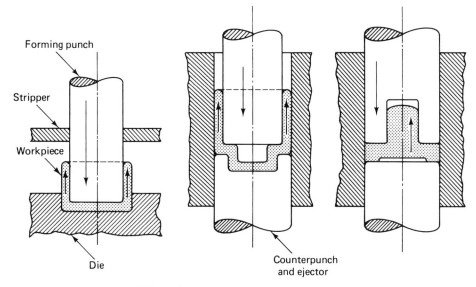

FIGURE 34-17 Backward extrusion

extrusion is frequently employed to make cup-shape parts, such as cans and shells. Bosses, projections, or recesses can be readily formed inside on the base of the part. The bore size of the die determines the outside diameter of the finished part; clearance between the punch and die determines the wall thickness. Collapsible tubes used for shaving cream, lotions, toothpaste, and paint pigments are made by this process, with an average production rate of approximately 60 tubes per minute.

In some applications, the sequence of operations may be extended by including other operations, such as piercing, coining, and coldheading. For some parts, the only secondary operation required after extruding the desired shape is trimming the excess stock.

Forward extrusion, also known as the *Hooker process,* is illustrated in Figure 34-18. The slugs used in this method may consist of short cylinders, small disks, thick washers, short lengths of tubing, or small cups. As the punch moves into the die, the impact forces the slug out through an opening in the end of the die. Unlike backward extrusion, the clearance between the punch and die walls is too small to permit an upward surge of metal. The flow of metal follows the path of least resistance, which is in the same direction as the motion of the punch. Forward extrusion may be used to form hollow or solid round and nonround parts. Straight, twisted, and ribbed rods may also be formed. Thin-wall tubing with one or both ends open may be extruded. Copper radiator tubing is an example of a product made by this process. Shafts may be produced with multiple stepped diameters. Hollow, tapered cylinders may be extruded by a single stroke of the punch, but the inside base configuration must remain plain because the end of the punch does not exert any forging action on the slug.

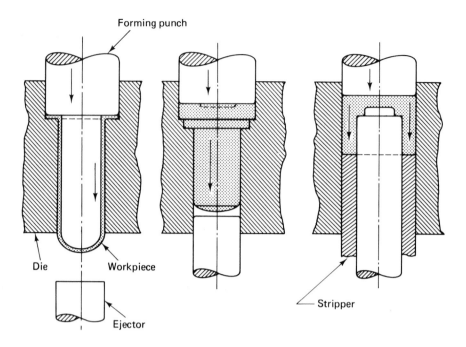

FIGURE 34-18 Forward extrusion

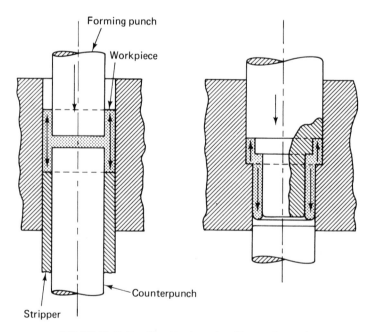

FIGURE 34-19 Combined backward and forward extrusion

The principle of *combination extrusion* is illustrated in Figure 34-19. Parts are produced by the simultaneous action of both forward and backward extrusion during a single impacting press stroke. Instead of flowing in only one direction through a single opening, the confined metal flows in two directions simultaneously. Solid or tubular shapes, or a combination of the two, can be produced in this way. After extruding, the punch is withdrawn and the completed part is ejected from the die.

Slugs are produced by various techniques, including sawing from bar stock, blanking from plate, shearing in a press, or, for some applications, a cast preform may be used. Close control of the initial slug size is necessary to obtain the proper volume to fit within the die cavity. The shape of the slug used in any one of the three extrusion methods depends largely upon the required shape of the finished part. For complex shapes, it is beneficial to preform the slug prior to extruding. Steel slugs, in some cases, are indented at the top and bottom for improved nesting in the die and positioning under the punch. No draft is required on the slugs.

Success in the cold extrusion of most metals is enhanced by careful attention to surface preparation of the slugs. Steel slugs, prior to extruding, are usually prepared by annealing followed by a series of cleaning and rinsing steps and an application of a zinc phosphate coating. While some aluminum alloys need no surface preparation, the surface on other, less extrudable types of aluminum alloys must first be prepared by etching followed by rinsing and coating. When necessary, the surfaces on copper and brass slugs are prepared by a sequence of cleaning, rinsing, pickling, and coating operations. Lubricants that react with one coating and adhere to the metal surface during the extrusion process help to reduce the effects of tool and workpiece friction.

Impact extrusion is often performed on the same machines that are used for sheet metal forming. The axis of such machines may be vertical or horizontal. Depending upon the workpiece requirements, both hydraulic and mechanical presses may be used. The slow squeezing action of a hydraulic press is especially well suited for loads over 2000 tons (1800 metric tons) and is particularly adaptable for cold extruding parts whose size requires a press with a long working stroke. Impact extrusion, however, requires faster types of machines, such as high-speed crank, eccentric, or toggle mechanical presses. Special Pitman slide presses are also used for impact extruding. Depending upon the ductility of the metal, extrusion pressures may be as high as 50 to 100 tons/in.^{2} (approximately 7000 to 14,000 kg/cm^2) of area.

Cold-impact-extruded shapes with straight and parallel walls may be conveniently produced and tapered walls are possible if tolerances are not excessively close. Parts can be extruded with multiple diameters having variations in wall thickness without causing structural defects. Hollow shapes and solids can have almost any cross section. Flanged, open-end, or closed-end tubes can be produced, and parts may include symmetrical bosses and lugs. Both internal and external ribs or flutes are also possible. Sharp corners should be avoided, particularly on internal shapes.

Commercial products are commonly impact-extruded in the following metals: aluminum, brass, copper, lead, tin, zinc, magnesium, titanium, various low-carbon and low-alloy compositions of steel, and stainless steel. All materials specified for impact-extruding should have the lowest yield strength compatible with the requirements for the finished part.

The allowable extrusion ratio, expressed as length-to-diameter, is limited by the type of material, the size of available extrusion equipment, and the extent of the cold work required. A practical limit for the backward extrusion of steel is 3:1. It is possible to produce aluminum alloy parts up to 16 in. (406.4 mm) in diameter using either the backward or the forward impact-extruding methods. Parts up to 5 ft (1.52 m) in length are commonly produced by backward extrusion. Depending upon the part configuration, it is sometimes also possible to produce parts to this same length by forward-extrusion methods. A ratio of 10:1 is common for extruding aluminum, and, under closely controlled conditions, a ratio as high as 17:1 has been obtained.

In backward extruding, the maximum ratio for copper is 5:1, while a typical extrusion ratio for magnesium is about 8:1. An increase in the length-to-diameter ratio is possible when impacting magnesium by the forward-extrusion method. Regardless of the method employed to impact-extrude magnesium, it cannot be considered a "cold" extrusion process because it is necessary to preheat both the slug and the tooling. Tubular shapes produced by the forward-extrusion method are limited to a maximum diameter of ⅜ in. (9.53 mm), with a maximum length of 10 in. (254 mm). Tubes are readily extruded with a wall thickness of 0.005 in. (0.13 mm).

It is impractical to state a definite tolerance for cold-impact extruding because the accuracy that can be obtained depends largely upon the size and shape of the section to be produced. In most cases, however, a tolerance of ±0.001 to 0.002 in. (0.03 to 0.05 mm) can be held on parts with diameters of less than 1 in. (25.4 mm). Tolerances on parts with diameters from 4 in. (101.6 mm) can usually be held to within ±0.005 in. (0.13 mm). The recommended minimum tolerance, for length dimensions, on long, slender parts up to 20 in. (508 mm), and longer, is about ±0.015 in. (0.38 mm).

In addition to the products noted previously, impact-extrusion methods are also used to produce an impressive variety of commercial products, such as lamp bases, lipstick cases, food and beverage cans, aerosol cans, Thermos cases, cartridge cases, flashlight housings, battery cases, seamless tubing, caulking or grease gun bodies, door-check cylinders, electric motor housings, fire extinguishers, and shielding cans used in radios, television sets, and for various electronic receivers. Impact extrusion is finding a growing market in the production of components for home appliances, such as washing machines, refrigerators, and air conditioners, for automotive and aircraft applications, construction and farm equipment, power lawn mowers, outboard motors, business machines, and in countless other fields.

Thread Rolling

Thread rolling is an economical, chipless, cold-forming process principally used to make external forms on cylindrical workpieces. The process has long been used by the fastener industry to produce smooth, precision threads at high production rates. More recently, a wide variety of other applications have been found for this process, particularly as a profitable method of forming other external cylindrical shapes, such as gear teeth, splines, and various forms of knurl patterns and serrations.

The process of thread rolling, shown schematically in Figure 34-20, begins with a plain cylindrical blank with a diameter that is sized partway between the major and minor diameters of the finished thread. The blank material must be sufficiently plastic to flow and to withstand the stress of cold working without disintegrating. The thread is formed by rotating and squeezing the blank between hardened steel dies whose working surfaces are the reverse of the thread form to be produced. The threads of the die penetrate the surface of the blank as it rolls between them. The displaced material forms the roots of the thread and, at the same time, is forced radially outward to form the crests. Thread-rolling dies may be either flat or cylindrical.

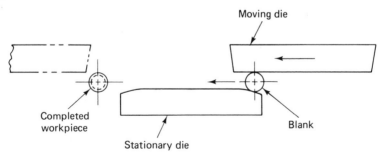

FIGURE 34-20 The principle of thread rolling-shown schematically

The equipment and tools used in producing rolled threads are usually dependent upon the nature of the part, the type and specifications of the thread, and the quantities required. Small as well as large quantities of stock may be rolled economically on hand-feed machines, since the setup is normally simple and quick. Very little time is consumed in rolling a thread.

FIGURE 34-21 This vertical model of a thread rolling machine utilizes a 3-die principle of self-centering action on the work. The machine set-up is fast and can be operated manually for job shop operations (Courtesy, Reed Rolled Thread Die Co.)

Most of the threads produced today are rolled on special thread-rolling machines or by using special attachments on automatic screw machines and automatic lathes. A typical thread rolling machine is illustrated in Figure 34-21. Machines of this kind use flat and cylindrical dies, while automatic screw machines use only cylindrical dies. In most instances, the entire length of thread is formed by an in-feed method without endwise feeding of the blanks or dies. Through feeding of the blanks is used on cylindrical die machines for continuous threading of long bars or for threading short, headless parts.

Flat dies, shown in Figure 34-22, are used in reciprocating-type thread rollers, including boltmaking machines. These machines are available in a number of sizes, each for a limited diameter range and with a specified length of die. Two dies are used, one stationary and one moving. The rolling faces of the dies are located opposite each other. A thread is rolled on the blank at a time during the forward stroke of the machine. There is no appreciable axial movement of the blank during rolling. The diameter of the finished thread is controlled by the diameter of the blank and the distance between the faces of the dies at the finish end of the stroke.

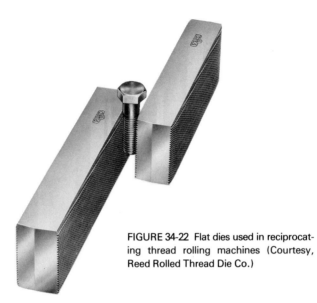

FIGURE 34-22 Flat dies used in reciprocating thread rolling machines (Courtesy, Reed Rolled Thread Die Co.)

Rotary planetary machines use dies as illustrated in Figure 34-23. The starting end of the segmented die is set so that the segment and rotary die will just contact the blank. The finishing end of the segment die is set closer to the axis of the rotary die so that the thread is fully formed when the blank rolls past the finish end of the segment die. One or several blanks may be rolled in this manner at one time, depending upon the setup. There is no appreciable axial movement of the blank during rolling.

FIGURE 34-23 Rotary Planetary Dies (Courtesy, Reed Rolled Thread Die Co.)

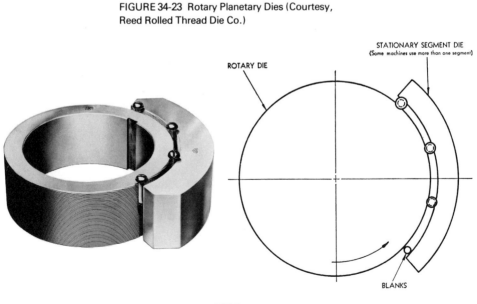

Cylindrical die machines have a wider range of speeds and feeds, and are widely used for in-feed rolling. Cylindrical die machines are available with two or three dies, as shown in Figure 34-24. As for flat and segmented dies, there is no appreciable axial movement of the blank during rolling. Since the dies are circular in shape, there are no limitations on the number of work revolutions provided for the rolling of a thread or the rate at which the dies feed into the work.

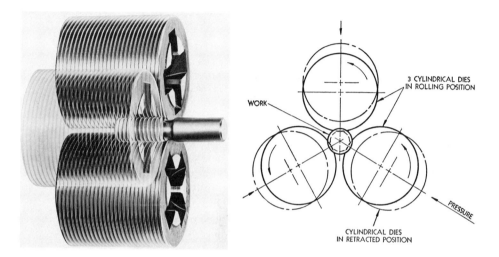

FIGURE 34-24 Three die types used in cylindrical die thread rolling machines (Courtesy, Reed Rolled Thread Die Co.)

Large forms, worm threads, gears of spur and helical design, straight and helical splines, and various knurl patterns and serrations can each be produced by roll forming. In most cases, thread rolling machines, presently in the field, can be used without modifications. A gear, worm, or spline, to be considered "rollable," must conform to the physical limits of the process with respect to the diametral pitch, tooth profile, and material adaptability. Large roll-formed shapes can be produced at extremely high rates and to precise limits of accuracy.

As a general rule, plain carbon steels, structural alloy steels, high-speed steels, nonleaded brasses, and copper constitute the majority of materials used in rolling threads and other external forms on parts. Selection of materials often depends upon an optimum combination of rollability and machinability of the material.

Rolling has long been conceded to be the fastest method of producing screw threads. Threads are completely formed in one pass of the dies.

The process is not economical for limited production runs, except for special applications on hand-feed machines. A practical workpiece hardness limit is Rockwell C40.

Figure 34-25 illustrates typical sample parts, which bring out the versatility of parts adaptable to this process.

FIGURE 34-25 (a) miscellaneous parts with straight gimlet point and special threads, knurling and other rolled forms (Courtesy, Reed Rolled Thread Die Co.); (b) various precision heat treated parts (Courtesy, Reed Rolled Thread Die Co.)

(a)

(b)

It is possible to roll-form internal threads without producing chips by using the special taps shown in Figure 34-26. Roll thread tapping is similar to that of conventional tapping operations except that a slight change in the size of the corresponding tap drill must be made. This innovative process has proved to be an effective method for producing deep blind holes, particularly in small thread sizes, where excessive tap breakage with conventional taps has been experienced.

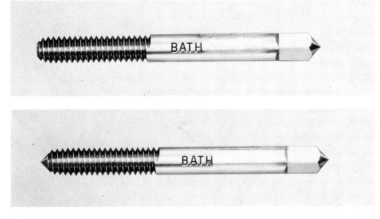

FIGURE 34-26 These taps cold-form or swage threads without cutting. They are suited for threading ductile materials such as aluminum, brass, copper, zinc, and, under proper conditions, steel (Courtesy, John Bath & Company, Inc.)

Chipless tapping eliminates the sometimes costly operation of chip removal from the threaded work. The fluteless design of these taps gives them added strength, which is especially important in the smaller sizes.

Roll thead taps are suitable for threading ductile materials such as aluminum, brass, copper, zinc, and, under proper conditions, leaded types of steel and stainless steel. Taps are available in sizes from No. 0 (0.0519 in. or 1.32 mm) to ¾ in. (19.05 mm) in diameter.

For most materials, spindle speeds may be at least doubled over those recommended for conventional cutting-type taps. Because of a tendency of the roll thread tap to displace a small amount of metal above the mouth of the hole, it is recommended that the hole be countersunk or chamfered prior to tapping.

Thread-rolling machines are usually equipped with automatic features for high-speed loading and unloading of the work blanks. Machines can be operated in batteries with several machines to each operator. Production rates vary with the nature of the work, hardness, kind of material, and type of equipment used. An approximate production rate for infeed rolling of ½ in. (12.7-mm) diameter stock, for example, can be as high as 400 in. (approximately 10 m) per minute. Diameters from 0.060 in. (1.53 mm) to 5 in. (127 mm) with 2 to 80 threads per inch are readily rolled.

In cold-forming operations, the surface finish left on the work is a close approximation of the surface finish of the dies. Because of the slight slipping and burnishing effect of the blank as it rolls against the ground and polished dies, the surface finish of thread-rolled parts is smoother than the dies. Rolling produces much the same finish on the work regardless of the properties of the material being rolled. A surface roughness of 4 to 32 μin. is obtainable.

The process is economical of materials and, because of the benefits of cold working, it results in parts with increased strength. When a thread is rolled, the fibers of the material are not severed as they are in the cut thread shown in Figure 34-27. Rolled threads are re-formed or forged in continuous unbroken lines that follow the contours of the threads, as also shown in Figure 34-27. Rolled threads resist stripping because shear failures must take place across rather than with the grain.

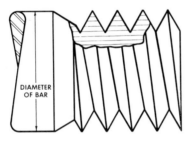 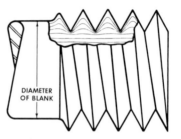

FIGURE 34-27 A comparison of a cut and a rolled thread: (a) cut threads; (b) rolled threads (Courtesy, Reed Rolled Thread Die Co.)

REVIEW QUESTIONS

34.1. What are the chief advantages of flexible die forming?

34.2. Explain why flexible die forming is regarding as a versatile process.

34.3. What are the chief disadvantages of flexible die forming?

34.4. What types of parts are particularly applicable to formation by the Guerin process?

34.5. What is the principal use of peen forming?

34.6. In what important way does peen forming differ from conventional press forming?

34.7. List some techniques metal fabricators employ to ensure successful results in peen forming.

34.8. Briefly explain how a desired workpiece shape is produced by the process of magnetic pulse forming.

34.9. What is the function of the mandrel or die used in magnetic pulse forming?

34.10. What are the principal advantages and limitations of magnetic pulse forming?

34.11. Under what conditions would one of the explosive forming processes be specified in preference to one of the conventional forming processes?

34.12. Explain the function of the mandrel in the electroforming process.

34.13. What factors determine which of the two kinds of mandrels to use for a given part?

34.14. In what major ways does the process of electroforming differ from electroplating?

34.15. What are the principal advantages of electroforming?

34.16. Briefly explain how the shape of a part is formed on a rotary swaging machine.

34.17. What are some of the principal advantages of swaging?

34.18. Explain the basic differences between a solid-die heading machine and an open-die machine.

34.19. What are some of the factors that affect the quality of the surface finish in the cold-heading process?

34.20. In what ways do the shape of parts commonly produced by cold heading differ from those produced by cold-impact extrusion?

34.21. Explain the basic difference between backward extrusion and forward extrusion.

34.22. Give some reasons why the various techniques of extrusion are widely employed for producing high-volume parts in a wide range of shapes.

34.23. What are the chief advantages of thread rolling?

34.24. Other than screw thread forms, what applications have been made for the thread-rolling process?

PROBLEMS

34.1. It is desired that the swaged steel parts described in Figure 34-P1 be limited to a maximum 40% reduction in area in one pass. Determine if this limitation is possible for each of the three parts with the sizes listed in the table.

34.2. Tube swaging without a mandrel results in an increase of wall thickness. Calculate the increase in wall thickness that will be produced on a tube with a wall thickness of 0.062 in. and an O.D. of 2 in. before swaging and an O.D. of 1.875 in. after swaging.

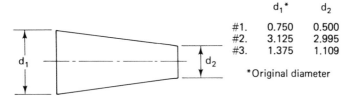

	d_1*	d_2
#1.	0.750	0.500
#2.	3.125	2.995
#3.	1.375	1.109

*Original diameter

All sizes are in inches.

FIGURE 34-P1

BIBLIOGRAPHY

Aluminum Extrusion Application Guide, EAG-7, New York: The Aluminum Association, 1976.

*BAUGHMAN, D. L., "Peen Forming," *Uncommon Metalworking Methods,* Cleveland, Ohio: Penton Publishing Co., 1971, p. 10.

*CASTELLANO, E. N., and H. WOELLMER, "Electroforming," *Uncommon Metalworking Methods,* Cleveland, Ohio: Penton Publishing Co., 1971, p. 29.

*"Cold Extruding," *Machine Design,* February 14, 1974.

"Cold Heading," *Metals Handbook,* Vol. 4, *Forming,* 8th edition, Metals Park, Ohio: American Society for Metals, pp. 465–475.

"Explosive Forming," *Metals Handbook,* Vol. 4, *Forming,* 8th edition, Metals Park, Ohio: American Society for Metals, pp. 250–254.

*PIERCE, R. D., "Magnetic Pulse Forming—A Unique Assembly Process," *Assembly Engineering,* Jan. 1968.

"Rotary Swaging of Bars and Tubes," *Metals Handbook,* Vol. 4, *Forming,* 8th edition, Metals Park, Ohio: American Society for Metals, pp. 333–346.

"Rubber Pad Forming," *Metals Handbook,* Vol. 4, *Forming,* 8th edition, Metals Park, Ohio: American Society for Metals, pp. 209–216.

SMITH, K. F., "Metal Forming by Electrical Energy Release—Trends and Applications," *Transactions of the Society of Automotive Engineers,* Paper 650190, 1966, 878 pp.

Source Book on Cold Forming, ASM Engineering Bookshelf Series, Metals Park, Ohio: American Society for Metals, 1975.

*STROHECKER, D. E., "Swaging Today," *The Tool and Manufacturing Engineer,* May 1968.

Thread Rolling with Flat Dies, Flat Die Catalog 1-110-A, Holden, Mass.: Reed Rolled Thread Die Co.

Upsetting, Catalog 215-B, Tiffin, Ohio: National Machinery Co.

*ZERNOW, L., "Explosive Forming—When and How It Pays Off," *Machine Design, Aug. 23, 1973.*

*Abstracted with permission.

35

CONVENTIONAL DRAWING AND RELATED PROCESSES

Drawing is a process for forming thin metal products such as cups, cones, boxes, tubular shapes, and shell-like parts. Most drawing operations start with a flat blank or a sheet of metal. There are a number of important drawing-related operations which will be explained. These are redrawing, sizing, ironing, striking, upsetting, bulging, expanding, necking, embossing, and coining.

DESCRIPTION OF THE PROCESSES

Drawing, as illustrated in Figure 35-1, consists of pressing or forcing the flat metal blank into a female die while stretching it to conform to a shape over a male die or punch. High compressive stresses act upon the metal, which, without the offsetting effect of a blank holder or a pressure plate, would result in a severely wrinkled workpiece.

Shallow drawing is generally understood to be a process of forming a cup no deeper than half its diameter, with little thinning of the metal. *Deep drawing* consists of producing a cup whose depth may exceed its diameter.

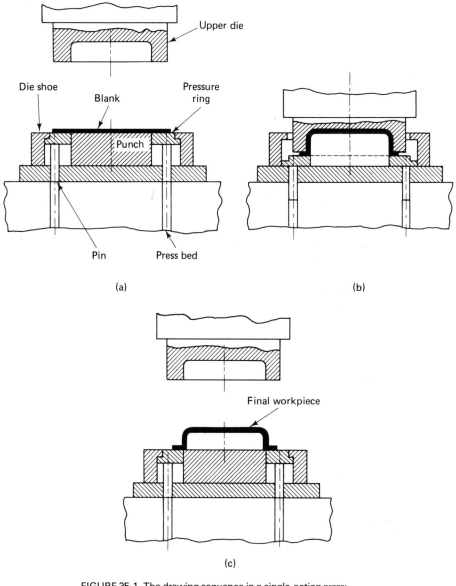

FIGURE 35-1 The drawing sequence in a single-action press:
(a) before drawing; (b) during drawing; (c) after drawing

Wall thinning, in this case, is usually more pronounced than for shallow-drawn cups. Parts made by deep drawing usually require several successive draws with progressive dies. One or more annealing operations may be required to reduce the work hardening by restoring the ductile grain structure.

RELATED OPERATIONS

Redrawing

Redrawing is a term that describes an additional drawing operation. Redrawing is required for shells that are too deep to be drawn in a single operation. In this case, a punch is used which pushes the cup into the die, causing a reduction in the diameter and an increase in the length of the cup.

Embossing

Embossing is generally understood to be a very shallow drawing operation. Theoretically, there is no change in metal thickness. Only moderate pressures are required. The depth of the draw is generally not more than one to three times the thickness of the metal. Embossing is commonly used to produce projections on workpieces, such as dowel buttons or bosses, which may later be used for welding or assembly purposes. Embossing is also a method used to make indented or raised identification numbers or letters, scales, or special surface designs on workpieces. Embossing can produce small projecting ribs, webs, or beads, which improve the stiffness of thin-gage metal parts. Embossing may consist of a distinctly separate operation, or it may be combined with blanking or piercing operations.

Sizing

Sizing is a term used to describe the further flattening or improvement of selected surfaces on previously drawn parts to closer limits of accuracy than are possible by conventional drawing methods. A sharp radius on a workpiece may be formed by sizing, for example. Sizing consists of squeezing the metal in a desired direction. There is little, if any, restriction to metal flow and the volume of metal forces the workpiece walls to bulge outward to conform to the contours of the die cavity. Figure 35-2 illustrates a method for bulging using rubber. Segmented internal die posts may also be used, in which various segments are held together by springs. The segments are pushed apart by the punch, which causes them to radially expand against the inner walls of the shell.

By varying the bulging technique, it is also possible to form a flange along the side wall or at the bottom of a flanged or open-end cup, as shown in Figure 35-3. The flange thickness is slightly less than double the wall thickness of the cup. The process is not expensive and often results in a savings in the machining cost normally required for matching fits. The O.D. tolerance of a 3 in. (76.2 mm) flange, for example, can be held to ± 0.008 in. (0.20 mm), and the contour may extend as much as 30% beyond the cup diameter.

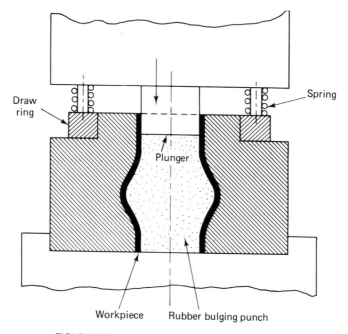

FIGURE 35-2 A method of forming a part using rubber

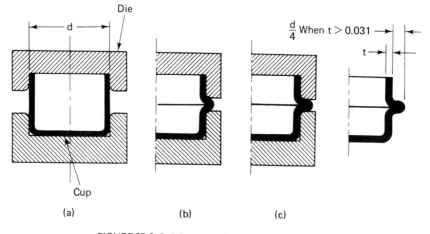

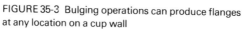

FIGURE 35-3 Bulging operations can produce flanges
at any location on a cup wall

Bulging

Bulging is often the preferred method of forming contours on teapots, water pitchers, kettles, watch cases, doorknobs, artificial limbs, parts of musical instruments, and in forming lamp bases and special lighting fixtures.

Expanding

Expanding, in a manner similar to bulging, is a method of enlarging portions of drawn workpieces in a press. Figure 35-4 illustrates a method of expanding the open end of a drawn shell by using a punch. In most cases, the workpiece is first annealed. After the cup has been positioned in the die, the punch moves downward and expands the top of the cup. The diameter of ductile metal shells of copper or low-carbon steel can be expanded as much as 30% in one press operation. Two or more operations, with intermediate annealing steps, are necessary when a diameter increase of more than 30% is required.

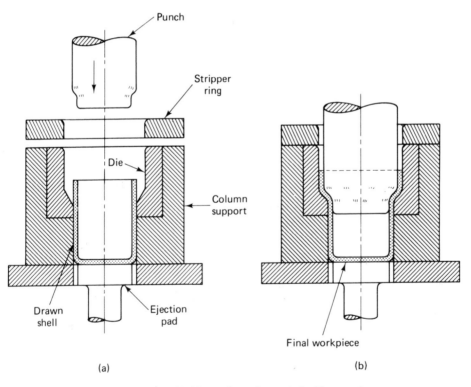

(a)   (b)

FIGURE 35-4 Expanding a drawn shell with a punch:
(a) before expanding; (b) after expanding

Necking

Necking is a die-reduction method by which the top of a cup may be made smaller than its body. Examples of such shapes are found on some electric coffee makers or on the carbon dioxide gas bottle shown in Figure 35-5. It is possible to reduce the top of the cup, if the material is not too thin, to about 20% of its diameter in one operation. Necking compresses the work metal, resulting in an increase in length and wall thickness.

FIGURE 35-5 Examples of necking on a carbon dioxide bottle

Nosing

Nosing, shown in Figure 35-6, reduces the open end of a drawn shell or tubing by tapering or rounding the end. Although it may be used for other applications, tubing is used chiefly in making ammunition.

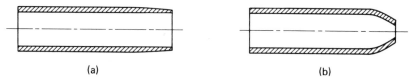

(a)

(b)

FIGURE 35-6 Nosing: (a) tubing tapered before nosing; (b) tubing after nosing

Coining

Coining, unlike embossing, requires high forming pressures, resulting in metal flow from one area of decreased thickness to another. Drawn parts, like some forgings (see Chapter 28), are often further processed by coining. Coining is generally the method selected to build up or to fill in corners on drawn parts, to obtain better tolerance control or to produce sharper definition of raised or indented surface detail than is possible to obtain by embossing. Top edges of cups are often rounded by coining. Other examples are in forming oil or gasket grooves, or for flattening.

TRIMMING DRAWN PARTS

It is necessary for metal fabricators to decide whether the shape of the drawn edge on a part to be produced will be acceptable. In most cases a somewhat irregular and oversize edge is formed because the flow of metal is rarely uniform throughout a drawn part. The necessity for trimming parts may often be omitted entirely by using a developed blank. The optimum shape of the blank is usually determined by making a series of tests using various shapes of developed blanks. The use of developed blanks often proves more economical than final trimming dies for the reason that blanking dies are generally less expensive to produce than trimming dies. In some cases, however, owing to the conditions imposed by an unusual part shape or tolerance requirement, or because of variations in work metal thickness, even parts drawn from a developed blank might require final trimming.

Trimming operations may be performed in various ways. The choice of a specific trimming method is influenced by the shape of the drawn part, the accuracy required, and the production quantity. Cylindrical or conical shape parts, for example, may be trimmed with a cutting tool on a lathe or by rotary shears combined with the drawing operation, by pinch trimming in a compound draw-and-trim die, or in a special trimming machine. Various other workpiece shapes can be trimmed in a press equipped with a special-purpose die or in a trimming machine.

MATERIALS FOR DRAWING OPERATIONS

The most suitable nonferrous metals for deep drawing are aluminum, brass, magnesium, and titanium. In addition, many excellent cold- and hot-rolled drawing quality steels are available. With some minor differences, techniques for drawing ferrous metals are essentially the same as for nonferrous metals. Drawn parts, particularly those which have been deep-drawn, may be entirely free of stress or fatigue areas. In fact, overall physical characteristics of drawn parts can be equal to, or better than, the raw material from which they are made. This is particularly true of alloys having inherent work-hardening characteristics, because cold working during the draw increases tensile and yield strengths. Drawn parts can be made from materials that can be heat-treated after drawing to obtain the desired strength.

Drawing is the process specified when thin metal parts must have no mechanical joints or seams, where dimensional stability is required for long runs, where uniformity and close tolerances are important, and where freedom from defects is necessary.

PRODUCT DESIGN FACTORS

Configurations: Round, oval, and square cross sections are the easiest to draw. Rectangular box shapes are generally the most difficult to draw and usually result in the most expensive tooling.

Tolerances: Tolerances on steel parts must be greater than those on aluminum parts because steel, unlike aluminum, does not readily cold-flow over the punch during the drawing process. Dimensional tolerances are normally a function of overall size and of the part configuration. Generally, a tolerance of $\pm \frac{1}{64}$ in. (0.397 mm) should be allowed for parts up to 6 in. (152.4 mm) and $\pm \frac{1}{32}$ in. (0.794 mm) above 6 in.

An allowance of 15 to 20% of nominal metal thickness should be provided for reduction in thickness of vertical walls. The depth to which a part may be drawn is limited by the inherent drawing characteristics and the thickness of the material, by the ratio of depth to width, and by the corner radii required. In general, the depth to which a part can be deep-drawn in a single operation depends upon the size of the smallest cross-section dimension. The "rule of thumb" for aluminum and brass parts is to allow a maximum depth-to-width ratio of 1:1 for a single draw.

Diameter and Depth of Drawn Shells: Cylindrical shells are most easily and economically produced when the diameter is equal to or greater than the depth, as shown in Figure 35-7. As the shell depth increases in proportion to the diameter, the number of operations and the related cost increase also. While the necessity for deep-drawn shells cannot always be avoided, a shallower shell can often be substituted, resulting in substantial savings.

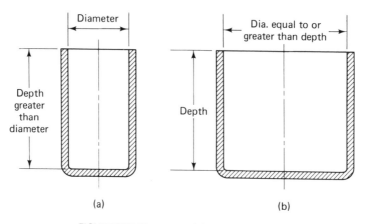

FIGURE 35-7 Diameter and depth of drawn shells:
(a) avoid if possible; (b) preferred

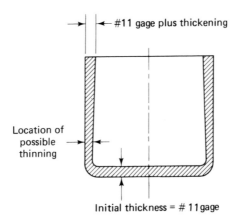

FIGURE 35-8 An example of a typical variation in wall thickness (sidewall thickness is exaggerated for emphasis)

Wall Thickness: The walls of drawn cylindrical shells normally thicken at the open end as the depth of the shell increases, as shown (exaggerated) in Figure 35-8. This fact should be kept in mind in order to avoid costly, restrictive gage tolerances. A recommended practice is to indicate the desired wall thickness as a nominal dimension in the material specifications.

Average Diameter: All drawn shells, because of the grain direction of the sheet metal stock, will tend to become "egg-shaped" at the open end. When dimensioning diameters of cylindrical shells, two basic characteristics must be considered:

1. The maximum out-of-roundness is measured by the T.I.R. (total indicator reading). The maximum out-of-roundness is equal to one-half of the T.I.R. and is a function of the variation of the radius. Total indicator reading is the total diametral variation normally measured while rotating the shell while using a dial indicator.

2. The average diameter is measured by at least three or more random diameters. The average diameter is found by averaging diameters A, B, and C, as shown in Figure 35-9. They should be selected consistent with the normal eccentricity of the part.

Draw Radii: When designing drawn shells, it is advisable to keep the draw radii equal to at least four times the stock thickness, as shown in Figure 35-10. If the radii need to be sharper, because of desired shell capacity or other design requirements, this can be readily accomplished but usually requires additional press operations.

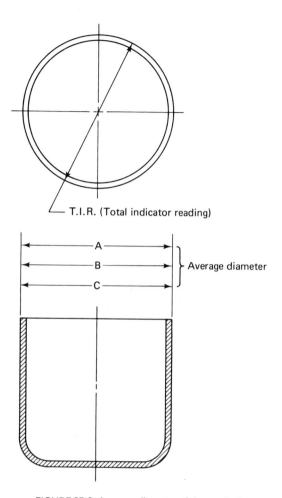

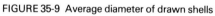

FIGURE 35-9 Average diameter of drawn shells

FIGURE 35-10 Draw radius: (a) avoid; (b) preferred

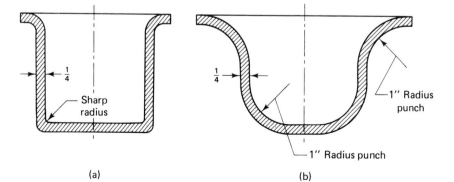

(a)   (b)

REVIEW QUESTIONS

35.1. Explain the basic differences between drawing and the process of backward extrusion discussed in Chapter 34.

35.2. Contrast the processes of shallow and deep drawing.

35.3. In what ways do drawing and redrawing differ?

35.4. What are the chief reasons why a product designer might specify one or more embossing operations on a given component?

35.5. Explain why a sizing operation is often necessary when producing some parts.

35.6. What is the purpose of bulging and expanding on some parts?

35.7. How do coining and sizing differ?

35.8. Explain how the necessity for a trimming operation may sometimes be eliminated by forming the desired parts from a "developed blank."

35.9. What are the major advantages of drawing?

PROBLEMS

35.1. Calculate the diameter of the blank for an aluminum hemispherical cup 0.012 in. thick by 3 in. diameter.

35.2. Calculate the diameter of the blank for the cup described in Problem 35-1 with the addition of a 6½ in. diameter flat lip.

35.3. Calculate the diameter of the blank for the part shown in Figure 35-P3.

FIGURE 35-P3

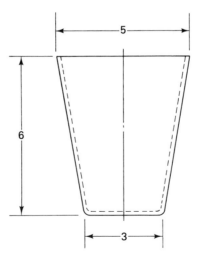

All sizes are in inches.

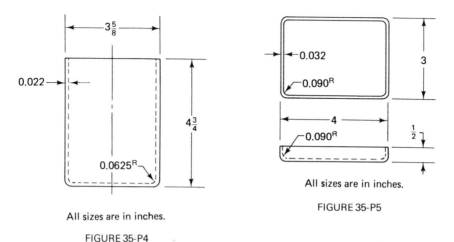

All sizes are in inches.

FIGURE 35-P4

FIGURE 35-P5

35.4. Calculate the drawing pressure required to form a circular cup shown in Figure 35-P4. The tensile strength is 60,000 psi and the material is deep-drawing steel stock.

35.5. Calculate the drawing pressure required to form a rectangular cover shown in Figure 35-P5. The tensile strength is 82,500 and the stock is B1112 steel.

35.6. The male member of the stamping tool is called the_____. The female member is called the_____.

35.7. Identify the three types of pressworking operations illustrated in Figure 35-P7.

35.8. In the drawing operation illustrated in Figure 35-P8 what is the percentage of reduction?

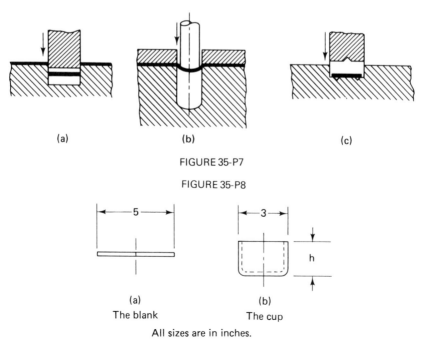

(a)   (b)   (c)

FIGURE 35-P7

FIGURE 35-P8

(a)   (b)
The blank   The cup

All sizes are in inches.

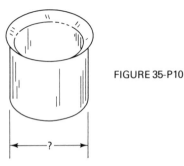

FIGURE 35-P10

35.9. With ideal conditions, the maximum reduction of mild steel from blank to cup is about 50%. If the part requires deeper drawing than this, the initial cupping must be followed by one or more redraw operations. Calculate the percentage of reduction for a drawn cup 6¾ in. diameter which is redrawn to 5 in. diameter.

35.10. If a blank diameter of 4 in. was used to produce the drawn cup shown in Figure 35-P10 to a reduction limit of 35%, what would be the diameter of the drawn cup?

35.11 Assuming the same materials, stock thickness and other conditions, which of the parts shown in Figure 35-P11 would you judge to be the most difficult to draw? Why?

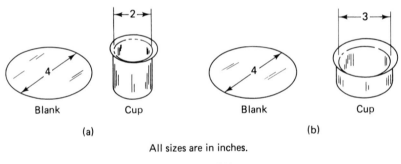

All sizes are in inches.

FIGURE 35-P11

35.12. Give three factors which limit the amount (percentage) of reduction in any one drawing?

BIBLIOGRAPHY

"Coining," *Metals Handbook,* Vol. 4, *Forming,* 8th edition, Metals Park, Ohio: American Society for Metals, pp. 78–88.

Die Design Handbook, American Society of Tool and Manufacturing Engineers, 2nd edition, New York: McGraw-Hill Book Company, 1965.

SACHS, G., *Principles and Methods of Sheet Metal Fabricating,* 2nd edition, New York: Van Nostrand Reinhold Company, 1966.

36

SPECIAL DRAWING
PROCESSES

The drawing processes in Chapter 35 applied to operations performed on *conventional* pressworking equipment. There are other drawing processes, such as stretch draw forming, radial draw forming, Marforming, hydroforming, and spinning, which are performed on *special* equipment. All drawing operations are alike in that they involve a plastic flow or adjustment of metal as the inner and outer shape of the workpiece is formed by the respective configurations of the male and female dies.

DESCRIPTION OF THE PROCESSES

Stretch Draw Forming

Stretch draw forming is the process of producing contoured parts by stretching metal sheets over a shaped form block. Stretch forming strains the metal beyond the elastic limit, to give the workpiece a permanent set. Two methods are used in stretch draw forming: the form-block method and the mating-die method. In the *form-block method,* shown in Figure 36-1, each end of the workpiece blank is securely held in tension by an adjustable gripper which is moved to stretch the blank over a form block. The desired shape of the workpiece is formed by the stretching action of the

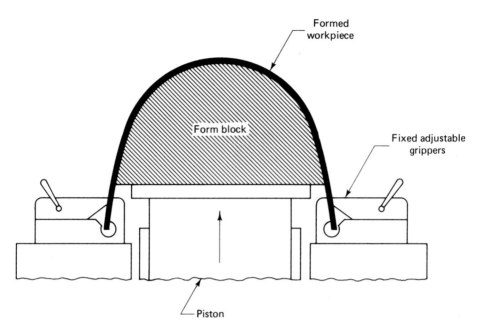

FIGURE 36-1 Stretch draw forming with a form block

form block as it moves hydraulically against the blank. The *mating-die method* is illustrated in Figure 36-2. The workpiece is held in tension by grippers which, as they move, preform the workpiece a predetermined amount (to approximately 2% elongation) over the lower die. The upper die then descends onto the blank, thus forming the workpiece shape by pressing the metal into both dies.

The stretch draw forming process is often the method used to produce parts for aircraft and aerospace applications, as well as for a wide variety of contoured panels for truck trailer and bus bodies and for the automotive industry.

FIGURE 36-2 Stretch draw forming with mating dies: (a) work blank is held in tension; (b) upper die moves down to form workpiece

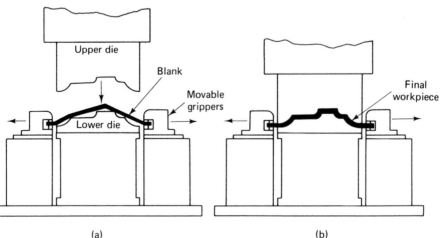

(a)   (b)

Radial Draw Forming

Radial draw forming is a method of contour-forming various shapes of rolled or extruded sections. As in stretch draw forming the metal blank is first stressed by pulling it to its yield point. In this semiplastic state, the part is easily made to conform to the shaped forming dies. Figure 36-3 shows how the ends of the workpiece are gripped by the jaws on a hydraulic cylinder. The metal is held in controlled tension while it is drawn over the forming die as the turntable rotates. At the same time, a compression shoe, shaped to mate with the profile of the die (with allowance for metal thickness), is pressed toward the die to confine the metal to the die face and assist with the forming. The combined action is necessary to maintain the shape of sections (such as Z's, L's, and I's) and channels during forming. The wiper shoe of the second hydraulic cylinder normally acts at the point where the uniform section of the part is tangent to the die curvature.

Contoured tubular and hollow shapes require the use of mandrels to maintain the cross-sectional shape during forming. These mandrels are usually laminated, flexible structures made to suit individual workpiece requirements.

Parts contoured by radial draw forming include industrial conveyor track, monorail guide beams, and various architectural shapes, as well as many forms of structural members for truck, bus, and aircraft applications.

When compared to conventional press forming methods, stretch draw forming and radial draw forming generally require simpler and less expensive dies. Most stretch draw forming can be done at room temperature, permitting the use of low-cost die materials. About 70% less forming tonnage is required. This is principally due to the fact that the parts are progressively

FIGURE 36-3 Schematic diagram of basic components and motions for radial draw forming

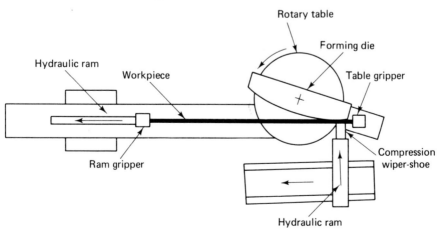

formed instead of being formed by a single blow and also because the material can be formed more easily in the plastic state. Materials such as steel, nickel, aluminum, and titanium alloys are commonly used for both high-production and short-run jobs in stretch draw forming. Savings in workpiece material costs are possible. In spite of the need for additional stock for gripping, the allowance for trimming is usually less than is required for conventional press forming methods. It is possible to obtain parts with uniformly improved tensile strength of up to 10%. There is less tendency for buckles and wrinkles to develop on workpieces. Strain marks, often a problem in conventional drawing and forming, may be prevented by using segmented or curved grippers, which equalize the amount of stretching. There is very little springback.

Stretch forming processes are not without limitations, however. There is a limitation to the shapes that can be formed. Shallow contours are best. It is practical to stretch-form only relatively symmetrical contours. An inside radius of at least 1 in. (25.4 mm) is recommended. Stretch forming elongates the material and reduces the thickness by as much as 7%.

Marform Process

The *Marform process*, illustrated in Figure 36-4, was developed by the Glenn L. Martin Company. It is a refinement of the low-cost tooling techniques used in the Guerin process (explained in Chapter 34), which permits some additional advantages. These are: parts may be drawn to greater shell depths, parts may be produced with wrinkle-free shrink flanges, and improved definition is possible on shallow shapes.

FIGURE 36-4 Tooling and set-up for the Marform process:
(a) before forming; (b) after forming

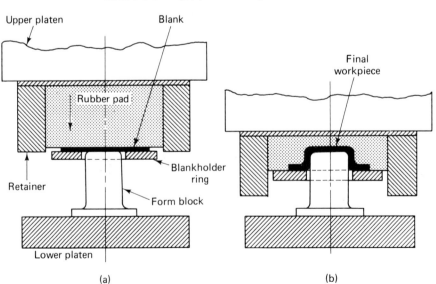

(a)   (b)

The production sequence on Marforming begins with positioning the metal work blank on the form block (or punch) and the stationary blank holder. The actual forming of the part begins as the downward-moving pad draws or envelops the work blank around the form block. After the stroke is completed, the shell is stripped from the form block by the blank holder and the cycle is repeated.

Because of a tendency for the solid rubber pads to become scored by repeated contact with the metal work blanks, fabricators often insert a loose thin sheet of rubber between the blank and the pad to prevent the sheet from being cemented to the pad. Urethane pads, when used, may consist of up to five layers or "wafers." Multiple wafers, as compared to solid pads, can be interchanged as they wear, thus extending the service life. There is also a tendency for the layers to deflect and conform more easily to the shape of the form block.

Specially designed presses, equipped with a cushion system on the bed, are currently available for Marforming. It is also practical to use standard 1000- to 5000-ton (900- to 4500-metric-ton)-capacity single-action hydraulic presses on which a Marforming unit can be installed.

Typical parts made by this process include flanged cylindrical and rectangular cups, spherical domes, shells with parallel or tapered walls, and a variety of unsymmetrical shapes.

Hydroforming

Hydroforming is a drawing process. Figure 36-5 illustrates the basic method of operation. The blank is positioned over the male die, called a *punch,* which is shaped to conform to the inner configuration of the desired sheet-metal workpiece. A flexible die, consisting of a rubber diaphragm or seal, is placed across the bottom of the pressure-forming chamber, which is filled with a hydraulic fluid. The drawing operation begins by positioning the blank on the draw ring. After the dome is lowered over the blank and locked into position, the preliminary hydraulic pressure is then applied. The punch is raised and pushed firmly into the blank. The pressure in the pressure chamber continues to increase, causing the metal blank to displace and uniformly flow and "wrap" around the punch to form the desired workpiece shape. The cycle is completed by releasing the forming pressure, raising the forming chamber, and stripping the workpiece from the punch.

Hydroforming differs from other flexible die-forming processes in that the shape of the workpieces is worked by *drawing* rather than formed by bending. The metal is displaced rather than stretched into shape. "Thinout," spot stresses, and springback are greatly reduced or often entirely eliminated.

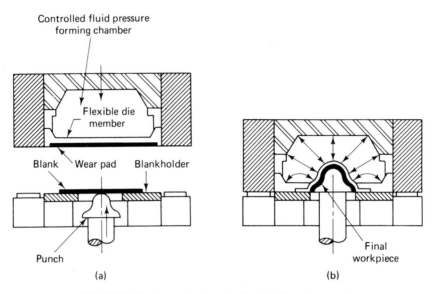

FIGURE 36-5 Tooling and set-up for the hydroforming process:
(a) before forming; (b) after forming

Special high-speed 15,000-psi (approximately 1050-kg/cm²) hydroforming presses are currently available which can draw a blank in production quantities up to 25 in. (635 mm) in diameter and form workpieces to a depth of 12 in. (304.8 mm). Depending upon the part configuration, the press can accomplish up to 900 automatic operations per hour, including loading and unloading. Other types of conventional hydroforming presses are available which are used principally for product development and prototype work. These presses can draw a blank up to 32 in. (812.8 mm) in diameter as deep as 12 in. (304.8 mm). Both special and conventional types of presses can form steel workpieces up to ⅜ in. (9.53 mm) thick and aluminum workpieces up to 1 in. (25.4 mm) thick.

Important advantages of hydroforming as compared to conventional drawing methods are the following:

1. Most parts can be formed in fewer drawing operations because more work is performed per operation.
2. There are significant savings in tooling costs. No female or matching dies are required.
3. Tooling can be rapidly changed because only a punch-and-draw ring need be handled.
4. Asymmetrical and irregular contours are formed as easily as symmetrical shapes.

580

5. Sharp detail is possible, particularly on inside radii.

6. Practically all sheet metals can be hydroformed—carbon steel, aluminum, copper, brass, precious metals, high-strength alloys, and so on.

7. Material is uniformly worked during the draw so that it retains superior mechanical and physical values.

8. Tolerances of ±0.005 in./in. (0.005 mm/mm) are practical.

Typical hydroformed parts are shown in Figure 36-6.

Spinning

Spinning is a process of forming seamless metal workpieces from a circular disk of sheet metal or from a length of tubing. Figure 36-7 illustrates representative shapes that may be produced by spinning. All parts produced by spinning are similar, in that they are symmetrical about a central axis.

There are two main methods of spinning: *manual spinning* and *shear spinning.* Shear spinning is also known by various proprietary trade names, such as Lodge and Shipley's Flo-Turn, Cincinnati Milacron's Hydrospin and Lear Siegler's Spin-Forge.

Most spinning operations are done at room temperature, although it is sometimes necessary to preheat work blanks to increase the ductility and reduce the strength of some hard-to-work metals. Any of the ductile metals and alloys that can be drawn by other processes can be spun.

Manual Spinning: This is performed on a speed lathe. In most cases, the operation consists of attaching a hardwood, fiber composition, or metal mandrel to the lathe spindle as shown in Figure 36-8a. The mandrel (also called a *form block*) is made to preestablished shape, conforming to the inside configuration of the desired workpiece. The work blank, in the form of a circular disk (also shown in Figure 36-8a), is centered and held between the mandrel and the block. As the assembly rotates, the operator presses a rounded tool or roller against the blank, causing a gradual change in shape, as shown in Figure 36-8b and c, until a final workpiece shape is developed, shown in Figure 36-8d, which conforms to the mandrel. Simple shapes can often be spun without a mandrel. In most cold-spinning operations, a lubricant, such as cup grease, soap, wax, or tallow, is applied to the work blank, which improves the drawing characteristics of the work material.

Hydraulically powered spinning lathes have replaced some manual spinning setups. Additional force is required for heavier-gage metals and for spinning complex parts. Some hydraulically powered lathes are available with either tracer controls or tape control and require less operator skill and give more uniform results.

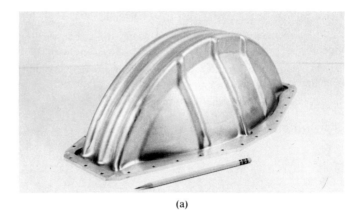

(a)

FIGURE 36-6 Typical hydroformed parts: (a) Gimbal cover—6061 aluminum; dip-brazed and heat treated to T6 (Courtesy, C. B. Knapp & Sons, Inc.)

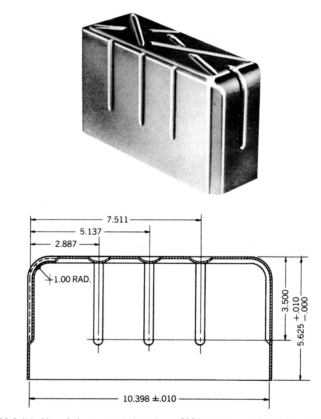

(b)

FIGURE 36-6 (b) Aircraft instrument, housing—6061 aluminum; 0.040 in. thick (Courtesy, C.B. Knapp & Sons, Inc.)

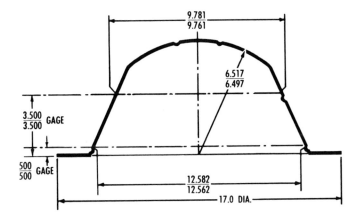

(c)

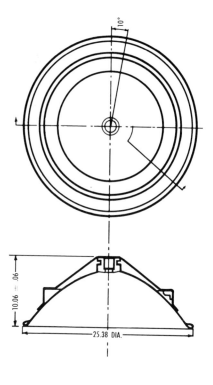

FIGURE 36-6(c) Ribbed bladder for aerospace fuel transfer—type 302 stainless steel; 0.007 in. thick (Courtesy, C.B. Knapp & Sons, Inc.)

(d)

FIGURE 36-6 (d) Parabolic reflectors assembly—6061 aluminum (Courtesy, C.B. Knapp & Sons, Inc.)

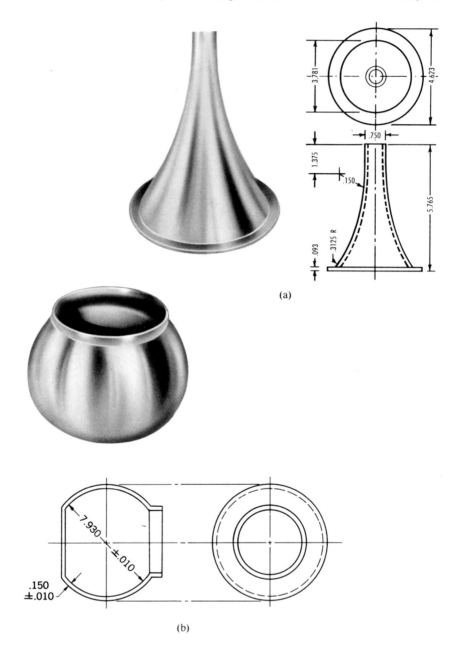

(a)

(b)

FIGURE 36-7 Representative shapes produced by spinning: (a) wave guide—6061 aluminum; (b) directional gyro laminated component spun simultaneously from three 0.050 in. thick aluminum blanks (Courtesy, C.B. Knapp & Sons, Inc.)

(a)

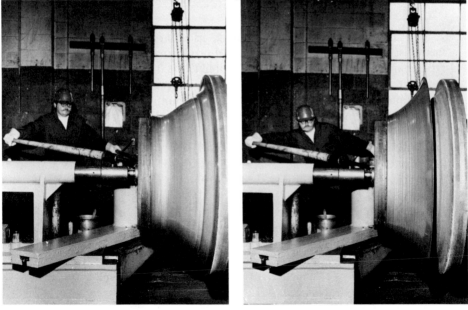

(b)   (c)

FIGURE 36-8 Manual spinning: (a) the circular blank is mounted in position against a mandrel on a lathe (Courtesy, Metal Spinners, Inc.); (b) and (c) the spinning technique shapes metal by forming it around a mandrel (Courtesy, Metal Spinners, Inc.);

(d)

FIGURE 36-8 (d) The formed product is removed from the metal spinning lathe (Courtesy, Metal Spinners, Inc.)

The advantages of manual spinning, as compared to other drawing methods, lie in the speed and economy of producing special parts or prototype samples in small lots, normally limited to 1000 pieces or less. Tooling costs are less and the investment in capital equipment is relatively small.

Aluminum parts as thin as 0.020 in. (0.51 mm) to about 12 in. (304.8 mm) in diameter may be spun by manual methods. It is not considered feasible to spin extremely large or deep parts from materials of very light gage. Aluminum stock as thick as $\frac{3}{16}$ to $\frac{1}{4}$ in. (4.763 to 6.35 mm) can be spun to form parts as large as 96 in. (2.43 m) in diameter. The maximum thickness limit for low-carbon steel is about $\frac{1}{8}$ in. (3.18 mm), and 72 in. (1.82 m) is the limiting diameter.

There is no substantial change in metal thickness, although parts may be intentionally thinned in selected areas by the operator as the spinning tool is manipulated against the rotating work blank. Considerable operator skill is required in manual spinning. Unless the metal is allowed to flow at the proper rate, there is a tendency for wrinkles, tool marks, and scratches to form on the workpiece.

When compared to other manufacturing processes, such as deep drawing, stamping, or machining, metal spinning cannot compete as a precision forming method. The recommended tolerance for manually spun parts ranges from $\pm \frac{1}{64}$ in. (0.397 mm) for small work to $\pm \frac{1}{16}$ in. (1.588 mm) on larger sizes. Parts should be designed with smooth curves with large radii and gradual tapered steps. In general, parts with a minimum radius of $\frac{1}{8}$ in. (3.18 mm) can be successfully manufactured, but a more generous radius simplifies the production problems and permits a better utilization of the unique advantages of this process.

Typical parts produced by manual spinning include an almost unlimited variety of odd-shaped domes, covers and trays, reflectors, tanks, floats, shields, bell-shaped musical instruments, components for the aircraft and aerospace industries, radar antennas, food-processing machinery, commercial laundry equipment, and various cooking utensils, bowls, pitchers, and kettles.

Shear Spinning: This is a method of forming complex shapes, such as cones with tapering walls, curvilinear shapes such as nose cones, and hemispherical and elliptical tank closures with either uniform or tapering walls. Tube spinning is a variation of shear spinning.

Basic shear spinning is also called *sine-law spinning,* since the final wall thickness of a spun cone can be determined by the sine-angle function. The process is illustrated in Figure 36-9 for the case of a 30°-angle cone. Since the sine of the 30° angle is 0.5 in. (12.7 mm), the thickness of the flat plate or starting stock will be reduced by 50%. The plate thickness, which lies in a plane parallel to the center line of the mandrel, does not change during spinning; only the normal wall thickness is reduced. Note also that the plate diameter does not change during spinning.

Since shear spinning always results in a substantial and predictable reduction in normal wall thickness, shear spinning machines are constructed more rigidly than conventional spinners. Special horizontal or vertical machines are required for shear spinning. Some are available that can spin workpieces as large as 240 in. (6 m) in diameter by 240 in. long. On some machines, it is possible to shear spin work as thick as $5\frac{1}{2}$ in. (139.7 mm). Unlike the low-cost laminated, solid, or collapsible mandrels used in manual spinning operations, shear-spinning mandrels must be solidly built to withstand the forces caused by increased eccentric loading. Also, they must be sufficiently hard to resist the deteriorating effects of wear. For low-quantity production of workpieces, mandrels are generally made of cast iron while tool steel is selected for high production. Tool steel or carbide rollers or tool rings, specially shaped to the requirements of the workpiece, are used. Tracer-controlled spinning machines are available which utilize precision templates to control the roller path. Such equipment reduces the degree of skill required of the operator, and lends the process to automation.

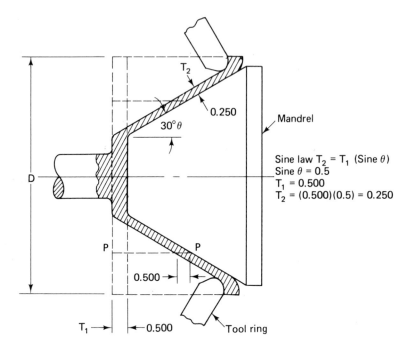

FIGURE 36-9 The sine law as applied to shear spinning cones
(Courtesy, *Materials Engineering,* February 1970)

Parts requiring greater total reduction than the material can accept in a single operation can be formed by multipass shear spinning. In this case, a preform is often used as the work blank. Preforms such as machined forgings or castings may be used. In some instances, preforms may be produced by manual spinning, hydroforming, explosive forming, or by conventional press forming.

Shear spinning is an effective method for manufacturing lightweight, high-strength cylindrical parts from steel alloys. Products made by shear spinning include rocket and missile cones, pressure bottles, fire extinguishers, jet-engine components, and harrow and truck wheel disks.

Finished parts have no seams or joints and often require little if any machining. Wall thickness on spun parts can be maintained to a tolerance as close as ±0.002 in. (0.05 mm).

Tube Spinning: These operations are also performed on shear-spinning machines. There are two methods: forward and backward, so named because of the direction of metal flow. In both methods, a tubular shape preform is required which may consist of a machined forging or a centrifugally cast tube or a welded or extruded tube.

Forward tube spinning is illustrated in Figure 36-10a. A preform is clamped to the mandrel at the tailstock end and the roller advances toward the headstock. In *backward* tube spinning (Figure 36-10b) the preform is not clamped by but slid over the mandrel to the headstock end. Since the preform is contained by the fixture at the headstock, the metal flow is directed backward toward the tailstock.

Spinning reduces the wall thickness of tubular shapes. It is possible by this method to produce one or more integral projecting rings or flanges at selected positions on a tube. In most cases, the process may be accomplished without heating the workpieces.

FIGURE 36-10 Tooling and set-up for forward and backward tube spinning: (a) forward tube spinning; (b) backward tube spinning

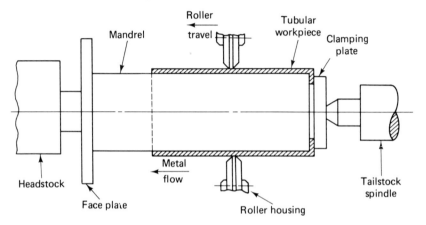

(a)

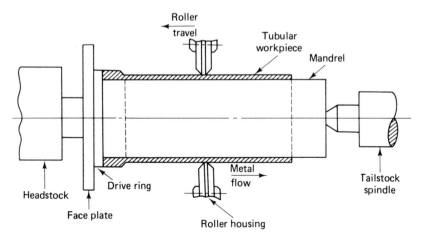

(b)

A surface finish of 15 to 62 μin. is possible, depending upon the configuration and position of the roller, the nature of the work metal, and the amount of reduction per pass.

REVIEW QUESTIONS

36.1. In what important way are all drawing processes alike?

36.2. In stretch draw forming, how do the form-block method and the mating-die method differ?

36.3. List the main advantages of radial draw forming over other conventional press-forming methods.

36.4. What are some of the chief limitations of stretch forming?

36.5. What are some of the benefits of Marforming?

36.6. In what significant ways does hydroforming differ from other flexible-die forming processes?

36.7. List some important design factors that must be considered when deciding the suitability of a part to be formed by spinning.

36.8. In what ways can the skilled operators aid in controlling the quality of the part they produce by manual spinning?

36.9. Compare the important features of mandrels used in manual spinning to the type used in shear spinning.

36.10. Explain why preforms are sometimes used in shear spinning.

BIBLIOGRAPHY

*BRAUER, E. H., "Contouring Parts by Stretch Contour Forming," *Machine Design,* Oct. 18, 1973.

"New Metal Spinning Techniques Form Large Shapes from Thick Metals," *Product Engineering,* May 1974, p. 48.

"Spinning," *Metals Handbook,* Vol. 4, *Forming,* 8th edition, Metals Park, Ohio: American Society for Metals, pp. 201–209.

*STEWART. J. D., "How to Process for Shear Spinning," *Materials Engineering,* Feb. 1970.

"Stretch Forming," *Metals Handbook,* Vol. 4, *Forming,* 8th edition, Metals Park, Ohio: American Society for Metals, pp. 239–244.

   *Abstracted with permission.

37

HEAT TREATMENT

Specifications for heat-treating processes are among the most important of those shown on an engineering drawing. Proper heat treatment is a powerful tool for developing the best possible properties that a material can possess. In general, heat treatment may be described as a combination of heating and cooling operations, timed and applied to a metal or an alloy in the solid state in a way that will produce desired properties. Principally, heat treatment is used to produce strengthening, but some heat-treating processes soften, toughen, or otherwise enhance properties.

Internally, a metal or alloy consists of one or more kinds of atoms packed together in orderly three-dimensional arrangements called *crystals.* The crystals, in turn, are bonded together in diverse ways which are described in terms of microstructure or grain structure. Any given structure can be altered to some extent by plastic deformation from compressive, tensile, or shear forces, but the available time–temperature treatments provide a greater variety of properties. Heat treatments are carefully controlled combinations of such variables as time, temperature, rate of temperature change, and furnace atmosphere. The selection of a specific treatment must be based upon knowledge of the properties desired in the finished part.

There is available today a multitude of metals and alloys designed for various purposes. There are also many different heat-treating processes.

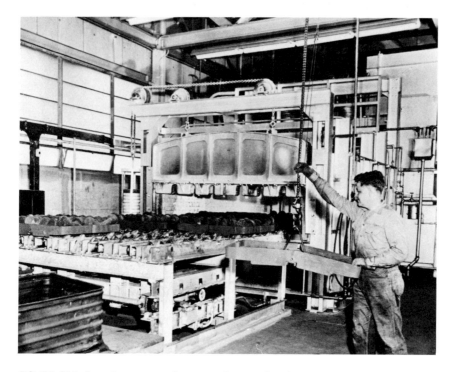

FIGURE 37-1 A pusher-type continuous cycle annealing furnace using trays which carry 500-lb loads of forgings (Courtesy, Forging Industry Association)

Not all the treatments can be used with each metal or alloy. In other words, the treatment selected must be one that is compatible with the specified material. Heat treatment cannot be selected independently of material. One is just as important as the other.

REASONS FOR HEAT TREATING

Ferrous Metals

Ferrous parts are heat-treated for several reasons: to relieve internal stresses, to change the microstructure by refining the grain size or producing uniform grain throughout a part, to alter the surface chemistry by adding or deleting elements, and to strengthen a metal part.

Nonferrous Metals

The reasons nonferrous parts are heat-treated are to relieve internal stresses resulting from forming, brazing, or welding; to obtain a more uniform structure in castings by allowing thorough diffusion to take place; to offset plastic strain inherent in a particular metal by bringing about recrystallization to provide new equiaxed and stress-free grains; and to strengthen by the age-hardening process.

The TTT Curve: The "Heat Treater's Road Map"

Figure 37-2 illustrates a TTT curve (time-temperature-transformation) for steel which is often referred to as the "heat treater's road map." The heat treater quenches steels to produce a particular microstructure. Each microstructure has its own set of properties and is selected according to the design requirements of the finished part. The TTT curve shows the heat treater what path to take and how severe the quench must be to obtain the necessary microstructure. The TTT curve (shown in Fig. 37-2) includes four different cooling paths (1–2, 1–3, 1–4, and 1–5). In each case, the steel is heated to about 1,333 °F approximately 723 °C. At this temperature, the steel is in a solid solution known as austenite.

Path 1-2 represents a rapid quench (about 60,000 °F/hr) from the austenite temperature to room temperature. At 550 °F, the austenite slowly begins to transform to another state, martensite. Below 325 °F, all the austenite has transformed to martensite, so that at room temperature, point 2, the quenched steel is fully martensite.

Path 1-3 yields a different structure. At 925 °F the austenite begins transformation to bainite. Transformation continues until 550 °F, where the remaining austenite transforms to martensite instead of bainite. Below 310 °F, transformation is completed and the result is a mixture of martensite and bainite.

Path 1-4 represents a delayed quench. At 925 °F, the austenite begins to transform to bainite, as before. But somewhere between 925 and 550 °F, the quench is halted. In this example, the steel is held at about 650 °F. During this period, all the austenite transforms into bainite. The final cooling yields a structure exclusively bainite. This sequence is called "austempering."

Path 1-5 is a slow-cooling annealing cycle. At about 1250 °F, the austenite begins to transform to pearlite rather than ferrite, so that at point 5, the structure is a combination of ferrite and pearlite, a soft annealed steel structure.

STRESS RELIEF, SOFTENING, AND GRAIN REFINEMENT

In most ferrous alloys the constituents that make up the microstructure are clustered into individual grains. Extensive cold and hot working of steel parts tends to increase grain size. Likewise, extensive machining and working

FIGURE 37-2 The TTT Curve

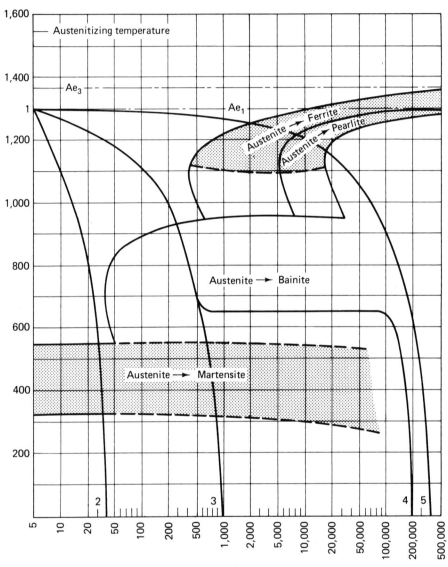

Transformation time (sec)

of steel introduce stresses that are stored away and may be released later in the form of metal displacement or distortion. Several heat treatments are designed to restore uniformity to the microstructure, to refine grain size, or to relieve stresses.

Annealing

Annealing is a simple cycle that has the effect of "starting off with a clean slate." Every alloy steel except stainless steel has an *upper critical temperature,* the temperature above which the alloy changes from its room temperature state to a form called *austenite.* Annealing consists of heating the material above its upper critical temperature and slowly cooling it back to room temperature at a controlled, predetermined rate. Annealed steel is in a state of maximum stability and is dead soft, fully ductile, free of internal stresses, and its structure is called *pearlite.* A number of annealing steps are commonly used. These are usually distinguished by their specific cooling rates.

Annealing is often specified for castings that, because they have abrupt changes in cross section, did not cool uniformly. In addition, forgings or parts made from rolled steel may be annealed to correct nonuniform structure. Finally, parts heat-treated in error sometimes may be annealed to "start over."

Severely worked parts that require subsequent heat treatment should be annealed. Annealed parts respond predictably to other heat treatments while unannealed metal may respond erratically.

Spheroidize annealing is a special annealing cycle frequently selected instead of plain annealing for high-carbon and alloy steels (usually, steels with over 0.50% carbon are considered high-carbon steels). Parts are held at about 1300 °F (705 °C) below their transition temperature for a long time, long enough for the carbides to spheroidize or come out of solution. (Actual time varies with the amount of carbon in the steel.) These spheroidize-annealed steels show improved machinability, so the process may be specified when a part requires a lot of machining. Often, extruded or headed parts are spheroidize-annealed before swaging, extruding, or heading to improve formability.

Normalizing

Normalizing is a process similar to annealing, but the material is allowed to air-cool after austenitizing. Normalizing is specified in place of annealing to refine grain size. As with annealing, the final structure is a soft mixture of free ferrite and pearlite.

The process consists of heating steel about 100 °F (37.8 °C) above the upper critical temperature and cooling it in still air. This process results in a homogeneous structure and imparts moderate hardness and strength. The process is ordinarily used with low- and medium-plain carbon and alloy steels. Most commercial steels are normalized after being rolled or cast.

Stress Relieving

Stress relieving removes internal stresses caused by machining and cold working but, unlike annealing, does not alter the microstructure. (An exception would be for some cold-worked steels whose structures may be changed by recrystallization.) Parts are heated below their critical temperature, so no structural change occurs. Parts that have been welded or heavily machined would be stress-relieved, especially if they are to be subsequently heat-treated and if they must be finished to close dimensional accuracy. Otherwise, stored internal stresses will be relieved during the final heat treatment and the parts will distort out of tolerance. Stress relieving is specified when the microstructure is not to be altered.

Specifications for annealing or stress relieving can include:

1. At what point during the manufacturing processing to perform the heat treatment (usually accomplished soon after massive working, welding, or machining).
2. Temperature and time.
3. Cooling rate.
4. Grain size (anneal and normalize only).

Alloy content, part size, and sometimes lot size affect most of these variables.

Hardening

Hardening is a familiar heat-treating term but is an oversimplification of an interaction among many mechanical properties. While annealing or stress relieving is often specified as an auxiliary treatment to assist manufacturing or to provide a uniform material for other heat treatments, the hardening or strengthening processes are specified to bring out desired properties in the finished part. These properties include hardness, toughness, ductility, fatigue strength, tensile strength, and wear resistance.

The principal methods of hardening metals and alloys are cold working by plastic deformation, precipitation hardening, quench hardening as applied to steel, and surface-hardening processes.

Cold Working

Cold working or *mechanical working* is sometimes the only practical way to obtain the desired hardness for a specific metal or alloy. When metals are deformed or cold-worked, there is an increase of hardness or strength. The metal remains in its stable hardened condition until it is heated to a temperature high enough to cause the occurrence of extensive recovery and recrystallization. Sheet stock may be hardened by the simple expedient of cold rolling and is available from the manufacturer in various degrees of desired hardness. Wire drawing increases hardness, and piano wire or wire used in bridge construction is readily available in different hardness values.

Precipitation

Precipitation or *age hardening* strengthens and improves the mechanical properties of a metal. The process is specified for certain types of stainless steels, special high-strength and heat-resistant alloys, or for some alloys of aluminum, copper, or magnesium.

Many alloys normally exist in two distinct phases at room temperature. (A *phase* is defined as a homogenous portion of the system throughout which the properties are uniform. A phase may exist as a vapor, liquid, or solid.) During precipitation hardening, the minor phase goes into solution at elevated temperatures, resulting in a stable single-phase solid solution. Holding the alloy to a suitable temperature for a period of time dissolves the hardening phase. Precipitation hardening can result from the controlled precipitation of the particles of the secondary crystalline phase inside the crystals of the primary phase. When the solid solution is suddenly cooled, as in quenching, there is insufficient time for material to precipitate out into the normal two phases, and the material is trapped in an unstable state as a supersaturated solid solution. The process of treating raw material in this way is called *solution treating.* Machining or cold working may be performed while the alloy is in its soft state. The completed part may be hardened after machining by reheating it to a slightly elevated temperature, an action that will precipitate out the second phase.

Aluminum alloys are hardened principally by precipitation hardening or by cold working. Copper, manganese, nickel, and silicon make precipitation-hardenable aluminum alloys. Precipitation hardening of some aluminum alloys may take place at room temperature after solution treatment, which may take up to 1 month for some alloys such as 2014 and 6061 to reach full hardness. The strength of alloys 7079 and 7075 continues to rise at a significant rate for years after quenching. The process can also take place at temperatures from 240 to 375 °F (approximately 115 to 190 °C) with times varying from 5 to 48 hours. Selection of time and temperature depends upon the alloy being hardened.

Copper and copper alloys are usually hardened by cold working and solid-solution alloying with zinc, tin, aluminum, silicon, manganese, and nickel. Precipitation-hardening methods may be applied to beryllium copper, aluminum bronze, chromium copper, zirconium copper, and silicon- and phosphorous-containing copper nickels.

Magnesium alloys respond to precipitation hardening. In this process, the heat-treating furnace is loaded at about 500°F (260°C) and then the temperature is slowly increased to the appropriate solution-treatment temperature. Often, temperatures vary from 3 to 24 hours.

Precipitation hardening is a gentle, low-temperature process and is ideal for critical, closely dimensioned parts. Parts can be hardened and strengthened, free from cracking or distortion.

Specifications for precipitation hardening can include:

1. Initial raw-material condition (often mills offer several solution treatments for a material).
2. When to perform the hardening.
3. Time and temperature for precipitation cycle (or aging).

Quench and Temper

Quench and temper, also known as the *harden-and-temper* hardening process, is the usual method of hardening a piece of steel. Steel (with carbon content of about 0.35% and over) can be heated above its critical temperature and rapidly cooled, or quenched. The resulting microstructure, called *martensite,* is hard, brittle, and full of internal stresses. (Hardness is directly related to the amount of carbon in the steel.) As-quenched steel is sometimes specified for parts which, in service, must support heavy loads or for maximum wear resistance, but as-quenched steel is usually too brittle to endure shock or impact.

Almost always, quenched martensite is tempered, that is, reheated below the critical temperature and held long enough to remove internal stresses and restore some ductility and toughness. Unfortunately, hardness is traded for toughness, so selecting tempering times and temperatures always involves a trade-off among mechanical properties. Furthermore, higher-alloy steels are subjected to greater internal stresses than the milder alloys, and may crack or distort when quenched. Thin walls, abrupt changes in cross section, and sharp corners are common sources of cracking. Asymmetrical parts made from higher-alloy steels (such as bushings with only one keyway) are prone to quench cracking.

Tempering or "drawing" must follow hardening of steel if it is to be useful and not brittle. As hardness is reduced, tensile strength decreases and ductility and toughness increase. In practice, the quenched workpiece is reheated to some temperature below the lower critical range, followed by any rate of cooling. Successful tempering, unlike annealing, requires close control. The final microstructure is tempered martensite. There is relatively little decrease in hardness between 300 and 400 °F (about 150 and 204 °C), that temperature range being suitable for stress relief. Around 600 °F (about 316 °C), the formation of tempered martensite is rapid.

Martempering and *Austempering* are special methods of quenching steel. Martempering is a less severe way of producing tempered martensite. In this process, the quench is slowed down just at the point where the martensite begins to form. Martensite formed under these less-severe conditions contains less internal stress, and thus may not crack or distort. The process takes longer and is more costly.

High-alloy steels (steels with high "hardenability") can readily be Martempered. Plain carbon steels also can be Martempered, but they require greater attention to section thickness since they do not have the depth of hardenability offered by alloyed steels.

Austempering is even less severe than Martempering. In this case the quench is completely stopped before martensite forms and the austenite is isothermally transformed, at this point, to another microstructure, called *bainite*. (Bainite is more ductile than martensite.) Austempered steel requires no tempering and is used as transformed. The part is stress-free and distortion-free.

Like Martempering, Austempering is time consuming, and each process requires careful furnace control. Austempering can be readily used on high-hardenability steels, or on carbon steels if the section thickness is not too large.

The composition of a steel dictates which of these processes may be used. It also indicates the severity of the required quench. Severe quenching increases the chance of cracking and distortion. Plain carbon steels require severe quenching to acquire a martensitic structure, so they are usually quenched in water. Alloy steels have a higher hardenability rating, so they may be quenched in oil, a less severe medium. High-alloy steels and many tool steels form martensite on air cooling and are almost completely free of distortion and cracking. Occasionally, a closely dimensioned part "grows" or distorts over a period of several hours to several days after quench-and-tempering or Martempering. This condition is caused by a delayed transformation of austenite trapped in the quenched material. Quenching is rapid, so small islands of austenite (the phase present above the critical and not normally in equilibrium at room temperature) cannot transform to martensite. Over a period of time, however, the "retained austenite" spontaneously transforms into martensite, a more stable condition. The transformation is accompanied by a slight expansion. Therefore, to prevent unexpected size changes, closely dimensioned parts are soaked at cryogenic temperatures immediately after tempering to complete the transformation of retained austenite to martensite under controlled conditions.

Many of the following variables are specified for the quench-and-temper processes:

1. Process cycle required (Martemper, Austemper).
2. Final hardness range.
3. Final microstructure.
4. Quenching medium (water, air, oil).
5. Temperature before quenching and time to "soak" at this temperature.
6. Initial temperature of quenching medium (optional).
7. Quenching cycle (optional).
8. Tempering time and temperature.
9. Minimum time elapsed between quench and temper cycles (optional).
10. NDT (nondestructive testing) for cracking.
11. Cooling rate after tempering.
12. Posttempering cold treatment, time, and temperature.
13. Minimum time elapsed between temper and cold-treatment cycles.
14. Maximum percent retained austenite.

Surface Hardening

Case hardening, differential hardening, or surface hardening produces a hard skin around the core of a ferrous part. Several processes can be used for case hardening. All but one, induction hardening, involve diffusion of another material into the metal surface.

Carburizing: The most common diffusion-type hardening process is *carburizing.* Briefly, the carbon diffuses into the metal at a high temperature (about $1700\,°F$, or $927\,°C$) and actually transforms the surface layer of the metal from a low-carbon steel to a high-carbon steel. Then, a simple quench and temper hardens the high-carbon skin (because hardness is a function of carbon content). The hardened case resists wear and scuffing, and provides higher fatigue resistance because of the beneficial compressive stresses at the surface. Tensile stresses at the surface are harmful to fatigue performance.

Specifications for carburizing must include core as well as case requirements. Closely controlled case properties usually require gas carburizing because the carbonaceous atmosphere can be monitored closely. The carbon atmosphere must completely surround the part for uniform case depth and hardness. Of course, carbon content gradually decreases from the surface of the case to the core, and so does hardness. The distinction between actual core and case, or "effective case," is sometimes defined as the point where hardness measures Rockwell C50. As little metal should be removed from the surface as possible after carburizing to prevent removing the best part of the hardened case: generally, no more than 20% of the carburized case should be removed after heat treatment.

Typical specifications for carburizing may include:

1. Case hardness on surface.
2. Effective case depth.
3. Carbide condition and restriction.
4. Hardness gradient across case.
5. Core hardness.
6. Core structure (fully martensitic, no austenite, etc.).
7. NDT (testing) for surface cracking.
8. Maximum stock removal after carburizing.

Variations on the basic carburizing cycle are designed to fit special needs:

Carbonitriding: Carbonitriding sounds like a combination of carburizing and nitriding. Actually, carbonitriding is essentially liquid carburizing, but it also involves some diffusion of nitrogen. The process is selected instead of carburizing, when high hardness with less distortion is required. But, because any steel that can be carburized can also be carbonitrided, the process introduces some of the benefits of nitriding (hard case, shallow case depths, lower processing temperatures, less distortion) to materials that cannot actually be nitrided.

Often carbonitriding is applied to low-cost, low-carbon steels where a wear-resistant surface is all that is required. Because the process is carried out at temperatures lower than those for conventional carburizing, carbonitriding is economical and allows more accurate distortion control. In addition, carbonitriding strengthens low-carbon steels subjected to light loads (where high-core properties are not required).

Vacuum Carburizing: This is a relatively new process where portions of the cycle are performed in a vacuum. The process, perhaps best described as partial-pressure carburizing, may produce a carburized case in less time than conventional carburizing. Of course, because of the vacuum requirement, equipment costs and processing costs may be higher than for conventional carburizing. Overall benefits of this process are yet to be established.

Carburizing can be followed by many other heat treatments. Carburizing followed by air-cooling, reaustenitizing, and then the conventional quench-

and-temper can produce finer martensitic grains and eliminate much of the retained austenite. In addition, a number of carefully selected tempering cycles can further refine the structure. Naturally, each added cycle increases processing costs and must be justified by improved properties.

Parts may be selectively carburized. Surfaces can be "stopped off" by plating. Only those surfaces exposed to the carbonaceous atmosphere are hardened. A sequence often employed to selectively harden is (1) carburize the entire part, but do not harden; and (2) machine the carburized case off surfaces that are to remain soft. Only those surfaces where the case is left will be hard when reheated and quenched.

Nitriding: This is not new, but recent improvements in furnace design and control have moved nitriding into the precision heat-treatment category. Nitriding is selected for precision parts where wear resistance and fatigue resistance are important.

Nitrogen, in the presence of certain catalysts and other elements, enters the surface of the metal and forms nitrides with some of the alloying metals. Few alloys can be nitrided: the "nitriding steels," some tool steels such as H-13, gray cast iron, and alloys such as the medium-carbon 4100 series. In general, special alloys high in aluminum and chromium content are used for nitriding.

Nitriding occurs at reasonably low temperatures (about 900 °F, or 483 °C) and quenching is not required, so distortion is minimized. Usually, material is first heat-treated for desired core properties (often quench-and-tempered, for example, to a hardness that will yield satisfactory toughness but remain machinable). After finish machining, the part is nitrided. Finish grinding may follow nitriding, if necessary. Nitriding does not affect the previously established core properties. Sometimes, stress relieving before nitriding, followed by semifinish machining, can eliminate distortion completely.

Gas nitriding, by far the most precise process, requires preliminary surface preparation. Parts must be clean and either etched or grit-blasted for a specific surface texture. Improperly treated parts nitride unevenly and slowly.

Compared to carburizing, nitriding has several inherent disadvantages. First, the range of materials that can be nitrided is limited. In addition, some nitriding alloys are expensive and difficult to machine. Also, the cycle times are long and pretreatment surface preparation can be costly. Finally, because nitriding results in a relatively thin case, only minimal machining can be done on parts after heat treatment.

Nitriding parts "grow" during heat treatment. Bars increase in length and bushings expand. Because so little case may be removed after heat treatment, most manufacturers anticipate this size change when they produce the part. Actually, with careful process control and preliminary testing, most parts can be designed to "grow" to size.

Parts to be nitrided may require the following specifications:

1. Core heat treatment (quench and temper, etc.).
2. Core hardness and condition.
3. At what point during the manufacturing sequence to treat core.
4. Nitride case depth.
5. Case hardness.
6. Maximum white layer permitted.
7. Surface pretreatment, size, or grit.
8. Cycle times, temperature, concentrations of ammonia.
9. Prenitriding stress reliefs.
10. When during the manufacturing cycle to stress relieve.
11. Which surfaces to finish after nitriding, and the maximum stock to remove.

Surfaces can be selectively nitrided. Special "stop-off" paint is available for masking surfaces from the nitrogen atmosphere. Also, surfaces to remain soft can be plated and later stripped.

Soft nitriding is a relatively new process. It involves only gaseous atmospheres and is more readily controlled for highly precise case depths and hardnesses. The process is basically a gas nitriding cycle, with the addition of carbon to the furnace atmosphere. The process can be performed on more alloys than conventional nitriding, and it still offers most of the advantages of nitriding. It produces a case similar to conventional nitriding, although the white layer may be more acceptable. Case depths are shallower than with conventional nitriding.

Boron Diffusion: This, done in certain high-carbon steels and tool steels, provides high surface hardness and good wear and oxidation resistance. The process, performed at low temperatures, is nearly distortion-free, but is also slow.

Materials such as bonded steel carbides, 1095 spring steel, H-13, and S-1 tool steel have been surface-hardened by boron diffusion. Some age-hardenable alloys can be boron-treated and then hardened throughout for maximum stability under high loads. The process is best suited for wear and tooling applications, especially in corrosive environments.

Case depth, usually around 0.007 in. (0.18 mm), is determined by time, temperature, and steel composition. The diffusion rate of the boron into the steel is decreased by alloying elements such as titanium, chromium, and vanadium.

Induction Hardening: This is basically a quench-and-temper process. Instead of heating the part throughout and quenching, only the surface skin is hardened and quenched. The result, of course, is a hardened outer skin with a soft core.

Induction coils provide the required heat in most cases, although a similar process called *flame hardening* uses an oxyacetylene torch for heating in less critical applications. Timing is important. The part must be heated to the correct depth and immediately quenched. As a result, abruptly changing cross sections and asymmetrical parts are often difficult to induction-harden uniformly. Sharp external edges heat too rapidly, and thin-walled cross sections may heat all the way through.

Induction hardening is specified for heavy-duty case-hardening applications. The case can be much deeper than a carburized case: the process is faster and requires less elaborate equipment. Frequently, heating coils must be custom-designed for a particular part, however, so tooling costs must be considered.

With induction hardening, selected surfaces of a part may be hardened for wear resistance, while others can remain soft for subsequent machining. Unfortunately, selectively hardened parts retain residual stresses near the hard/soft zone, so this zone should be located a safe distance away from sharp edges, grooves, or other stress raisers to avoid cracking.

Cracking is the greatest concern to the designer of induction-hardened parts. Alloys with high hardenability are more sensitive to heat transfer into the core of the part and may be more likely to crack. Occasionally, a pretreatment such as normalizing may prevent cracking. Careful material selection, proper heat-treatment cycle, and—most important—careful part design can prevent cracking.

Some *variables often specified for induction hardening* are:

1. Case depth.
2. Case hardness.
3. Case structure.
4. Pretreatment, and at what point to perform them during manufacturing sequence.
5. Core hardness.
6. Core treatments.
7. Post treatments.
8. NDT for cracking.
9. Hardness gradients desired between selectively hardened and soft areas.
10. Heating times and temperatures.
11. Quenching medium.
12. Preheating, if required.

DESIGN FOR HEAT TREATING

In all critical applications, the selection of material and the heat treatment should be carried out with the help of a metallurgist or experienced heat treater. Together, the metallurgist and the designer work to piece together pertinent design parameters, manufacturing requirements, and cost restrictions to select the best combination of commercially available material and heat treatment. An informed designer can communicate with the metallurgist and know *why* a particular process was selected, *how* it is performed, and—most important—*how* this information is conveyed to the heat treater.

Many handbooks advise how to select steel with the correct hardenability, how to determine proper case depths, when to avoid asymmetrical parts, or how to design for heat treating. However, a number of subtle design requirements are often neglected.

The surface condition of the raw material is often ignored. Most mill barstock has a skin of decarburized steel (often called "bark" or "decarb") that must be removed before heat treatment. On quench-and-tempered alloy steels, the decarburized zone does not harden thoroughly. All surfaces to be hardened must be machined below the decarburized region. Likewise, decarb must be removed wherever a hardened case is required. When a part is to be nitrided, the decarb must be removed everywhere—even where a hard case is not needed, because nitriding over a decarburized zone creates a very brittle layer that peels off under slight loading.

Threaded parts shown in Figure 37-3 and parts with sharp edges require special attention if they are to be case-hardened. At the sharpest edge the part is case-hardened throughout. In some instances, the case cannot withstand shock and loading without the support of a ductile and tough core. Nitrided threads experience brittle failure when they are used frequently. For critical parts, threads should be protected or masked from case hardening. Also, thin-walled parts, of thin cross sections of abruptly changing wall, as shown in Figure 37-4, may be inadvertently case-hardened throughout. Typically, a minimum 30% of the cross section should be core.

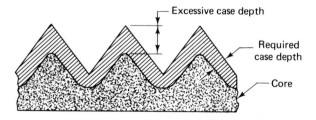

FIGURE 37-3 Case hardened threaded parts require special attention

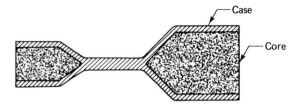

FIGURE 37-4 Thin cross sections with abruptly changing wall configurations present problems when case hardening

Steel harder than Rockwell C35 to C40 is difficult to machine by standard metal-cutting processes. Hardened parts are best machined by EDM, ECM, or abrasive processes such as grinding and lapping. Because these processes are slow, finishing after heat treatment should be reserved for correcting distorted surfaces, correcting errors due to growth (or shrinkage), touching up surface finishes, or bringing ultracritical dimensions into tolerance.

REVIEW QUESTIONS

37.1. Define "heat treatment."

37.2. What is the purpose of heat treating?

37.3. Name the class of alloys that accounts for nearly 90% of all heat-treating operations.

37.4. Explain the purpose of a TTT curve.

37.5. What is the effect of extensive machining operations upon a steel part in production?

37.6. List the steps involved in annealing and explain why the process is sometimes necessary for steel parts.

37.7. In what ways do normalizing and annealing differ?

37.8. Explain why it is often necessary to stress-relieve weldments or heavily machined parts.

37.9. What is the effect of cold working upon a metal workpiece?

37.10. Explain why precipitation hardening is considered an ideal process for hardening highly accurate parts.

37.11. What is the purpose of "quenching and tempering"?

37.12. Explain the basis for deciding which of the two tempering processes to use—Martempering or Austempering.

37.13. What are the advantages of carburizing?

37.14. Under what conditions would carbonitriding be selected in preference to carburizing?

606

37.15. Describe the process of "selectively carburizing."

37.16. What are the principal reasons for nitriding?

37.17. Why is only minimal machining recommended following nitriding?

37.18. How does the depth of case in induction hardening compare to the carburizing case?

37.19. Regarding quality control, what is the greatest potential concern when parts are induction-hardened?

37.20. Tell why it is necessary to remove the decarburized steel "skin" on stock before heat treating.

37.21. What precautions should heat treaters take when case-hardening parts with threads? Why?

37.22. Under what conditions would finishing after heat treatment be justified?

PROBLEMS

37.1. Explain the advantages of the following procedure. When extruding alloys that will be heat treated such as aluminum 6061, common practice is to extrude the slug in the 0 temper, solution treat the preform to the T-4 temper and then extrude to the final size. After sizing, the part can be aged to the T-6 temper if required.

37.2. Bearing surfaces on large machine tools or teeth on large gears are often flame hardened. (a) Describe the process of flame hardening and list the advantages. (b) What depth of hardness can be expected? (c) What would be the hardness at the surface?

37.3. Warping and cracking are serious problems to be encountered in nearly every hardening operation. List some important factors relating to production configuration and material which may influence these heat treating problems.

37.4. Explain the effect of carbon upon the hardenability of steel.

37.5. Explain the causes of work hardening.

37.6. In order to harden steel by heating and quenching, two conditions must be maintained. What are these conditions?

BIBLIOGRAPHY

*BITTENCE, J. C., "The Basics of Heat Treating," *Machine Design,* Part I, Jan. 24, 1974, and Part II, Feb. 7, 1974.

"Do's and Don'ts for Heat Treaters," *Production's Manufacturing Planbook,* Feb. 17, 1975, 126 pp.

"A Guide to Surface Hardening," *American Machinist,* Sept. 3, 1973, p. 53.

"Heat Treating," *Metals Handbook,* Vol. 2, *Heat Treating, Cleaning and Finishing,* Metals Park, Ohio: American Society for Metals, pp. 1–271.

*MISKA, K. H., "Engineer's Guide to Heat Treatment," *Materials Engineering,* Dec. 1971.

"Move to Electric Furnace Predicted," *American Machinist,* Nov. 12, 1973, p. 51.

Source Book on Heat Treating, ASM Engineering Bookshelf Series, Vols. 1 and 2, Metals Park, Ohio: American Society for Metals, 1975.

WILSON, R., *Metallurgy and Heat Treatment of Tool Steels,* Metals Park, Ohio: American Society for Metals, 1975.

*Abstracted with permission.

38

SURFACE FINISH
AND MICROFINISHING

Product designers constantly strive to design machinery that can run faster, last longer, and operate more precisely than ever. Modern development of high-speed machines has resulted in higher loadings and increased speeds of moving parts. Bearings, seals, shafts, machine ways, and gears, for example, must be accurate—both dimensionally and geometrically. Unfortunately, most manufacturing processes produce parts with surfaces that are either unsatisfactory from the standpoint of geometrical perfection or quality of surface texture.

As industry tries harder to approach perfection, interest has focused more closely than ever before on the microfinishing processes covered in this chapter—*honing, lapping,* and *superfinishing.* Each process was designed to generate a particular geometrical surface and to correct specific irregularities and so must be applied carefully to a given production sequence. Also, each process is a final operation in the machining sequence for a precision part and is usually preceded by conventional grinding. This chapter begins by explaining how industry controls and measures the precise degree of smoothness and roughness of a finished surface.

SURFACE FINISH

Control

In most cases, *surface finish control* starts in the drafting room. The designer has the responsibility of specifying a surface that will give the maximum performance and service life at the lowest cost. In selecting a required surface finish for a particular part, the designer must base his/her decision on past experience with similar parts, on field service data, or on engineering tests. Such factors as size and function of the parts, type of loading, speed and direction of movement, operating conditions, physical characteristics of various materials as they contact one another or whether they are subjected to stress reversals, type and amount of lubricant, contaminants, and temperature influence the choice.

There are two principal reasons for surface-finish control: (1) to reduce friction, and (2) to control wear. When a film of lubricant must be maintained between two moving parts, the surface irregularities must be small enough so they will not penetrate the oil film under the most severe operating conditions. Bearings, journals, cylinder bores, piston pins, bushings, pad bearings, helical and worm gears, seal surfaces, and machine ways are examples where this condition must be fulfilled.

Surface finish is also important to the wear service of certain parts that are subject to dry friction, such as machine-tool bits, threading dies, stamping dies, rolls, clutch plates, and brake drums.

Smooth finishes are essential on certain high-precision parts. In mechanisms such as injectors and high-pressure cylinders, smoothness and lack of waviness are essential to accuracy and pressure-retaining ability. Smooth finishes are also required on such items as micrometer anvils, gages, and gage blocks.

Often, surface finish must be controlled for the purpose of increasing the fatigue strength of highly stressed members which are subjected to load reversals. A smooth surface eliminates the sharp irregularities which are the greatest potential source of fatigue cracks.

Smoothness is often essential for eye appeal of the finished product. The surface finish on dies used for extruding and on those used in precision casting is carefully controlled to accomplish this objective.

For parts such as gears, surface-finish control may be necessary to ensure quiet operations. In other cases, however, where a boundary lubrication condition exists or where surfaces may not be compatible, as in two extremely hard surfaces running together, a slightly roughened surface will usually assist in lubrication.

A specific degree of surface roughness is also required in order to accommodate wear-in of certain parts. Most new moving parts do not attain a condition of complete lubrication as a result of imperfect geometry, running clearances, and thermal distortions. Therefore, the surfaces must *wear in* by a process of actual removal of metal. The surface finish must be a compromise between sufficient roughness for proper wear-in and sufficient smoothness for expected service life. Too smooth a surface will produce too slow an initial wear. In fact, the surfaces may never wear in, and improper clearances may result in local *hot spots* and high oil consumption. If the surface is too rough, the initial wear particles are large. These large particles act as abrasives, and wear continues at a high rate.

Measurement

To meet the requirements for effective control of surface quality under diversified conditions, a system for accurately describing a surface has been evolved. A standard, Surface Texture, ANSI B46.1-1962 (R-1971), published by the American National Standards Institute, deals with the height, width, and direction of surface irregularities, these being considered of practical importance in specific applications. The standard does not indicate specific surface roughness, smoothness, or waviness, or what type of lay is necessary for any specific purpose.

Most of the following definitions are condensed from this standard.

Surface: A *surface* of an object is confined by the boundary which separates that object from another object, substance, or space.

Microinch: A *microinch* is 1 millionth of an inch (0.000001 in.). For written specifications or reference to surface roughness requirements, microinches may be abbreviated as μin. or nm (nanometer) in S.I. units.

Surface Texture: This is the overall condition of a surface. It consists of repetitive or random deviations from the nominal surface which form the pattern of the surface. Surface texture includes roughness, waviness, lay, and flaws.

Roughness: This is the surface texture evidenced by the tiny nicks and gouges on a workpiece resulting from the action of the cutting tool employed, such as a lathe tool, milling cutter, or grinding wheel. Fine irregularities may also be generated by a casting process or by a shearing action. Roughness is sometimes referred to as a *primary* texture, while waviness (explained in a following paragraph) may be thought of as a *secondary* texture.

Roughness Height: This is rated as the arithmetical average deviation expressed in microinches or micrometers, measured normal to an imaginary center line, running through the roughness profile. Roughness height is schematically illustrated in Figure 38-1.

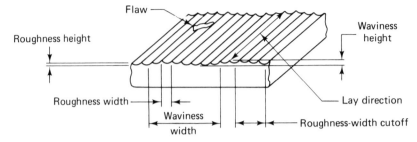

FIGURE 38-1 Surface characteristics (Courtesy, ANSI B46.1-1962)

Roughness Width: Roughness width is the distance parallel to the normal surface between successive peaks or ridges that constitute the predominant pattern of the roughness. Roughness width, also shown in Figure 38-1, is rated in inches.

Roughness-Width Cutoff: Roughness-width cutoff, shown in Figure 38-1, is also rated in inches. The cutoff width must always be greater than the roughness width in order to obtain the total roughness-height rating. Careful control must be exercised so that the primary surface effects of the production process (roughness) may be studied separately and without interference from the surface effects of the machine inefficiencies. It is for this reason that cutoff or sampling length facilities are built into most surface-quality measuring devices. Roughness-width cutoff values, according to both British and American standards, are 0.003 in. (0.08 mm), 0.010 in. (0.25 mm), 0.030 in. (0.76 mm), 0.100 in. (2.54 mm), and 1.0 in. (25.4 mm). Instruments with the cutoff feature generally are equipped with a choice of three or more cutoff values.

Arithmetic Average (AA): A close approximation of the *arithmetic average* roughness-height can be calculated from the profile chart of the surface, as shown in Figure 38-2. Averaging from a mean centerline may also be automatically performed by electronic instruments using appropriate circuitry through a meter or chart recorder. In a critical analysis of a surface, it is useless to state merely the AA value, because it tells little about the physical appearance of the surface.

Root Mean Square (rms): This method, also shown in Figure 38-2, was formerly an American standard. In 1955 it became obsolete, but it is still encountered occasionally. Its numerical value is about 11% higher than that of AA.

Roughness-Height Values: Figure 38-3 shows a range of typical surface *roughness-height values* obtained by common production methods. The cost of producing a surface increases as the roughness-height value increases.

612

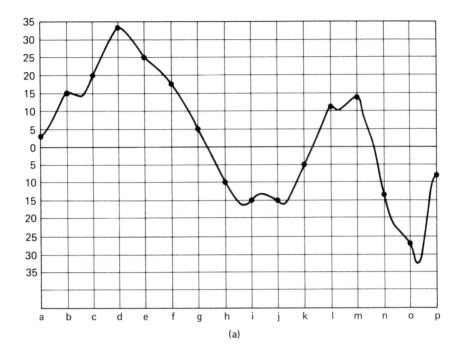

(a)

a	3	a^2	9
b	15	b^2	225
c	20	c^2	400
d	33	d^2	1089
e	25	e^2	625
f	18	f^2	324
g	5	g^2	25
h	10	h^2	100
i	15	i^2	225
j	15	j^2	225
k	5	k^2	25
l	11	l^2	121
m	14	m^2	196
n	13	n^2	169
o	27	o^2	729
p	8	p^2	64

TOTALS 234   4551

(b)

Arithmetical average = $\dfrac{237}{16}$ = 14.8 μ in.

Root-mean-square average = $\sqrt{\dfrac{4551}{16}}$ = 16.9 μ in.

(c)

FIGURE 38-2 Using a profile chart to calculate the approximate values of arithmetic average and root mean square average

Roughness Height Rating Micrometres, μm (microinches, μin)

Process	50 (2000)	25 (1000)	12.5 (500)	6.3 (250)	3.2 (125)	1.6 (63)	0.80 (32)	0.40 (16)	0.20 (8)	0.10 (4)	0.05 (2)	0.025 (1)	0.012 (0.5)

Flame cutting
Snagging
Sawing
Planing, Shaping

Drilling
Chemical milling
Elect. discharge mach.
Milling

Broaching
Reaming
Electron beam
Laser
Electro-chemical
Boring, turning
Barrel finishing

Electrolytic grinding
Roller burnishing
Grinding
Honing

Electro-polish
Polishing
Lapping
Superfinishing

Sand casting
Hot rolling
Forging
Perm. mold casting

Investment casting
Extruding
Cold rolling, drawing
Die casting

The ranges shown above are typical of the processes listed.

Higher or lower values may be obtained under special conditions.

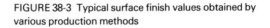

KEY ▨ Average application ▨ Less frequent application

FIGURE 38-3 Typical surface finish values obtained by various production methods

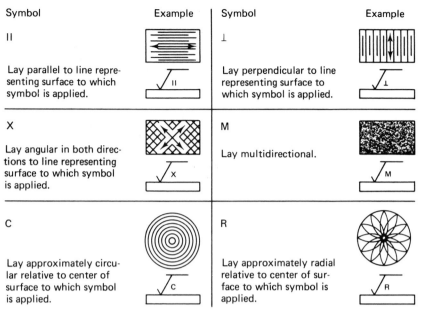

Symbol	Example	Symbol	Example
‖ Lay parallel to line representing surface to which symbol is applied.		⊥ Lay perpendicular to line representing surface to which symbol is applied.	
X Lay angular in both directions to line representing surface to which symbol is applied.		M Lay multidirectional.	
C Lay approximately circular relative to center of surface to which symbol is applied.		R Lay approximately radial relative to center of surface to which symbol is applied.	

FIGURE 38-4 Lay symbols (Courtesy, ANSI B46.1-1962)

Waviness: This is the usually widely-spaced component of surface texture and is generally of wider spacing than the roughness-width cutoff. Waviness may result from such factors as machine or work deflections, vibrations, chatter, heat treatment, or warping strains. Roughness may be considered as being superimposed on a wavy surface. Distances representing waviness width and waviness height are also shown in Figure 38-1.

Lay: This is the direction of the prominent surface pattern, ordinarily determined by the production method used. Lay direction is also shown in Figure 38-1. Lay symbols shall be specified as shown in Figure 38-4.

Flaws: *Flaws* are irregularities that occur at one place or at relatively infrequent or widely varying intervals in a surface. Flaws include such defects as cracks, blowholes, checks, ridges, and scratches. Unless otherwise specified, the effect of flaws shall not be included in the roughness-height measurements.

Surface Symbol: The *surface symbol,* shown in Figure 38-5, is used to designate the characteristics of surface texture on a drawing of a production part. The symbol is always placed in the standard upright position, as shown in Figure 38-6, never at an angle or upside down. The symbol is generally omitted on views of parts when the finish quality of a surface is not important. Generally speaking, the ideal finish is the roughest one that will do the job satisfactorily.

615

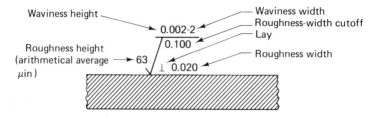

FIGURE 38-5 The surface symbol (Courtesy, ANSI B46.1-1962)

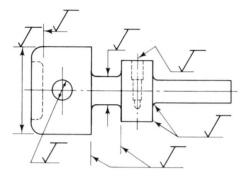

FIGURE 38-6 How to apply the surface symbol to representative surfaces on the drawing on a part (Courtesy, ANSI B46.1-1962)

Evaluation

There are three general methods by which the surface texture and the surface geometry may be explored and evaluated: electronic, optical, and visual or tactual.

Electronic: There are two types of *electronic* instruments which measure actual surface texture: *averaging* (or velocity type) and *profiling* (or displacement type). Averaging or tracer-type instruments employ a stylus that is drawn across the surface to be measured. The vertical motion of the tracer is amplified electrically and is impressed on a recorder to draw the profile of the surface or is fed into an averaging meter to give a number (AA) representing the roughness value of the surface. Figure 38-7 illustrates a profilometer, which is used in routine shop inspection applications to check the roughness of a surface. The only reading that such an instrument can provide is the average roughness height of a surface.

Profiling equipment is used principally in laboratories for research and development applications. Considerable skill is required to operate the equipment and to analyze and interpret the data.

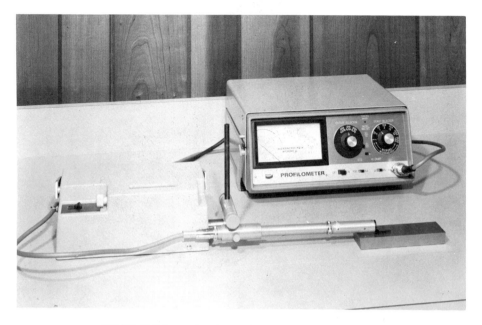

FIGURE 38-7 Inspecting the roughness of a surface with
a profilometer (Courtesy, Norton Company)

Optical: Optical or *area systems* use optical methods for surface evaluation. Equipment ranges from exploration of the surface with simple microscopes or three-dimensional microtopography to highly sophisticated techniques such as interferometry. Area systems inspect all the surface, not simply one line across it. The surface texture in this process is clearly distinguished from the surface geometry. Because there is no stylus, the surface is not mechanically contacted, and thus there can be no damage to the workpiece surface. Another important advantage of optical inspection methods is that the biasing effect of the stylus radius is eliminated.

Visual or Tactual: The *visual* or *tactual* is the simplest and most straightforward method of surface measurement. It is also the least accurate. Figure 38-8 shows a commercial set of master precision reference specimens with 15 replicated surfaces, ranging in roughness from 2 to 125 μin. in height. Comparators of this type are readily available with various surface-finish profiles representing high-fidelity duplicates of actual surfaces. A complete range in finish from 2 to 1000 μin. is available. The scales, used with or without a magnifier, are placed adjacent to the workpiece under examination and the surfaces are compared visibly or tactually by drawing the tip of the fingernail across each at right angles to the tool marks. The fingernail touch or "feel" will be the same when both finishes are identical.

FIGURE 38-8 A set of eight precision reference standards which are exact duplicates of actual machined surfaces (Courtesy, GAR Electroforming, Div. of MITE Corporation)

HONING

Expanded research into the process has enabled *honing* to compete with some of the other abrading processes used for fast stock removal or to obtain precision and fine finish. Honing was originally used only as a method for finishing and polishing. Recent engineering improvements in the basic honing machines, together with refinements in tooling, abrasives, and honing techniques have combined to greatly increase the utilization of honing as an important production process.

The most common honing applications relate to removing stock from internal and external cylindrical surfaces. Honing is a mechanical means of stock removal that uses spring-loaded abrasive stones as the cutting tool. The stones may be composed of aluminum oxide, silicon oxide, or in some cases diamond grains held together by a vitrified or organic bond. Each abrasive grit protruding through the bond contacts the part being honed and acts as a tiny cutter to shear off minute chips. Of particular advantage is the "self-sharpening" aspect of the abrasive grits. They work by breaking out of the bond when they become dull, allowing other sharp, fresh grains to contact the workpiece surface. Abrasive stones are spaced at regular intervals around the periphery of a mandrel. The number of stones in the toolhead,

618

their length and width, are determined by the size and nature of the work to be performed. Figure 38-9 illustrates a stub-type tool used for internal honing.

Unlike internal grinding, in which there is only a line contact of a boring tool, there is a relatively large area of the honing stones in contact with the workpiece. A slow, combined reciprocating and rotating action is transmitted to the stone by a driving shaft on the toolhead. Contact with the surface being finished is maintained through a positive but variable-pressure mechanism. Although the amount of stock removed by each cutting edge is small, the combined action of the numerous cutting edges working simultaneously provides a relatively fast and accurate means of stock removal.

Manual honing operations may be performed in the shop by rotating the tool in a drill press, lathe, portable electric drill, or by similar means. Honing in this way requires skilled workers who must either build special fixtures for holding and reciprocating or stroking the work or must duplicate this action manually. The results of portable honing in limited production or in salvage work of this kind cannot usually compare with the precise methods of production honing achieved on specialized equipment.

FIGURE 38-9 Multi-spindle high production automotive brake drum honing (Courtesy, Barnes Drill Co.)

FIGURE 38-10 A vertical-type honing machine with a 25 in. stroke used for honing a 4⅞ in. diameter cylinder liner (Courtesy, Barnes Drill Co.)

The type of microhoning machine required for a particular operation is principally determined by the length of the part to be honed. The shape of the part or availability of space facilities may also be important determining factors. Both horizontal and vertical honing machines (see Figure 38-10) are produced. Engine cylinders and other parts with short bores are generally honed on vertical machines, while long holes in parts, such as cannons or rifle barrels, are honed in horizontal machines.

620

Internal honing is by far the most common application of the process. The process is characterized by a unique motion in which the tool floats and aligns itself in the bore while rotating and reciprocating within the securely held workpiece. Occasionally when honing small bores, the conditions may be reversed so that the tool is rigidly held while the workpiece floats freely to seek its own center. The rotational and reciprocating speeds are independently adjustable and, in fact, may be set at an odd ratio to one another. In this way, the individual grains of the abrasive stick are free to act on the workpiece in an ever-changing path.

External cylindrical surfaces of piston rods, piston pins, spindles, shafts, and other similar parts may also be honed. In this process, a rigid sleeve with an internal cone is used to enclose the tool body with its abrasive stones. This hollow toolhead is placed over the cylindrical part to be honed. Finish size is achieved in vertical machines in the same manner as for internal tools. In horizontal applications, either the tool or the work may be rotated.

Honing produces geometrically accurate cylindrical forms by correcting various inaccuracies remaining from previous operations, such as high spots, chatter marks, out-of-roundness, taper, or deviations in axial straightness. Honing is an effective process for controlling both the I.D. and O.D. size of workpieces. Tolerances within 0.0001 in. (0.003 mm) are easily maintained. Bore-to-bore sizes may be held to a variation of between 0.0001 to 0.0003 in. (0.003 to 0.008 mm) on machines equipped with automatic size-control devices.

It should be emphasized that the honing tool follows the neutral axis on the workpiece initially established by a preceding operation. Honing can only improve the feature size and shape. It cannot correct errors of hole location and alignment nor can it improve concentricity with other diameters.

Honing produces a characteristic cross-hatch matte finish or lay pattern made up of "hills and valleys." This unique finish has a functional use for the bulk of parts processed by honing as a load-bearing surface. Each minute scratch serves as an oil reservoir for lubricants, thus diminishing the possibility of wear on a workpiece in service by minimizing friction and heat. The surface generated by honing is free of torn, smeared, or "burned" metal.

Almost any material can be honed. While cast iron and steel are materials most commonly honed, the process may also be used for finishing titanium, copper, bronze, carbides, glass, ceramics, and certain plastics. Workpiece sizes are limited by the available equipment, but a range in diameters from 0.060 in. (1.52 mm) up to 30 in. (762 mm) is most common. As a general rule, bores of almost any length-to-diameter ratio can be honed. Successful bores on workpieces as large as 36 in. (914.4 mm) in diameter have been honed.

While the honing process is not limited by workpiece material hardness, the stock removal rate is affected by hardness. The amount of stock removed for hard materials is less than that for softer materials. The actual rate of stock removal is dependent upon many variables, including physical characteristics of the work material other than hardness and other controllable factors within the honing process itself. It is not practical to state definite removal rates, owing to the influence of so many variables.

Operations such as boring, reaming, or grinding normally precede honing. The amount of stock allowance depends upon the accuracy of the previous operation in regard to surface condition and roundness and straightness of the bore. On high-production honing jobs, for example, it is consistent with good practice to leave not more than 0.002 to 0.005 in. (0.05 to 0.13 mm) per surface for stock removal. When sizing the bore of long lengths of seamless steel tubing, as much as 0.060 in. (1.52 mm) and more of stock can be economically honed, thus eliminating the necessity of a previous boring operation. In this example, the honing process would usually be divided into roughing and finishing operations in order to obtain the maximum rate of stock removal. The most economical range of stock removal for most parts is from 0.001 to 0.025 in. (0.03 to 0.64 mm).

It is possible to hone bores having undercuts, shoulders, oil grooves, keyways, ports, and cross holes. In general, blind-end holes are not practical, because of the difficulty in honing the sides flush with the bottom. Also, it is difficult to supply adequate honing fluid to the work area. One design technique that improves the finishing of blind-end holes is to include a specification for a small relief at the blind end. In some cases, a relief no larger than ⅛ in. (3.18 mm) in width will alleviate the situation. Tapered cylindrical surfaces may be honed using special equipment.

The degree of finish can be accurately controlled by varying the speed of rotation, changing the grit of the abrasive stones, and altering the coolant mixture. Any desired finish between 4 and 10 μin can be duplicated from part to part and from job to job. Because of the slow rate of stone travel across the workpiece and the slight amount of cutting pressure needed, not enough heat is generated to deform the workpiece.

LAPPING

The process of *lapping* is performed on workpieces by manual or by machine methods, principally to increase accuracy. Other important advantages which are obtained automatically as a part of the process are the correction of minor surface imperfections, improvement of surface finish, and achieving a close fit between mating surfaces. Lapping is a gentle, final operation commonly used to microfinish flat or cylindrical surfaces, but the process is also adaptable to spherical or specially formed surfaces.

The major difference between lapping and honing is in the type of tooling employed and in the abrading motions of the workpiece as it acts against the abrasive medium. All lapping methods are done at low speed, so slow, in fact, that there is no sparking. Lapping is not considered a stock removal process. Standard practice is to provide only 0.0005 in. (0.013 mm) of stock allowance for lapping, and preferably less.

Hand lapping of individual pieces can be done on flat or on external or internal cylindrical surfaces. Highly skilled operators are necessary to obtain consistent and accurate results. Figure 38-11 shows a hand-lapping operation. Hand lapping is commonly employed as a toolroom method of finishing carbide die parts and other exceptionally hard materials. For *flat work,* a serrated lapping block or plate charged with a fine-grain loose abrasive compound with a suitable oil- or water-base vehicle is used. Silicon carbide, fused alumina, or boron carbide are excellent abrasives for metal lapping. Most laps are made of soft cast iron or of a material softer than the workpiece.

The process of hand lapping consists of manually rubbing the work with an ever-changing motion over the accurately finished surface of the lapping plate. The rate of stock removal and the quality of the ultimate finish obtained depend primarily upon the pressure exerted by the operator. A soft metal ring lap may be used on *external cylindrical surfaces* to correct errors

FIGURE 38-11 Operator hand lapping a workpiece on a charged cast iron plate (Courtesy, Surface Finishes, Inc.)

of roundness and taper on plug gages and other precision parts. A tolerance of 0.00005 in. (0.0013 mm) may be achieved. The workpiece is mounted on a lathe and the adjustable ring is carefully slid back and forth along the length of the rotating workpiece. Size adjustments are made as needed to the lap as the operation progresses. The lapping compound is periodically fed between the lap and the workpiece through a slot in the ring. *Internal cylindrical surfaces* on workpieces may be finished by rotating a lap in a lathe or drill press and sliding the work back and forth over the part at a relatively rapid rate. Various designs of commercially available laps are used, including adjustable types with replaceable perforated sleeves or solid laps which are grooved with a right- and a left-hand helix along their length. The loose abrasive compound is stored in the holes or grooves and is periodically released to the workpiece surface.

Today's versatile lapping machines can consistently turn out precision work at an acceptable production rate. Mechanical lapping processes include a number of machines and methods, each characterized by relatively light work pressures, moderate speeds, and the generation of little or no heat during the process. Standard machines can handle multiple parts in sizes from 0.060 to 32 in. (1.52 to 812.8 mm) in diameter.

Single-lap machines are used to process single or multiple parts when only one surface needs lapping. Lapping can be performed either with cast iron laps using loose abrasives or with bonded abrasive wheel laps, as shown in Figure 38-12. The work may be manually held against the lap, or workholders may be used to loosely hold the work and guide it into the proper motion. Skilled operators using single-lap machines can often produce surfaces on parts to a flatness within 0.0005 to 0.00001 in. (0.013 to 0.0003 mm). Parallelism cannot be held on workpieces processed on single-lap machines.

One model of a single-lapping machine will handle 800 ½-in. (12.7 mm)-diameter parts per load of three $9^{11}\!/_{16}$ in. (246.06 mm) diameter parts at a time. It is commonly used to produce precision finishes on components used in tool, gage, and die manufacture or for valve seats, compressor disks, and so on. A loose abrasive grain mixed with a lubricant is used. The parts are confined within cast iron conditioning rings, as shown in Figure 38-13. This process can produce flat parallel surfaces to tolerances within 0.00001 in. (0.0003 mm). Surfaces on cylindrical parts or parts on which two parallel surfaces must be lapped are done on vertical spindle machines equipped with two laps.

Another type of lapping machine used principally for commercial production work can lap soft materials as well as hard materials. As in the previous lapping method, parts are also held and guided in workholders but are lapped between two bonded abrasive wheels. This method can produce lapped surfaces on parts typically parallel to within 0.005 in. (0.013 mm) and to sizes within a tolerance of 0.0001 in. (0.003 mm). The practice of

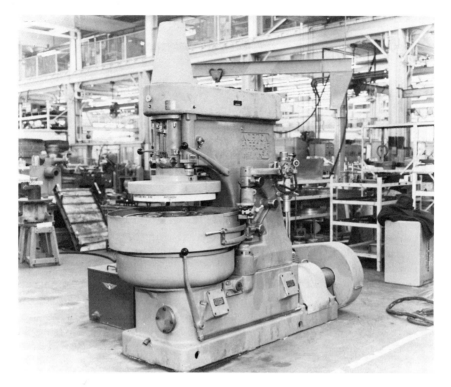

FIGURE 38-12 This high production bonded abrasive lapping machine performs parallel face lapping (Courtesy, The Warner & Swazey Co.)

FIGURE 38-13 A single lap machine for production parts up to 9 11/16 in. diameter. Note the conditioning rings used to contain the various parts as they are lapped on the lap plate (Courtesy, Crane Packing Company)

lapping soft materials with loose abrasives is not generally recommended because of the tendency of the grit to lodge in the surface of the part being lapped.

High-production parts, such as pistons, valve tappets, shafts of all kinds, and various injectors and plungers, can be lapped by a centerless process. Centerless lapping machines operate much in the same way as centerless grinders. The wheels used generally have a 22 in. (558.8 mm) face width, to provide adequate contact with the work. In order to achieve the finest finish obtainable, about 1 to 2 μin., a sequence of at least three successive lapping operations is necessary. A series of progressively finer bonded abrasive wheels must be used. Tolerances up to 0.000025 in. (0.0006 mm) for roundness and straightness and 0.00005 in. (0.0013 mm) for size can be held.

Spherical surfaces can be lapped by using any of several well-established mechanical methods. In one method, concave or convex laps shaped to correspond to the surface to be finished are used in a machine similar to a drill press. A rotating crank, held in the chuck, imparts a unique gyroscopic motion to the lap as it contacts and slides over the stationary workpiece. There are two economical methods employed for lapping balls for ball bearings. Both produce very close dimensional tolerances to extremely smooth finishes. Special-purpose high-production machines called the "multigroove lapper" and the "single-groove lapper" are used for this process.

Matched-piece lapping is a "wearing-in" method by which pairs of mating parts may be equalized by rubbing one against the other. Loose abrasive compound mixed with a lubricant is applied to the contact surfaces of various parts, such as gears and worms, plungers and cylinders, valves and valve seats, pin and hole sets, shafts and bearings, and tooling components. In this way, unwanted small irregularities in surface geometry caused by heat treatment, tool chatter, cutter marks, or scratches left by grinding or honing may be eliminated, thus increasing the service life of moving parts. Matched-piece lapping is commonly employed by manufacturers as a method to form tight, leakproof seals on engine heads and blocks or on pump components. Mating parts processed in this manner are stocked in pairs.

Flat surfaces on some individual parts are finished by manipulating them by hand using fine-grit abrasive paper. A rotary lapping plate may be used or the operation may simply consist of manually rubbing the workpiece against a sheet of abrasive paper that is placed on a flat surface. The main purpose is to produce a bright reflective surface generally on softer materials. No significant amount of stock is removed.

Practically all materials may be lapped, including steels, stainless steel, cast iron, brass, bronze, aluminum, magnesium, plastics, quartz, glass, and ceramics. Figure 38-14 illustrates a wide variety of lapped parts.

FIGURE 38-14 Typical examples of lapped parts
(Courtesy, Crane Packing Company)

SUPERFINISHING

Superfinishing is a proprietary name given to a microfinishing process that produces a controlled surface condition on parts which is unobtainable by any other method. Superfinishing produces the ultimate in the refinement of metal surfaces. The early principles of superfinishing were developed by Chrysler Corporation engineers early in 1934 for the improvement of automobile wheel bearings. Superfinishing is an abrading process in which the cutting medium for cylindrical work is a loosely bonded abrasive stick or stone. An abrasive cup wheel is used for flat or spherical work. The process consists of removing fragmented or smear metal from the surface of a dimensionally finished part formed by a previous operation, notably by turning or grinding, but possibly by honing or lapping. Dimensional changes are principally limited to the removal of high spots. Superfinished parts are bright and reflective with an undisturbed crystalline structure.

When two mating parts are in service, high spots or fragmented metal particles extending from one or both contact surfaces wear down rapidly and soon result in a loose fit. These minute surface projections are also objectionable, because they tend to penetrate the thin film of lubricant used to separate the moving parts in service, thus contributing to wear. The major purpose of superfinishing is to produce a surface on a workpiece capable of sustaining an even distribution of a load by improving the geo-metrical accuracy. The wear life of parts microfinished to maximum smoothness is extended considerably.

Superfinishing is not primarily a sizing operation. Although as much as 0.001 to 0.002 in. (0.03 to 0.05 mm) of stock may be efficiently removed in some production applications, the process becomes most economical when stock removal can be limited to not more than 0.0002 in. (0.005 mm). Owing to the many variables, it is not practical to state absolute values for stock removal rates. There are, however, tables available in some manu-facturers' catalogs listing approximate times for production superfinishing operations, depending upon such factors as workpiece material, required surface finish, and number of spindles on machines.

Figure 38-15 illustrates an automatic superfinisher for cylindrical sur-faces. During superfinishing operations a flood of low-viscosity oil, often a light mineral oil, is applied in the work area between the abrasive

FIGURE 38-15 This superfinisher model is an automatic modular machine for producing con-trolled surface finishes on short-type or cylindrical parts. The machine has an automatic load-unload system. Work can be held between centers or with a chuck and tailstock (Courtesy, Giddings & Lewis-Bickford Machine Company)

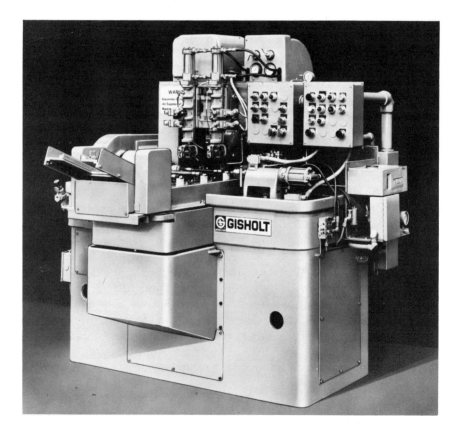

stones and the workpiece. The lubricant–coolant carries away the abraded particles and keeps the work at a uniformly cool temperature.

In this process, there is a large area of abrasive stone in contact with the work. On cylindrical surfaces, for example, the width of the abrasive stone may be two-thirds of the diameter of the part to be finished and often the same length. Only a relatively short time is required to complete surface improvement, since most of the finishing involves removal of the minute projections. Parts may be superfinished to a smoothness of 3 μ in. as rapidly as 15 to 50 seconds, depending upon the condition of the original surface. However, an additional 2 or 3 minutes may be required to develop sufficient cutting action to obtain an improved surface finish of 2 μin. or better.

Stones are virtually "self-dressing." Aluminum oxide is used principally for superfinishing steels and silicon carbide abrasives are generally used for cast iron and nonferrous metals. Vitrified bonded stones are most often used in preference to the slower cutting action associated with shellac or resin bonds.

Figure 38-16 illustrates the multidirectional motions required for superfinishing cylindrical work. The abrasive stone is applied with light but constant pressure and at the same time reciprocates a distance of usually $\frac{3}{16}$ in. (4.763 mm) at a rate up to 700 complete cycles per minute. On short workpieces, where the stone is the same length as the work surface, traversing the stone is not necessary. The stone is originally dressed to the radius of the work surface. After the stone has been dressed, it is not dressed again, since the workpiece surface will dress the stone to the desired "master" contour during the finishing operation. In use, after the initial roughness of the unfinished workpiece is cut away, the stone becomes dull and glazed until a condition is reached where the viscosity of the lubricant will almost completely separate the stone and the work. At this point, the cutting ceases. Tapered work, both internal and external, may be superfinished using slight modifications of this process.

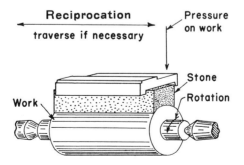

FIGURE 38-16 The motions of superfinishing for cylindrical work

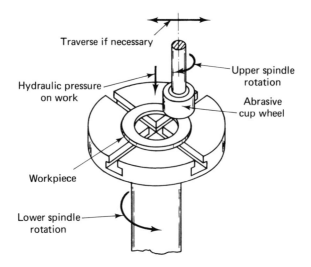

Traverse if necessary

Upper spindle rotation

Hydraulic pressure on work

Abrasive cup wheel

Workpiece

Lower spindle rotation

FIGURE 38-17 The motions of superfinishing for flat work

Both flat and spherical operations may be performed on the same type of superfinishing machine using vertically opposed upper and lower spindles. Figure 38-17 illustrates the motions for a typical setup. The upper spindle has a spring or hydraulically loaded quill on which a cup-shaped stone is mounted. The lower spindle contains a circular table which carries the work. As the lower spindle and the work revolve, the end face of the cup-shaped stone in the rotating upper spindle is brought into contact with the work. Some degree of offset is given the cup stone with relation to the work position, so that the path of any one grit is not often, if ever, repeated. A portion of the diameter of the cup wheel overhangs and cuts free of the workpiece surface, creating a self-dressing action for the wheel face. When the ends of the two spindles are exactly parallel, the combined effort of the revolutions of both stone and work results in a very true, flat surface, often within 0.0002 in. (0.005 mm) of perfect flatness. A spherical shape may be generated by adjusting the upper spindle to some angle with the lower. No oscillating motion is used.

There are five basic factors that can affect the results obtained by the superfinishing process:

1. The grit size and grade of the abrasive stone.
2. The surface speed of the workpiece (about 50 fpm).
3. The pressure of the stone (from 10 to 40 psi or about 0.7 to 2.8 kg/cm^2).
4. The viscosity of the stone lubricant.
5. The reciprocation speed.

Typical work applications for cylindrical shapes include calender rolls in paper mills, computer memory drums, sewing machine parts, and a vast number of automotive applications on components, such as cylinders, brake drums, bearings, pistons, piston rods and pins, axle shafts, clutch

630

plates, tappet bodies, guide pins, and over 50 transmission parts. Figure 38-18 shows some typical superfinished cylindrical parts. There are a wide variety of product examples on which one or more flat surfaces have been superfinished. Such parts include seal faces on pump housings, packing glands, thrust collars, disk and hub assemblies for turbine converters, and bearing races. One typical example of a surface on which a part is spherically superfinished is the end of valve lifter bodies.

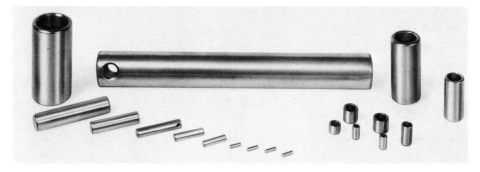

FIGURE 38-18 Examples of superfinished parts

REVIEW QUESTIONS

38.1. Give two principal reasons for surface-finish control.

38.2. Explain why a smooth surface finish on a workpiece can often ensure against the development of fatigue cracks.

38.3. Why would a smooth finish on an extruding die result in an extrusion with a corresponding smooth finish?

38.4. Explain the term "wearing in."

38.5. Referring to Figure 38-3, compare the range (average application) of surface-roughness-height values of (a) drilling, grinding, and polishing (b) sand casting, forging, and die casting.

38.6. Give the term associated with surface texture which denotes the marks caused on a workpiece by tool chatter.

38.7. Explain why visual or tactual methods of surface measurement are less accurate than electronic or optical methods.

38.8. Briefly explain the process of honing.

38.9. What is the purpose of honing?

38.10. Explain why honing cannot be used as a method to improve errors of hole location or alignment on a workpiece.

38.11. Explain how the characteristic matte finish produced by honing may be used to an advantage on some parts.

38.12. Prepare a sketch that illustrates how a small relief at the end of a blind-end hole can facilitate honing conditions.

38.13. What is the chief purpose of lapping?

38.14. Explain the difference between the type of abrasive cutting tool used for honing and that used for lapping.

38.15. What is the chief purpose of superfinishing?

38.16. Which process—honing, lapping, and superfinishing—is classified as a stock removal or sizing operation?

38.17. Explain why, in superfinishing, there is no need to redress a stone once it has been dressed to the required workpiece configuration.

38.18. Compare the relative surface finish obtained in honing, lapping, and superfinishing.

BIBLIOGRAPHY

ARMAREGO, E. J. A. and R. H. BROWN, *The Machining of Metals,* Englewood Cliffs, N.J.: Prentice-Hall, Inc., 1969, pp. 378–381.

BROADSTON, J. R., *Tolerance vs. Surface Quality: The Case for a Change,* ME72DE015, Dearborn, Mich.: Society of Manufacturing Engineers, 1972.

*BUSCH, T., "Let's Clean Up the Mess in Surface Measurement," *The Tool and Manufacturing Engineer,* Jan. 1968.

DOYLE, L. E., *Manufacturing Processes and Materials for Engineers,* Englewood Cliffs, N.J.: Prentice-Hall, Inc., 1961, pp. 681-690.

HINES, C. R., *Machine Tools and Processes for Engineers,* New York: McGraw-Hill Book Company, 1971, pp. 425–477.

"Honing," *Metals Handbook,* Vol. 3, *Machining,* 8th edition, Metals Park, Ohio: American Society for Metals, pp. 288–297.

"Lapping," *Metals Handbook,* Vol. 3, *Machining,* 8th edition, Metals Park, Ohio: American Society for Metals, pp. 298–310.

*MCCONNELL, B. R., "Hone to Size—Superfinish to Shape," *The Tool and Manufacturing Engineer,* Nov. 1966.

"Surface Finish and Surface Integrity," *Machining Data Handbook,* Second edition, Dearborn, Mich.: Society of Manufacturing Engineers, 1972, pp. 795-848.

"What Cost Surface Roughness?" *American Machinist,* Nov. 26, 1973, p. 39.

*Abstracted with permission.

39

SURFACE-FINISHING PROCESSES

Mass surface-finishing processes include *barrel* finishing, *vibratory finishing, spindle finishing,* and *orbital* or *centrifugal finishing.* Although not strictly considered mass-finishing processes because in most cases the parts are hand-held, other important surface-finishing processes explained in this chapter include *abrasive belt finishing, polishing,* and *buffing.*

MASS-FINISHING

Precisely controlled results can be consistently obtained by certain modifications of the mechanical grinding and honing processes. Each of the mass-finishing processes to follow have combined to make hand-finishing operations such as deburring, descaling, radii forming, and general surface refinement obsolete.

In addition to other similarities, these mass-finishing processes each use a mixture of abrasive grain media, principally aluminum oxide or silicon carbide, together with special compounds and water. In barrel finishing, the media may consist of random-shaped abrasive grains or nuggets in coarse sizes ranging from a grit size of 00 (2 in. or 50.8 mm) down to 20. When used as cutting additives, as grain on a "carrier" such as soft steel, the grit size may vary from 80 to 600. The size of the medium selected for a given

633

job is determined by the nature of the work to be performed, the work-piece material, and the type of machine. Other media used in barrel finishing are steel balls or slugs, granite, gravel, sand, or sawdust. Compounds are added to keep the media functioning efficiently or to vary the cutting properties and also to prevent rusting and pitting and other forms of corrosion on the workpiece.

Although not generally classified with mass-finishing processes, other methods, such as abrasive belt finishing, pressure blasting, polishing, buffing, and power brushing, are also relatively high-speed processes and, in most cases, are readily adaptable to semiautomatic production finishing machines.

Barrel Finishing

Some authorities make a distinction between *barrel finishing* and *tumbling,* in that the first process denotes a more precise operation. This is probably true, although common practice makes no such difference.

A typical barrel-finishing machine is shown in Figure 39-1. This model is a horizontally mounted eight-sided unit with a watertight loading door.

FIGURE 39-1 A barrel finishing machine with a canvas unloading boot and portable screening unit (Courtesy, Queen Products Division-King Seeley Thermos Co.)

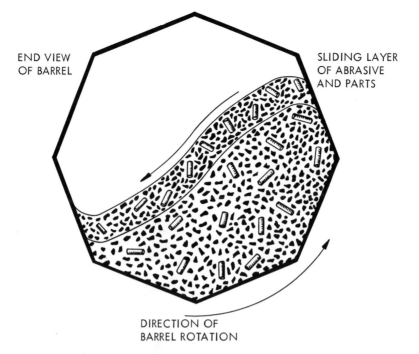

END VIEW
OF BARREL

SLIDING LAYER
OF ABRASIVE
AND PARTS

DIRECTION OF
BARREL ROTATION

FIGURE 39-2 The basic action of barrel finishing is a sliding movement of the upper layer of the work load as the barrel rotates (Courtesy, Norton Company)

The barrel is loaded within 50 to 60% of its rated capacity with parts, media, and compound. As the barrel turns, parts and media are carried up the side until gravity causes them to slide down again. This action is schematically illustrated in Figure 39-2. All the work is performed during this slide. In this brief description are most of the parameters for the operation of the barrel. Barrel finishing is relatively slow. During the time that a part of the load is being carried up to the start of the slide, no work is being performed. The barrel can be only partially filled; if it were full, there would be no work at all.

The action also dictates the top speed of the barrel (i.e., it is going too fast when the work begins to cascade, with consequent nicking and marring of the work). The optimum operating condition is achieved when the barrel is rotated at a speed just below the cascade point. As a consequence, tumbling operations are relatively slow. Much of the time the machines operate without supervision. For instance, it is quite feasible to load the barrel shortly before the finish of the workday, to run during a period of time when there is at most a watchman on the premises, and to be shut off at a predetermined time, either by an automatic timer or by a janitor or other employee.

Obviously, it is not possible to examine work in process in a barrel without first stopping the barrel. Once a time cycle has been set up, however, this requirement does not cause any problems.

635

The barrel tends to abrade edges and exposed surfaces much more than recesses and the insides of bores. Large corner radii can be formed on parts without difficulty, and large burrs can be removed without rollover. A distinct finish pattern is developed on barrel-finished workpieces caused by the sliding–rubbing action of the media, as shown on the parts illustrated in Figure 39-3. Since the barrel is closed, it is possible to build up thick suds consisting of the water and the mixture, which facilitates very high finishes, as for plating.

FIGURE 39-3 Three parts shown before and after barrel finishing (Courtesy, Norton Company)

Tumbling barrels are relatively uncomplicated machines, easy to repair and needing little maintenance. For a given capacity, compared with a vibrator (to be explained later), the initial cost is about one-third, and the power consumption is about one-half. On the other hand, the cycles are much longer.

The barrel is, of course, a batch-type operation; it is usually not compatible with in-line processing. However, this is not a true gage of its productivity. The capacity of the barrel in comparison to the size of the parts being processed can make it a very productive unit. Barrels come in a range of sizes from a fraction of a cubic foot to as large as 40 ft³ (1.3 m³). Rotational speeds of barrels range from 50 to 200 fpm. Low surface speeds are usually employed for burnishing, while faster surface speeds are used for heavier stock removal operations.

One variation of the basic barrel finishing technique is the compartmented barrel, actually a series of smaller barrels mounted on a common pair of rolling rails. As long as the parts take the same general range of rotation, this is a convenient means of finishing small lots of parts. It is also possible to have a compartmented barrel rotating on a single axle.

Slide honing, with small medium-size barrels mounted on a common frame, is another variation of the same principle and achieves similar results. It is particularly advantageous for long, slender, and fragile parts.

Vibratory Finishing

Vibratory finishing in its simplest denominator is the finishing of parts in a tub whose vibrating motion creates an abrasive action in the loose abrasive. The action removes burrs, rounds, corners, finishes bores and other concealed surfaces, or accomplishes the other purposes of loose-grain finishing. It is faster and more aggressive than barrel finishing; the cycles are in terms of minutes rather than hours. In fact, one authority refers to it as a "coarse grinding action," although there are those who feel that it will do just as fine a finishing job, although in a different surface pattern, as the barrel.

Figure 39-4 shows one model of a vibratory finishing machine which has an open-top U-shaped tub which tilts 10° for discharge of media and processed parts. The basic finishing action of this machine is shown schematically in Figure 39-5. Since the tub remains upright at all times, it may be loaded to about 90% of capacity. As it vibrates, the mass of media, parts, and compound usually move in an elliptical path, the plane of which depends upon the form of the tub. The abrasive is in constant motion against the work, deburring, rounding corners, and working on all the surfaces to which it has access.

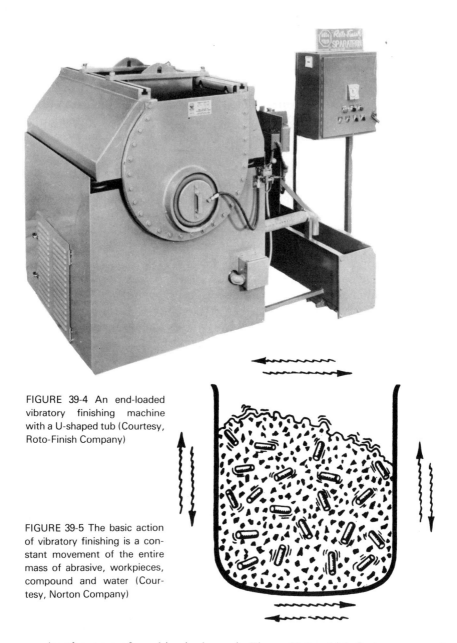

FIGURE 39-4 An end-loaded vibratory finishing machine with a U-shaped tub (Courtesy, Roto-Finish Company)

FIGURE 39-5 The basic action of vibratory finishing is a constant movement of the entire mass of abrasive, workpieces, compound and water (Courtesy, Norton Company)

Another type of machine is shown in Figure 39-6, which features a round bowl. Parts travel in a circular path and tend to space themselves at regular intervals. The vibratory action is rapid and constant. The tub on most vibratory machines is mounted on springs and is vibrated by eccentric or magnetic mechanisms.

In contrast to the barrel, which produces a relatively long scratch pattern, the vibrator produces a short choppy pattern. The work, of course, can be examined at any time simply by picking it up out of the tub.

The vibrating mechanism of the tub, although not a complicated piece of machinery, is subject to a greater degree of wear than the simple barrel. On the other hand, the vibrator lends itself quite well to in-line processing, with parts moving into the vibrator from another process, and then on to another.

In the original vibrators, the entire contents of the load were dumped at the end of each run. This has been changed now, and parts are readily separated from media and compound, often by so simple a mechanism as an appropriately placed screen of proper size, which allows the parts to be vibrated out of the mass while the media falls back through to continue its work.

Most applications for vibratory finishing are for relatively small parts. Anything that can be loaded into the machine, as a generalization, can be finished.

Vibratory finishing has an advantage over barrel finishing for parts that are likely to get tangled up; in the vibrator, parts and media tend to retain their distances. This is a continuous and unbroken path; hence, the parts do not as easily get entangled with each other. Barrel action results in a random sliding motion of the mass of the parts and media.

FIGURE 39-6 A vibratory finishing machine with a round bowl (Courtesy, Queen Products Division-King-Seeley Thermos Co.)

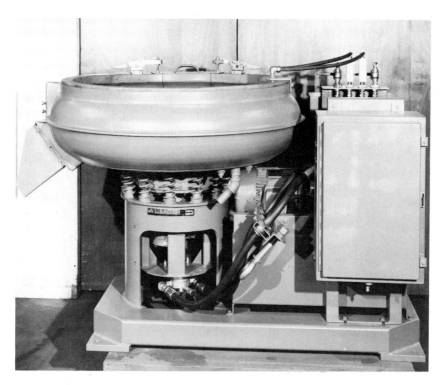

Still another variation is the fixtured vibrator for parts that cannot tolerate the possibility of impingement or for those that are quite large in relation to the size of the barrel. In either case, the work is held stationery while the media and compound vibrate through and around it. This variation reduces total finishing to roughly one-fourth of the conventional time.

The machines are available in a range of sizes from $\frac{1}{8}$ ft.^{3} (0.0038 m^3) to 70 ft.^{3} (2.1 m^3) capacity. Depending on the machine, they may be equipped with amplitude settings adjustable from $\frac{1}{64}$ in. to $\frac{1}{4}$ in. (0.397 to 6.35 mm) and with variable frequency controls ranging up to 3600 vpm (vibrations per minute).

Spindle Finishing

Spindle finishing is a loose-grain process that has been termed the "form-fitting grinding wheel." The work is chucked on one or more spindles and lowered into a tub containing an abrasive grain. The spindles with the attached parts are slowly rotated to expose all surfaces to a high-velocity abrasive stream. During processing, the tub spins up to 1200 fpm in a direction opposite to the rotation of the spindle. Control of the finishing process is achieved through variation of the following major elements of the cycle:

1. Speed of rotation of the tub.
2. Speed of rotation of the part on the spindle.
3. Depth of submersion of the part.
4. Angle of the spindle.
5. Length of the cycle.
6. Type and size of abrasive.

These elements are all interrelated. If the tub speed is too fast, for example, the grain will have a peening action rather than a cutting action and will roll over the burrs rather than cutting them and rounding the edges. There is a definite relationship between tub speed and work speed, although it is very likely to be different for different types of parts. This constant, from such evidence as is available, varies more by part geometry than by size, but is something to be worked out by trial and error rather than something that can be derived from a table. The part must be covered by abrasive at all times. The abrasive is thrown up in the shape of a wave by the spindle. The depression back of the wave is called the *cavitation.* The higher the tub speed, the deeper the cavitation, and consequently the deeper the part must be submerged.

The spindle angle determines the degree to which the interior of the part is worked on. When vertical, it confines the action to the outside of the part. The more acute the angle, the more action there is on the inside of the part. The effect of length of cycle upon the workpiece is obvious.

The abrasive used in this operation is aluminum oxide grain wetted down with water, and with charges of detergent periodically pumped in. The water flows onto the top of the tub in a steady stream and out the bottom in the same manner. Its only purpose is to keep the grain wet. The detergent softens the water for more effective flushing of the sludge and the spent abrasive.

This process is practical for use on fragile parts ordinarily fixtured to prevent impingement, and where close control of deburring and edge breaking is imperative. Part geometry is not a problem, as long as the part can be chucked. Figure 39-7 shows two production parts as they appear before and after spindle finishing.

(a)

(b)

FIGURE 39-7 Two parts shown before and after spindle finishing: (a) idler gear finished in 2 minutes (Courtesy, Queen Products Division—King-Seeley Thermos Co.); (b) cam finished in 10 seconds (Courtesy, Queen Products Division—King-Seeley Thermos Co.)

641

PROCESS SELECTION FACTORS

Product Applications: Stampings, castings, forgings, screw machine products, and other machined parts in a variety of sizes, shapes, and weights can be mass-finished. Delicate or large parts may be barrel-finished in specially designed fixture-type equipment.

Cost: Most mass-finishing processes require relatively low-cost equipment when compared to grinding operations, for example. Production costs may often be reduced by mass-finishing processes. By changing from wheel buffing to barrel finishing, for example, a manufacturer of automobile hardware found that finishing costs were reduced over 90%. Other substantial cost savings have been reported because of fewer part rejects as well as in labor savings made possible due to a reduction in hand-finishing operations.

Work Volume: Equipment is now available which is replacing time and labor-consuming batch-type operations. Higher levels of productivity are now possible through the use of automated loading and unloading systems and in improved materials-handling systems.

Workpiece Materials: Practically all kinds of materials in soft or hardened condition can be mass-finished, including ferrous and nonferrous metals, plastics, rubber, metal, and wood.

Process Effect on Workpiece: Recent technological developments make mass-finishing methods feasible on high-precision machine parts without changing their dimensional characteristics.

Types of Finishing Operations: Perhaps the most common type of mass-finishing operations pertain to deburring fins, flash, and other sharp edges. Other important applications are in forming corner radii, blending parting lines on castings or forgings, or in removing tool and chatter marks on machined parts. Burnishing, descaling, and degreasing also account for a large proportion of mass-finishing work.

Porosity: Porosity of metal parts is reduced by barrel finishing, which provides better surfaces on workpieces preparatory to plating.

Workpiece Uniformity: Uniformity is virtually impossible to obtain by hand-finishing methods. The fine cutting action of mass-finishing consistently makes the parts completely uniform.

Workpiece Finish: The nature of the final finish on a part is affected by the condition of the prior finish (i.e., a cast surface or milled surface). As an example, one manufacturer reported an improvement in surface roughness of 6 μin. was obtained on the exterior of steel bushings, which measured 10 μin. before finishing. The superior finish was obtained after tumbling for 3 hours.

Corner Radii: A producer of aircraft engine parts reports that mass-finished parts are consistently obtained with corner radii of 0.008 to 0.010 in. (0.20 to 0.25 mm).

Work-Hardening Effect: The tumbling action of the parts in contact with the media in the barrel combine to impart an increased hardness to brass, steel, and other metal workpieces.

ABRASIVE BELT FINISHING

The Process

Even when the process was principally limited to flat grinding, abrasive belt finishing was an important, low-cost, relatively fast finishing process. Now with the development of better resin-bonded belts having increased flexibility and improved joints coupled with improvements in machinery, the versatility of the process has reached even greater acceptance on the production line. *Abrasive belt finishing* is now an important size and surfacing process for the precision finishing of flat, concave, and convex surfaces.

Stock removal is accomplished by the abrasive grains on a moving belt as they continuously pass over the work area. In this way, burrs, high spots, the coarse texture on cast parts, parting lines, or machining marks are refined or totally removed from a workpiece.

Most production belt-grinding and polishing operations use a lubricant that is applied at frequent intervals. The function of a lubricant is to serve as a coolant and to assist in stock removal by preventing the belt from clogging with grinding debris. Mineral oil and emulsions of soluble oil and water are commonly used as lubricant. Water is frequently used as a coolant in wet-abrasive belt grinding. Aluminum alloys are finished dry. Belt life is extended by using a series of coated belts of varying degrees of abrasive fineness, usually roughing, polishing, and fine polishing. The final surface produced is superior to that obtained by milling or turning. In spite of the term "*fine polishing,*" however, the resulting surface on the workpiece is always characterized by fine scratches. Any abrasive belt or setup wheel coated with abrasive grain will leave grit lines on the workpiece regardless of the grit size employed.

Machines

Most belt-finishing machines consist of a motor-driven contact wheel and idler arrangement over which an endless coated-abrasive-tensioned belt rides.

Several belt-finishing setups are illustrated in Figure 39-8. *Flat finishing* requires a sturdy support platten as in Figure 39-8a. *Contoured finishing,* shown in Figure 39-8b, can be performed by rotating the part against a flat contact wheel, against a formed contact wheel, or offhand, entirely against the moving belt without support. Tubing and rolls may be finished in centerless abrasive belt machines, as in Figure 39-8c.

FIGURE 39-8 Variations of belt finishing set-ups: (a) flat finishing; (b) contoured finishing;

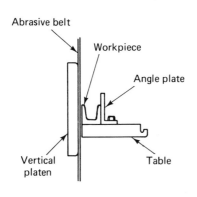
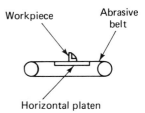

(a)

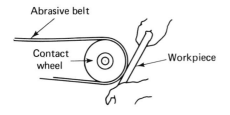
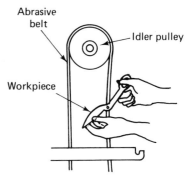

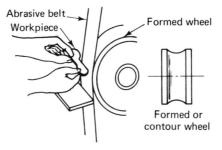

(b)

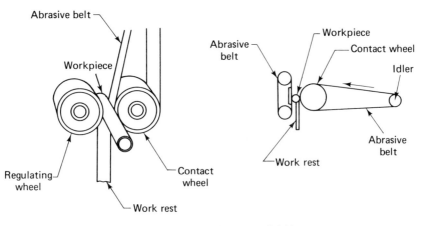

FIGURE 39-8(c) Center line finishing

Simple workholding fixtures that decrease the time lost in part handling are often used in place of hand-held methods. Fixtures are particularly advantageous for generating end radii on parts and for sizing and finishing flat surfaces in desired locations on certain parts. Fixtures are often used to simplify parts handling in repetitive operations on large production runs. Wide sheet and coil stock are finished by abrasive belt machines in a smooth and continuous motion.

Product Applications

Typical parts commonly processed by belt finishing include bicycle parts, golf-club heads, parts for fountain pens, jet-engine turbine blades, hand tools, including screwdrivers, hammerheads, hatchets, axes, and wrenches, and sole plates for electric irons, plumbing supplies, and so on. Belt finishing is often the process selected to remove welding beads and various other unwanted projections on machine parts.

Design Factors

Users report a tremendous advantage in coated abrasive belts over older, previously used finishing methods, particularly for the finishing of contoured parts. The rapid movement of the belt (up to 5500 fpm for heavy stock removal and 7500 fpm for polishing operations) makes possible an extremely high production rate. Also, new belts may be very quickly changed and substituted for worn belts on most machines. While it is possible to remove ⅛ in. (3.18 mm) stock and more on some of the heavy-duty machines, the least amount of cut returns the greatest production advantage in terms of finishing time and belt life.

As a general rule, the recommended stock allowance for sizing operations for ferrous materials should be about $\frac{1}{32}$ in. (0.794 mm) and up to $\frac{1}{16}$ in. (1.588 mm) for ferrous materials. Polishing operations require less stock

645

allowance, usually not more than 0.015 to 0.020 in. (0.38 to 0.51 mm). Under controlled conditions, flat surfaces on cast iron and soft steel parts may be sized to tolerances of ±0.002 in. (0.05 mm) on flatness and parallelism. In centerless grinding operations, a tolerance of ±0.0005 in. (0.013 mm) can be achieved. A surface finish of 10 μin. may be readily obtained using a fine-grit abrasive belt.

Possible Limitations

Unless careful control is exercised, sharp fins or projections will slit and tear belts, especially when used for heavy stock removal applications. Also, somewhat fragile edges of belts tend to crease and break down when surfaces are ground adjacent to shoulders. The process is not effective for finishing recesses, undercuts, or interior corners. Unlike the one-step finish that is produced by abrasive machining, for example, a fine finish in this process requires a progression of at least two belt changes and sometimes more. Finally, in offhand belt finishing, the process becomes somewhat slow, owing to repetitive work-handling motions.

POLISHING

The Process

Polishing, or *flexible grinding,* is an intermediate, dimensionless step in the formation of a finished surface. It is generally preceded by grinding with a solid abrasive wheel and followed by buffing. A polished surface is accomplished by the cutting action of millions of small abrasive grains adhering to an endless coated belt or flexible wheel as they wear away the metal. The complete polishing sequence usually involves several steps, first to remove the initial scratches and defects and then to gradually impart the final surface condition. Polishing operations may be minimized by constantly monitoring the preceding machining operations so as to produce the finest practical finish on the workpiece prior to polishing. Adequate surface preparation can dramatically reduce polishing costs. The general practice is to polish the work with a succession of wheels set up with different grain sizes. The progression is from a coarse abrasive grit to a fine grit, followed by buffing. The fact remains that the finer the surface finish, the longer it takes to accomplish.

Machines

Polishing may be done on an abrasive belt machine, shown in Figure 39-9, or on various kinds of bench or floor-stand-type machines equipped with a polishing or setup wheel, as shown in Figure 39-10. Because most abrasive belts and polishing wheels are flexible, they will readily conform to

FIGURE 39-9 Fine abrasive belts are used to polish edge surfaces of spoons to a smooth finish (Courtesy, International Silver Company)

FIGURE 39-10 Production polishing operations (Courtesy, Norton Company)

the shape of a contoured workpiece when necessary. Polishing operations on endless belt machines differ from grinding and sizing operations, in that an almost negligible amount of stock is actually removed. The quality of the resultant polished surface is principally regulated by the degree of fineness in the abrasive belt grit.

In some companies, the belt has largely displaced the setup wheel for polishing applications. Some of the reasons are: (1) a belt may have from three to four times the abrasive surface of a setup wheel; (2) various workpiece shapes can be accommodated without changing the contact wheel because the coated abrasive belt can flex when run on a pleated cloth wheel; and (3) setup wheel inventory can be practically eliminated. Another reason for using the belt in preference to the wheel is that due to its length, a belt cuts cooler. This becomes important when loading or clogging of the abrasive surface with metal is a serious factor. Setup wheels will continue to be used for complex shapes that belts cannot handle—narrow grooves are an example.

Polishing wheels are made of cloth or canvas glued or hand-sewn to make up the desired width, or from solid felt, leather, sheepskin, leather-covered wood, or rubber. The abrasive grain is chiefly aluminum oxide or silicon carbide. The abrasive may be impregnated or "charged" into certain types of wheels during bonding, or it may be mixed with a solid-stick-type carrying medium (called polishing compound) and applied to the face of the rotating wheel at suitable intervals. The required glossy finish is obtained only after the workpiece has been held in contact with several different wheels, each coated with a decreasing grit size. For most polishing operations, the wheel or belt surface speed is in the range 5000 to 7500 fpm.

Product Applications

In addition to the many product applications already listed under *belt finishing,* manufacturers of cutlery and small hand tools are particularly dependent upon polishing for finishing, as shown in Figure 39-11. No other process is said to give equally good results. Extensive polishing operations are normally required for internal work on tools and dies. Many companies regularly insert polishing operations into their normal production sequence for such work as "satin finishing," deburring, and for cleaning up irregularly shaped parts prior to plating or buffing.

Possible Limitations

In some cases, semiautomatic polishing machines may be substituted for some work to reduce handling time and to make the process more economical. Rotary polishing machines, for example, are high-production units that greatly speed up some polishing operations. On such machines, parts are

(a)

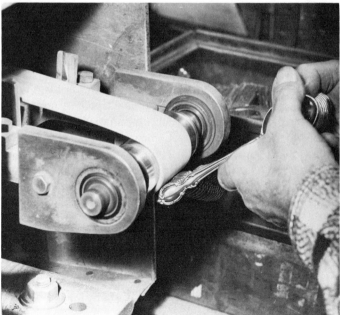
(b)

FIGURE 39-11 Production polishing operations (a) (Courtesy, Norton Company); (b) (Courtesy, International Silver Company)

held in fixtures mounted on a turntable. The parts are automatically rotated into contact with one or more revolving polishing wheels. In spite of such advances, however, labor costs tend to be high in comparison to other steps in processing certain parts.

BUFFING

The Process

Buffing generally follows polishing and is usually the final operation that is performed on a workpiece. In buffing, the rubbing action is more gentle than the vigorous and aggressive cutting action employed in polishing. Buffing removes negligible amounts of material. A buffed surface is formed in two distinct steps: (1) cutting down and (2) coloring. During the initial finishing stage of cutting down, minute scratches left by polishing and other surface irregularities are reduced or entirely eliminated. It is during the final stage of buffing (coloring) that the ultimate reflective, highly lustrous surface is produced.

Hand-held or offhand methods of producing a luster by bringing the workpiece into contact with a revolving wheel or coated abrasive belt, as shown in Figure 39-12, are virtually identical to those used in polishing. Various types of finishing machinery such as automatic straight-line or rotary conveyors are designed to incorporate both polishing and buffing operations.

Buffing wheels, called "buffs," are often fabricated into a number of piles from a series of individual fabric disks of various kinds. Some buffs are hardened and stiffened by special treatments to promote faster cutting action. Loose disks of sheepskin are used in buffing hardware and in highlighting fine coloring of both ferrous and nonferrous metals. Buffing-wheel speeds are in the range 6500 to 8000 fpm.

The principal abrasive used for buffing compounds on aluminum, copper, brass, and for zinc-alloy die castings is Tripoli. Aluminum oxide is commonly used for hard metals while chromium oxide is used to obtain the highest color buffing of stainless steel, chromium, and nickel plate. Rouge is the preferred compound for high coloring on copper, brass, gold, and silver. Crocus is used as the abrasive ingredient in coloring compounds employed in the cutlery trade. Thermosetting plastics such as Bakelite, Catelin, Beetle, and Plaskon may be buffed without difficulty. The procedure requires the use of a soft polishing wheel with a fine grade of greaseless compound to the roughened areas. The polished areas may then be buffed using pumice as the cutting agent. Moderate wheel speeds (4000 to 6000 fpm) are recommended. Thermoplastics and plastic laminates may also be buffed, but careful experimentation under conditions of low speeds and light pressures are strongly recommended.

FIGURE 39-12 A lusterous finish is applied to a spoon handle on a high-speed revolving buffing wheel (Courtesy, International Silver Company)

The buffing compounds may be manually applied to metal products in the same manner as is outlined for polishing (i.e., solid bar compound). The compound may also be supplied to the wheel face in liquid form by using a patented airless-spraying system. Liquid compounds are highly suitable for machine or automatic buffing.

Product Applications

The decorative mirror-like finish obtained by buffing is applied to a wide selection of metal products, including objects used on mobile homes, automobiles, motorcycles, boats, bicycles, as well as sporting items, tools, store fixtures, commercial and residential hardware, and household utensils and appliances. Buffing may be specified both prior to and following plating.

REVIEW QUESTIONS

39.1. What two abrasive grain medias are most commonly used in mass-finishing processes?

39.2. In barrel finishing, cutting efficiency is the greatest when the barrel is loaded within 50 to 60% of its rated capacity. Why?

39.3. What is the purpose of barrel finishing?

39.4. Which process is fastest—barrel finishing or vibratory finishing? Why?

39.5. Compare the surface texture produced by barrel finishing to that obtained in vibratory finishing.

39.6. In spindle finishing, how are the workpieces positioned?

39.7. Why is spindle finishing a particularly appropriate process to use for deburring *fragile* parts?

39.8. What procedure is used to extend the belt life in the belt-finishing process?

39.9. How would you describe the appearance of a workpiece that has been belt-finished?

39.10. Explain how the use of fixtures may be advantageous in belt finishing.

39.11. In what important ways do conventional belt finishing and polishing differ?

39.12. For what special applications are setup wheels particularly appropriate? Why?

39.13. What is the purpose of buffing?

39.14. What is the major difference between buffing and polishing?

39.15. Describe the two buffing steps—cutting down and coloring.

39.16. Explain why only moderate wheel speeds are recommended for buffing most types of plastics.

BIBLIOGRAPHY

"Abrasive Belt Grinding," *Metals Handbook,* Vol. 3, *Machining,* 8th edition, Metals Park, Ohio: American Society for Metals, pp. 277–279.

"Brushed Embossed Finish Produced in a Single Pass," *Modern Metals,* Nov. 1974, p. 69.

DOHERTY, C. B., "Surfaces Produced by Abrasive Blasting of Steel," *Materials Performance,* Nov. 1974, p. 12.

"Finishing: The Final Opportunity," *Production's Manufacturing Planbook,* Feb. 17, 1975, p. 141.

"Guide to Abrasive Belts," *Metal Progress,* June 1974, p. 223.

*McLAUGHLIN, J. K., "Abrasive Belt Finishing of Brass and Copper Parts," *Metalworking,* Aug. 1963.

"Mechanical Finishing," *Metals Handbook,* Vol. 2, *Heat Treating, Cleaning and Finishing,* 8th edition, Metals Park, Ohio: American Society for Metals, pp. 371–410.

MURPHY, J. A., *Surface Preparation and Finishes for Metals,* New York: McGraw-Hill Book Company, 1971.

Use of Abrasive Grain in Mass Finishing, Booklet 7, Cleveland, Ohio: Abrasive Grain Association.

*Abstracted with permission.

40

SURFACE CLEANING PROCESSES

Most products require a final cleaning treament to remove surface contamination that remains after processing. In some cases, a cleaning operation is necessary to improve the appearance of the article; in other cases, workpiece surfaces are cleaned so that their adhesion may be improved prior to the application of a final finish or coating such as those discussed in Chapter 41. Standard procedure in some companies is to clean or protect parts at one or more intervals during the course of processing to facilitate certain production operations and inspections and to protect against rusting. Most cleaning operations are generally time-consuming and consequently add to the cost of producing a product.

Surface cleaning processes may be divided into two general categories: *mechanical,* such as impacting grit against the surface or by ultrasonic methods or power brushing, or *chemical,* such as by alkaline or acid cleaning methods.

653

MECHANICAL

Impacting

The Process: Impacting consists of a number of various well-established processes for blasting the surface of a metal part with a material traveling at a relatively high velocity. In each case, the process applies a finish and deburrs or otherwise cleans and prepares a surface. The specific term used to distinguish one impacting process from another is governed by the nature of the material that is hurled against the workpiece. In abrasive cleaning, for example, a stream of abrasive particles is directed against the work. Sand blasting employs silica sand particles while shot blasting cleans by using metal shot particles or steel grit. Impacting should not be confused with peen forming, explained in Chapter 34. Peen forming is used primarily to alter mechanical properties or to impart a compressive skin over the surface of a metal part. Nevertheless, a secondary benefit of peening is surface cleaning.

Various methods of impacting are commonly employed to remove burnt sand from castings or scale from forging or heat-treated parts. Impacting is also used to remove slag, oxides, and discoloration after welding and as a method for removing paint and rust when reconditioning used parts.

Grit Acceleration: There are two main methods of accelerating the grit: (1) an air-blast nozzle (dry or liquid), and (2) a centrifugal wheel. Large parts may be cleaned when necessary by a high-velocity blast of air and dry abrasive grit manually directed onto selected surfaces through a nozzle held by an operator, as shown in Figure 40-1. The operator wears suitable protective clothing, which includes an air-supplied hood or, in some cases, an air-conditioned hood. The work is performed in a special tightly enclosed cleaning room or chamber which is usually equipped with a turntable for positioning the work. Smaller parts may be impact-cleaned by loading them into an abrasive pressure cabinet, as shown in Figure 40-2; the parts are cleaned by directing a blast of compressed air and dry abrasive grit against the workpieces. The cabinet has access holes fitted with protective gloves with gauntlets into which the operator extends his hands. In this way, the parts may be handled without danger with one hand leaving the operator's other hand free to direct the blast nozzle. A viewing port is provided for visual inspection, and a foot treadle is used to control the on-and-off pressure action of the grit particles. Air-blast cleaning methods are principally used for low-volume production and for parts with difficult-to-clean areas, such as deep holes or partially blocked recessed areas.

Another method of surface cleaning, *liquid blast cleaning,* involves air blasting water in which very fine abrasive particles are suspended. Addi-

FIGURE 40-1 Shot blasting in a special cleaning room (Courtesy, Norton Company)

FIGURE 40-2 An abrasive pressure cabinet (Courtesy, Alexander Saunders & Co., Inc.)

tives are generally used in the slurry to facilitate suspension of the abrasive grit and to protect the workpieces from the effects of corrosion. The abrasive slurry is pumped at low pressure into a nozzle or gun and is accelerated by a jet of compressed air. Velocities as high as 80 to 90 psi (about 5.6 to 6.3 kg/cm^2) are used.

Equipment for liquid blast cleaning is similar to that used in dry blast impacting. Abrasives commonly used for cleaning and for light deburring operations include silica sand, quartz, and aluminum oxide. The method is often selected in cases where a fine decorative matte finish is desired either over the entire surface area of a product or, through the use of masking techniques, over selected areas. The use of glass beads has proved to be an effective media for producing a distinct bright surface finish somewhat resembling that obtained by polishing. *Hydroblasting* is a form of liquid blast cleaning, except that sand is used instead of abrasive particles. The principal advantage of liquid blast cleaning methods over dry blast cleaning operations is that the air is not filled with dust.

Centrifugal wheel cleaning is an airless blast cleaning method in which a dry abrasive is fed to a position near the center of a rapidly revolving wheel. The abrasive is propelled by vanes on the wheel which throw the grit onto the work in a fan-shaped pattern by a centrifugal action. Airless-type cleaning is widely used for high-volume production requirements. Because it requires no compressed air equipment, the power requirement is decreased. As a consequence, centrifugal-wheel impacting cleaning is a highly efficient method.

Possible Limitations: Impacting, particularly abrasive blasting, is not a recommended method for cleaning parts if sharp edges or corners must be maintained. It is difficult to prevent rounding of the corners. Also, some other process should be selected for cleaning parts with inaccessible recesses or pockets.

Ultrasonic Cleaning

The Process: Ultrasonic cleaning is rapidly replacing time-consuming hand methods such as brushing on production lines in many companies. For certain applications, ultrasonics can increase cleaning speeds as much as 40 or 50 times and still do a better job. The process consists of immersing parts in a tank containing cleaning fluid that is charged with ultrasonic energy. A phenomenon known as *cavitation* cleans workpieces by a controlled action that explodes the dirt off the contaminated surfaces. The phenomenon is explained as follows. When high-frequency sound waves of sufficient power level are propagated in a cleaning solution, the liquid reacts by exploding many thousands of times per second into millions of submicroscopic vacuum bubbles. These cavitation bubbles work by releasing and violently scrubbing the dirt off the workpiece. The cleaning action may be compared to the effect caused by millions of tiny, rapidly agitating brushes. Except for a low sizzling sound caused by the vapor bubbles rippling through the cleaning solution, the cleaning action is otherwise quiet. The ultrasonic sound waves are inaudible to the average person.

The process can be used to remove all types of soil (contamination) except those that are an integral part of the metal surface, such as rust, scale, or tarnish. Ultrasonic cleaning is particularly applicable to removing residual soils caused by compounds used in metal drawing, grinding, lapping, polishing, or buffing and other organic or inorganic soils that include finely divided particulate soil or smut.

Ultrasonic cleaning is fast and thorough. Most cleaning operations can be performed within 3 or 4 minutes. The countless impacts produced by the exploding bubbles penetrate even the smallest holes and quickly remove the most minute particles of soil.

Equipment: Energy for ultrasonic cleaning equipment is produced by high-frequency generators (20 to 500 kHz) and fed to transducers. The transducers, located in the liquid-filled cleaning tank, convert the electrical energy into sound energy. Some tanks contain as many as 12 or more transducers, which are available for modular mounting into existing installations. Also available is ultrasonic cleaning equipment featuring stainless-steel tanks with capacities as large as 100 gallons (378.5 liters) or more. One model of an ultrasonic and vapor degreaser is shown in Figure 40-3.

A wide selection of off-the-shelf ultrasonic equipment is currently available. Many companies offer component accessories consisting of special baskets, racks, drain boards, process timers, heaters, signals for evaluating peak performance, and so on. Some basic equipment may be readily assembled into completely automated custom systems consisting of pre-programmed or automatic work-handling units for loading and unloading parts. One manufacturer has recently installed an automatic six-stage ultrasonic cleaning system that can process over 2100 lb (945 kg) of automobile engine parts per hour.

Importance of Cleaners: Cleaners must be selected with extreme care to avoid the possibility of a potential chemical reaction between the work and certain chemical cleaners. High temperatures created by imploding vapor bubbles as well as a tendency of some solvents to penetrate the neutral film or boundary layer on the work surface contribute toward attack of the base metal. Protection for the base metal is generally assured by using neutral or alkaline cleaners rather than acid. It is through the use of specialized cleaners that ultrasonics is an economical and efficient cleaning technique.

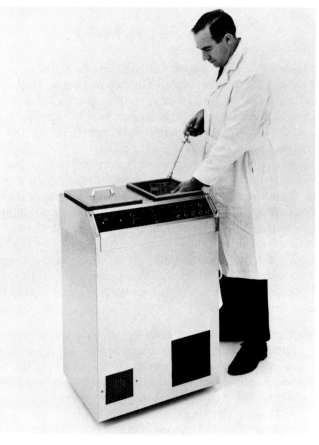

FIGURE 40-3 An ultrasonic and vapor degreaser
(Courtesy, Tronic Corporation)

The Cleaning Sequence: In general, the following steps are included: (1) a prerinse to remove heavy greases and oils, (2) ultrasonic cleaning in a proprietary formulation of some alkaline water solution, (3) spray rinsing with clear water to remove the particles dislodged by ultrasonic cleaning, (4) preserving parts by submerging them in water-displacement oil at room temperature, and, lastly, (5) draining the parts and reclaiming the oil.

Product Applications: Typical applications for the process include the cleaning of precision machine parts; aircraft, marine, and automotive engine parts; electronic, computer, and business machine parts; hydraulic, magnetic, and optical components; glassware; and so on. An outstanding advantage of ultrasonic cleaning is made possible because cavitation readily occurs in crevices and isolated areas normally too difficult to reach by most other cleaning processes. As a consequence, it is an ideal process for cleaning assemblies such as switches and relays, gas, water, and electric meters, parking meters, small gear trains, circuit boards, laboratory apparatus, and so on. Other users of commercial ultrasonic cleaning equipment are hospitals, atomic energy plants, scientific laboratories, and various maintenance facilities. In addition, the process has been widely adopted by dentists, doctors, opticians, jewelers, and in watch repair shops.

Power Brushing

The Process: Power brushing with high-speed rotary brushes is a well-accepted production process for metal finishing. Although most power-finishing operations were traditionally performed by hand with the operator holding the work against the rotating brushing wheel, new technologies have been developed in automatic brushing equipment which are finding wide acceptance. A three-brush setup with a rotating workpiece fixture is illustrated in Figure 40-4. In most cases, however, automatic equipment is restricted to fairly simple shapes.

Except for removing slight projections such as burrs, particles, or high spots, little or no metal is removed. A surface finish of 5 μin. is possible under controlled conditions.

FIGURE 40-4 Parts may be finished by power brushing using special semi-automatic equipment such as this (Courtesy, Anderson Corporation)

Because most power brushes are naturally flexible, they are able to conform to many contoured surfaces on workpieces, with the added advantage of being able to work in difficult-to-reach areas.

Contact Pressure: It is important to control the pressure applied to the brush against the workpiece surface because of the effect of the quality of the surface and upon the ultimate brush life. Only the sharp tips of the brush filaments actually do the work. Uniform contact pressure may be monitored through the use of an ammeter, which measures the load on the drive motor.

Types of Power Brushes: Power brushes are available in a wide range of materials, shapes, and sizes. Brushes with bristles or filaments of oil-tempered steel, stainless steel, and nonferrous wire such as brass, bronze, beryllium, copper, nickel silver, and aluminum are used for the bulk of metal-finishing applications. The form of the individual wire bristles may be crimped, twist-tufted as shown in Figure 40-5, or straight. Another important brush-fill material is Tampico, a natural vegetable fiber, which is used for wet-scrubbing and washing applications as well as for deburring and edge blending. Other brush materials include a nylon monofilament in which an abrasive grit such as aluminum oxide or silicon carbide is securely

FIGURE 40-5 A twist-tufted brush being used to finish a steel workpiece (Courtesy, Anderson Corporation)

FIGURE 40-6 Various shapes of encapsulated wire brushes. A leading advantage of the elastomer encapsulation is in preventing inefficient bristle separation and flare-out (Courtesy, Anderson Corporation)

encapsulated. Brush bristles are also made of synthetic fibers which have proved to be long-wearing and resistant to acid and alkali solutions. Figure 40-6 illustrates a relatively new development, the "encapsulated brush."

The various types of power brushes may be classified as follows:

RADIAL OR WHEEL: Available in a wide range of diameters and brush-face widths with a filament of steel or nonferrous wire, natural fibers, and a variety of synthetics.

CUP BRUSHES: Available in different diameters and wire sizes and materials. Wire may be crimped or twist-tufted. Cup brushes are designed for high-speed air and electrical tools, drill presses, and pipeline machinery.

END BRUSHES: Available in a wide range of sizes and shapes as well as wire diameters for a variety of applications. End brushes are used for light-to heavy-duty work in confined areas. Used on high-speed portable air or electric tools.

TUBE BRUSHES: Available in a complete range of oil-tempered steel or stainless-steel wire sizes and diameters. Tube brushes are used for cleaning internal surfaces.

STRIP BRUSHES: Available in a wide range of fill materials and trim lengths. Strip brushes are mounted on a hub to make an open-face rotary brush of varying lengths. They are used for many cleaning jobs in metal working, rubber, food, textile, paper, and chemical plants.

Product Applications: Industrial power brushes are employed to perform thousands of important production functions. Whether the brushing action is gentle or aggressive, flexible or rigid, the versatility of the process is most impressive. The following list describes the various categories of work commonly performed by power brushing:

DEBURRING: One of the most widespread uses is to eliminate rough edges, burrs, and fins on cut, stamped, or cast metal. Used on all types of materials and suitable for irregularly shaped parts. Both wheel and cup-type brushes are used.

EDGE BLENDING: Developing a rounded corner at a juncture of two surfaces. Most irregularities can be blended so that the line of demarcation cannot be detected.

SURFACE FINISHING: Brushes remove small raised metal particles on workpieces without changing dimensions. Many different surface appearances can be achieved.

CLEANING: Removal of surface film or soil, dirt, rust, scale, welds, paint, and so on, can be "dry"-cleaned or in conjunction with water or a chemical solution.

ROUGHENING: Power brushes may be used to create a surface for adhesion purposes on leather, plastics, rubber, and metals before bonding, painting, or coating.

CHEMICAL

Chemical cleaning operations are widely used to remove various kinds of surface contaminants from parts at some stage in their processing or prior to a final prefinish or coating process. Cleaning operations may be divided into two main job categories: *precleaning* and *cleaning*. In precleaning, the solid and caked deposits are removed by solvents and emulsions. In cleaning, the oxides, scales, and chemically embedded foreign material are removed by an alkaline cleaning or a pickling process.

Precleaning

Solvent Cleaning: Soils such as oils, grease, wax, asphaltic materials, resins, and gums are dissolved without difficulty by various kinds of solvent cleaners. Depending upon the required conditions, precleaning may be performed with either hot or cold solutions. The process of precleaned parts may be as simple as immersing them into a tank of petroleum (mineral spirits) or in a chlorinated solvent, commonly trichlorethylene or per-chlorethylene. In most cases, the parts are successively dipped in a series of tanks to help offset the buildup of contamination in any one tank. Chlorinated solvents are also manually applied to the parts by pressure spraying or by automatic means in elaborate mechanically operated conveyorized systems. Dipping and spraying operations are generally followed by one or two water rinses.

Vapor Degreasing: A typical model of a vapor degreaser is shown in Figure 40-7. In this process, the chlorinated solvent is vaporized at the bottom of a tank equipped with heating coils. A cold water jacket surrounds the upper portion of the tank. Except for heavily caked parts, the grease and oil is readily dissolved and runs off when the solvent vapor condenses on the suspended workpieces placed inside the tank. A subsequent drying operation is generally necessary except in ultrasonic vapor degreasing systems.

FIGURE 40-7 A vapor degreaser designed for bench-top mounting (Courtesy, Tronic Corporation)

There are a variety of other proprietary solvents available from many sources for removing particular types of soil. Problems such as the removal of fingerprints or water spots on some parts often require special-purpose solvents. In some cases, the addition of rust or corrosion inhibitors to solvents may be economically justified. Except for solvent emulsions, no solvent-type cleaner will remove water-soluble contaminants. These must be removed before or after degreasing.

Emulsion Cleaning: This process combines the detergent power of a soap with the solvency power of solvent cleaners. For this reason, emulsion cleaners are particularly effective for the removal of gross contaminants such as caked greases, buffing compounds, or water-soluble contaminants that cannot usually be removed in a single cleaning treatment. Emulsion cleaners consist of a mixture of a soap and an organic solvent such as kerosene with a small amount of water. The solution may be used at room temperature or, in other cases, raised to some suitable elevated temperature. Precleaning is accomplished by dipping and agitating the parts or by pressure spraying followed by a series of thorough rinsing and drying steps. The residue from an emulsion cleaner is a thin film of oil. This film offers temporary rust protection to the workpiece following the cleaning operation but prohibits the use of emulsion cleaners for final cleaning. If the parts are to be electroplated, for example, alkaline cleaning is necessary.

Cleaning

Alkaline Cleaning: Water and inexpensive cleaning compounds are used in this low-cost process to produce an alkali. Caustic soda, trisodium phosphate, or sodium metasilicate are commonly used agents to which a soap or wetting agent is added to increase emulsification and to ensure thorough cleaning. Some alkaline cleaners also contain rust inhibitors. Close control of the pH value in the solution must be exercised to ensure optimum cleaning results. Parts are cleaned by immersing or soaking them in heated tanks and are usually agitated to expedite the cleaning action. Power or spray washing methods may also be used, during which the parts may be automatically carried through the spray by conveyors in specially enclosed booths. Alkaline cleaning of aluminum, zinc, brass, or tin parts is not recommended because of the injurious chemical attack on the metal by the alkali.

Pickling: The two most commonly used acids in pickling are sulfuric acid and hydrochloric acid. Dilute water solutions of nitric, muriatic, and hydrofluoric acids are also used to a lesser extent for special applications. Most acids are used at concentrations of 10 to 25% in solutions at temperatures of 150 to 185 °F (about 66 to 85 °C). While pickling is widely used to remove oxides and scale from metal parts, it will not clean dirt or oil left on the part by a previous operation. As a consequence, the process must be used following some other cleaning method, such as alkaline cleaning, for example. Special inhibitors are generally added to pickling solutions to minimize base-metal attack by the acid and yet permit efficient removal of the oxides. Parts may be dipped or sprayed and, in most cases, pickling requires only a few minutes.

All traces of residue acid must be thoroughly removed from the parts by high-pressure spraying and rinsing. Any traces of acid on the part will prevent adhesion of paint or other subsequent coatings. As a final step, to prevent rusting, parts are frequently dipped in a slightly alkaline bath mixture.

REVIEW QUESTIONS

40.1. Give some reasons why workpieces may require a cleaning treatment before use.

40.2. Essentially what does the process of impacting accomplish?

40.3. What are the principal differences between impact cleaning (with compressed air and abrasive grit) and barrel finishing?

40.4. In what way does hydroblasting differ from the liquid blasting cleaning process?

40.5. List some limitations of abrasive blasting.

40.6. What are the major advantages of ultrasonic cleaning?

40.7. Most power brushes used in metal finishing are naturally flexible. Why is this considered to be an advantage?

40.8. Give two reasons why it is important to control the contact pressure of the moving brush against the workpiece surface being cleaned by power brushing.

40.9. There are two main categories in chemical cleaning: precleaning and cleaning. Explain these terms.

40.10. Which of the precleaning methods would be most appropriate to specify for the removal of heavily caked grease?

40.11. Explain why alkaline cleaning of aluminum, zinc, brass, or tin workpieces is not recommended.

40.12. What is the purpose of a "wetting agent"?

BIBLIOGRAPHY

"Checking Out Degreasing Solvents," *American Machinist,* Aug. 5, 1975, p. 38.

"Cleaning and Finishing of Nonferrous Metals," *Metals Handbook,* Vol. 2, *Heat Treating, Cleaning and Finishing,* 8th edition, Metals Park, Ohio: American Society for Metals, pp. 611–663.

*"Four Case Histories Show the Effectiveness of Ultrasonic Cleaning," *Materials Engineering,* Nov. 1972.

"Heat Treating, Cleaning and Finishing," *Metals Handbook,* Vol. 2, *Forming,* 8th edition, Metals Park, Ohio: American Society for Metals.

"Metal Cleaning," *Metals Handbook,* Vol. 2, *Heat Treating, Cleaning and Finishing,* 8th edition, Metals Park, Ohio: American Society for Metals, pp. 307–370.

*SANDFORD, J. E., "Cleaning with Ultrasonics," Special Report 583, *American Machinist,* Feb. 14, 1966.

SARICA, D., "Abrasive Blast Speeds Deburring," *American Machinist,* Jan. 1975, p. 35.

SPRING, S., "Chemicals Used in Industrial Cleaning," *Metal Finishing,* Dec. 1974, p. 33.

SPRING, S.., *Preparation of Metals for Painting,* New York: Van Nostrand Reinhold Company, 1965, 318 pp.

"Ultrasonics Versus Degreasing Die Castings," *Metal Progress,* May 1974, pp. 105–115.

*Abstracted with permission.

41

FINISHES AND COATINGS

Many of the chapters in this book have dealt with various ways of manufacturing parts by producing surfaces finished by mechanical, chemical, electrical, or other means. The term "finish" in this case denotes the condition or the surface characteristics of the final finish on the metal part as a function of the manner by which the surface was originally produced (i.e., milling, turning, grinding, EDM, etc). It is important to clearly differentiate the term "finish" as just described from the same term used in this chapter. The term "finish" as used in this chapter refers to a surface layer of material which is applied to the original metal surface of the part by one means or another. Finish may be characterized by full matte or high gloss, ranging down to a dead flat finish.

Originally, finishes and coatings were deposited on metal surfaces principally for protective reasons such as resistance to abrasion, weathering, and corrosion or simply to improve the appearance of a part or product. Often, two or more purposes were served by a single process so that an applied coating, for example, was both protective and decorative. The usefulness of today's finishes and coatings has now been extended to include *functional* applications. These are electrical insulation, temperature control and fire retardation, lubricity, control of marine fouling, and sound deadening. In some cases, the role of the underlying metal part has been reduced to that of a substrate to carry the coating.

The subject of finishes and coatings can be conveniently subdivided into five principal headings: *precoated metals, organic finishes, inorganic finishes, metallic coatings,* and *conversion coatings.*

PRECOATED METALS

While finishes and coatings are often applied as a final step in the production process, more and more parts are initially being fabricated from prefinished or *precoated* materials. The chief advantage of using prefinished materials is that they are not affected by subsequent manufacturing operations such as roll or press-brake forming, blanking, stamping, piercing, or deep drawing. Armco Steel Corp. has developed an enameling steel called I-F, which consists of vacuum decarburized steel sheets precoated with porcelain enamel. Parts such as household washing machine components may be fabricated from this material, because the material is not subject to damage by severe forming and drawing operations. Precoated vinyl-clad aluminum has been used for some time for refrigerator and freezer parts by a major producer. Other prepainted materials, also known as "coil-coated" metals, are available which require only minor alterations in production procedures normally used for uncoated materials.

ORGANIC FINISHES

An *organic coating* forms a film over a substrate. The film coheres to itself and adheres to the surface over which it is applied. All surfaces must be free of foreign matter such as grease, oil, sand, dirt, or other impurities in order to assure maximum adhesion of the coating. It is sometimes desirable to roughen the surface before coating and between successive coats by wire brushing, sandblasting, or by some other suitable method. Organic finishes usually require a primer coat followed by one or more finish coats. An organic coating forms a decorative and a protective barrier on the base metal to which it is applied. Organic finishes are applied by roller coating, silk screen, spraying, flow coating, dipping, tumbling, electrocoating, or by the centrifugal process. Many of the foregoing methods of applying finishes can be done either by hand or by automatic means.

Oil Paint

External surfaces of products are often decorated and protected with one or more coats of an *oil paint*. Oil paint is a mixture of a solid pigment and a liquid vehicle. The finely divided metallic pigments, usually lead, titanium, or aluminum, have three functions: (1) to provide the necessary decorative color, (2) to completely hide the metal surface, and (3) for protective value. The liquid vehicle may consist of a natural or a synthetic resin, thinner, and a drying oil. Because of the attendant production difficulties in parts handling associated with the relatively long drying time, only large structures usually are oil-painted.

Alkyd Paint

Alkyds are widely used in industry today for protecting steel surfaces as well as for all-purpose protection for wood. Low-cost alkyd resin coatings are resistant to weathering and moisture, stable to the effects of sunlight, and retain a gloss finish. Urea and/or melamine with alkyd is used as a finish for refrigerators, washing machines, or venetian blinds. A common primer for an automobile consists of urea with alkyd. Most alkyds are little affected by greases or by many chemicals and, when exposed to the weather, to moderate temperatures. Alkyds are used for drapery hardware, curtain rods, metal containers, swimming-pool sides, indoor and outdoor furniture, baseboard heating jackets, metal awnings, barbecue equipment, and similar products.

Epoxy Paint

One of the most important families of synthetic resins developed in recent years are the *epoxies*. Epoxy resins offer a combination of toughness, flexibility, adhesiveness, and chemical resistance. In common with other paints containing drying oils, epoxy paints are not suitable for water immersion or as a topcoat for exterior applications. The ability to adhere to almost any type of surface is the prime reason for the wide use of epoxy coatings. Epoxies are excellent primers. In addition to being exceptionally corrosion-resistant, epoxies are tough, abrasion-resistant, and flexible. Epoxy coatings do not require special application techniques. They are popular as an electrical insulation film. They have high dielectric strength over a wide temperature range, up to 1200 volts/mil in heavier thicknesses.

Silicone Paint

Because of their thermal stability, *silicone paints* are used in formulating heat-resisting coatings. Applications range from protective coatings on mufflers, stoves, space heaters, and appliances to industrial equipment and maintenance. Silicone paints are rated as excellent in resistance to moisture, oxidation, chalking, or fading. To offset the disadvantage that silicone resin films tend to be somewhat brittle and relatively poor in adherence to metal surfaces, they are combined with alkyd, phenolic, or amine resins. In comparison to oil-base or alkyd paints, silicone paint is somewhat more expensive.

Varnishes and Enamels

For many years, *varnishes* were traditionally produced from rosin, fossil resins, or vegetable origins. More recently, these have been largely replaced by the newer oil-modified alkyds combined with synthetic resins. When used without a pigment, varnishes produce a clear film which fails to hide the surface to which it is applied. The addition of a pigment to a varnish produces an *enamel* that resembles paint in that it obscures the base material. By regulating the texture of the pigment, paint manufacturers can produce enamels in flat, air-drying, and bake-drying types in semigloss or gloss finishes. Drying accelerators are usually added to varnishes and enamels to hasten the drying process. When compared to oil paint, varnishes and enamels are somewhat more flexible and tougher and less affected by moisture.

One of the first synthetic resins to find wide use in organic coatings was oil-modified phenolic. These air-dried varnish films are dried by solvent evaporation and oxygen absorption in the same manner as oil-base and alkyd paints. Generally, phenolic coatings have better chemical and water resistance than oil-base and alkyd coatings. However, phenolics tend to darken slightly with age and to chalk more rapidly. They adhere well to metals, are fast-drying, and withstand high temperatures for short periods of time. Phenolic varnish finishes are used commonly in marine and exterior applications and find their widest uses in products subjected to exposure of chemicals, alkalis, and moisture.

Polyesters

Unlike the solvent-type coatings, *polyester* or *oil-free alkyd enamels* contain 100% solids with no solvent to evaporate. Polyesters are applied in thick films. Polyester coatings have good resistance to weathering, excellent resistance to acids and solvents, but have poor resistance to alkalis. They are being used for a variety of exterior and indoor uses.

Lacquers

Another type of organic finish available as a clear film is unpigmented *lacquer.* (Pigmented lacquers act like oil paint or enamels to hide the surface on the part to which the finish is being applied.) Lacquers are composed of nitrocellulose or other plastic resins, plasticizers with or without pigments, and a fast-evaporating solvent. Modern lacquers provide a means of obtaining a hard, quick-drying finish superior in mar and abrasion resistance to most enamels. A surface cleaning operation is always essential before applying lacquers. Because of the poor adherence of lacquer to metal surfaces, it is necessary to apply a nonlacquer prime coat to the surface for best results. As a general rule, when compared to paint or enamel coverage for a given surface, more coats of lacquer are required to provide an equivalent film coating thickness. Lacquer is available in a flat, semigloss, or gloss finish.

Acrylics

Widely used as a finish for appliances and automobiles are the fast-drying and durable *acrylics.* Generally, acrylic lacquers are solutions of substituted polyacrylic and acrylate resins together with plasticizing resins. Coating formulations containing acrylate and methacrylate resins generally are baked and can be used for white enamels having excellent resistance to chemical fumes and to acid and alkali solutions. The acrylates in conjunction with vinyl chloride polymers have been used for clear coatings on a variety of metals. Acrylics are available in a wide range of colors. Properties such as resistance to stains, chalking, abrasion, and permanence of color and gloss are rated as outstanding. Typical uses include coatings on exterior building products (siding, rain-carrying equipment, etc.), appliance housings, cabinets, lighting fixtures, office furniture, and similar products. For some applications, acrylic coatings have replaced porcelain enamel.

Cellulose

Two important members of this large class of solvent-type resins are *nitrocellulose lacquers* and *ethyl cellulose lacquers.* Because of the excellent solubility of nitrocellulose lacquer, various resins, such as alkyds, phenolics, or vinyls, can be incorporated to increase adhesion, resistance to moisture, acids, and alkalis, and to improve the general finish durability. Ethyl cellulose is soluble in many esters, ketones, and aromatic hydrocarbons but is more limited in its compatability with resins. When compared to nitrocellulose lacquers, ethyl cellulose is more resistant to heat, light, and electrical breakdown and is less flammable. A hot-melt coating of cellulose is often applied to finished ordnance hardware.

Vinyls

Another group of synthetic organic coatings comprise a variety of *vinyl* coatings. Some of the original vinyl materials were introduced as long ago as the 1930s. Vinyls have outstanding durability and excellent resistance to acids, alkalis, chemicals, and salt water. Vinyl coatings are applied to components used in household appliances, business machines, automobiles, and electrical equipment. Materials used in manufacturing closures, containers, tanks, metal furniture, and shelving are often vinyl-coated. Adhesion of the vinyl to metal substrates is poor unless special primers are used.

Polyurethane

Several types have been developed, including a one-component moisture-cured formulation, an oil-modified *polyurethane,* and a two-component polyol-cured formulation. Most of the various types of coatings are highly resistant to weathering and chemical attack, coupled with the ability to readily adhere to a metal surface. Polyurethane coatings provide a tough, flexible, protective film. Although the physical properties vary considerably from one type to another, most of them display a high order of abrasion resistance. In one typical product application, polyurethane has been used successfully to protect automotive valve covers.

Nylon

Polyamide coatings, including *nylon,* offer many advantages for commercial and industrial applications. They produce a uniform or semimatte finish in a wide range of colors. They are highly resistant to impact, abrasion, and wear and have excellent bearing qualities. In some cases, nylon coatings are replacing chromium plating, paint, and enamel on such products as handles, electronic components, metal indoor furniture, instrument panels, and on decorative parts of buildings.

Some of the outstanding characteristics of nylon coatings are nontoxic and noncontaminating properties and low water absorption when subjected to steam sterilization.

Polypropylene

An organic coating widely used as a lining on metal tanks, pipes, containers, and so on, is *polypropylene.* Plastic coatings have outstanding resistance to corrosive elements such as acids, alkalis, and salts and are resistant to damage from impact and abrasion. In addition, polypropylene has excellent flexibility and adhesion properties. Its post-forming ability makes it suitable for application to wire goods such as baskets, strainers, and other similar products.

Fluorocarbons

Teflon (PTFE) is perhaps the best-known *fluorocarbon* coating. An outstanding feature of fluorocarbon coatings is the fact they remain stable over a wide range of temperatures from -245 to $155\,°F$ (or about -154 to $68\,°C$). They are unexcelled in their combination of properties, including formability, color retention, resistance to solvents, and chalking. A typical fluorocarbon coating remains inert chemically, except on exposure to molten alkali metals, fluorine, and chlorine trifluoride at high temperatures and pressures. Fluoropolymers are premium coatings and are costly when compared to other organic coatings, possibly as much as seven or eight times the cost of some coating types. However, when the best in exterior durability is needed, the high cost usually can be justified. Typical applications include coatings on hand saws, snow shovels, power saw blades, putty knives, molds, valves, and linings.

Phenolics

A modified vinyl primer coat is usually necessary to obtain a satisfactory bonding for *phenolic* finishes. Organic coatings of this type are relatively fast drying and have good stain-resisting characteristics. They are highly durable and resist the effects of moisture and most chemicals and solvents. Uses include protective linings for beverage can interiors, food and chemical processing equipment, tank cars, and various metal containers. Phenolics are also used as an automobile primer.

INORGANIC FINISHES

Two inorganic coatings are widely used as a metal coating. They are the vitreous coatings known as *porcelain enamel* and *ceramic.*

Porcelain Enamels

Embracing a wide variety of compositions, *porcelain enamel* consists essentially of a glass matrix (usually alkali borosilicate) in which may be suspended crystalline opacifiers and coloring pigments known as "ceramic oxides." A porcelain finish is characterized by an easy to clean, smooth, hard, lustrous or matte-finished surface. Porcelain enamels are resistant to attack by chemicals, heat, or mechanical abrasion.

Porcelain can be applied to most metals and alloys provided that the metal remains solid at the baking or firing temperatures and does not oxide excessively during firing. The more commonly coated metals are the specially processed "enameling irons and steels," cold-rolled mild steel, hot-rolled steel plate, aluminized steel, stainless steel, and cast iron. In addition, porcelain enamel can also be applied to gold, silver, platinum, copper, aluminum, and superalloys.

While a full spectrum of scratch-free colors, including black and white, can be produced, some colors, such as burgundy, purple, and certain shades of red and orange, are more difficult and expensive to control than others during application. Metallic lusters such as copper, gold, and platinum can be produced but are usually restricted to surfaces having small patterns or designs.

A full range of finishes from high gloss to semigloss and full matte may be obtained. Various special textures are available, including glass-smooth, pebble-rough, stippled, grained or marbleized, and other decorative effects.

Traditionally, conventional porcelain enamel coatings on steel have comprised up to three layers. Special one and two-coat formulations are now available which cut costs considerably. Porcelain enamel finishes vary in thickness with the type of underlying metal, the application process used, and the intended use of the product being coated. A one-coat finish on steel can be in the order of 0.003 to 0.005 in. (0.08 to 0.13 mm) thick while a coating on cast iron may range from 0.025 to 0.070 in. (0.64 to 1.78 mm).

Two methods are used to fuse a porcelain-enameled coating to a metal. One process, called "wet porcelain enameling," consists of applying a mixture of glass powders suspended in a water vehicle in a continuous layer on the surface of a metal. The liquid mixture is called *slip*. After application, the water is removed by drying and the resulting film of dry particles is then fused to the metal by firing at a suitable temperature. A second method, called the "dry process," is used primarily to coat heavy sections of cast iron and steel parts. The base metal is first heated. The porcelain enamel in the form of a dry powder is then sprinkled on the hot surface. Complete fusion is assured by reheating the coated piece. A high degree of skill is required in this process to consistently form a good bond between the porcelain and the underlying metal.

Parts that have been coated with porcelain enamel have high resistance to alkaline solutions and acids at room temperatures. They are nonporous, easily cleaned, and color retentive. They are highly resistant to atmospheric corrosion. The upper temperature limit for porcelain coatings on aluminum is about 600 °F (approximately 316 °C) to about 1100 °F (approximately 594 °C) on steel. Porcelain-coated parts are brittle and are not generally considered repairable.

The biggest users of porcelain-enameled finishes are manufacturers of kitchen ranges and heating equipment, followed by makers of washers and dryers, refrigerators, water heaters, and dishwashers. Porcelain-enameled panels in steel sheet 0.0359 to 0.0598 in. (0.91 to 1.52 mm) thick in a wide range of colors are available from some suppliers.

Ceramic Coatings

Another type of vitreous coating is a hard, glass-like material called *ceramic.* When appreciable quantities of refractory compounds such as alumina, chromic oxide, or silica are incorporated to increase heat resistance, these special porcelain enamels are termed *ceramic coatings.* There are two types of commercially available coatings important in the protection of refractory metals: the silicides and the aluminides. Both are based upon the principle that a protective coating is formed when the oxides of an oxidation-resistant metal are diffused into the surface of the substrate. Some of the silicides offer protection up to 3500 °F (approximately 1928 °C). Aluminides are generally restricted to below 3000 °F (approximately 1650 °C). Ceramic coatings are generally applied in powder form by advanced flame or plasma spray techniques.

Ceramic coatings are used in applications where it is required to protect metal surfaces from the devastating effects of corrosion at high temperatures. The strength and rigidity of metal parts are enhanced by ceramic coatings. Other important advantages obtained by diffusing refractory particles into surfaces of metal parts is an increased resistance to chemical attack and mechanical abrasion. At this stage of development there remain several problems associated with occasional part failures due to small, undetected flaws in coatings. Many coatings are not impermeable.

Typical applications of ceramic coatings are found in jet aircraft engines, space vehicles, and in the chemical, textile, steel, and data-processing industries.

METALLIC COATINGS

Metal may be deposited upon a base metal or to a nonmetal by *electroplating, dipping, immersion, diffusion, vapor,* or by *metallizing.* Essentially, metallic coatings are applied to objects to protect against corrosion and to serve as a decorative finish. In special cases, however, metallic coatings are applied for protection against wear or to increase the dimensions of a part by building up the size with a metallic layer.

Electroplating

In *electroplating,* an adherent coating of metal is electrodeposited on a surface of an object. It may help to understand electroplating by thinking of it as a cold-casting process. While the process may be carried on in a number of different ways, the basic operation consists of passing an electric current from an anode to the object to be plated through the medium of an electrolyte containing dissolved salts of the plating metal in solution. Figure 41-1 shows a basic electroplating circuit. A direct electric current is required for all plating operations. Other salts, acids, or alkalis are generally added

FIGURE 41-1 Schematic diagram of a basic electroplating circuit

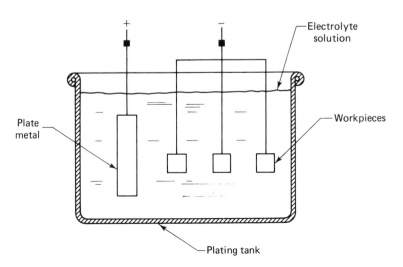

to the electrolyte solution to improve the action. The anode is a piece of pure metal of the type to be deposited. The object to be plated is the cathode. Both the anode and the cathode are suspended in the plating solution in the tank by special racks. When a direct current is passed through this circuit, metallic ions migrate to the cathode and, upon losing their charge, are deposited as a metal coating upon the workpiece. Tanks are usually lined with rubber, polyvinyl chloride, or vinyl resin as protection against attack by the plating solution.

The rate of metal deposition on an object varies with the current density; the higher the current (expressed in amperage per square foot), the faster the plate will be applied. Also, the thickness of the metal deposit may be increased by raising the current density and prolonging the plating time. *Throwing power* is a trade term used in electroplating. It denotes a capability to deposit a metal in remote areas and recesses on workpieces. In spite of the attendant dangers to personnel and equipment, cyanide solutions are generally used to plate parts of complex geometry because of their superior throwing power. The throwing power of some acids and alkaline solutions is good for most straightforward workpiece shapes. By contrast, the throwing power of chromium solutions is rated as poor.

Careful surface preparation for firm adhesion of the plate is extremely vital. Plating does not hide surface defects. All pinholes and scratches on the object must be removed prior to plating. The surface must be entirely cleaned of foreign materials. Impurities remaining on the workpiece surface will greatly diminish the forces of attraction as they interact between the atoms of the plated metal object and the plate. Voids occurring at the interface between the metal coating and the underlying metal will result in an unacceptable porous surface which greatly affects the protective value of the plated surface.

A process known as "periodic reverse-current electroplating" is often employed to plate a relatively rough-base material. This novel process results in a finish that resembles a polished surface. The plating cycle involves a sequence consisting of normal plating for a short period of time followed by a condition known as "deplating," which is accomplished by reversing the current. During this short interval some of the metal coating previously applied is permitted to deplate. The plating and deplating cycles continue until the desired aggregate thickness of thin coatings has built up. The principal advantage of the process is that denser and more homogenous deposits are produced on surfaces than is possible with conventional continuous direct-current methods.

Electroplating is a process by which metals such as cadmium, chromium, copper, nickel, silver, tin, and zinc may be deposited on the same base metal or on a different metal applied over another electroplate, or on nonmetals. Alloys such as brass, bronze, lead, and solder, as well as copper-zinc,

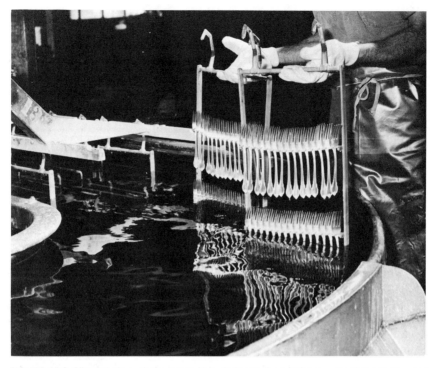

FIGURE 41-2 Silver plating—forks rotate in a plating tank containing a cyanide solution and bars of pure silver. The amount of silver on each piece is determined by the amount of electric current and the length of time the pieces remain in the tank (Courtesy, International Silver Company)

nickel-molybdenum, tungsten-cobalt, tungsten-iron, and so on, are also electrodeposited. Precious metals other than silver are electroplated, including gold, platinum, palladium, and rhodium.

Plastic-parts plating is considered to be one of the fastest-growing businesses in recent years. Heavy users of ornamental as well as functional plated plastics products include automobile, television, radio, boating, and appliance manufacturers. Surfaces of plastic parts are prepared for plating in various ways by roughening methods such as deglazing or etching. Most plastics parts are first preplated with a thin layer of copper, which provides an adherent base coating (somewhat like the function of a primer in an organic finish), which can later be plated with nickel or chromium. Figure 41-2 shows a rack of forks being removed from a silverplating tank.

Dipping

Parts may be coated by a process known as *hot dipping* in tanks containing molten aluminum, tin, zinc (also called *galvanizing*), or in an alloy of lead and tin. Dipping is a common method employed to form a thin protective coating on ferrous sheet and wire stock or on simple shapes of parts formed by forging, casting, or by welding. Only uncomplicated shapes can be dipped, because the molten metal tends to cling in recesses and pockets after the part is extracted from the molten bath. The speed of withdrawal helps to maintain a consistent coating thickness. In spite of this apparent disadvantage, corrosion-resistant coatings can usually be applied to base metals at less cost than by electrocoating.

One well-known commercial hot-dipped product is called Terne plate. It consists of an alloy of lead and about 20 to 25% tin. Terneplated steel sheets are used in roofing. As in electroplating, the success of the final coating is highly dependent upon the thoroughness of the surface preparation.

Immersion

Another method of applying a thin protective coating to a base metal is by the process of *immersion*. Parts may be immersion-coated with a thin film of gold, platinum, nickel, silver, tin, or zinc. The process is characterized by its simplicity. Instead of using a molten metal as in dipping, a watery solution containing ions of the coating metal is used. The solution is held at room temperature and to facilitate the process, the immersed parts are usually mildly agitated. Unlike electroplating, no electric current is used. This low-cost process is often used instead of electroplating when a uniform coating is needed. In spite of the fact that only a thin, almost a superficial coating, is usually deposited, there is an excellent adhesion of the deposit on the base metal. Immersion coating of zinc is often applied to aluminum and its alloys subsequent to electroplating. Immersion-coating methods are adaptable to bulk plating procedures.

Diffusion Coatings

Protective metallic coatings may be applied to a wide range of ferrous metals by placing the heated parts in intimate contact with a more corrosion-resistant metal in a powder, solid, liquid, or gaseous form. Essentially, *diffusion-coating* processes may be classified as surface-alloying heat treatments. There are more than a dozen different processes that fall within the category of diffusion coatings. The procedures for applying the various coatings vary from process to process. The principal advantages of diffusion-coated parts include an ability to withstand high temperature oxidation, corrosion resistance, and the production of hard and wear-resistant surfaces.

Selected examples of diffusion coatings follow. In *Calorizing,* a powdered aluminum compound or an aluminum chloride vapor is diffused onto the surface of a low-alloy steel part, forming a coating that resists oxidation at temperatures up to 1400°F (approximately 760°C). Calorized parts are widely used in furnaces, steam superheaters, and so on. *Chromizing* diffuses a chromium-containing powder compound to steel and iron parts, resulting in surfaces that are highly resistant to wear, abrasion, and corrosion. Chromized metals are used for tools and for aircraft, railroad and automotive parts. *Sheridizing* is a process of alloying zinc with ferrous metals for parts that in service must resist atmospheric corrosion. The process is used for coating bolts and small castings. *Siliconizing* introduces silicon carbide and chlorine into low-carbon and low-sulfur steels. Parts heat-treated in this way are used as pump shafts, cylinder liners, valves, valve guides and fittings, and so on.

Silicides and *metal additives* are introduced into the surface of refractory metals such as columbium, molybdenum, tantalum, and tungsten as a fluid or vapor at high temperature in a controlled-atmosphere furnace. Surface alloying treatments of this kind are used to reduce oxidation of metals for short periods of time at temperatures of up to 3000 °F (approximately 1650 °C). Important users of components processed in this manner are aerospace manufacturers.

Vapor-Deposited Coatings

There are four principal processes in this category of nonelectric coating: *vacuum-metallized* or *evaporated, ion-sputtered, chemical vapor plating,* and *glow discharge.* Practically any material can be vapor-coated, since the workpieces do not have to conduct electricity. Accordingly, substrate materials consisting of metal, glass, ceramics, plastics, textile fabrics, or paper may be coated. Essentially, any metal, but also many nonmetals, may be used as a coating material.

In the *vacuum-metallized* or *evaporated* process, the metallizing material, usually aluminum, is placed with the objects to be coated in a high-vacuum chamber and heated. Vapors of the metallizing material heated by tungsten filaments form the required coating as it condenses on the exposed surfaces of the slowly rotating workpieces. A controlled deposit of high-purity film ranging from 0.002 to 0.0000001 in. (0.05 to 0.0000025 mm) can be maintained. While the deposition process is misleadingly fast, often requiring less than 10 seconds, the total processing time must also include other necessary functions, such as loading and unloading the material and preparing the vacuum chamber before and after the deposit. A protective coating of lacquer is often applied to the coated surface as a final finish after metallizing.

Product applications include decorative coatings on costume jewelry, toys, household goods, functional coatings on automotive trim, instrument panels, hardware, microelectronic circuits, electrode surfaces on ceramics and piezoelectrics and optical surfaces, and for corrosion and thermal protection.

In the *ion-sputtered* coating process, the metallizing material which is used as the cathode is placed in the vacuum chamber with the workpieces.

A high voltage is applied to the cathode, causing the cathode metal to disintegrate, with the result that the ions bombard the workpiece surfaces. The "sputtered" atoms condense and diffuse on the workpiece to form a uniform coating. As in the evaporated process, the film thickness ranges between 0.002 and 0.0000001 in. An extremely high purity of film coating can be achieved. In addition to the product applications listed for the vacuum-metallizing process, ion-sputtered vapor coatings are also used for phonograph-recording masters and for surgical gauze.

The *chemical vapor plating* process is principally used to form tungsten shapes, to produce active semiconductor devices, and as a method of depositing oxidation-resistant coatings (silicides) or wear-resistant coatings (carbides and borides) on parts. This process can also be used to apply a uniform coating of various other refractory metals over complicated surfaces. Deposits may approach theoretical density.

Glow discharge is a process used to deposit a coating of organic polymers upon a variety of underlying materials. A pinhole-free film may be formed on leather and fabric materials with equal ease to that employed in forming a protective film over steel. Insulating films for capacitors are also formed by this process.

Metallizing

Dense and strongly adherent metal coatings may be applied to a variety of metals and nonmetals. There are two principal methods: *wire metallizing* (or *flame spraying*) and *plasma spraying.*

The principal uses of *wire metallizing* are in corrosion protection, repair work for building up worn parts or salvaging mismachined parts, and as a coating for improving wearing surfaces. In addition, plasma spraying may also be used as a protective coating on parts used in high-temperature applications to reduce thermal shock and to build up a heat barrier. Recently, powder-sprayed coatings in the form of a solid-film lubricant composite material have been successfully applied to parts to provide low frictional surfaces. Powdered metal coatings are also used as electrical conductors or dielectrics, and as primers or bonding coats on metal surfaces.

The wire-metallizing process consists of melting the material to be sprayed and simultaneously projecting the molten droplets in a high-velocity air jet to the surface of the base material. The apparatus used may consist of a spray pistol, shown in Figure 41-3, resembling that used in paint spraying. The pistol may be hand-held for manual spraying or mounted on a lathe tool post or on special equipment for automatic application. The raw material is automatically fed into the pistol in the form of wire and the necessary heat is obtained either from an electric arc produced from a transformer rectifier or from a propane or oxygen-acetylene source. Although

FIGURE 41-3 A metallic coating is applied to a rotating workpiece mounted on a lathe (Courtesy, METCO. Inc.)

the deposits are applied to the workpiece in a molten state, the base material undergoes only a relatively small rise in temperature. Because of this, sprayed metal coatings may be applied to plastics, wood, paper, and leather.

Adequate surface preparation is necessary for an effective bond. Substrates must be thoroughly cleaned followed by grit blasting to roughen the surfaces. The deposits produced by wire metallizing possess a high degree of bond strength. The interparticle cohesions of the flattened molten droplets on the workpiece surface is extremely high, all the particles being firmly bonded to each other. The network of interlocking plates gives a coating that is highly resistant to chipping, spalling, or cracking even when machined by conventional methods.

There is almost no limit to the maximum deposit thickness. Deposits in excess of ½ in. (12.7 mm) may be applied without difficulty, but most applications seldom require a coating greater than 1/8 in. (3.18 mm). Most metals and alloys may be readily applied to other materials by the wire metallizing process. These include carbon steel, stainless steel, brass, bronze, molybdenum, babbit, copper, zinc, aluminum, nickel, monel, tin, and cadmium. Mixed deposits of metals are possible. Metallizing has been successfully used as a method of building up wear-resistant coatings on brake disks, clutch pads, and on electrical contact surfaces. Other wear-resistant applications include mixed deposits of steel and bronze to bearing surfaces on machine components and on machine-tool slideways.

Although *plasma spray* is still a relatively new process, the type and quality of coating obtained cannot be matched by any other thermal process. In fact, some high-temperature materials cannot be sprayed by any other method. Figure 41-4 shows a coating being applied on a workpiece. Various coating materials, such as nickel-chromium alloys, tungsten-carbide, ceramics, cermets, and many other materials, may be used. Over 150 different metals, thermosetting and thermoplastic materials, and ceramics can be plasma-sprayed. Although temperatures can reach 30,000 °F (about 16,700 °C), most powdered materials can be applied below 15,000 °F (or about 8350 °C). In operation, the coating powder is blown at a high velocity across a plasma formed by a gas, such as nitrogen, argon, or hydrogen, inside a spray gun. As the particles of powder melt, they impinge with great force on the substrate to form a continuous coating on the workpiece. Powder can be fed at various rates, depending upon requirements. The apparatus generally includes a power supply with built-in controls to regulate the power and control the powder flow, the plasma spray gun, a powder feeder, and auxiliary equipment such as gas and water hoses and switches. Coatings vary from 0.0015 to 0.030 in. (0.38 to 0.76 mm) thick. The surface of the part to be coated is subjected to heat only to the extent of the molten powder. This permits powder deposition on temperature-sensitive wood and paper.

FIGURE 41-4 The service life of gages may be increased 20 times over by applying a spray coating of tungsten-carbide to worn, undersize gages

CONVERSION COATINGS

Conversion coatings or chemical finishes convert the surface of a metal to a compound of the metal. The principal types are: *chromate, phosphate, oxide,* and *anodized coatings.* In the case of anodic coatings, the process is electrochemical rather than chemical.

According to the process selected, conversion coatings can offer a pleasing, silvery white or brilliant, glossy surface, or produce special films that can be used to fulfill a specific function while they are principally used to form a base for subsequent finishes. They may also serve as a final finish. While conversion coatings find their widest use as a base for organic coatings, they also have some protective and decorative appeal.

Chromate Coatings

The precise nature of most *chromate*-bath compositions is proprietary, but all of them contain two basic ingredients, hexavalent chromium ions and a mineral acid. Some baths also contain one or more organic acids.

Chromate conversion coatings are used extensively on nonferrous materials such as aluminum and magnesium, and on cadmium, tin, or zinc-coated materials. Because of the nonporous structure, chromate coatings provide high corrosion protection and, owing to their low electrical resistance, they are often used for applications requiring good electrical conductivity. Chromate coatings are particularly appropriate when used as a treatment to improve paint bonding. Because of the availability of a clear film finish or a wide variety of attractive colors, they may also be used on some products as a final decorative finish. Chromate coatings are usually applied by the immersion process or by spray, brush, or roller techniques. In addition, promising electrolytic methods have recently been developed. Chromate coatings are generally very thin, usually not more than 0.00002 in. (0.0005 mm) thick. Processing times, including cleaning, are relatively short. Most chromate solutions operate at or near room temperature.

Phosphate Coatings

There are four types of *phosphate* coatings: *iron, zinc, lead,* and *manganese phosphate.* The choice depends upon such factors as thickness of coating, which is usually 0.0001 to 0.003 in. (0.003 to 0.08 mm), and the amount of corrosion protection desired. Iron phosphates are unique in that they prepare steel surfaces for painting while simultaneously acting as a cleaner. Zinc phosphates provide a high-quality base for oils as well as for organic coatings. They are applied over steel, iron, and cadmium or zinc-plated parts to give improved adhesion for subsequent finishes, particularly when extensive outside exposure is expected. Iron and zinc phosphate coat-

683

ings prolong the life of any subsequent finish. Manganese phosphate coatings are used for lubricating and as friction improvement on metal surfaces. Items such as internal combustion parts and gears are coated with this material to aid in "break-in" and to prevent galling. Lead phosphate coatings provide corrosion-resistant protection to iron and steel parts. Phosphate coatings are usually applied either by immersion or by spraying. A thorough cleaning before phosphating is very important.

Phosphate conversion coatings have good wear resistance as well as excellent resistance to humidity and weathering. Coatings vary in shade from an iridescent green to a light green, depending upon coating thickness. Dyed finishes in other colors, such as red, blue, gray, and black, are available as an aid in component identification. Phosphate coatings are also used to facilitate cold metal drawing and reduce friction in operations such as embossing, stamping, stretch drawing, and heading.

Typical applications for phosphate coatings as a paint base include pre-painting of auto bodies, hoods, fenders, refrigerator and freezer cabinets, air-conditioner housings, fan housings and blades, metal furniture, and similar products.

Oxide Coatings

Next to phosphate coatings, *black oxide* coatings are the most widely used conversion coatings for iron and steel. Oxide coatings are principally applied to metals for decorative effects, although abrasion and chip resistance are other advantages. Oxide coatings are produced by causing a chemical action to take place between the exposed metal and the hot oxiding solutions or gases. Oxide conversion coatings are used on iron, steel, stainless steel, zinc, cadmium, copper, aluminum and titanium. The appearance of the coating depends chiefly upon the condition of the metal surface to which it is applied. Attractive, lustrous finishes are obtained on a highly buffed surface while dull-matte coatings result from etched, brushed, or sandblasted surfaces.

A common coating for steel is composed of the black oxide of iron, Fe_3O_4. Of the several methods for producing the black oxide finish, the aqueous alkali-nitrate method is the most widely used. In this method, the parts are immersed in a highly alkaline solution containing strong oxidizers and other rectifying chemicals which react with the iron in the steel to form the black oxide.

There are a number of oxide conversion coatings for protecting and decorating aluminum and magnesium. These coatings provide an acceptable degree of corrosion resistance and serve as an excellent paint base. Oxide coatings for copper and its alloys as well as for brass are also available. Also readily available are colors for chemically treated copper and brass, which include gray, green, red, blue, brown, and yellow. Decorative blackened surfaces may be produced on zinc, zinc sheet, galvanized coatings, and zinc-base alloys with various proprietary oxide treatments. Oxide conversion coatings are applied to titanium to facilitate forming and drawing. An oxide coating also increases the wear resistance of the metal by reducing its tendency to gall and seize.

Anodized Coatings

Anodizing is a well-established electrochemical process for applying a permanent decorative and protective oxide coating on aluminum, magnesium alloys, zinc die castings, wrought zinc, and galvanized steel. In this process, the workpiece is made the anode and is immersed in an electrolytic circuit. An integral oxide coating is formed on the base metal over a short period of time during the anodizing operation. The method produces a nonflaking and adherent surface oxide on parts.

Various thicknesses of anodic coatings can be formed by controlling the electrolyte and operating conditions. For example, by using a 3 to 10% solution of chromic acid, a coating thickness from 0.00005 to 0.0002 in. (0.0013 to 0.005 mm) can be produced on aluminum. When a 12 to 15% solution of sulfuric acid is used on aluminum, a heavier coating, ranging from 0.0001 to 0.003 in. (0.003 to 0.08 mm), can be obtained. There are a number of proprietary anodizing solutions available which are recommended for coating magnesium and zinc.

Coatings may be clear or colored. Colors are obtained by using impregnated dyes or by establishing a chemical reaction to certain constituents added to the solution. A variety of finishes can be obtained, including high-dielectric coatings, hard, abrasion-resistant coatings, and architectural finishes of bronze, gray, or black.

Degreasing and cleaning operations are especially necessary when anodizing magnesium and zinc products. When compared to anodic coatings on aluminum, magnesium coatings are much softer and less dense. It is not possible to impart effective corrosion protection to magnesium by anodizing. In most cases, additional treatments, such as the application of inorganic sealants or organic finishes, are usually given to anodized magnesium parts. An anodized coating on zinc gives superior surface protection, because of the electrochemical formation of a barrier-layer type of bond with the base metal.

Parts may be formed from preanodized aluminum sheet without damaging the anodic film to any significant degree. Important products formed by using this material in this way include aircraft, automobile and electronic components, appliance parts, food-industry equipment, mobile-home components, furniture, sporting goods, and jewelry.

REVIEW QUESTIONS

41.1. What is the main advantage of using precoated metals?

41.2. What is the function of an organic coating?

41.3. What is the chief disadvantage of oil paint when used for finishing high-volume production parts?

41.4. What are the properties of an epoxy paint that make it superior to oil and alkyd paints for many applications?

41.5. What are some of the important limitations of epoxy paints?

41.6. What is a major application of silicone paint?

41.7. What is a major application of phenolic varnish?

41.8. In what way does polyester differ from oil or alkyd paint?

41.9. What procedure must be followed when applying lacquer to a metal surface in order that it properly adhere?

41.10. Give five uses of acrylics and explain those properties which are most applicable to the particular application listed.

41.11. What are the important properties that make vinyls a popular coating for many products?

41.12. What are the chief reasons for specifying a polyurethane coating for a product?

41.13. What important properties make polypropylene an outstanding coating for many applications?

41.14. In spite of the increased cost, under what conditions would the use of fluorocarbons be justified?

41.15. Give some important uses of phenolic coatings.

41.16. List some principal properties of porcelain enamels.

41.17. Prepare a brief outline showing the steps involved for each of the methods employed in fusing a porcelain enamel coating to a metal.

41.18. Give some examples showing some common applications of porcelain-enamel finishes.

41.19. What are the two types of ceramic coatings?

41.20. Under what conditions would the specification of a ceramic coating be specified?

41.21. Prepare a brief outline showing the steps involved in electroplating a workpiece.

41.22. Define the term "throwing power."

41.23. Explain why the surfaces of products to be electroplated must be carefully prepared beforehand.

41.24. What is the purpose of dipping?

41.25. Why would the process of immersion be preferred to electroplating in some cases?

41.26. What are the chief advantages of diffusion coatings?

41.27. Give the principal uses of wire metallizing and of plasma spraying.

41.28. Explain why the wire metallizing process results in a coating highly resistant to chipping, spalling, and so on.

41.29. In what form is the raw material applied to the workpiece in the plasma spraying process?

41.30. What are the principal applications for conversion coatings?

41.31. Describe the main characteristics of an anodized surface.

BIBLIOGRAPHY

BLOCHER, J. M., JR., J. H. OXLEY, and C. F. POWELL, *Vapor Deposition,* New York: The Electrochemical Society, Inc., and John Wiley & Sons, Inc., 724 pp.

*"Conversion Coatings," Manual 246, *Materials Engineering,* Mar. 1974.

DOWNEY, B. L., *There Is No Mystery in Specialty Coating,* FC72-826, Dearborn, Mich.: Society of Manufacturing Engineers, 1972.

"Electrodeposited Coatings, *Materials Engineering,* Sept. 1974, p. 312.

"Electroplating," *Metals Handbook,* Vol. 2, *Heat Treating, Cleaning and Finishing,* Metals Park, Ohio: American Society for Metals, pp. 409–488.

*"Finishes and Coatings, Porcelain Enamels, Diffusion Coatings," Section 7, *Materials Engineering,* Materials Selector Issue, mid-Sept. 1973.

GRAHAM, A. K., *Electroplating Engineering Handbook,* Metals Park, Ohio: American Society for Metals, 1971.

*LAVOIE, F. J., "Coatings That Fight Fire and Heat," *Machine Design,* July 27, 1972.

"Metallic Coating Processes Other Than Plating," *Metals Handbook,* Vol. 2, *Heat Treating, Cleaning and Finishing,* Metals Park, Ohio: American Society for Metals, pp. 489–530.

*MOCK, J. A., "Guide to Organic Coatings," *Materials Engineering,* Aug. 1972.

MOHLER, J. B., *Electroplating and Related Processes,* New York: Chemical Publishing Co., Inc., 1969, 311 pp.

MOHLER, J. B., "High Speed Electroplating," *Metal Finishing,* July 1974, p. 29.

"Nonmetallic Coating Processes," *Metals Handbook,* Vol. 23, *Heat Treating, Cleaning and Finishing,* Metals Park, Ohio: American Society for Metals, pp. 531-598.

*Abstracted with permission.

42

WELDING PROCESSES

Welding is one of the major ways to fabricate metal parts. It consists of permanently fastening together two or more pieces into a single homogeneous part by the application of heat, pressure, or both. While there are more than over 40 separate welding processes, only a relatively few processes are industrially important. The remaining processes are used only in specialized fields of application or are rarely used at all.

Three types of welding processes are most important today: *arc welding, gas welding,* and *resistance welding.* Each of these processes will be discussed in some detail in this chapter. They are similar, in that metal parts are joined by fusion in what is essentially a localized small-scale casting operation. The workpieces are melted along a common edge or surface so that the molten metal—and often a filler metal, also—is allowed to form a common pool or puddle. The weldment is formed by the permanent joining of the various pieces after the puddle solidifies.

Other processes, such as plasma-arc, electroslag, laser, electron-beam, friction, and thermit welding, although gaining in usefulness and general acceptance, generally continue to be used in specialized operations and, although they are commercially important welding processes, are not widely used.

SELECTING A WELDING PROCESS

The designer must carefully consider a great many factors pertaining to welding as well as the influence of other factors that do not directly relate to welding operations. Simple, straightforward weldments of mild steel are rarely troublesome to weld. Complex shapes, on the other hand, demand compatible choices in equipment and metals and in developing a suitable welding sequence. It is the designer who often has the responsibility of assuring such compatibility. The following checklist summarizes important factors to be considered when selecting an appropriate process for a particular welding application.

1. The design of the welded parts has a profound effect upon the fatigue strength of the weldment. Will the in-service requirements affect the choice of the joint and the base metal?

2. Some parts, such as valve bodies and small pressure vessels, may be welded by mass-production methods. Other applications, such as structural fabrications like ships, bridges, and cranes, require portable welding equipment. What is the nature of the product? Are the parts small enough to be moved to the welding equipment or are they so large to require that the welding equipment be brought to them?

3. The designer tries to use the methods already in use in his plant. What equipment and skills are available?

4. If the order is large, new equipment may be necessary. The number of parts to be made usually helps the designer decide whether or not to weld by manual methods, semiautomatically, or automatically. Does the size or volume of the job or its potential justify the purchase of new equipment?

5. Will the thickness of the metal parts restrict the choice of the welding method? For example, thin sheet cannot be welded with electroslag nor can 2 in. (50.8 mm) thick plate be spot-welded.

6. Difficulties in welding ferrous and nonferrous metals often increase, for a variety of metallurgical reasons, as their alloy elements increase. Will the metal require special treatment to meet the specified welding quality?

7. Does the material and the process suit the application? The metal must be compatible with the welding process. For example, mild steel welds readily by many processes, but only a few processes are suitable for welding aluminum.

8. It is impractical to friction-weld long butt joints on plate stock. Will the shape or type of the joint limit the choice of the welding method?

9. Access is important. Hard-to-reach joints require special high-cost automatic equipment. Is the joint accessible to weld?

10. It is usually easiest to weld "downhand" (a condition when welding is performed from the upper side of the joint and the face of the weld is horizontal), which makes for faster welding.

689

11. Will special equipment for preparing, holding, positioning, or moving the weldment be required?

12. Can the problems of on-site welding (such as a bridge, for example) be overcome? On-site welding often means more difficult working conditions, poorer fit-up, and so on.

13. Can welds meet inspection standards?

14. Can the methods be mechanized or the design changed for automatic welding? Would such costs be justified? Can assemblies be standardized for faster welding? The designer tries to find ways to weld entire fabrications by automatic methods.

15. Can parts of the job be subcontracted by outside vendors? Subcontracting is usually justified when specialized equipment, such as electron beam equipment, is required for certain applications.

16. Will the cost eliminate the profits? Economy must not be overlooked.

ARC WELDING PROCESSES

The major industrial welding process is *arc welding,* where heat is generated by an electric arc struck between a consumable welding electrode, or rod, and the workpiece. Intense heat is generated by the arc. Melting and subsequent solidification of the weld metal occur very rapidly and largely because of this speed, arc welding has been universally accepted in industry.

The earliest attempts at arc welding employed carbon electrodes, but the welds tended to be brittle because the weld puddle was often contaminated by particles of carbon. Carbon was soon replaced by metal electrodes, which are the dominant types in use today.

Some forms of carbon arc welding are still used occasionally since recently improved techniques minimize the possibility of contamination. In addition, because carbon produces a particularly hot arc, it is often used for brazing and for welding materials that soak up heat readily. In general, the carbon arc technique is not a major metal-fabrication method. Descriptions of the most commonly used metal electrode processes follow.

Shielded Metal-Arc Welding (SMAW)

Molten weld metal absorbs atmospheric gases, chiefly oxygen and hydrogen. These gases reduce the strength and promote brittleness in the welded joint. *Shielded metal-arc welding* employs a chemically coated electrode which decomposes and vaporizes with the heat of the electric arc. Ingredients in the vaporized coating form a cloud of protective gas over the weld puddle to shield the arc from the atmosphere and prevent the molten metal from reoxidizing. The filler metal is obtained from the electrode. The process is illustrated in Figure 42-1.

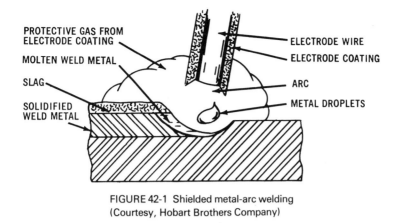

FIGURE 42-1 Shielded metal-arc welding
(Courtesy, Hobart Brothers Company)

The electrodes used in SMAW are stick-like rods 14 to 18 in. (355 to 457 mm) long. These are held by a clamp-type holder which the welding operator manipulates manually. The process is often thus referred to as "stick" or manual welding. The process welds all nonferrous metals. It is normally used for small-lot production, general maintenance and repair, and for on-site erection of large structures. After welding, the slag or molten flux must be removed to avoid possible flux corrosion. Figure 42-2 shows the major components required for shielded metal-arc welding.

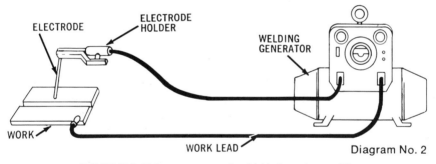

FIGURE 42-2 Major components for shielded metal-arc welding
(Courtesy, Hobart Brothers Company)

Flux-Cored Arc Welding (FCAW)

Flux-cored electrode wire, like the stick electrodes used in SMAW, are consumable; that is, they melt in the heat of the electric arc and contribute metal to the weld. In the *flux-cored arc welding* process the shielding may be obtained from a chemical flux supplied from the hollow core within the tubular wire electrode. This variation is called "self-shielding." A second variation, called "external shielding," requires a cloud of gas, normally CO_2 for steels, to be additionally supplied to the weld zone from an external source. Figure 42-3 shows the basic principle of these two process variations.

691

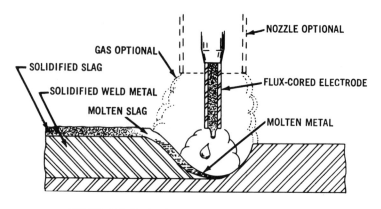

FIGURE 42-3 Basic principles of flux-cored arc welding
(Courtesy, Hobart Brothers Company)

There are several advantages of FCAW; regardless of which variation is used, the arc and the weld pool are visible to the welder and, depending upon the size of the wire electrode used, "all-position" welding is possible. Also, there are no restrictions governing the type of joint that can be welded. A wide range of material thicknesses can be welded, starting at $\frac{1}{16}$ in. (1.588 mm) and up, depending upon the welding-wire diameter. Electrode wires are available on spools and coils packed in special containers to protect them from moisture. FCAW was developed to provide a faster, automated alternative to stick welding (SMAW) for common grades of steel.

Gas Metal-Arc Welding (GMAW)

This process is schematically illustrated in Figure 42-4. It employs an electric arc generated by a continuously fed consumable electrode wire externally shielded by a flow of inert or special active gas. The composition of the electrode wire must be selected to match the requirements of the metal being welded. The shielding gases for welding must be specified as "welding grade," which ensures a specific purity level and moisture content.

FIGURE 42-4 Basic principles of gas metal-arc welding
(Courtesy, Hobart Brothers Company)

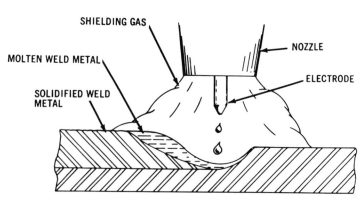

There are three variations of *gas metal arc welding,* depending upon the type of shielding gas employed and the nature of the type of metal transfer. One variation, formerly referred to as MIG, uses a pure inert gas of argon, helium, or a mixture of argon and helium. Solid, bare wire in sizes from 0.035 in. (0.89 mm) to ⅛ in. (3.18 mm) in diameter is used. MIG is most commonly used to weld nonferrous metals such as aluminum, magnesium, and copper, but it can also be used for production welding on many other metals, including titanium. A second variation is the microwire process, which uses a small-diameter filler wire. A low-cost shielding gas of carbon dioxide or a mixture of carbon dioxide and argon makes the process economical. Like the MIG process, a neat weld is produced. There is virtually no slag or spatter to remove. In addition, welding is possible in all positions. The microwire process is used for welding most steels, particularly at thinner gages than is possible with other arc welding processes. The third variation of GMAW is CO_2 gas metal-arc welding, which employs a larger-diameter filler wire requiring a special electrode wire feeder. A constant-voltage machine is used to produce high-speed welds, principally on low- and medium-carbon steels.

Gas Tungsten-Arc Welding (GTAW)

Figure 42-5 shows the basic principles of this process. Metals are welded by using the intense heat of an electric arc created in the space between a single tungsten electrode and the work. Contamination of the weld metal by the oxygen and nitrogen in the air is eliminated by an inert gas or an inert gas mixture which displaces the air around the weld zone. Argon or helium gas or a mixture is used for shielding. Stick electrodes used in this process consist of 3 to 24 in. (approximately 76 to 610 mm) lengths of tungsten or tungsten alloys. Because of the high melting point [6170°F (3412°C)] of the tungsten electrode, they are almost nonconsumable. In practice, the

FIGURE 42-5 Basic principles of gas tungsten-arc welding (Courtesy, Hobart Brothers Company)

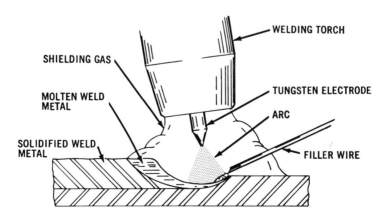

SHIELDING GAS

WELDING TORCH

MOLTEN WELD METAL

TUNGSTEN ELECTRODE

ARC

SOLIDIFIED WELD METAL

FILLER WIRE

electrode tip does not contact the metal workpiece. Accidental contact with the weld puddle will cause contamination, resulting in a sputtering arc. The intense heat generated by the arc keeps the weld puddle fluid. Filler material is normally used except when welding very thick stock. The filler metal may be added to the liquid puddle manually or by automatic means. High-quality welds in nonferrous metals may be produced.

Other advantages of GTAW are: (1) little if any weld cleaning is necessary, (2) there is no accumulation of weld spatter or slag, (3) the arc and the weld pool are visible to the welder, (4) welding is possible in all positions, and (5) the process is suitable for welding workpieces in a wide range of thicknesses. Figure 42-6 shows the major equipment requirements for gas tungsten-arc welding.

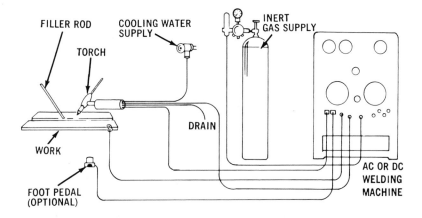

FIGURE 42-6 Major components for gas tungsten-arc welding (Courtesy, Hobart Brothers Company)

Submerged Arc Welding

Instead of gas shielding, this process uses a blanket of fusible granular material applied directly on the weld seam ahead of the electrode to shield the arc from the effects of contamination. Figure 42-7 illustrates the basic principles of this automatic process. A bare wire metal electrode is continuously fed from a coil to the weld zone. Heat generated by an electric arc forms a weld puddle under the protecting blanket of flux. The flux operates in much the same way as a chemically coated electrode used in the shielded metal-arc-welding process; it prevents sparks and spatter and helps to clean and purify the weld deposit. Some of the flux in the weld zone is melted by the heat of the arc, and upon cooling, forms a layer of easily removed slag. The upper layer of flux is unaffected by the heat and can be reclaimed and

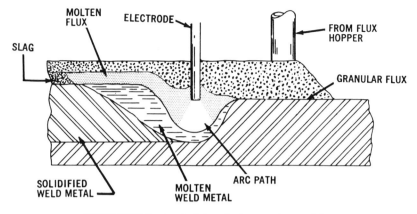

FIGURE 42-7 Basic principles of submerged-arc welding
(Courtesy, Hobart Brothers Company)

reused. Electrode wire sizes are available from $\frac{1}{16}$ in. (1.588 mm) to $\frac{1}{4}$ in. (6.35 mm) in diameter. Filler metal may be obtained from the electrode or from a supplementary welding rod. Submerged arc welding is used for a wide range of steels in various thicknesses from No. 16 gage to $\frac{1}{2}$ in. (1.588 to 12.7 mm). No edge preparation is required. Advantages of this process include high-speed, good-appearing welds with deep penetration. The process can be utilized only on flat or nearly flat and horizontal surfaces where the flux will not spill off the workpiece.

OXYACETYLENE GAS WELDING (OAW)

Commercial gases used in gas welding include acetylene, hydrogen, propane, butane, and natural and manufactured illuminating gases. The most common form of gas welding is oxyacetylene welding (OAW). Commercially pure acetylene gas and oxygen are combined in the proper proportions to produce a concentrated high-temperature flame of about 5300 °F (approximately 2930 °C). Edges of the base metal are joined without the need of pressure by fusion of the melted metal. Ordinarily, a slight gap exists between the pieces being joined, so filler material in the form of a wire or rod is added. Flux is usually employed in the welding of nonferrous metals and for some steels and cast iron. It is applied by dipping the heated rod into a paste solution. Flux helps to dissolve and remove impurities, in addition to giving some protection from atmospheric contamination.

Three adjustments of the flame are schematically illustrated in Figure 42-8. Each adjustment has a practical significance. The *neutral flame,* the most commonly used, is obtained when approximately equal volumes of oxygen and acetylene are used. A neutral flame is used for a wide variety of

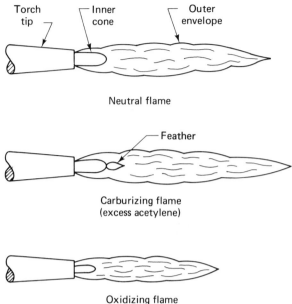

Torch tip — Inner cone — Outer envelope

Neutral flame

— Feather

Carburizing flame
(excess acetylene)

Oxidizing flame
(excess oxygen)

FIGURE 42-8 Types of oxyacetylene flames

welding and cutting operations. Two zones, composed of an inner cone and an outer cone, can be seen in the illustration. The highest temperature occurs at the tip of the pale blue inner cone; the outer cone serves to provide a reducing atmosphere, which protects the molten metal.

The *oxidizing flame* has an excess of oxygen over the ratio required for a neutral flame. In this case, the inner cone is purple in color and is shorter than the corresponding cone in a neutral flame. The oxidizing flame is principally restricted to work applications of braze welding and fusion welding of Monel metal and nickel or for welding certain alloy steels. For the third adjustment, the *carburizing flame,* there is an excess of acetylene over the ratio required for a neutral flame. A third zone, known as the "excess acetylene feather," whitish in color, can be observed between the inner cone and the outer envelope. A carburizing flame is used for welding steel with low-alloy-type welding rods.

While some oxyacetylene welding operations can be automated, the basic equipment shown in Figure 42-9 may be used as well with some variations for most manual welding operations. Essentially, a cylinder of oxygen and one of acetylene is needed along with two regulators, two lengths of hose, a welding torch with an assortment of welding tips and nozzles, and hose fittings. In addition, accessories are needed, including a spark lighter for igniting the torch, protective glasses, welding rods, flux, and welding torch wrenches. Oxyacetylene equipment is versatile. In addition to welding, it can be used for brazing, soldering, cutting, and for heat treating.

696

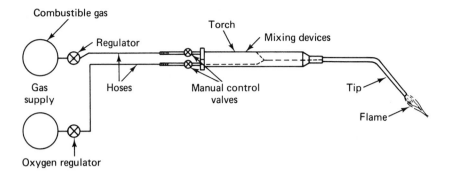

FIGURE 42-9 Basic equipment for oxyacetylene welding

Unlike arc welding processes, where the heat source may be quickly applied and confined to a relatively small area in the weld zone, large areas of the work are heated in gas welding. As a result, metal parts may buckle and distort, particularly on thin workpieces. Aggravated by prolonged heat, there is often a loss of corrosion resistance on pieces being joined by gas welding. Also, it is difficult to prevent contamination at the weld joint caused by the exposure of the heated metal parts to the various gases in the flame and in the atmosphere.

Practically all metals may be gas-welded, including steel, wrought and cast iron, brass, bronze, copper, aluminum, and many other alloys. It is also possible to gas-weld dissimilar metals, such as steel and iron or steel and brass.

Welding skills for this process are relatively easy to master and the equipment is comparatively inexpensive. Compared to other modern welding methods the process is slow, so gas welding is normally confined to job and repair shops for repair and maintenance work rather than as a mass-production technique. Automobile repairmen and shop handymen, for example, are major users of gas welding. Unlike arc welding, no electrical power source is required for gas welding. Oxyacetylene equipment is portable and can easily be transported for on-site work.

RESISTANCE WELDING

There are five variations of resistance welding: *spot welding, projection welding, seam welding, flash butt welding,* and *percussion welding.*

Resistance welding is widely used in mass-production work particularly in the joining of thin-gage metals up to a practical thickness limit of about ⅛ in. (3.18 mm). However, a thin piece may be readily welded to another piece thicker than ⅛ in. As in arc welding, resistance welding employs electricity but no arc is generated. Instead, heat is created from resistance losses as high-amperage current is sent across the joint between two firmly held mating metal pieces. The process does not require shielding or flux.

697

The workpieces are in sufficiently close contact to each other prior to welding as to discourage air from contaminating the molten weld metal at the joint. Methods of applying pressure during heat application on the fitted pieces range from purely manual action to mechanical, pneumatic, and hydraulic means.

Resistance welding normally is confined to assembly-line types of application since the tooling and equipment required are most easily justified under these conditions. The process, for example, is widely used on automobile production lines or for building containers, appliances, small metal structures, and so on. Resistance welding is also used as a "mill" process to turn out structural shapes such as I-beams or other shapes built up from ordinary plate or angles.

Spot Welding

One of the simplest variations of the resistance welding processes is spot welding. *Spot welding* offers a quick and inexpensive method of joining two or more metal pieces of flat stock or wire. The size and shape of the welds are limited only by the size and contour of the electrodes. Weld nuggets are usually circular. Spot-welding can often be used instead of riveting on some products. Figure 42-10 shows a typical layout for a spot welder. The process consists of the following steps: (1) position the work and squeeze between the electrodes; (2) apply a low-voltage current to the electrodes; (3) hold until proper welding temperature is achieved; (4) release current, continue pressure; and (5) release pressure and remove work.

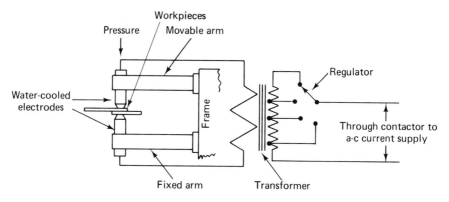

FIGURE 42-10 A typical arrangement for spot welding

High-quality welds require reasonably clean surfaces free of scale, paint, or excessive impurities. Practically all metals may be spot-welded, although there are some difficulties in welding tin, lead, or zinc. Figure 42-11 shows a typical spot/projection welder. Only one weld is made at a time.

FIGURE 42-11 A spot/projection welder
(Courtesy, Taylor-Winfield Corporation)

Projection Welding

Projection welding closely resembles spot welding. There is, however, one important difference. In projection welding at least one of the pieces to be joined receives a small embossment or projection, which is placed there by a previous operation or by a punch press. Except for the electrodes, spot-welding machines may be used for projection welding. Electrodes for projection welding are generally larger than those used for spot welding. High-quality welds are produced by the combined forging action of heat and pressure, which flattens out the projections and tightly forces the components together. Resistance welding is the only welding process that uses pressure while heat is applied. Unlike spot welding, which is restricted to one weld at a time, multiple welds may be simultaneously produced, depending upon the locations of the projections and the capacity of the machine.

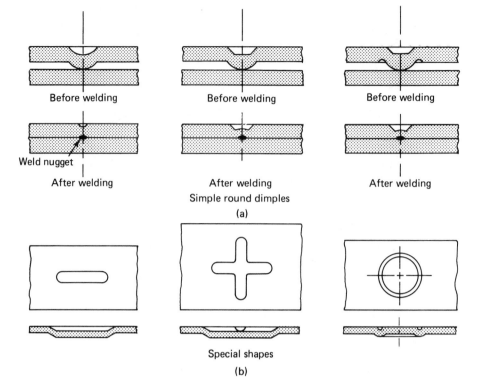

Before welding

Before welding

Before welding

After welding

Weld nugget

After welding
Simple round dimples
(a)

After welding

Special shapes
(b)

FIGURE 42-12 Typical embossed shapes for projection welding

The shape, form, size, and location of the projections vary according to the nature and form of the pieces, stock thickness, design of the joints, and the manufacturing process used to form the individual pieces to be welded. Typical projection shapes are shown in Figure 42-12. The following recommendations apply to projection design:

1. Round shapes are preferred because they are best both from functional and manufacturing standpoints.

2. As a general rule, projections are formed only on one of the pieces; only in special cases is it necessary to prepare projections on both pieces.

3. Where the pieces are of different hardness, projections are made on the harder piece.

4. Where there are two different gages, the projections are made on the thicker piece.

5. With materials of dissimilar electrical characteristics, the projections are located on the piece of higher conductivity.

6. Allow about three to five times the stock thickness between projection centers.

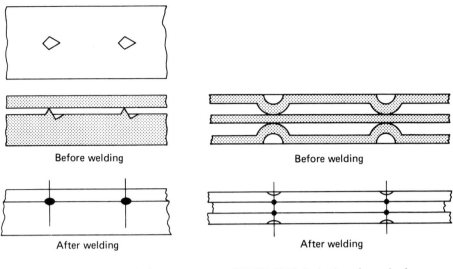

Before welding

Before welding

After welding

After welding

FIGURE 42-13 Swaged or staked projections

FIGURE 42-14 Projections for projection welding three components

For heavy stock or where invisible welds are desired, the projection may be swaged or staked as shown in Figure 42-13. In special cases, three pieces may be projection-welded. One common method, shown in Figure 42-14, is to emboss both the outer components, leaving the middle component flat. Another method is shown in Figure 42-15. In this special case, the middle component receives the alternately spaced projections.

Projection welding is frequently used to form cross-joints composed of two flat parts welded together so they form an angle, usually at 90°. A variety of commercial pins, bolts, nuts, and other types of fasteners also lend themselves to projection welding. Projection welding may be used to join dissimilar metals such as sheet steel and brass, for example. However, best results with the least problems are obtained with low-carbon, stainless, or alloy steels.

FIGURE 42-15 Projections for projection welding three components

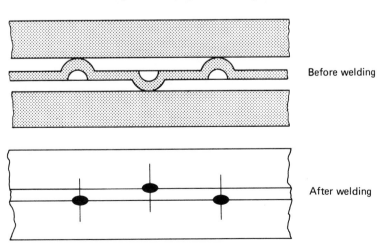

Before welding

After welding

Seam Welding

Seam welding resembles spot welding except that two rotating rollers are used for electrodes instead of stationary rod-shaped electrical contacts. A continuous welded seam is produced on the outer surfaces of two overlapping edges held firmly together by the pressure of the copper-alloy electrodes. The welded joint is formed by a series of spot welds located in a line along closely spaced intervals (called "stitch welding") or overlapping one another in a continuous seam. Figure 42-16 shows a typical seam-welding machine. Figure 42-17 illustrates common types of welded seams.

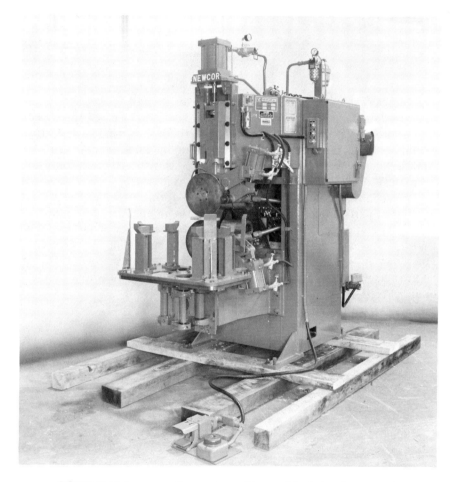

FIGURE 42-16 A seam-welding machine with a specially designed pantograph fixture for welding halves of automobile fuel tanks (Courtesy, Hall, Inc.)

702

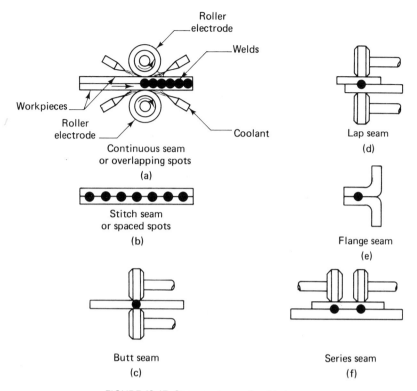

FIGURE 42-17 Common types of welded seams

In operation, the work is fed between the revolving electrodes while the current is automatically turned on and off. The timing schedule of the "on" and "off" heat patterns forms the individual spot welds, which accumulate to form the desired seam. In some cases, pieces being welded are "prespotted" or "tack-welded" along the intended path of the seam. Prespotting is done to assist the operator in preventing the parts from slipping out of alignment before being welded. A coolant (usually water) is normally used to prevent overheating of the work metal and to conserve the electrodes. For satisfactory results, the surfaces of the components to be welded must be clean and free from impurities, paint, oil, and so on.

Seam welding is most commonly done on flat or curved metal sheets and on plates of low-carbon, stainless, or alloy steels, but other materials, such as aluminum, brass, and titanium, may also be welded. Perhaps the most common application of seam welding is in making pipe. This process is shown schematically in Figure 42-18. Other product applications include mufflers, barrels, gasolene tanks, appliance cabinets, and metal containers.

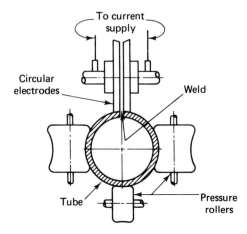

FIGURE 42-18 Continuous seam welding of pipe

Flash Butt Welding

Flash butt welding is shown schematically in Figure 42-19. This resistance welding process consists of firmly clamping two pieces in alignment in fixtures and pressing them together in light contact. A current is applied to cause flashing. When the proper temperature is achieved and as the metal reaches a plastic state, the pieces are simultaneously pushed together. The sudden upsetting action forces out the impurities at the joint in the form of flash and forges together the two workpieces. The metal in the weld consists entirely of parent material which is sound and virtually free of oxide particles. (An oxide is formed by the chemical action that takes place between the metal and the active oxygen.) The ragged and sharp projecting ridge or fin extending entirely outside and around the welded joint may be easily removed.

FIGURE 42-19 Basic principles of flash butt welding

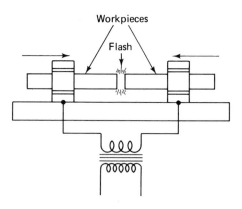

For successful results pieces to be flash-butt-welded should have about the same area at the mating surfaces. The ends of the pieces need no special preparation. For steel parts, currents ranging from 2000 to 5000 amperes/in.² are used during the flashing period, but during the upsetting cycle, between 2000 and 50,000 amperes is required. The forging pressure is measured in pounds per square inch. Depending upon the type of alloy and the cross-sectional area of the pieces being welded, the pressure can vary from 5000 to as high as 25,000 psi (351 to 1758 kg cm²). Flash butt welding is fast. A ¼ in. (6.35 mm)-diameter steel rod, for example, can be welded in about 10 to 12 seconds. It takes about 35 to 40 seconds to weld a ¾ in. (19.05 mm) diameter steel rod.

Various methods are used in welding machines to apply the power to push the pieces together. Some machines have manually operated drives while others use more complicated mechanical or hydraulic means. Figure 42-20 shows a typical flash-butt resistance welding machine.

Dissimilar metals may be end-welded by prolonging the flashing period until the fusion temperature of each metal is reached. Most of the common metals may be flash-butt-welded, including cast iron; low-carbon, stainless, and alloy steels; and aluminum, copper, and nickel alloys. Sheets, strips, bars, small structural shapes, pipes, tubes, rings, special fittings, and forgings may be joined in this way.

FIGURE 42-20 A typical flash butt welding operation for welding an automobile wheel rim (Courtesy, Hall, Inc.)

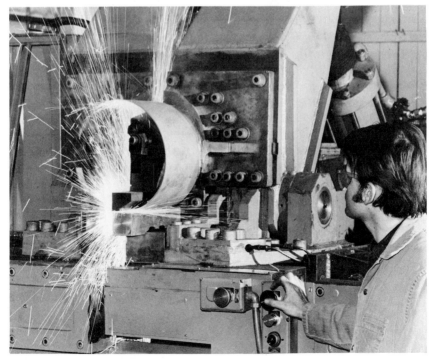

Percussion Welding

Percussion welding, the last of the five resistance welding variations, uses the heating effect of a high-intensity electric arc in combination with a percussion blow to forge-weld metal pieces together. Figure 42-21 shows the general arrangement of percussion welding equipment. In operation, the two pieces are first loaded in an aligned position in clamping fixtures. The ends are spaced slightly apart. A high-voltage discharge of electrical energy is applied, causing an intense arc between the heating surfaces of the workpieces. The arc is flashed for only about 0.001 second. During this short interval the surface of each part is sufficiently heated, causing the metal to soften to a depth of a few thousandths of an inch. A trigger controlling a spring-loaded clamp is released, causing one piece to suddenly impact under high force against the other. The weld is formed by the fusion of the thin layer of molten metal which forms on the end of each mating component.

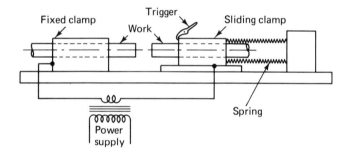

FIGURE 42-21 General arrangement of equipment for percussion welding

Unlike parts that have been joined by flash welding, there is no upset or flash at the joint. Dissimilar metals may be readily welded. Excellent results have been obtained in welding copper to Nichrome, for example, or welding Stellite tips to tools. Percussion welding is restricted to wire, rods, or tubes of relatively small diameter, with a mating surface area of less than $\frac{1}{2}$ in.^{2} (3.2 cm^2). In this process, mating surfaces must be reasonably smooth and parallel for sound welds.

OTHER WELDING PROCESSES

Plasma Arc Welding

Recent developments have placed this process in contention with other well-established welding processes because of its advantages in certain welding and cutting applications. *Plasma arc welding* resembles gas tungsten arc welding except in the way the shielding gas is supplied to the weld area. The difference is that an extremely hot shielding gas, sometimes in the form of a special "plasma" gas, is blown directly through the arc instead of forming a protective cloud-like atmosphere around it.

The training of operators on plasma arc welding equipment requires a much shorter time than for GTAW. There is less operator skill required. Compared to GTAW there is less emphasis in the training on maintaining close control over variations in the standoff distance between the electrode and the workpiece. Because of the protected position of the electrode within the torch, electrode contamination is eliminated in plasma welding. The plasma arc is superior to ordinary GTAW for welding thin stock such as foil and thin- to medium-gage materials and in joining dissimilar metals such as stainless steel to mild steel, for example. The highly concentrated arc gives deep penetration and a narrow weld bead. In spite of the extremely hot arc, the heat-affected zone adjacent to the weld area is narrow, with a minimum amount of distortion on the weldment.

Several factors have contributed to broadening the application base for plasma arc cutting. Obnoxious fumes formerly associated with plasma cutting operations have been virtually removed, leading to eliminating the need for costly pollution control systems. Also, lightweight tools and portable power units are currently available which permit on-site plasma cutting and welding operations. Another factor that has accelerated the trend to plasma cutting has been the coupling of the process to numerical control methods, which has increased the cutting output significantly.

In cutting applications, the plasma arc cuts cleanly, leaving little, if any, slag on most materials. Plasma cutting requires no preheating before starting the cut. Most nonferrous metals up to 1 in. (25.4 mm) thick and mild steel up to ½ in. (12.7 mm) thick can be readily cut.

Electron Beam Welding

Some of the unique advantages of *electron beam* (EB) welding include: (1) high-quality welds can be made at high speeds; (2) the fusion and heat-affected zone is extremely narrow; (3) a wider range of materials (including ceramics) than any other fusion process can be welded; (4) welds may be as shallow as 0.001 in. (0.03 mm) or as deep as 6 in. (152.4 mm); and (5) precise control is possible—welds can be made between components in remote locations.

In some respects, an EB welder works in a manner similar to the workings of a television set. Electrons in a picture tube are emitted by a heated tungsten filament. An electron optical system focuses a small diameter beam which is actuated so rapidly by a deflection system that a picture is produced on a fluorescent screen. While an electron beam welder can multiply by several thousand times the beam intensity of a television picture tube, it has very similar operating features and is almost as simple to operate. The weld setting and all other necessary variables on most EB equipment can be changed by simple knob adjustments at the console while the operator continuously inspects the weld joint through the optical viewing system. The weld quality depends upon such variables as the material to be welded, welding speed, shielding gas (if required), and various joint characteristics, including the design, surface preparation, and fit-up.

There are three basic types of EB welders: hard vacuum, soft vacuum, and nonvacuum, shown schematically in Figure 42-22. The three classifications are based upon whether or not the workpiece must be welded in a vacuum chamber. Each type offers specific advantages and disadvantages.

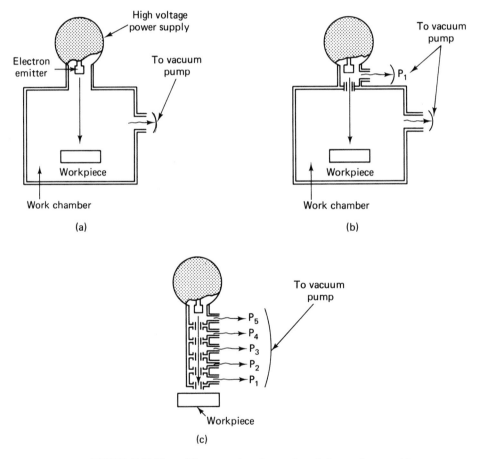

FIGURE 42-22 Three different modes of operation of electron beam welders: (a) hard vacuum; (b) soft vacuum; (c) nonvacuum

FIGURE 42-23 A small component electron beam welder (Courtesy, Electron Beam Welding, Inc.)

The *hard vacuum welder* is the most powerful machine of the three. The electrons can be projected a distance of about 3 ft (approximately 0.91 m), with a maximum penetration in steel of up to 6 in. (152.4 mm). Welding machines of this type are generally restricted to limited production runs because of the time required to pump down or evacuate the vacuum chamber during the period while the workpieces are loaded and unloaded. Increased production runs are possible with *soft* (or *partial*) *vacuum welders,* because of less time needed to evacuate the chamber. The design of the machine includes separate pumps, one for the emitting chamber and one for the workpiece chamber. Soft vacuum welders can project an electron beam about 2 ft (609.6 mm) with a maximum penetration in steel of abut 4 in. (101.6 mm). The third type of welder, *nonvacuum* (or *out-of-vacuum*), can be continuously operated without shutdown. The electron beam passes through several pumping chambers, each at a progressively lower vacuum pressure until the emerging electrode beam at the output nozzle reaches atmospheric pressure. Since a workpiece chamber is not required, pump-down time is eliminated, resulting in significantly increased production rates. The maximum penetration in steel for nonvacuum welders is reduced to about 1½ in. (38.1 mm). A small-component electron beam welder is shown in Figure 42-23.

Workpieces welded by either the hard vacuum or the soft vacuum process have high-quality joints entirely free of contamination. EB welding is fast. Material as thick as ⅛ in. (3.18 mm) can be butt-welded with a horizontal electron beam welder at speeds up to 300 to 350 ipm. Final, leaktight closure welds may be readily made under a vacuum on containers of various kinds. Vacuum welding is extensively used in the manufacture of high-intensity lamps and vacuum-insulated cryogenic equipment, and for envelope brazing and hot rolling. One leading automobile producer welds distributer cams and plates at the rate of 1100 per hour in a soft vacuum welder. Other automotive EB welding applications include weldments of flywheels, ball joints, steering column tubes, and frame members. The aerospace industry has been a major user of electron beam welding for some time. Although most EB welding is performed in at least a soft vacuum, uses for nonvacuum electron beam welding continue to grow significantly.

Laser Welding

Like the electron beam, the *laser* is a well-collimated beam of intense energy. The light energy of the laser beam causes thermal energy which rapidly heats the surfaces of the metal parts to their melting points without any significant vaporization taking place. The principle of laser operation is explained and illustrated in some detail in Chapter 25.

The key advantages of laser welding are uniquely well suited to the electronics industry. Examples of important applications are shown in Figure 42-24. In spite of the fact that laser welding applications have not been widely adopted in other industries, there are some technically effective applications which are of current interest. A partial listing of nonelectronic uses include laser welding operations on orthodontic dental braces, strain gages, automotive parts, sensitive watch components, gyro housings, and welding stainless-steel components on delicate instruments.

FIGURE 42-24 Applications of laser welding: (a) continuous contour welding of end plugs onto a turbine vane assembly. These welds are made at the rate of 30 in. per minute (Courtesy, Holobeam Laser, Inc.)

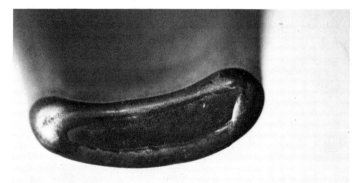

(a)

(b)

(c)

FIGURE 42-24(b) Seam welding relay cans (Courtesy, Holobeam Laser, Inc.); (c) hermetic welding of fully loaded battery containers at the rate of one every 0.9 in. per second (Courtesy, GTE-Sylvania)

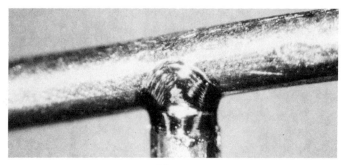

(d)

(e)

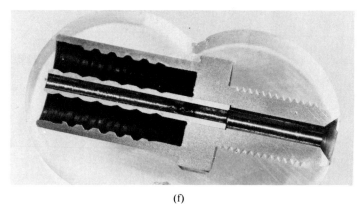

(f)

FIGURE 42-24(d) Spotwelding on 0.050 in. diam. platinum wire (Courtesy, Holobeam Laser, Inc.); (e) a main spring is welded to a coiled assembly. The materials are free-machining brass and stainless steel (Courtesy, Holobeam Laser, Inc.); (f) long standoff welding of a stainless steel hose coupling at a cycle time of 1.5 seconds. The narrow cone of the laser beam focused by a 5 in. focal length easily passes through the limited throat opening producing a weld at the bottom of the 1-in. deep base welding the serrated O.D. tubing into the outer casing (Courtesy, GTE-Sylvania)

(g)

FIGURE 42-24(g) Precision spot welding of a clock motor (Courtesy, GTE-Sylvania).

One leading advantage of the process is that laser welds can be made without difficulty through any atmosphere and through transparent materials. Also, there is a direct access to the components being welded because laser beam welding does not require a vacuum atmosphere. Welding of wire interconnections in butt, lap, cross, and tee configurations is readily accomplished. Repair welds may be made *inside* vacuum tubes. Transistor and transducer connections may be easily made, and integrated circuits may be attached to thin films by laser welding. There is no contamination, because the welding head never contacts the workpiece, and no welding pressure is required. The beam may be precisely focused at any desired position on the workpiece by use of relatively simple optics. Welds may be positioned at off-axis or at right angles and in areas on components impossible to reach by conventional processes. Well-defined spots as small as 0.005 in. (0.13 mm) can be produced with extreme accuracy. Because of a very small heat-affected zone, heat-sensitive components and assemblies may be readily welded without damage to adjacent parts. Laser welding is adaptable to practically all ferrous and nonferrous metals, and dissimilar metals may be readily joined.

A leading disadvantage of high-power laser systems is that the equipment is very bulky in size and relatively expensive. Components to be welded must have close-fitting joints. Laser welding is most efficient in stock less than 0.040 in. (1.02 mm) thick. In general, laser welding is seldom practical except in work applications not possible by conventional welding. In spite of the highly specialized nature of laser welding, laser systems are fast becoming competitive.

The type of laser to use depends upon the nature of the welding requirements, workpiece material, thickness to be welded, and the speed required. A solid-state laser, particularly one with a long pulsed mode, may be the first choice on some products for producing local spot welds. Seam welds require continuous-wave operation or a very high repetition pulsed mode. Lasers such as CO_2 and YAG are generally used for seam welds. Because of the relatively low power output, the argon laser is restricted to work applications in foil thicknesses.

Diffusion Bonding

In this process, heat and pressure are applied to carefully cleaned and mated metal surfaces so that they actually "grow" together by atomic diffusion. The word "diffusion" refers to the transmigration of atoms. The bond line often cannot be discerned in a properly made diffusion weld. *Diffusion bonding* is sometimes referred to as the "perfect" or "ideal" welding process because of its simplicity and applicability to an almost unlimited range of materials.

The concept of diffusion bonding is not new—it is one of the oldest welding processes known to mankind. The maximum advantages of the process are still to be fully developed and exploited. It is principally used on small work for welding materials particularly difficult to join by conventional techniques.

Hydraulic presses resembling those used in forging are used to heat and press the components. They are equipped to pressurize from three directions: from the top in the conventional manner and from the side and from the end. The presses range in capacities from 300 to 6000 tons (approximately 270 to 5400 metric tons). While specially equipped presses of this type add considerably to the cost, they need only one-tenth the capacity of those required for forging. Diffusion-bonded components are initially cut to near or final design size so that the bonded weldment requires only minimal secondary operations. A forged part, on the other hand, requires extensive machining of the surplus material. Extensive lead time is usually required to change forging dies when compensating for design changes. Ordinarily, tooling for diffusion bonding can easily be changed over a shorter time period. Recent advancements have been made which permit the bonding of assemblies up to 8 or 9 ft (approximately 2.43 to 2.75 m) in diameter.

One of the major difficulties encountered in diffusion bonding is the removal of oxide and the contaminating layers present on practically all metals exposed to natural or industrial environments. Also, the opposing surfaces must be mated in size to within a few angstroms of each other in order to achieve a satisfactory metal bond. (A 16-μin. surface finish has a typical peak-to-trough height of 10,000 angstroms.) The intimacy of actual contact of the two mating surfaces of parts with a 16-μin. finish, for example, would be limited to a relatively small percentage of the total area available.

The diffusion bonding process requires a relatively long, time-consuming thermal cycle, as well as requiring precise metallurgical control. It is not classified as a mass-production process. The least difficult condition in diffusion bonding is when the same or similar materials are being bonded together. With dissimilar materials, difficulties due to time/temperature/pressure requirements are frequently encountered.

When necessary, components may be shielded from contamination during bonding by using an inert or reducing gas, by protecting the joint by welding or using a flux in the form of a paste. As in EB welding, the process can be performed if necessary in a protective vacuum environment.

A recent development in diffusion bonding involves electromagnetic impact bonding, sometimes called "thermomagnetic bonding." The technique shows considerable promise in increasing the welding speed and in reducing costs particularly for producing socket-type joints. One advantage is that the process does not require a press. In practice, the components are first preheated to a plastic-range temperature slightly below their melting point. The mating surfaces are then bonded by forging them together by application of a high-pressure surge of electromagnetic force. The time period required in this process is in fractions of a second compared to several minutes normally needed for press equipment. Parts ranging from ¼ in. (6.35 mm) to 6 in. (152.4 mm) in diameter have been welded in this way. It is expected that the process will accommodate work up to 14 in. (355.6 mm) as larger equipment becomes available.

There are ample examples available to show that diffusion bonding techniques are being used more and more to fabricate components formerly produced as complex castings or forgings. Other applications include fabricated products made possible by the bonding of dissimilar metals.

Among the principal advantages of the process are: (1) there is little or no change in physical or metallurgical properties; (2) heat treating operations can be incorporated during the bonding cycle; (3) many different bonds can be made during one operation, particularly on hollow components; (4) high-quality, leakproof joints may be produced; (5) precut components may be near final size, thus reducing machining requirements; and (6) the process is particularly well suited for welding dissimilar metals and ceramics.

Electroslag Welding

Another process gaining wider acceptance is *electroslag welding,* whose chief advantage lies in its ability to rapidly produce welds in relatively thick materials. This process is essentially a continuous casting operation, in that filler wire is fed into a molten weld puddle held in place by water-cooled dams that usually are moved along the joint as the weld metal solidifies. The weld seam must be maintained in an upright position so that the molten pool does not spill out of the dams.

A layer of molten slag floats on the weld metal to protect it from oxidation. Heat for the process is generated in the slag by resistance heating caused by an electric current flowing between two filler wires. The electroslag process is used primarily in shipbuilding and in other heavy-plate work or as a technique in the production and repair of large machine parts.

Thermit Welding

This process is principally used for maintenance and repair work or for forming parts too large and heavy to be feasible by other methods. Sometimes it is possible to fabricate very large and complicated cast or forged parts by using *thermit welding* to assemble various irregular shapes of pre-forged or precast sections. Unlike most other welding processes, the preparation of the components requires no unusually careful fit-up procedures. Joints on parts to be thermit welded may be flame-cut, sawed, or rough-machined, or the mating surfaces may be welded in their as-cast or as-forged condition.

The process consists of casting molten steel to form a weld bead in the joint between the abutting surfaces of the adjoining parts. In practice, the pieces to be welded are first aligned and supported in a mold in the proper assembled position and then, in a manner somewhat resembling the investment casting process, wax is used to establish the ultimate shape of the weld bead in the gap or space between the adjoining parts. Wax is also built up to form the riser and the runner system. Refractory sand is packed around the zone on the parts to be welded. The mold is then heated and the wax is permitted to run out. Filler material consisting of finely divided metallic aluminum powder mixed with iron oxide is ignited by a mixture of magnesium powder. As in casting, the mold cavity is filled with molten metal, the joint being formed by the chemical reaction obtained by reducing the oxide with the aluminum. (After cooling, the mold is broken away and the weld is mechanically finished to the required shape.) The reaction is very rapid, requiring only about 30 seconds to reach a temperature in excess of 4500 °F (2484 °C).

An important advantage is there is no practical size limitation to the process. Also, no pressure is required. A wide range of materials, both ferrous and nonferrous, can be thermit-welded.

Inertia Welding

A major advantage of *inertia welding* is that dissimilar metals may be joined even when they have quite different melting points. In some cases, it would be extremely difficult, if not impossible, to weld certain materials by any other method. Successful welds have been made using practically every combination of metals, including most of the steels, copper alloys, brass, bronze, aluminum, titanium, various refractory alloys, and many sintered alloys. Examples of successful combinations are high-speed steel to carbon steel, aluminum to stainless steel, and copper to aluminum. One material unsuitable for this process is cast iron, because of its brittleness and also because the free graphite in iron acts as a lubricant to severely limit frictional heating. Other materials not recommended are free-machining steel containing lead and more than 0.13% sulfur, and copper alloys having a high lead content. Inertia welding can be performed very quickly; most welds require only about 1 second.

The process consists of pressing one end of a spinning rod or tube against the end of a second part held stationary in a chuck or fixture. During rotation, sufficient heat is developed to cause the material to melt at the interface. At this point, rotation is stopped and an increase in axial thrust is applied, forcing the ends of the two workpieces together. The kinetic energy stored in the rotary mass is rapidly transformed into frictional heat at the interface, causing the ends of two workpieces to become plastic. A very narrow heat-affected zone is produced, and even when dissimilar metals with widely differing melting points are welded, no significant melting takes place. Instead, the permanent bond formed in inertia welding is the result of mechanical mixing of a thin layer of molecules on each side of the interface during the plastic state. For some metals the depth of the mix across the interface is only about 0.002 in. (0.05 mm). In most cases, the newly forged metals are joined with a strength equal to or greater than the parent metal. There are no inclusions, gas pockets, or similar defects—these have been extruded from the joint together with the impurities that form the flash. Each side of the bonded joint is backed by a mass of cold metal that cools the weld rapidly by conduction.

A sufficient amount of stock allowance must be provided to compensate for mismatch on the workpieces. The concentricity of the joined rods or tubes can usually be held to 0.010 to 0.030 in. (0.25 to 0.76 mm) depending upon the diameters. Allowance must also be made to offset the decrease in length of each component. This loss will typically be in the range $\frac{3}{16}$ to $\frac{5}{16}$ in. (4.763 to 7.938 mm).

No fluxes, rods, or filler material are used in forming the joint. Flash rings around the joint are usually relatively smooth and are removed from the products only when considered objectionable. Welding machines are available to accommodate solid bars up to 4 in. (101.6 mm) in diameter and tubing up to 14 in. (355.6 mm) O.D. One automated model attains 1200 parts per hour.

Brazing

Welding and brazing both use heat and filler materials. They both can be performed on a production basis, but except for a few other vague similarities, there the resemblance ends. It is important to understand the nature of the *differences* between welding and brazing and to know when to use each process.

Unlike most welding processes, brazing does *not* fuse the base metals. Only the filler metal is molten. While brazing temperature is in excess of 800 °F (427 °C), it is invariably lower than the melting points of the base metal and always lower than welding temperature. In brazing, the joint is formed by flowing a molten filler metal between the mating surfaces. Only enough heat to melt the filler material is required and when the filler material and the base metal cool, a permanent metallurgical bond is produced. Brazing alloy is distributed to the joint by capillary action. A close fit-up of joints is usually required because capillary action functions best in narrow spaces. Generally, clearances of 0.002 to 0.004 in. (0.05 to 0.10 mm) are suitable. Figure 42-25 illustrates some typical joint designs for brazing.

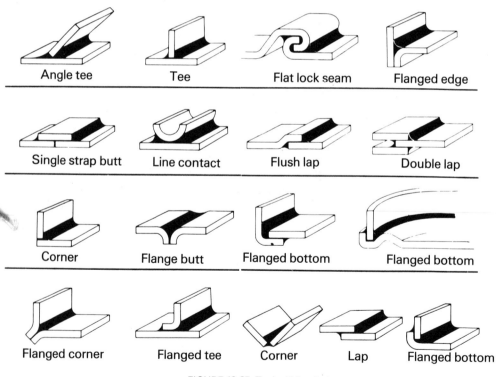

FIGURE 42-25 Typical joint designs

718

Strong brazed joints depend upon clean surfaces. In practice, the various mating surfaces on the pieces to be joined are first thoroughly cleaned of contaminants such as oxides, shop dirt, oil, grease, paint, drawing lubricants, or other impurities. Cleaning may be accomplished either by mechanical or chemical processes. The components are next manually placed in the desired assembly position by "gravity positioning" or "jigging" them on the welding table. When feasible, to facilitate more accurate alignment of parts prior to brazing, parts may be preassembled by spot welding, expanding, pressing, staking, crimping, seaming, riveting, or by using threaded fasteners. In other cases, special jigs or fixtures may be needed to align and temporarily hold irregularly shaped parts. Jigs and fixtures are also used as a faster, more convenient means of holding parts while increasing brazing production. Next, a flux is applied to all surfaces to be brazed. The purpose of the flux in brazing as in welding is to dissolve the oxides prior to heating the metal workpieces and to prevent the formation of oxides while the parts are being heated. Flux is available in paste, liquid, or powder form. Nonferrous aluminum, copper, or silver alloys are commonly used as a filler material metal for brazing.

In *torch brazing,* the heat is broadly applied to the base metal by an oxyacetylene, oxyhydrogen, or other gas-flame supply. In *furnace brazing* the work is placed in a controlled-atmosphere furnace and brought up to a suitable temperature. Filler metal must be properly positioned on the joint and prevented from moving about. Gravity is usually sufficient to maintain the filler material in place. *Dip brazing* consists of immersing preassembled parts in a bath of molten brazing metal. The process is usually limited to joining small parts, such as wire connections. In *induction brazing,* normally a batch process, the joint is rapidly heated by a high-frequency induction current applied directly to the joint by an electric coil. Filler metal is usually applied to the joint prior to brazing.

Brazing filler-metal alloys are available in the form of sheet, wire, rods, strips, bars, or rings and washers of various sizes. Filler material can be flattened or otherwise altered or preformed in shape to conform to the required joint configurations. In some instances, it may be advantageous to tack-weld the filler metal to the parts before brazing.

There are several important advantages of brazing. A properly designed joint will be leaktight. Brazing does not impede electrical conductivity. The filler metal is thinly distributed over a relatively large area. Because of the low heat requirement, ductile joints are obtained with little or no brittleness or stress in the joint area. Also, metals have less tendency to warp or distort than in welding. Dissimilar metals may be readily joined without regard to possible heat damage to metals. Brazed joints are corrosion-resistant because practically all of the filler material is contained within the joint. Brazed joints are strong, have a good appearance, and normally require no finishing. The principal limitation of brazing is workpiece size. Brazing is generally unsuitable for the joining of large assemblies.

There has been a definite increase in recent years in the automation of brazing processes. Currently available are rotary indexing and in-line endless belt machines to load, assemble, and, in some cases, to unload the parts automatically. In addition, equipment is now available with the capability of automatically applying the filler metal and dispensing the flux to the joint area.

REVIEW QUESTIONS

42.1. Prepare an outline with major headings representing the three main types of welding processes. Show the various processes that fall within these headings.

42.2. Explain the role "fusion" plays in welding.

42.3. List some important factors that should be considered when justifying the purchase of new welding equipment.

42.4. Explain what is meant by the statement, "The material and the process must suit the welding application."

42.5. Define the term "downhand" welding.

42.6. Give a principal reason why metal electrodes are used almost exclusively rather than carbon electrodes in arc welding.

42.7. Explain the justification for *shielded* welding.

42.8. Compare the two variations of shielding methods associated with FCAW.

42.9. List the major advantages of FCAW.

42.10. Explain why a welder must prevent the electrode tip from contacting the workpiece in the GTAW process.

42.11. What are the major differences between submerged arc welding and GMAW?

42.12. Explain why the submerged arc welding process is necessarily restricted to flat or nearly flat horizontal surfaces.

42.13. Explain the function of the *flux* used in oxyacetylene welding.

42.14. Give a reason why the oxidizing flame is principally restricted to the welding of brass, bronze, nickel and monel metal.

42.15. Why would a carburizing or reducing flame be particularly desirable for the welding of high-alloy steels?

42.16. Explain the reasons why workpieces have an increased tendency to distort and buckle in oxyacetylene welding.

42.17. Give a leading reason why gas welding is not ordinarily classified as a production welding process.

42.18. Explain why resistance welding processes do not require a flux.

42.19. Explain why spot welding is uniquely well suited to mass-production techniques.

42.20. Give the major differences between projection and spot welding processes.

42.21. Contrast the process of seam welding to spot welding.

42.22. Give a reason why "prespotting" or "tack" welding operations might be justified prior to seam welding.

42.23. Prepare a brief listing of the steps involved in flash butt welding.

42.24. Explain the major differences between percussion welding and flash butt welding.

42.25. How does the appearance of the final joint area of pieces joined by percussion welding differ from the joint area on parts joined by flash butt welding?

42.26. In what ways is plasma arc welding superior to gas tungsten arc welding?

42.27. EB welding requires particularly careful surface preparation and fit-up of components to be welded. Explain why this is so.

42.28. Of the three basic types of EB welders, which method provides the deepest workpiece penetration? Which method results in the fastest production rate?

42.29. Prepare a list of the principal advantages and disadvantages of laser welding.

42.30. Explain why diffusion bonding is sometimes called the "perfect" or "ideal" welding process.

42.31. Give some important reasons why electromagnetic impact bonding, if developed to its full potential, would be preferred to diffusion bonding.

42.32. Assume that a large section of a cast-iron flywheel rim has been broken. Prepare a sketch with suitable notation showing how the casting might be repaired using the thermit welding process.

42.33. Contrast the major steps involved in electroslag welding to those used in thermit welding.

42.34. For what type of work is electroslag welding best suited?

42.35. Prepare a sketch illustrating the process of inertia welding the ends of a 1 in. (25.4 mm)-diameter solid bar. Label fully.

42.36. For what general workpiece shape is inertia welding best suited?

42.37. In what important ways do brazing and welding differ? Under what conditions would brazing be preferred over welding?

42.38. Why is the joint clearance so important in brazing?

42.39. Prepare a list of the important advantages of brazing.

42.40. What is the principal disadvantage of brazing?

42.41. How does torch brazing differ from furnace brazing?

BIBLIOGRAPHY

AMERICAN SOCIETY OF TOOL AND MANUFACTURING ENGINEERS, *Tool Engineers Handbook,* 3rd edition, New York: McGraw-Hill Book Company, 1976.

"Bonding and Joining Technology," Bulletins 5918 (01), 5925 (01), 5925 (03), and 5978 (03), Springfield, Va.: Technology Utilization Office, NASA, 1975.

*"Choosing The Right Welding Method," *Welding Design & Fabrication,* Oct. 1975.

*KOHN, R., "Beginner's Guide to Welding," *Machine Design,* May 13, 1971.

LINDBERG, R. A., and N. R. BRATON, *Welding and Other Joining Processes,* Boston: Allyn and Bacon, Inc., 1976.

LITTLE, R. L., *Welding and Welding Technology,* New York: McGraw-Hill Book Company, 1954.

PATTEE, H. E., *Plasma Welding and Cleaning,* Columbus, Ohio: Battelle Memorial Institute, 1974.

PHILLIPS, A. L., Editor, *Welding Handbook,* 6th edition, New York: American Welding Society, 1968.

Pocket Welding Guide, Troy, Ohio: Hobart Brothers Company, 1973.

*Projection Welding Solves a Multitude of Problems," *Product Engineering,* June 1974.

"Welding and Brazing," *Metals Handbook,* 8th edition, Vol. 6, Metals Park, Ohio: American Society for Metals.

The Welding Encyclopedia, 17th edition, Chicago: Welding Engineer Publishing Co., 1974.

"Welding Iron Castings," Committee Report, New York: American Welding Society, 1965.

Welding Processes, Troy, Ohio: Hobart Brothers Company, 1975.

*Abstracted with permission.

43

PLASTICS—MATERIALS AND PROCESSES

Evidence showing the importance of plastics may be illustrated by the following example: In 1970, the Stanford Research Institute reported that the total contribution of the plastics industry to the U.S. gross national product was 1.9%. By the year 2000 it is estimated that the figure will increase to 7.2%. Substituting plastics for metals has been a common trend in recent years, involving thousands of products—often, but not always, in non-load-bearing applications that do not demand appreciable dimensional stability.

Historically, the beginnings of plastics as a material can be traced back to the discovery of hard rubber by Charles Goodyear in 1839, followed by the development of celluloid by John Wesley Hyatt in 1868. In 1909, a Belgian-American chemist, Leo Hendrik Baekeland, mixed phenol and formaldehyde and produced a synthetic substance that he named Bakelite. A patent was granted in 1909 for what has been established as the first plastic injection-molding machine. Two years later the first plastics extruder was patented. A huge array of new resins and compounds are currently available, along with some highly sophisticated plastics-processing machinery to handle these materials.

MATERIALS

A *plastic* is a synthetic material of high molecular weight composed of repeating organic chemical units. There are over three dozen *families* of plastics, and numerous grades of each are available to suit different product requirements. Naturally, no one material can completely meet each requirement and a compromise is often necessary. Special properties, applications, and processing techniques vary among members of each family. For example, one supplier has formulated over 400 different types of polyethylene, each slightly different from the other. Differences between members of a family of plastics are analogous to the differences between various grades of steel. Engineers should have a general acquaintance with the various families of plastics and with the principle methods of forming them. Material selection in metals or plastics almost always involves determining which properties are essential and working out the best possible balance of properties. Just as many of the technical aspects of metals are deferred to a metallurgist, details involving the chemical compositions of plastics are left to the polymer scientist.

Like metals, which are usually either hardenable or not, plastics are usually either *thermosetting* or *thermoplastic*. These terms may be considered the "heads" of the many families of plastics in current use. In some cases, it has been possible for plastics experts to make new materials by taking some of the best qualities from existing ones and combining them, regardless of "thermoplastic" or "thermoset" status. In response to demanding product requirements, new plastics have been developed that cannot really be categorized as belonging to either of the major families. A number of resins now available, although technically classified as thermoplastics, behave like thermosets. These materials are often referred to as "unclassed plastics."

Just why are thermoplastics and thermosets so different? Thermoplastics such as polyethylene, vinyl, or styrene can be softened and reshaped as often as desired by the application of heat. Thermoplastic materials are easily molded or extruded and, in fact, their excellent processibility is the major reason they are used so widely. Among these materials wide varieties of properties are available, from hard and brittle polystyrene to nylons with high moduli, high elongations, and impressively high tensile strengths.

If heat resistance or maximum rigidity is needed, the choice is usually a thermoset. By definition, no flow or melting takes place when a thermoset plastic is heated. Usually, a prepolymer such as epoxy, phenolic, or unsaturated polyester is used which itself is thermoplastic, and during fabrication, the resin thermosets. The wide differences in heat resistance and mechanical properties are due to the different molecular structures of thermoplastics and thermosets. This is the key to why these materials behave as they do.

724

Of major importance to the engineer are the "high performance" or "engineering" plastics. These are characterized by how well they rank on the scale of measurable characteristics—tensile strength, heat-deflection temperature, flammability, hardness, dielectric strength, and so on. Some will rate higher than another—none rates highest in all characteristics. High-performance plastics are recommended for rigorous mechanical applications —machinery parts, electrical or mechanical components, in punishing or hostile environments, and for structures that must withstand continuous and/or impact loads. Among materials in the "engineering plastics" category are nylons, acetals, fluoroplastics, acrylics, and phenolics.

In general, thermoplastics offer higher impact strength, easier processing, and better adaptability to complex designs than do thermosets. However, most thermosets have better dimensional stability, heat resistance, chemical resistance, and electrical resistance than do the thermoplastics. Thermosets are used principally to increase dimensional stability or other properties or are selected for reasons of economy.

The brief summaries that follow show the general properties and characteristics of the most common families of thermoplastics and thermosets used in industrial and consumer products.

Thermoplastics

ABS (acrylonitrile-butadiene-styrene): Very tough, yet hard and rigid; fair chemical resistance; low water absorption, hence good dimensional stability; high abrasion resistance; easily electroplated.

Acetal: Very strong, stiff engineering plastic with exceptional dimensional stability and resistance to vibration fatigue; low coefficient of friction; high resistance to abrasion and chemicals; retains most properties when immersed in hot water.

Acrylic: High optical clarity; excellent resistance to outdoor weathering; hard, glossy surface; excellent electrical properties, fair chemical resistance; available in brilliant, transparent colors.

Cellulosics: Family of tough, hard materials, cellulose acetate, propionate, butyrate, and ethyl cellulose. Property ranges are broad because of compounding; available with various degrees of weather, moisture, and chemical resistance; poor to fair dimensional stability; brilliant colors.

Fluoroplastics: Large family (PTFEM, FEP, PFA, ETFE, and PVF_2) of materials characterized by excellent electrical and chemical resistance, low friction, and outstanding stability at high temperatures; strength is low to moderate; cost is high.

Nylon: Outstanding toughness and wear resistance; low coefficient of friction; excellent electrical properties and chemical resistance; generally poor dimensional stability (varies among different types, however).

Phenylene Oxide: Excellent dimensional stability (very low moisture absorption) superior mechanical and electrical properties over a wide temperature range; resists most chemicals but is attacked by some hydrocarbons.

Polycarbonate: Highest impact resistance of any rigid, transparent plastic; excellent outdoor stability and resistance to creep under load; fair chemical resistance; some aromatic solvents cause stress cracking.

Polyester: Excellent dimensional stability, electrical properties, toughness, and chemical resistance, except to strong acids or bases; not suitable for outdoor use or for continuous service in water over 125 °F (51.7 °C).

Polyethylene: Wide variety of grades; low-, medium- and high-density formulations. LD (low-density) types are flexible and tough. MD and HD types are stronger, harder, and more rigid; all are lightweight, easy to process, low-cost materials; poor dimensional stability and heat resistance; excellent chemical resistance and electrical properties.

Polymide: Outstanding resistance to heat—500 °F (260.2 °C) in continuous use, 900 °F (482.6 °C) intermittent—and to heat aging. High impact strength and wear resistance; low coefficient of thermal expansion; difficult to process by conventional methods; high cost.

Polyphenylene Sulfide: Outstanding chemical and heat resistance 450 °F (232.4 °C) in continuous use; excellent low-temperature strength; inert to most chemicals over a wide temperature range; inherently flame-retardant; requires high processing temperature.

Polypropylene: Outstanding resistance to flex and stress cracking; excellent chemical resistance and electrical properties; good impact strength above 14° (−10 °C); good thermal stability below 225 °F (107.3 °C); light weight, low cost, can be electroplated.

Polystyrene: Low cost, easy to process, rigid, crystal-clear, brittle material; low heat resistance, poor outdoor stability; often modified to improve heat or impact resistance.

Polysulfone: Highest heat-deflection temperature of melt-processible thermoplastics; requires high processing temperature; tough, strong, and stiff; excellent electrical properties, even at high temperature; can be electroplated; high cost.

Polyurethane: Tough, extremely abrasion- and impact-resistant material; good electrical properties and chemical resistance; can be made in solid moldings or flexible foams; UV exposure produces brittleness, lower properties, and yellowing; also made in thermoset formulations.

Polyvinyl Chloride: Many formulations available, but most are classed as either rigid or flexible; rigid grades are hard, tough, and have excellent electrical properties, outdoor stability, and resistance to moisture; flexible grades (containing plasticizer) are easier to process but have lower properties; heat resistance is low to moderate for most types of PVC; low cost.

Thermosets

Alkyds: Good dimensional stability; very good dielectric properties; resistance to 300 °F (149 °C) in continuous use. High cost.

Epoxies: Excellent strength and toughness; outstanding adhesion to many materials; good resistance to many acids, alkalis and solvents; versatility and ease of processing. High cost.

Melamines: Very high hardness; resistant to detergents, water, and staining; permanency of color and molded in designs; fair dimensional stability; low impact strength.

Phenolics: Resistant to 300 °F (149 °C) with some formulations up to 600 °F (315.8 °C) over an extended period; outstanding resistance to deformation over load and over wide temperature range; chemically resistant to common solvents, weak acids, and many detergents. Low cost and ease in processing; some color limitations and low impact strength.

Urethanes: Good dielectric properties; very good wear resistance and toughness; retain properties over a wide temperature range.

Table 43-1 summarizes typical uses for the most commonly employed plastic families.

TABLE 43-1

General part requirement and/or shape	Typical candidate plastics
Structural, mechanical	Acetal
Gears, cams, pistons, rollers, valves,	Nylon
pump impellers, fan blades, rotors	Phenolic (reinforced)
	Polycarbonate

General part requirement and/or shape	Typical candidate plastics
	Polyester (TP)
	Polyphenylene sulfide
Light-duty mechanical and decorative	ABS
Knobs, handles, battery cases, cable	Celluosics
clamps, trim moldings, camera cases,	Phenolic
eyeglass frames, auto steering wheels,	Phenylene oxide
handles for hand tools	Polycarbonate
	Polyester (TP and TS)
	Polyethylene
	Polypropylene
	Polystyrene
Small housings and hollow shapes	ABS
Telephone and flashlight cases, sports	Acrylic
helmets, headlamp bezels, housings for	Cellulosics
office machines, power tools, pumps,	Phenolic
and small appliances	Polyethylene
	Polypropylene
	Polystyrene
	PVC
Large housings and hollow shapes	ABS*
Boat hulls, shrouds for motorcycles	HD polyethylene*
and agricultural equipment, housings	Phenylene oxide*
for large appliances, pressure vessels,	Polyester* (TP)
tubs, ducts	Polyester/glass (TS)
	Polypropylene*
	Polystyrene*
	Polyurethane*
	PVC*
Optical or transparent parts	Acrylic
Optical lenses and fibers, tail-light	Cellulose Butyrate
lenses, safety glazing, refrigerator	Polycarbonate
shelves	Polystyrene
Parts for wear applications	Acetal
Gears, bearings, wear strips, tracks,	Fluoroplastics
chute liners, roll covers, industrial tires	Nylon
	Phenolic
	Polyester (TP)
	Polyimide
	Polyurethane
	UHMW polyethylene

TP, thermoplastic; TS, thermoset.
* Often in structural foam.

PROCESSES

Injection Molding

The principal method of forming thermoplastic materials is *injection molding,* and with some material and machine modifications, the method may also be used to form thermosetting plastics. A typical machine consists of two sections mounted on a common base, as shown in Figure 43-1. One section clamps and holds the mold halves together under pressure during the injection of the material into the mold. The other section, the plasticizing and injection unit, includes the feed hopper and a hydraulic cylinder which forces the rotating screw forward to inject the plastic material into the mold. Also included is a motor which rotates the screw and a heated barrel that encloses the screw.

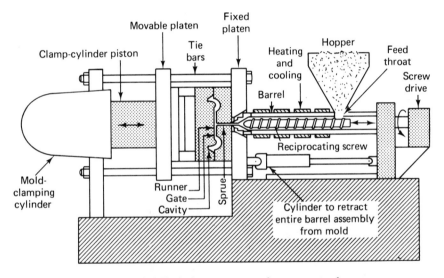

FIGURE 43-1 Typical arrangements of components of an injection molding machine

Practically all injection-molding machines operate on an automatic cycle. In operation, the raw material in granular or pellet form is fed from the hopper by gravity into the feed throat of the barrel. It is then moved forward by the action of the rotating screw or, in some machines, by the action of a plunger. As it is conveyed along the barrel, the plastic picks up conductive heat from the heating element of the barrel and frictional heat from the rotation of the screw. As the material changes from a granular to a semiviscous consistency, it forces the screw backward in the barrel against a preset hydraulic pressure. This back pressure is an important processing variable.

729

The screw stops turning when the proper amount of material has reached the nozzle end of the barrel as sensed by a vernier-set limit switch. The material at the nozzle end—the charge—corresponds to the exact volume of material needed to fill the sprue, runners, and cavities of the mold. Finally, as the screw is moved forward by hydraulic pressure on the plastic, the hot plastic melt is forced through the nozzle of the barrel, through the sprue of the mold, and into the runner system, gates, and the mold cavity. When the plastic material in the mold solidifies, the mold halves open and the finished parts tumble out. Heavy clamping pressure is necessary to keep the mold halves together in preventing the plastic material from escaping. While tonnage on some machines can reach or exceed 4000 tons (3600 metric tons), the majority of machines in use are in the 500- to 1200-ton (450- to 1080-metric-ton) range. Processed in quantities up to millions, parts leave the machine completely finished, trimmed, and ready for immediate use. As in die casting, multicavity molds can be designed to produce only a single part or many parts in one cycle in injection molding. For example, a 60-cavity mold on a 6-second production cycle would result in 36,000 parts per hour.

A mold made for one material cannot ordinarily be used with another material. Mold shrinkage of various plastic materials varies from a few thousandths per inch (approximately 0.08 mm) to as much as 0.05 in./in (0.05 mm/mm). Figure 43-2 shows a volume production technique developed using modular molds.

FIGURE 43-2 The modular mold shown above and to the right is a six cavity cup mold that runs at 30 shots per minute. Each module can be readily serviced by removing four bolts holding it to the base plate. Each core/cavity stack is like a single cavity mold mounted to a runner system on the stationary side and an injection system on the moving side (Courtesy, Husky Injection Molding System)

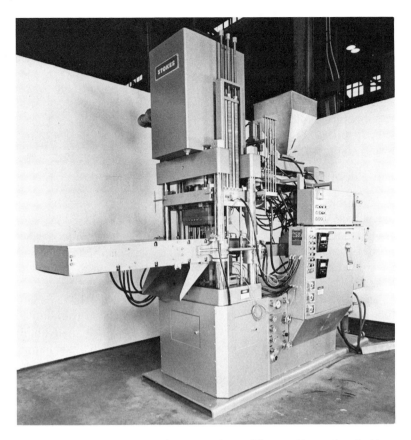

FIGURE 43-3 A 150-ton automatic compression molding machine equipped with a uni-screw preformer and a preheater (Courtesy, Stokes-Pennwalt Corporation)

Compression Molding

This process is largely restricted to molding thermosetting resins because of the need to cool the mold before removing the material. It consists of squeezing a material such as phenolic into the desired shape in a mold by the application of heat and pressure.

A typical *compression molding* press is shown in Figure 43-3. Thermosetting resins are rarely used alone. In practice, the powder or preform shape is mixed with various fillers such as wood flour, chopped canvas, glass fibers, cotton, alpha-cellulose, asbestos, mica, or saw dust. Other additives may include pigments and lubricants. The material is loaded directly into the open but heated mold cavity. Sufficient heat and pressure are applied to cause the plastic to fill all the portions of the cavity as it becomes liquid.

Then while the mold halves are held closed for a predetermined period of time, the thermosetting material crosslinks and permanently hardens. (*Note:* Crosslinking, or the joining of polymer chains, is comparable to "vulcanizing" in rubber processing.) Once cooled, the finished part is ejected. Depending upon the material the curing time may range from 30 to 120 seconds.

Compression molding requires rigid and substantial presses. Die halves must be in precise alignment. Platen pressure must build up quickly and uniformly to the precise density required.

Transfer Molding

A variation of compression molding, the *transfer molding* process, begins by placing the raw material into a preheating chamber. The temperature of the material is raised until it is in a semifluid condition. A ram then forces the liquid stock through runners and gates into the mold cavities. The main difference between transfer molding and compression molding is that the plastic material is preheated before it goes into the mold. Thus, the charge is able to flow more easily than the powder in conventional compression systems and completely penetrate the mold-surrounding metal inserts, if any are used, filling small, deep holes. A principal advantage of the process is the problem of precuring is avoided. This can happen in compression molding if the mold is too hot and the charge begins to cure before all of it is melted from powder to liquid. Both the compression molding and transfer molding processes can produce intricate parts with wide variations in section thickness to a consistently precise size and with an excellent finish.

Extrusion

When a component is too large for other processes, such as injection, compression, or transfer molding, or when it may be too intricate with undercuts, internal threads, or other features that could never emerge from a mold cavity, *extruded* plastic shapes are often used. A plastic extruder is shown in Figure 43-4. Almost all types of cross-sectional shapes are possible. Rods, tubing, bars, and sheets may be formed by extruding in addition to the typical applications shown in Figure 43-5. Conventional metalworking equipment can be used to fabricate extruded components, which may later form the finished product by assembling them by cementing or bonding. Plastic insulation may be applied to wire by the extrusion process.

In extruding, powdered raw material, usually thermoplastic, is fed into a hopper and carried along by a conveyer screw through a heating chamber, where it becomes a viscous fluid. In a manner similar to extruding metal or the way toothpaste emerges from a tube opening, the material is forced

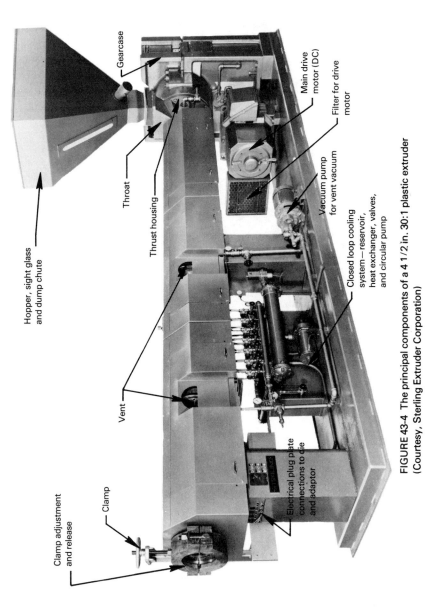

FIGURE 43-4 The principal components of a 4 1/2 in. 30:1 plastic extruder (Courtesy, Sterling Extruder Corporation)

FIGURE 43-5 Typical extruded shapes (Courtesy, Crane Plastics)

through a heated die. Material emerging from the die rests on a moving conveyor and, once cooled by air or a spray of water, the extruded material retains the shape of the die opening. Rigid plastic materials are cut to convenient lengths, while flexible plastics are usually coiled.

Blow Molding

This process begins with the preparation of a soft, extruded thermoplastic preformed tube called a "parison." The parison is formed over a core pin with one end closed like a test tube. The core pin with its attached parison is inserted between the open halves of the *blow mold* cavity. As the mold halves close, air pressure inflates the thin-walled parison and forces it outward against the mold sides. When the thermoplastic material has achieved the desired part shape in the mold, the hydraulically operated mold halves open and the hollowware part is removed. Except for a small fin of excess material in the form of flash, which may be manually cut off by the machine operator, the thin-walled part is ready for use.

Products include a wide range of blown-ware container shapes for the huge beverage and food markets. Other important nonfood uses of blow-

FIGURE 43-6 A blow molding machine. Machines of this kind can produce bottles from 11 in. high with a 42 oz. capacity to much smaller units holding only 5 c.c.'s (Courtesy, Rocheleau Tool & Die Co.)

molded containers are for packaging cosmetics, pharmaceuticals, artist's paint and powder, certain chemicals, detergents, and glue. Blow-molded products are also used as liners for steel drums, lighting globes, floats, and for various automotive parts. While the bulk of food container sizes are in the 12- to 30-oz (approximately 340- to 850-g) capacity range, equipment is now available for producing containers for a variety of applications in sizes of 64 oz (1811.2 g) and larger, with a nominal wall thickness of 20 mils. Machines range from those with single molds to highly sophisticated automatic models having molds with up to 12 cavities capable of producing parts in an 18- to 20-second cycle. Figure 43-6 shows a model of a single-cavity blow-molding machine.

Rotational Molding

The molds used in this relatively new process are generally made of cast aluminum, electroformed copper-nickel, or sheet steel. Depending upon the intricacy of the hollow parts to be produced, the molds may consist of two or more pieces. Unlike the open molds used in the slush casting process for casting metal parts (see Chapter 10), *rotational molds* are completely enclosed when assembled. Rotational molding equipment is available for small lot, batch-type production as well as multispindle machines for continuous, automatically indexed operations for increased production. Thermoplastics and, to some extent, thermosets have been successfully used in rotational molding. There is virtually no scrap, as just enough raw material is fed into the molds to make the part with the desired wall thickness.

In practice, a predetermined quantity of raw plastic material in the form of a powder or a liquid is poured into a mold, which is then placed in an oven. During the time when the molds are heated to the desired temperature, they are biaxially rotated, causing the liquid plastic to coat the inner mold walls. When the deposit has developed to a sufficient thickness, usually between $\frac{1}{32}$ in. (0.794 mm) but sometimes as thick as 1 in. (25.4 mm), the molds are removed from the oven. Following cooling, the molds are parted and the finished product is removed.

Except for limitations imposed by oven sizes or rotary mechanisms, there is no theoretical limit on the size of the part that can be rotational-molded. Commercial products made in this way include hollow-type toys, novelty items such as coin banks, parts for games, and beach balls, and various objects used for Halloween, Christmas, or Easter decorations, artificial fruit, and candles. Parts such as floats, liners, special containers, gasoline tanks, squeeze bulbs, bicycle seats, automobile sun visors, and armrests are also widely used rotational-molded products.

Thermoforming

This process consists of heating a thermoplastic sheet to its softening temperature and forcing the hot and flexible material against the contours of the mold. When held to the shape of the mold and allowed to cool, the plastic part retains the mold's shape and detail. Sheet plastics can often be used to an advantage when lightweight parts are required. Typical *thermoformed* products include windshields, luggage, signs, displays, toys, various covers and panels, guards, housings, cases, carriers, liners, trays, and tote boxes.

There are three basic sheet forming methods: vacuum forming, air blowing, and mechanical forming. In each case, tooling costs are relatively low.

Depending upon such factors as the complexity of the required part and the depth of draw, there are many variations of the basic vacuum forming process. In its simplest form, *vacuum forming* consists of clamping a sheet of thermoplastic material to a frame, heating it to its softening point, and then lowering it over a die or a form. As air is drawn through holes in the form, producing a partial vacuum, atmospheric pressure forces the sheet into intimate contact with the contours of the form. After the plastic material has been air-cooled sufficiently to retain its "set," the vacuum is discontinued and the part shape is removed from the form. Figure 43-7 shows a typical thermoforming machine.

A second thermoforming process, *air blowing,* employs air pressure to form thin-gage containers and other parts. The heated plastic sheet is inserted between the upper and lower mold halves. The mold is then tightly closed and sealed. Finally, compressed air is rapidly blown against the trapped sheet, forcing the part to retreat into the female mold. There are several variations of air blowing, ranging from a method that resembles blow molding to other techniques, which differ principally in the manner of applying the air pressure.

Another variation of thermoforming is *mechanical forming,* which consists of stretching a heated plastic sheet over a mold and forming the part

FIGURE 43-7 A typical thermoforming machine
(Courtesy, Comet International Inc.)

shape without vacuum or air pressure. The process strongly resembles sheet-metal forming, in that matched mold halves are used to press the sheet into the desired shape. To assure conformity of the plastic sheet to the mold contour, the mold is vented to allow trapped air to escape. Raised lettering and grained surfaces can be readily produced on mechanically formed plastic parts because of the ability of the process to reproduce intricate mold detail.

Laminates

Industrial *laminated* thermosetting products, whether in sheet, tube, or rod form, consist of plies of fibrous reinforcing sheet material such as cellulose, paper, asbestos paper, cotton fabric, glass fabric, or nylon. These materials are impregnated or coated with a thermoset resin and consolidated under high pressure and temperature into material with the desired properties. The principal resin used in cloth-or paper-reinforced laminated products are the phenolics. Depending upon the service conditions of the required product, other resins such as melamines, silicones, epoxies, and polyesters may be also used.

When possible laminated plastic shapes are cut or machined from sheet, rod, or tube to the desired size and later assembled for end use. Individual layers of irregularly shaped products, such as protective headgear, boat hulls and panels, or automobile bodies, may be precut and then shaped by molding under moderate pressure and temperature. The molded-laminated method is used to produce shapes that would be uneconomical to fabricate from flat, thick laminates. The properties of laminated products can be varied to a considerable degree by altering the resins, the sheet materials, and the important factors of pressure and temperature in the manufacturing processes. As a result, the desired properties—electrical, physical, thermal, chemical and mechanical—can be emphasized. In most cases, however, improvement of a particular property is at the expense of some other property.

Composite sheet laminates of copper, aluminum, nickel, and steel are widely used for some applications. For example, copper-clad sheets laminated on one or both sides are widely used for printed-circuit or multilayer boards. Nonmetallic laminates include elastomers, vulcanized fiber, asbestos, and cork. Composites can also be produced in rods and tubes.

STAMPING PLASTICS

A recently perfected technique using conventional metal-forming presses to *stamp* plastic parts is now an accepted forming process. The full utilization of the process was made possible by the formulation of materials known as AZDEL (G.R.T.L. Company, Southfield, Michigan) or STX (Fibers Division,

Allied Chemical Corp., Morristown, N.J.). The material consists of a glass fiber and particle-reinforced thermoplastic, highly resistant to heat distortion. Parts made from this material are strong, lightweight, and corrosion-resistant.

There are four basic steps involved in forming the part: the blank is heated to a temperature of about 420 °F (232.4 °C), and automatically transferred into the metal stamping press, where the part is formed and ejected. An infrared oven is used for heating the blanks. The oven has five stations: one loading, three heating, and one unloading. The blank moves through the various stations over a carefully timed sequence of 16 seconds for each movement. A fast-rise, fast-decay heat source is desirable to heat the material uniformly and quickly and to prevent heat degradation. A mechanical arm with insulating fluoroplastic-tipped fingers grips the blank at the unloading station of the oven and transfers it into heated matched metal dies maintained at a temperature of 212 to 302 °F (100 to 150 °C).

In operation, an adjustable limit switch and timer is used to signal the clutch and brake on the press to stop the ram at bottom dead center. At this point, the heated thermoplastic sheet absorbs practically all the energy of the downward-moving press. The press remains in this position to properly form the part for about 10 seconds and then retracts. The formed part is lifted from the male die by ejector pins and is pushed out of the press. Molding pressures are in the 1000- to 3000-psi (approximately 70- to 211-kg cm²) range and, during forming, the blank may flow as much as 2 in. (50.8 mm). While modified mechanical presses can be used for this process, no conversions are necessary for hydraulic presses.

There is a growing list of product applications for stamped plastic parts, particularly in the automotive field. Typical parts include tail lamp assemblies, fans and fan guards, rocker panels, battery trays, distributor caps, and sunroofs. Other commercial nonautomotive applications of stamped plastic parts are beginning to take their place on the market.

MACHINING PLASTICS

Plastic materials may be satisfactorily machined using conventional wood-working and metal-working tools, often with no, or only slight, modifications. There are two major considerations when machining plastics: how to control the heat generated by the cutting tools, and how to maintain a suitable finish on machined surfaces.

Heat Effects

Plastics have a greater heat sensitivity than metals. When worked with cutting tools, most thermoplastics soften and become gummy. In metal-cutting operations, the heat generated by the action of the cutting tool as it works against the material is usually carried away by the chips. On operations such as sawing, milling, turning, boring, or drilling, the burned plastic resin builds up on the cutting edges and often interferes with the performance of the cutting tool. Thermoplastic materials quickly heat up and clog the cutting area. While this condition is less troublesome for thermosetting plastics, considerable difficulty is usually encountered in maintaining a keen cutting edge when machining filled materials because of the abrasive action of the filler.

Several remedies have been developed to improve machining conditions for plastics. In most cases, coolants such as compressed air, plain water, or various mixtures of water-soluble oil and water may be used to an advantage. The front and side clearance angles on cutting tools should be increased to prevent rubbing and heating the plastic. In actual practice, the optimum speed, feed, and depth of cut is usually determined by the type of plastic and by the kind of cutting tool used. In general, most thermoplastics may be successfully machined at relatively high surface speeds up to 800 sfpm and thermosets up to 3000 sfpm. Slow feeds of 0.002 to 0.005 in. (0.05 to 0.13 mm) are recommended with a light depth of cut. Stellite and carbide-tipped tools will increase production and tool life. Cutting tools should be kept sharp to reduce frictional heat. Methods of holding and positioning parts during subsequent machining operations are usually carefully reviewed prior to manufacturing to prevent the possibility of work-holding damage to plastic parts.

Surface Finish

The lustrous surface on most molded plastic products, particularly phenolic or urea materials, is destroyed by machining. Whenever possible, plastic products are designed to be used in their "as-molded" condition without the necessity for machining operations. If machining cannot be avoided, the designer should try to specify machining operations only in locations not visible when the product is in use. Most molded plastic parts require some hand-finishing operations, such as the removal of flash and gates. Often, excess material may be removed by scraping, cutting with a knife, sanding, or by filing. The trimming of flash from a molded plastic part with filler material often cuts through a protective resin film on the surface and exposes a ridge of raw filler. For this reason, designers attempt to place parting lines on molded plastic products in the least objectionable position. The finish on some plastic articles may be improved by polishing or buffing using only light pressure. Minor projections and small fins on some parts may be successfully removed by barrel tumbling in wet or dry pumice compounds.

REVIEW QUESTIONS

43.1. Explain what is meant by the term "family" in reference to plastics.

43.2. In what ways do thermoplastics and thermosetting plastics differ?

43.3. Explain the major reasons why injection molding is so well suited for processing thermoplastic materials.

43.4. Explain why the same mold usually cannot be used for more than one plastic material when molding a product.

43.5. Explain why injection-molding equipment is not ordinarily used for molding thermosetting materials.

43.6. Explain the reason compression molding is adaptable to processing thermosetting materials.

43.7. What is the main advantage of transfer molding?

43.8. Would thermoplastic or thermosetting materials be more suitable for extruding? Why?

43.9. In what important ways do blow molding and injection molding differ?

43.10. Describe the method for producing plastic bottles.

43.11. Prepare a brief outline describing the steps involved in producing a part by rotational molding.

43.12. Give the leading reason why a designer might specifically specify the thermoforming process for producing a particular part.

43.13. What is a plastic laminate?

43.14. Discuss some of the important difficulties in machining plastics.

BIBLIOGRAPHY

The ABC's of Modern Plastics, New York: Union Carbide Corp., 1972.

Basic Plastics, Indianapolis, Ind.: Howard W. Sams Co., Inc., 1974.

BEACH, N. E., *Plastic Laminate Materials,* Dover, N.J.: Plastec, Picatinny Arsenal.

BECK, R. D., *Plastic Product Design,* New York: Van Nostrand Reinhold Company, 1974.

BIKALES, N. M., Editor, *Molding of Plastics,* New York: John Wiley & Sons, Inc., Interscience Division, 1971.

Cutting Costs in Short Run Plastics Injection Molding, Morgan Industries, Inc., Chicago, Ill., 1974.

GLANVILL, A. B., *The Plastics Engineer's Data Book,* New York: Industrial Press, Inc., 1973.

Machine Design, Reference Issue, Metals, Plastics, Elastomers and Other Engineering Materials, Mar. 4, 1976.

Machining the Engineering Plastics, Detroit, Mich.: Cadillac Plastics and Chemical Co., (Div. Dayco Corp.), 1975.

MILBY, R. V., *Plastics Technology,* New York: McGraw-Hill Book Company, 1973.

Modern Plastics Encyclopedia, New York: McGraw-Hill Book Company, 1976–1977.

Plastics Fabrication Manual, Reading, Pa.: Polymer Corp., 1975.

The Plastics Industry in the Year 2000, Menlo Park, Calif.: Stanford Research Institute, Apr. 1973.

"Plastics—Your Future Feedstock," *American Machinist,* Special Report 677, May 15, 1975.

A Ready Reference for Plastics, Boonton, N.J.: Boonton Molding Co., Inc., 1976.

The Story of the Plastics Industry, New York, N.Y.: The Society of the Plastics Industry, Inc., 1975.

*Abstracted with permission.

APPENDIX

USEFUL FORMULAS

MILLING OPERATIONS

To Calculate Feed:

$$F = f_t nN$$

where: F = feed (ipm)
          f = feed per tooth (in.)
          n = number of teeth on cutter
          N = revolutions per minute of cutter (rpm)

To Calculate RPM:

$$N = \frac{sfpm \times 12}{\pi D}$$

where: N = revolutions per minute of cutter (rpm)
    $sfpm$ = surface feet per minute of cutter
        D = cutter diameter (in.)
        S = cutting speed (sfpm)

To Calculate Cutting Speed:

$$S = \frac{\pi DN}{12}$$

743

where: S = cutting speed (sfpm)
          D = cutter diameter (in.)
          N = revolutions per minute of cutter (rpm)

To Calculate Time Required Per Cut:

$$t = \frac{L + \Delta L}{F}$$

where: t = time (min./cut)
          $t*$ = when the diameter of the cutter is equal to or only slightly larger
                 than the width of the work surface
          L = length of cut (in.)
          ΔL = one-half the cutter diameter (in.)
          F = feet (ipm)

Or,

$$t = \frac{L + \sqrt{dD - d^2}}{F}$$

where: d = depth of cut (in.)
          D = cutter diameter (in.)

To Calculate Stock Removal Rate:

$$V = Fwd \qquad \text{or} \qquad V = F_t TNdw$$

where: V = stock removed (in.^{3}/min.)
          F = feed (ipm)
          w = width of cut (in.)
          d = depth of cut (in.)
          F_t = feed per tooth (in.)
          T = number of teeth of cutter
          N = revolutions per minute (rpm)

To Calculate Horsepower Required at the Cutter:

$$HP_c = VHP_u$$

where: HP_c = horsepower required at the cutter
          V = stock removed (in.^{3}/min.)
          HP_u = horsepower required to cut various hardnesses of materials at the
                   rate of 1 in.^{3}/min

To Calculate Horsepower Required at the Motor:

$$HP_m = \frac{HP_c}{\eta} + HP_i$$

where: HP_m = horsepower required at the motor
HP_c = horsepower required at the cutter
HP_i = idle horsepower
η = mechanical efficiency factor of the machine
(usually 0.9 for direct belt drives and 0.7 to 0.8 for gear drives)

To Calculate the Length of the Tool Path:

Peripheral Milling:
Up milling

$$L = \frac{\pi}{180°} \ R\cos^{-1}\frac{(R-d)}{R} + \frac{F_r}{2\pi R} \ \sqrt{2Rd - d^2}$$

Down milling

$$L = \frac{\pi}{180°} \ R\cos^{-1}\frac{(R-d)}{R} - \frac{F_r}{2\pi R} \ \sqrt{2Rd - d^2}$$

where: L = length of tooth path (in.)
R = cutting radius (in.)
d = depth of cut (in.)
F_r = feed per revolution

Face milling (When the ω is nearly equal to D)

$$L = \frac{\pi}{90°} \ \sqrt{r^2 + R^2} \ \frac{(1 - r^2)}{4R^2} + \cos^{-1} \ \frac{R - \frac{w}{2}}{R}$$

To Calculate the Cutting Ratio (see Figure A-1):

$$C_r = \frac{L_c}{L_s} = \frac{d}{d_o}$$

where: C_r = cutting ratio
L_c = length of chip (in.)
L_s = length of path of cutting tool while forming the chip (in.)
d_o = thickness of the chip (in.)

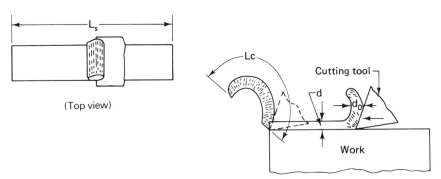

(Top view)

FIGURE A-1

Determining the "cutting ratio".

LATHE OPERATIONS

To Calculate Horsepower:

$$HP = HP_u \, 12 \, CVFd$$

where: HP = horsepower
   HP_u = horsepower required to cut various hardnesses of materials at the
   rate of 1 in.^{3}/min
   C = feed correction factor
   V = cutting speed (ipr)
   F = feed (ipm)
   d = depth of cut (in.)

To Calculate RPM:

$$rpm = \frac{sfpm \, 12}{\pi \, d}$$

where: rpm = revolutions per minute
   d = stock diameter (in.)

To Calculate Cutting Time:

$$T = \frac{L}{FN}$$

where: T = cutting time (min.)
   L = length of cut plus an additional distance of $\frac{1}{16}$ to $\frac{1}{4}$ in. for tool
   to enter and clear the cut
   F = feed (ipr)
   N = revolutions per minute (rpm)

To Calculate the Rate of Metal Removal:

$$Q = 12 \, VFd$$

where: Q = rate of metal removal (in.^{3}/min.)
   V = cutting speed (rpm)
   F = feed (ipr)
   d = depth of cut (in.)

BROACHING OPERATIONS (see Table A-1)

To Calculate the Power Required for Broach Pull:

Surface broaching

$$F = NLRC$$

746

where: F = broach pull (lbs)

          N = number of teeth engaged

          L = effective length of tooth (in.)

          R = rise per tooth or chip thickness (in.)

          C = broaching constant of metal being broached

Round hole broaching:

$$F = N \pi DRC$$

where: F = broach pull (lbs)

          N = number of teeth engaged

          D = hole diameter (in.)

          R = rise per tooth or chip thickness (in.)

          C = broaching constant of metal being broached

Splined hole broaching:

$$F = NSWRC$$

where: N = number of teeth engaged on workpiece

          S = number of splines

          W = spline width (in.)

          R = rise per tooth or chip thickness (in.)

          C = broaching constant of metal being broached

TABLE A-1 Broaching constants of metals

Metal	Constant
*Aluminum	50,000
*Copper	250,000
Cast iron, bronze	350,000
Malleable iron	400,000
Mild steel	450,000
Alloy steel	600,000

*Annealed.

POWDER METALLURGY (see Table A-2)

To Calculate Tonnage Capacity:

$$L = M_p A$$

where: L = press load (tons)

          M_p = molding pressure (tons/in.²)

          A = projected area (in.²)

To Calculate Die Fill Depth:

$$D_f = tC_r$$

where: D_f = die fill depth (in.)
 t = part thickness (in.)
 C_r = compression ratio

TABLE A-2 Tonnage requirements and compression ratios for various materials

Type of material or part	Tons per sq. inch	Compression ratio
Aluminum	5 to 20	1.5 to 1.9:1
Brass	30 to 50	2.4 to 2.6:1
Bronze	15 to 20	2.5 to 2.7:1
Carbon	10 to 12	3.0:1
Copper-Graphite brushes	25 to 30	2.0 to 3.0:1
Carbides	10 to 30	2.0 to 3.0:1
Alumina	8 to 10	2.5:1
Sterites	3 to 5	2.8:1
Ferrites	8 to 12	3.0:1
Iron bearings	15 to 25	2.2:1
Iron parts:		
Low density	25 to 30	2.0 to 2.4:1
Medium density	30 to 40	2.1 to 2.5:1
High density	35 to 60	2.4 to 2.8:1
Iron powder cores	10 to 50	1.5 to 3.5:1
Tungsten	5 to 10	2.5:1
Tantalum	5 to 10	2.5:1

The above tonnage requirements and compression ratios are only approximations and will vary with changes in chemical, metallurgical, and sieve characteristics; with the amount of binder or die lubricants used; and with mixing procedures.

(Note: To convert tons per in.^{2} to kg per mm^2, multiply tons by 1.41).

PRESSWORKING (see Table A-3)

To Calculate Press Size Requirements Without Shear (Piercing and Blanking):

$$L = \frac{ltS_s}{2000} + F_s$$

where: L = load on the press (cutting force-tons)
 (Note: this value is sometimes reduced by 30% to compensate for partial shear)
 l = length of perimeter of cut (in.)
 t = stock thickness (in.)
 S_s = shear strength of the stock (psi)
 F_s = safety factor (usually = 2)

748

EVALUATING ECONOMY OF PRESS BRAKE
OPERATIONS vs PUNCH PRESS OPERATIONS

$$X = \frac{S_B - S_A}{P_A - P_B}$$

where: X = the minimum lot size for punch brake operations that in less
quantity may be produced more economically by press brake operations

S_B = setup time per 100 pieces for a punch press (min.)
S_A = setup time per 100 pieces for a press brake (min.)
P_A = production time per piece for a press brake (min.)
P_B = production time per piece for a punch press (min.)

TABLE A-3 Shear strengths of various steels and nonferrous metals at room temperature

Metal	Shear strength, psi
Carbon steels:	
0.10% C	35,000 to 43,000
0.20% C	44,000 to 55,000
0.30% C	52,000 to 67,000
High-strength	
low-alloy steels	45,000 to 63,700
Silicon steels	60,000 to 70,000
Stainless steels	57,000 to 129,000
Nonferrous metals	
Aluminum alloys	7,000 to 46,000
Copper and bronze	22,000 to 70,000
Lead alloys	1,825 to 5,870
Magnesium alloys	17,000 to 29,000
Nickel alloys	35,000 to 116,000
Tin alloys	2,900 to 11,100
Titanium alloys	60,000 to 70,000
Zinc alloys	14,000 to 38,000

To Calculate Bending Pressure: (see Table A-4)

$$F = \frac{KLSt^2}{W}$$

where: F = bending force required (tons)
K = die opening factor (varies from 1.20 for a die opening of 16 times
metal thickness to 1.33 for a die opening of 8 times metal thickness)
L = length of bent part (in.)
S = ultimate tensile strength of material (tons/in²)
t = stock thickness (in.)
W = width of V-channel or U-ing of lower die (in.)

749

To Calculate Bend Allowance:

$$L = 2 \pi (r + 0.4t) \frac{\theta}{360°}$$

where: L = arc length of neutral axis (in.)
r = inside radius of bend (in.)
t = stock thickness (in.)
θ = angle of bend (degrees)

TABLE A-4 Ultimate strengths for various materials

Metal	Tons/in.^{2}
Aluminum and alloys	6.5 - 38.0
Brass	19.0 - 38.0
Bronze	31.5 - 47.0
Copper	16.0 - 25.0
Steel	22.0 - 40.0
Tin	1.1 - 1.4
Zinc	9.7 - 13.5

SWAGING

To Calculate Reduction in Area:

$$R = 100 \left(1 - \frac{d_2}{d_1}\right)$$

where: R = reduction in area (%)
d_2 = original O.D. before swaging (in.)
d_1 = O.D. after swaging (in.)

To calculate increase in wall thickness:

$$t = \frac{(d_1)(t_1)}{d_2}$$

where: t = wall thickness after swaging (in.)
d_1 = original O.D. before swaging (in.)
d_2 = O.D. after swaging (in.)
t_1 = original wall thickness before swaging (in.)

To Calculate the Required Blank Diameter for Symmetrical Shells: (see Figure A-2 (a), (b), (c), and (d).

750

To Calculate Drawing Pressure For:

Circular cups

$$P = \pi\,dtS\left(\frac{D}{d} - C\right)$$

where: P = pressure (lbs.)
D = blank diameter (in.)
d = shell diameter (in.)
t = stock thickness (in.)
S = tensile strength of material (psi)
C = constant = 0.6 to 0.7 to compensate for friction or bending

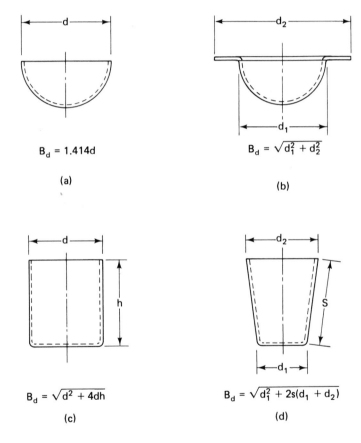

$B_d = 1.414d$

(a)

$B_d = \sqrt{d_1^2 + d_2^2}$

(b)

$B_d = \sqrt{d^2 + 4dh}$

(c)

$B_d = \sqrt{d_1^2 + 2s(d_1 + d_2)}$

(d)

Rectangular shells:

$$P = tS\,(2\pi r C_1 + LC_2)$$

751

where: r = inside corner radius of the shell (in.)
         L = perimeter of shell (in.)
         *C_1 = constant = 0.5 for a very shallow shell and up to about 2 for a shell having a depth of 5 or 6r
         *C_2 = constant =0.2 for each draw radius with ample clearance and no holding pressure up to a maximum of 1 for stock clamped too tightly to flow

TABLE A-5 Weights of metals

Metals	Weight lbs/in³
Aluminum	0.098
Beryllium-Copper	0.298
Brass, 64-40	0.310
Brass, 70-30	0.315
Bronze-Phosphorus	0.311
Bronze-Tobin	0.291
Copper	0.325
Duralumin	0.102
German silver	0.306
Iron, wrought	0.285
Iron, cast	0.260
Iron, ingot	0.285
Lead	0.409
Magnesium	0.063
Monel metal	0.320
Nickel	0.320
Ni-Al-Br	0.302
Steel, cast	0.283
Steel, carbon	0.283
Steel, alloy	0.283
Steel 18-8	0.285
Steel, 13% chromium	0.283
Steel, 14% tungsten	0.312
Steel, 22% tungsten	0.321
A-286	0.288
Discaloy	0.287
N155	0.296
Inconel	0.307
M308	0.304
M252	0.298
Hastelloy "A"	0.318
Hastelloy "B"	0.334
Hastelloy "C"	0.323
Hastelloy "D"	0.282
Titanium (pure)	0.163
Titanium RC 130B	0.167

*Values for C_1 and C_2 are approximate.

INDEX

INDEX